Marine Geomorphometry

Marine Geomorphometry

Special Issue Editors

Vanessa L. Lucieer
Vincent Lecours
Margaret F.J. Dolan

MDPI • Basel • Beijing • Wuhan • Barcelona • Belgrade

MDPI

Special Issue Editors

Vanessa L. Lucieer
University of Tasmania
Australia

Vincent Lecours
University of Florida
USA

Margaret F.J. Dolan
Geological Survey of Norway
Norway

Editorial Office
MDPI
St. Alban-Anlage 66
4052 Basel, Switzerland

This is a reprint of articles from the Special Issue published online in the open access journal *Geosciences* (ISSN 2076-3263) from 2017 to 2018 (available at: https://www.mdpi.com/journal/geosciences/special_issues/marine_geomorphometry).

For citation purposes, cite each article independently as indicated on the article page online and as indicated below:

LastName, A.A.; LastName, B.B.; LastName, C.C. Article Title. *Journal Name* **Year**, *Article Number*, *Page Range*.

ISBN 978-3-03897-954-8 (Pbk)
ISBN 978-3-03897-955-5 (PDF)

Cover image courtesy of Kartverket/MAREANO.

Contents

About the Special Issue Editors

Vanessa L. Lucieer, Senior Scientist. Vanessa L. Lucieer is a senior scientist at the Institute for Marine and Antarctic Studies at the University of Tasmania, Australia. Her multidisciplinary team collaborates closely with government and industry partners to apply the latest ocean mapping technologies to solve real-world problems. Her research interests include the application of seafloor and water column acoustics from shipboard and autonomous underwater vehicles for studying, mapping, and monitoring marine ecosystems working across the fields of marine ecology, geology, oceanography, geomatics, and ocean technology. Dr. Lucieer is the founding member of Seamap Australia, a leading member of the International Multibeam Backscatter and Water-Column Working Group, Seabed 2030 South and West Pacific Committee Member, and is on the Steering Committee of AusSeabed. Dr. Lucieer is committed to training the next generation of marine spatial scientists who endeavor to continue mapping the world's oceans.

Vincent Lecours, Assistant Professor of Marine Remote Sensing and Geospatial Analysis. Dr. Lecours is an assistant professor of Marine Remote Sensing & Geospatial Analysis at the University of Florida. He completed a B.Sc. in applied geomatics from the Université de Sherbrooke and received his Ph.D. in geography from Memorial University, where he studied methods associated with the mapping of deep-water habitats. Dr. Lecours conducts cross-disciplinary research using geospatial technologies and spatial sciences. His research program bridges the spatial sciences with the marine sciences by studying ways to improve marine habitat mapping methods through a better integration of spatial concepts such as spatial scale, autocorrelation, and spatial data quality in the habitat mapping workflow. He also studies how those spatial concepts affect the marine geomorphometry workflow, and how marine geomorphometry can assist in mapping marine habitats. His research is aimed at developing best practices in the application of geomatics-based marine habitat mapping to ecological and management questions.

Margaret F.J. Dolan, Research Scientist, Geological Survey of Norway. Dr. Dolan is a Research Scientist in the Marine Geology Group at the Geological Survey of Norway, where she is engaged in all aspects of seabed mapping from the coastal zone to the deep sea, with a focus on habitat mapping. Dr. Dolan holds a Ph.D. in Earth and Ocean Science from the National University of Ireland, Galway, where she conducted research on deep-sea habitat mapping using remotely operated vehicles. Her research interests center on the development of methods for seabed geological and habitat mapping, combining the use of mapping technology, quantitative spatial analysis, and predictive modelling. Marine geomorphometry is a core theme throughout Dr. Dolan's multi-disciplinary work, which is underpinned by bathymetry data from various sources.

Preface to "Marine Geomorphometry"

Preface by Tom Hengl (OpenGeoHub Foundation; https://opengeohub.org/people/tom-hengl) and Peter Guth (US Naval Academy) Tom Hengl is the former vice-chair of the International Society for Geomorphometry (2011–2015) and Peter Guth is the current chair of the International Society for Geomorphometry (since 2015).

Geomorphometry is the science of digital relief analysis, and is often identified only with terrestrial applications, which focus on land surface processes such as water flow, snow movement, solar radiation modeling, soil erosion and deposition, and vegetation distribution mapping. Many well-known global digital elevation and/or surface models stop where water bodies start. However, the underwater landscapes and processes can also be characterized using geomorphometry. In fact, most of the planet is under water! Understanding seabed complexity and its associated ecology is of importance to our economies and everyday lives: it is the world's largest supply of oil, diamonds, and rare earth metals. It is also still the main sink for all our liquid/sewage waste. The world's seas are also home to the largest biological habitat on Earth, and 70% of the oxygen in the atmosphere is produced by marine plants. We owe so much to having a healthy and functioning ocean. Unfortunately, for decades we knew little about the sea bed topography. In fact, it is often said that we have "a more complete understanding of the Moon or Mars than we do of our own seabed—much of the Earth's ocean floor is a mystery". Marine geomorphometrists are researchers who try to improve this situation by diligently collecting pieces of data and increasing our knowledge of marine topography, resources, and biodiversity. In essence, marine and land geomorphometry have so many differences that, for years, the communities developed independently. The three biggest differences between marine and terrestrial geomorphometry are:

1. Data acquisition methods in marine geomorphometry are most often based on acoustic sonar systems, while terrestrial acquisition methods are based on visible light, LiDAR, and radar sensors. Hence, different background knowledge of physics and surface properties is required.

2. Surveying of the seabed is an order of magnitude more complex and more expensive than land surface topography. It takes much more patience and diligence to produce seabed digital terrain models (DTMs).

3. Hydrological flow modeling equations from terrestrial geomorphometry do not apply directly to seabed topography.

Applications of marine and terrestrial geomorphometry are also quite different. Producing consistent (fixed accuracy and spatial resolution) topographic models of seabed is more complex in comparison to terrestrial applications, for two main reasons:

1. The vertical accuracy of sonar systems decreases with sea surface height, such that most complete DTMs of seabed topography can be inconsistent. For that reason, it is currently difficult to produce bathymetric DTMs covering large areas with a constant pixel size. This can be compared with the terrestrial DTMs, which lose some consistency in shifting sand areas and around poles (permafrost), but in general are very consistent.

2. Terrestrial survey imagery is multispectral, that is, it covers fairly diverse parts of spectra/different wavelengths so that one can compare signals and produce a more complete image of topography, vegetation, soil moisture, mineralogy, etc. Marine geomorphometry can often rely only on the topographic data and backscatter imagery. Overall, it can be said that marine geomorphometry is an order of magnitude more challenging and more complex than terrestrial geomorphometry.

A 2018 article in BBC News claimed that only 15% of the Earth's ocean has been mapped (probably implying only bathymetric mapping, i.e., to produce DTMs of seabed). However, it is not only about the coverage. Many ocean areas can be mapped at coarse spatial resolutions, but this still does not make them usable for many applications. Thanks to missions and products such as the German Space Agency's TANDEM-x (https://download.geoservice.dlr.de/TDM90/), Japan Aerospace Exploration Agency's AW3D30 (https://www.eorc.jaxa.jp/ALOS/en/aw3d30/index.htm), and NASA's DEM (https://earthdata.nasa.gov/community/community-data-system-programs/measures-projects/nasadem), the land surface has been mapped at unprecedented detail and at spatial resolutions of 30 m or better. We have much less data on sea bottom topography. Even though we now seem to know much more about sea bottom relief, our planet's oceans remain largely unexplored. To increase the consistency, spatial detail, and usability of global marine elevation models, the Nippon Foundation-GEBCO has initiated the Seabed 2030 Project (https://seabed2030.gebco.net/). Seabed 2030 "aims to bring together all available bathymetric data to produce the definitive map of the world ocean floor by 2030 and make it available to all" (https://doi.org/10.3390/geosciences8020063). While this sounds like an excellent initiative, we also need data products and applications before 2030.

There are also many common points between marine and terrestrial geomorphometry. A great deal of the current sea bed used to be above sea level, and hence many of the terrestrial processes that happened in the past can be tracked by mapping the seabed topography. Also, many of the terrestrial wetland conservation/modeling projects, such as those focusing on modeling world mangroves and river deltas, require topographic information from both above and below the water's surface. This is where marine and terrestrial geomorphometry need each other. So, in essence, marine geomorphometry will remain connected with terrestrial applications and there is certainly many reasons to host joint marine and terrestrial geomorphometry meetings and workshops (which I am personally, as a terrestrial geomorphometrist, happy to continue supporting).

The authors of this book try to fill some of the gaps between what we know about seabed topography and process models and the importance of such data for both exploitation and nature conservation projects. They advance the marine geomorphometry field forward in terms of technological and methodological innovation, data availability, and applications. The Special Issue has resulted in 18 accepted papers, which is excellent news for our field. We are happy to see that the Special Issue has now been converted to a full-volume book covering various sub-fields of marine geomorphometry (including seabed geomorphology, habitat mapping, natural hazards, engineering, and seabed mining) and that its authors also have a clear vision of how to proceed (https://doi.org/10.3390/geosciences8120477).

Tom Hengl and Peter Guth

geosciences

MDPI

Editorial

Charting the Course for Future Developments in Marine Geomorphometry: An Introduction to the Special Issue

Vanessa Lucieer [1,*], Vincent Lecours [2] and Margaret F. J. Dolan [3]

1   Institute for Marine and Antarctic Studies, University of Tasmania, Tasmania 7000, Australia
2   Fisheries & Aquatic Sciences | Geomatics, School of Forest Resources & Conservation, University of Florida, Gainesville, FL 32653, USA; vlecours@ufl.edu
3   Geological Survey of Norway (NGU), Postal Box 6315 Torgarden, NO-7491 Trondheim, Norway; margaret.dolan@ngu.no
*   Correspondence: vanessa.lucieer@utas.edu.au; Tel.: +61-3-6226-6931

Received: 7 December 2018; Accepted: 10 December 2018; Published: 13 December 2018

Abstract: The use of spatial analytical techniques for describing and classifying seafloor terrain has become increasingly widespread in recent years, facilitated by a combination of improved mapping technologies and computer power and the common use of Geographic Information Systems. Considering that the seafloor represents 71% of the surface of our planet, this is an important step towards understanding the Earth in its entirety. Bathymetric mapping systems, spanning a variety of sensors, have now developed to a point where the data they provide are able to capture seabed morphology at multiple scales, opening up the possibility of linking these data to oceanic, geological, and ecological processes. Applications of marine geomorphometry have now moved beyond the simple adoption of techniques developed for terrestrial studies. Whilst some former challenges have been largely resolved, we find new challenges constantly emerging from novel technology and applications. As increasing volumes of bathymetric data are acquired across the entire ocean floor at scales relevant to marine geosciences, resource assessment, and biodiversity evaluation, the scientific community needs to balance the influx of high-resolution data with robust quantitative processing and analysis techniques. This will allow marine geomorphometry to become more widely recognized as a sub-discipline of geomorphometry as well as to begin to tread its own path to meet the specific challenges that are associated with seabed mapping. This special issue brings together a collection of research articles that reflect the types of studies that are helping to chart the course for the future of marine geomorphometry.

Keywords: bathymetry; digital terrain analysis; geomorphometry; geomorphology; habitat mapping; marine remote sensing

1. Introduction

Geomorphometry (or digital terrain analysis, digital terrain modelling) is the science of quantitative surface analysis [1,2] that evolved from mathematics, Earth sciences, and computer science [3]. Geomorphometry developed its roots in geomorphology, but its branches reached out to a variety of end-user disciplines, such as the environmental sciences, space exploration, and civil engineering [4]. The widespread integration of geomorphometric tools into Geographic Information Systems (GIS) made geomorphometric analyses accessible to a wide range of end-users, many of whom are not necessarily aware of the science underpinning the tools [5]. Over the last decade, efforts have been made (e.g., [6–13]) to bridge the gap between the discipline of geomorphometry, which has traditionally focused on terrestrial and planetary applications, and the marine sciences.

Those efforts have resulted in the broader geomorphometry community becoming more aware of the challenges specific to the analysis of seafloor bathymetry data. Over the same period, the marine sciences community has become more aware of the field of geomorphometry, although some of its concepts, tools, and applications are still perhaps more widely recognized than others.

This special issue on Marine Geomorphometry is timely, as mutual recognition by the geomorphometry and marine sciences communities is higher than ever. The need to work together to solve emerging issues in quantitative seafloor analysis is also increasingly being acknowledged by both communities. This special issue explores existing and emerging trends where marine science applications and geomorphometry meet. As technology takes us deeper into the oceans and reveals its landscapes at increasingly higher resolution, there is much to learn about the seafloor. At the same time, it is important for the scientific community to show some restraint and critically assess the methods applied to analyze and explain these seafloor environments. Through this special issue, the community is demonstrating this restraint by questioning, assessing, and discussing the spatial geomorphometric techniques that they are applying to their seafloor data. The 17 papers in this issue address the five fundamental steps for implementing geomorphometric analysis, and these steps allow us to expose many of the important lessons learned to address various challenges. By continuing research along these lines, we will ensure that, as new characterization methods are developed, they are valid, repeatable, and robust to classification ontologies.

2. Five Steps to Implement Geomorphometric Analyses

The complete geomorphometry workflow involves five main steps [3]: sampling a surface, generating a digital terrain model (DTM) from the sampled surface, preprocessing the DTM (e.g., correcting for errors) for subsequent analyses, deriving terrain attributes and/or extracting terrain features from the DTM, and using and explaining those attributes and features in a given context. Early applications of marine geomorphometry often only focused on the analysis of the digital bathymetric model (DBM) and its application, disregarding the importance of the first three steps [7]. Over the last few years, however, end-user awareness about the impacts of the earlier steps of the geomorphometry workflow on applications has significantly increased [10]. The articles in this special issue highlight this trend, and Sections 2.1–2.5 summarize the contributions to this special issue according to each of those five steps.

2.1. Sampling the Depth of the Seafloor

While acoustic remote sensing technologies remain the main tools used to sample the depth and composition of the seafloor (Figure 1), the challenges that are associated with using them in very shallow waters have long made the coastal environment one of the most difficult in which to collect depth information. However, optical remote sensing technologies, such as bathymetric lidar and multispectral satellite imagery, are slowly gaining traction in coastal applications due to recent developments in hardware and processing methods. For instance, Walbridge et al. [14] used a 3 m resolution lidar dataset of the Buck Island Reef National Monument, in the U.S. Virgin Islands, to classify the seafloor into nine geomorphic classes. Linklater et al. [15] empirically derived depth estimates from 2-m resolution WorldView-2 and 2.4-m resolution Quickbird satellite images that were corrected for atmospheric effects and sun glint. Those depth estimates were then combined with existing acoustic data from deeper waters to develop a seamless, high-resolution DBM of the shelf around Lord Howe Island (Southwest Pacific Ocean), from which geomorphometric analyses were performed. The use of optical remote sensing in marine geomorphometry is likely to increase in the next few years as both empirical [16] and physical [17] approaches for bathymetric derivation become more widely available in user-friendly tools (e.g., Traganos et al., [18]) and new techniques (e.g., satellite-derived photogrammetric bathymetry, see [19]) are developed.

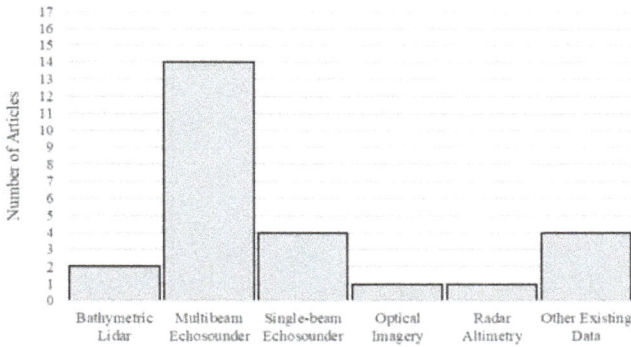

Figure 1. The techniques used in the special issue to sample seafloor depths. Some articles combined multiple techniques. The category 'other existing data' includes, for example, navigational charts.

The use of radar altimetry to estimate depth has declined over the years due to its inability to capture seafloor morphology at scales relevant to many applications. Recognizing the poor state of global single-resolution ocean depth maps and the critical role that such knowledge plays in understanding our planet, the International Hydrographic Organization-Intergovernmental Oceanographic Commission (IHO-IOC) General Bathymetric Chart of the Oceans (GEBCO; information available at www.gebco.net) framework and the Nippon Foundation have joined forces to establish the Nippon Foundation-GEBCO Seabed 2030 Project. This represents an international effort with the objective of facilitating the complete mapping of the world's oceans by 2030. The concept paper by Mayer et al. [20] outlines the ambitious Seabed 2030 initiative and the possibilities that these global data will bring to users worldwide and that will only be seen in the years to come. Methods to process these data will indeed require the support of the Global Data Assembly and Coordination Centre (GDACC), who will likely turn to the research community for efficient and robust spatial-data-processing methods to extract common and valuable variables of interest to the international marine community. This special issue provides a relevant summary of these methods (cf. Section 2.4) and the applications of these to multidisciplinary research (cf. Section 2.5).

In managing the oceans, it is widely recognized that one must first acknowledge their nature as a system, and in the large marine ecosystem paradigm, our perception of it is influenced by the scales at which they are examined, which are directly dependent on the sampling technique. Figure 2 shows that applications presented in this special issue looked at geosystems and ecosystems at a wide range of spatial scales. Finer-resolution data were most often produced by optical means (i.e., lidar and satellite imagery) and by acoustic means in shallower waters, while broader-resolution data were produced by acoustic methods in deeper waters and by using datasets with sparse coverage (cf. Section 2.2).

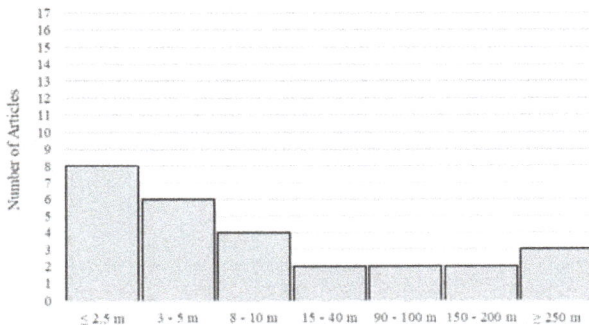

Figure 2. The range of resolutions presented or discussed in the articles.

2.2. Generating a Digital Bathymetric Model

Methods to generate DBMs have not significantly changed in the last few years. For example, many studies, including [21], still use the CUBE (Combined Uncertainty and Bathymetric Estimator) algorithm [22] to grid multibeam echosounder data. Given the costs that are associated with collecting bathymetric data over large areas of seafloor, there has also been a growing interest in using data fusion to combine existing data from multiple sources. Zimmermann and Prescott [23] accomplished the feat of combining 18 million data points from more than 200 individual sources to produce the best bathymetric model to date of the Eastern Bering Sea, enabling them to study 29 canyons in the area and to confirm the legendary status of some pinnacles. Bourguignon et al. [24] used single-beam echosounders data and chart data to produce a 200-m resolution DBM by interpolating more than 150,000 points.

In some cases, the costs of data collection are not the only impediments to the production of DBMs, but our inability to travel through time may be: Goswami et al. [25] presented an innovative approach to produce a DBM representing a reconstruction of paleobathymetry from 94 Ma, with implications for paleoclimate studies, among others. This new bathymetric model at $0.1° \times 0.1°$ resolution improves upon present global paleoclimate simulation model layers that are developed from bathtub-like, flat, featureless ocean bathymetry models, which are neither realistic nor suitable. This approach represents an important step forward for this type of application.

2.3. Preprocessing

Unlike in terrestrial applications of geomorphometry, for which DTMs need to be hydrologically corrected (e.g., by removing sinks), the preparation of DBMs for marine applications mainly consists in correcting errors and artefacts that could not be accounted for during the processing of raw data to generate the DBM or filling in data gaps to facilitate analyses and reduce potential edge effects. For instance, Porskamp et al. [26] used Delaunay triangulation to stitch multiple datasets and fill any holes in the final product.

While simple methods to correct for different types of artefacts in DBMs are still lacking, the awareness of artefacts and their potential impacts on applications is now regularly acknowledged and reported on (e.g., Ryabchuck et al. [27]), which used to be very uncommon [28,29]. In this special issue, Hughes Clarke [30] addresses the main factors that affect data quality in bathymetric data collected using multibeam echosounders. Multibeam acoustic swath systems are the common instrument of choice for a full-coverage bathymetric survey. Within each swath of data, the variables of distance, azimuth, and elevation angles will influence significantly the quality of the data. This variability will translate through to the DBM and subsequent users, if unfamiliar with the original acquisition geometry, may potentially misinterpret such variability as real attributes on the seabed, particularly if the artefacts are at the same scale as the morphologic features of interest [31]. Hughes Clarke [30] warns of the uncertainty that can arise with the ever-increasing ambition of higher-resolution data and cautions that relief close to either the resolution limit or the scale of artefacts increases the risk of over-interpretation by morphological studies.

2.4. Analysing the Digital Bathymetric Model

There has not been much change in terms of general geomorphometry (which focuses on the derivation of terrain attributes) since the reviews by Lecours et al. [7,10]. As identified in Figure 3, slope remains the most commonly used terrain attribute, followed by measures of curvature, rugosity, and topographic position. Tools to automatically compute and analyze those measures are, however, increasingly being developed and made available to the broader community (see examples in [7]). In this special issue, Walbridge et al. [14] offers a review of such tools and toolboxes and presents the most recent developments to their Benthic Terrain Modeler (BTM) toolbox.

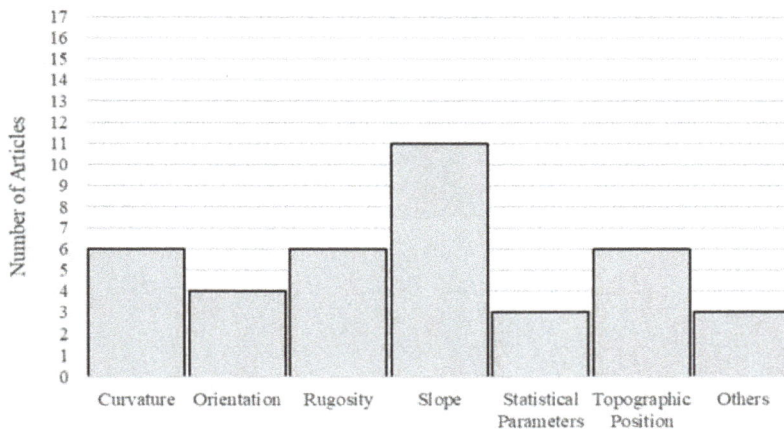

Figure 3. The categories of terrain attributes used in the articles of the special issue.

Specific geomorphometry, i.e., the branch of geomorphometry that deals with the extraction of terrain objects/features, is also well-represented in this special issue. Di Stefano and Mayer [21] developed a scale-based model for extracting and quantifying characteristics of submarine landforms (mainly sand dunes, ripples, mega ripples, and coral reefs) from high-resolution digital bathymetry. Their approach follows a two-part procedure wherein the first part the model extracts terrain features based on differential geometry principles and the second part evaluates the models for their relationships to scale-dependency, simulating their sensitivity to variation in the input parameters. Diesing and Thorsnes [32] present a methodology that combines image segmentation and random forest spatial prediction with the aim to derive maps of cold-water coral carbonate mounds with associated, spatially explicit measures of confidence. This approach is successful in mapping the presence and absence of carbonate mounds with high accuracy and confidence and shows promise for more widespread application. The variables used to facilitate carbonate mound detection include curvature, roughness, length, width, and bathymetric position index, demonstrating how general geomorphometry underpins further applied analysis and modelling. Finally, Masetti et al. [33] adapted the geomorphons concept introduced by Jasiewicz and Stepinski [34] for terrestrial and planetary settings to make it more meaningful for the study of marine bedforms. The identified "bathymorphons", a term used by the authors, provide a robust and flexible way to segment acoustic seafloor data based on principles of topographic openness, pattern recognition, texture classification, object similarity, and multi-modality.

2.5. Applications

In line with the review by Lecours et al. [10], the two main applications of marine geomorphometry remain in the general fields of geomorphology and habitat mapping (Figure 4). While Figure 4 shows a stronger representation of geomorphology, we note that the publisher for this special issue might have influenced the relative proportions of submissions from each field. Gardner [35] studied the Mendocino Channel, a deep-water sinuous channel, and quantitatively described its morphology and structural maintenance. The author asserts that the formation, maintenance, and modification of the Mendocino Channel have occurred through a combination of significant and numerous earthquakes and wave loading resuspension by storms forming turbidity currents. Ryabchuk et al. [27] used both a multibeam echosounder and a sub-bottom profiler to identify and map submerged glacial and post-glacial geomorphological features, enabling them to interpret the sedimentation regimes of two post-glacial basins in the Gulf of Finland. The geomorphological analysis has led to the identification of Late Pleistocene sediment and more modern bottom relief, which together indicated the occurrence of a deep-water level fall in the Early Holocene and multiple water-level fluctuations during this period. Also, in this special issue, Gafeira et al. [36] introduced a semi-automated approach to spatially delineate pockmarks of different shapes and sizes in different geological settings. Their approach proved to be less subjective and faster than traditional methods, such as manual expert identification and delineation. Sánchez-Guillamón et al. [37] used morphometry and size to classify deep seafloor mounds, such as domes and volcanoes, in the Canary Basin and proposed a growth model of those mounds informed by their geomorphometric characteristics.

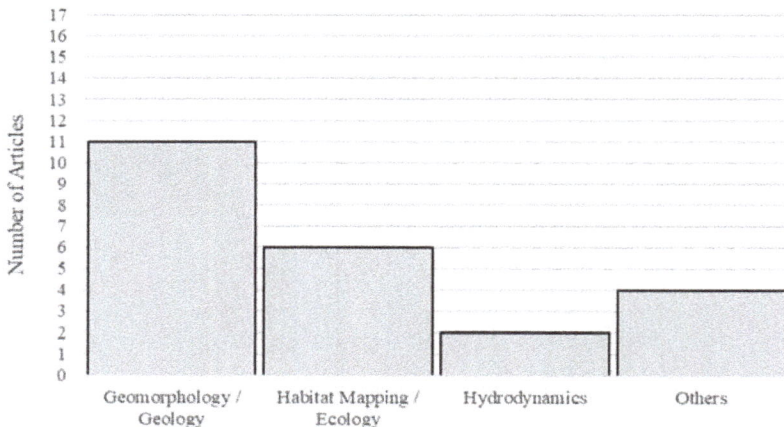

Figure 4. The categories of applications that were presented in the special issue. Some articles had multiple applications; for instance, when the geomorphology was interpreted and then used to map habitats.

It is also noteworthy that many studies have both a geomorphological focus and a habitat-mapping focus that complement each other. For instance, Greene et al. [38] studied deep-water sand wave fields in the San Juan Archipelago of the Salish Sea, which form habitat for Pacific sand lances and sand-eels. Of note, their interpretation of the features and habitats also considers the complex hydrodynamics of the area. Linklater et al. [15] examined and compared reef morphology around the subtropical island shelves of Lord Howe Island and Balls Pyramid in the Southwest Pacific Ocean for the first time. Diverse accretionary and erosional geomorphic features were mapped, with highlights including fossil reef systems dominating the shelves in 25–50 m water depth. A geomorphological analysis was used to provide insight into the geological and ecological processes that have influenced the formation of these shelves around the two islands. Bourguignon et al. [24] examined the use of seabed geomorphology and sedimentology to study the influence of sedimentary regimes on physical marine habitat distribution. They then used that information to define potential fishing grounds and predict fishing activities. The results are to be used for marine spatial planning on the Eastern Brazilian Shelf. Picard et al. [39] supported this theme with a study of hydrodynamics patterns by documenting the use of semi-automated methods to map and quantify the form and density of pockmark fields in one of the regions with the highest concentration of those features in the world: the Northwest Australian continental shelf. Whilst regional bi-directionality of pockmark scours corresponded to the modelled tidal flow, localized scattering around banks suggested turbulence regimes. The geomorphological analysis of these data proposed that pockmark scours can act as a proxy for bottom currents, which could help to inform modelling of benthic biodiversity patterns.

3. Discussion

Geomorphometric analysis continues to evolve across each of the five themes mentioned in this paper. There are several questions in seafloor quantitative characterization research that will occupy our attention for decades to come and for which this special issue may progress discussion. There is a complex interplay where new developments in this field will ebb between improved data collection and data-processing technique to create higher-resolution and more accurate DBMs and workflows with improved big data processing algorithms to handle larger and more complex automated methodologies for feature extraction. As the paper by Mayer et al. [20] rightly points out, new approaches to seafloor mapping will particularly enhance efficiency and coverage. However, as Hughes-Clarke [30] identifies, analysts unfamiliar with acquisition geometry may potentially misinterpret variability in the data as geomorphometric features, and similarly, sparse depth soundings can lead to a false impression of flat seabed terrain. Bathymetric coverage of the seabed at various resolutions builds up the quest for robust methods for the production and analysis of multiscale DBMs, which perhaps will become the next major demand for marine geomorphometry. Methods for multiscale grid structure applied to bathymetric data have been recently explored by Maleika et al. [40], whilst options for the generation of multiresolution surfaces are now available in some multibeam processing software [41]. Further down the line, data end-users now have the option to merge datasets at multiple resolutions and use these directly in an analysis through data management solutions, such as the ESRI® Mosaic Dataset. It is essential for the future integrity of marine geomorphometry that these various types of multiresolution DBMs are produced and analyzed with due regard for the additional complexities of multiresolution surfaces, supported by adequate documentation to make the methods transparent and verifiable. It seems likely that terrain analysis methods focused on an analysis of distance rather than pixels will become more applicable in providing suitable outputs from multiresolution surfaces.

Seafloor mapping is inherently a multidisciplinary task—a mix of hydrography, computer science, engineering, physics, and mathematics—that also delivers valuable data to many more disciplines, such as marine geology, oceanography, biology, habitat and species prediction modelling, remote sensing, and hydrographic surveying. New applications for marine geomorphometry will continue to be discovered as high-resolution data and marine geomorphometry becomes valued by even more applications, such as seafloor energy harvesting, marine archeology, and deep-sea resource assessment. Any new application areas will bring with them new challenges to marine geomorphometric analysis, which can best be met through a strong partnership between those advancing marine remote sensing and those developing geospatial techniques. We hope that this special issue identifies a breadth of perspectives and integrates ideas that will help to further establish the discipline of marine geomorphometry and provide the conduit to solve these future challenges.

Author Contributions: V.L. (Vanessa Lucieer), V.L. (Vincent Lecours), and M.F.J.D. conceived and designed the special review. V.L. (Vanessa Lucieer), V.L. (Vincent Lecours), and M.F.J.D. equally contributed to the writing of this paper.

Funding: This work was supported by the National Environmental Science Program (NESP) Marine Biodiversity Hub at the University of Tasmania (Lucieer), the University of Florida (Lecours) and the Geological Survey of Norway (Dolan).

Acknowledgments: We wish to thank all the authors and co-authors who published in this special issue, and the reviewers that have contributed to the success of this collection of high-quality and broad impact research.

Conflicts of Interest: The authors declare no conflict of interest.

References

1. Pike, R.J. Geomorphometry: Progress, practice, and prospect. *Z. Geomorphol.* **1995**, *101*, 221–238.
2. Rasemann, S.J.; Schmidt, J.; Schrott, L.; Dikau, R. Geomorphometry in Mountain Terrain. In *Geographic Information Science and Mountain Geomorphology*; Springer Science & Business Media: Berlin, Germany, 2004; pp. 101–146.

3. Pike, R.J.; Evans, I.S.; Hengl, T. *Geomorphometry: A Brief Guide, Geomorphometry—Concepts, Software, Applications*; Hengl, T., Reuter, H.I., Eds.; Elsevier: Amsterdam, The Netherlands, 2009; Volume 33, pp. 3–30.

4. Bishop, M.P.; James, L.A.; Shroder, J.F., Jr.; Walsh, S.J. Geospatial technologies and digital geomorphological mapping: Concepts, issues and research. *Geomorphology* **2012**, *137*, 5–26. [CrossRef]

5. Bishop, M.P.; Shroder, J.F., Jr. GIScience and mountain geomorphology: Overview, feedbacks, and research directions. In *Geographic Information Science and Mountain Geomorphology*; Springer Science & Business Media: Berlin, Germany, 2004.

6. Bouchet, P.J.; Meeuwig, J.J.; Salgado Kent, C.P.; Letessier, T.B.; Jenner, C.K. Topographic determinants of mobile vertebrate predator hotspots: current knowledge and future directions. *Biol. Rev.* **2015**, *90*, 699–728. [CrossRef] [PubMed]

7. Lecours, V.; Dolan, M.F.J.; Micallef, A.; Lucieer, V.L. A review of marine geomorphometry, the quantitative study of the seafloor. *Hydrol. Earth Syst. Sci.* **2016**, *20*, 3207–3244. [CrossRef]

8. Lecours, V.; Lucieer, V.; Dolan, M.F.J.; Micallef, A. An ocean of possibilities: Applications and challenges of marine geomorphometry. In *Geomorphometry for Geosciences*; Jasiewicz, J., Zwoliński, Z., Mitasova, H., Hengl, T., Eds.; Adam Mickiewicz University in Poznań—Institute of Geoecology and Geoinformation: Poznan, Poland; pp. 23–26.

9. Lecours, V.; Dolan, M.; Micallef, A.; Lucieer, V. Geomorphometry in marine habitat mapping: Lessons learned from the past 10 years of applications. In Proceedings of the 15th International Symposium GeoHab, Winchester, UK, 2–6 May 2016; 2016.

10. Lecours, V.; Lucieer, L.; Dolan, M.F.J.; Micallef, A. Recent and future trends in marine geomorphometry. In Proceedings of the Geomorphometry 2018, Boulder, CO, USA, 13–17 August 2018; pp. 1–4.

11. Lundblad, E.R.; Wright, D.J.; Miller, J.; Larkin, E.M.; Rinehart, R.; Naar, D.F.; Donahue, B.T.; Anderson, S.M.; Battista, T. A Benthic Terrain Classification Scheme for American Samoa. *Mar. Geodesy* **2006**, *29*, 98–111. [CrossRef]

12. Micallef, A.; Lecours, V.; Dolan, M.F.J.; Lucieer, V.L. Marine geomorphometry: Overview and opportunities. In Proceedings of the EGU General Assembly 2016, Vienna, Austria, 17–22 April 2016.

13. Wilson, M.; O'Connell, B.; Brown, C.; Guinan, J.C.; Grehan, A.J. Multiscale terrain analysis of multibeam bathymetry data for habitat mapping on the continental slope. *Mar. Geodesy* **2007**, *30*, 3–35. [CrossRef]

14. Walbridge, S.; Slocum, N.; Pobuda, M.; Wright, D. Unified Geomorphological Analysis Workflows with Benthic Terrain Modeler. *Geosciences* **2018**, *8*, 94. [CrossRef]

15. Linklater, M.; Hamylton, S.; Brooke, B.; Nichol, S.; Jordan, A.; Woodroffe, C. Development of a Seamless, High-Resolution Bathymetric Model to Compare Reef Morphology around the Subtropical Island Shelves of Lord Howe Island and Balls Pyramid, Southwest Pacific Ocean. *Geosciences* **2018**, *8*, 11. [CrossRef]

16. Stumpf, R.P. Retrospective and future studies of coastal water clarity and sediment loads. In Proceedings of the Third Thematic Conference on Remote Sensing for Marine and Coastal Environments, Seattle, WA, USA, 18–20 September 1995; pp. 376–377.

17. Lyzenga, D.R.; Malinas, N.P.; Tanis, F.J. Multispectral bathymetry using a simple physically based algorithm. *IEEE Trans. Geosci. Remote Sens.* **2006**, *44*, 2251–2259. [CrossRef]

18. Traganos, D.; Reinartz, P. Mapping Mediterranean seagrasses with Sentinel-2 imagery. *Mar. Pollut. Bull.* **2018**, *134*, 197–209. [CrossRef]

19. Hodúl, M.; Bird, S.; Knudby, A.; Chénier, R. Satellite derived photogrammetric bathymetry. *ISPRS J. Photogramm. Remote Sens.* **2018**, *142*, 268–277. [CrossRef]

20. Mayer, L.; Jakobsson, M.; Allen, G.; Dorschel, B.; Falconer, R.; Ferrini, V.; Lamarche, G.; Snaith, H.; Weatherall, P. The Nippon Foundation—GEBCO Seabed 2030 Project: The Quest to See the World's Oceans Completely Mapped by 2030. *Geosciences* **2018**, *8*, 63. [CrossRef]

21. Di Stefano, M.; Mayer, L. An Automatic Procedure for the Quantitative Characterization of Submarine Bedforms. *Geosciences* **2018**, *8*, 28. [CrossRef]

22. Calder, B.R. Automatic statistical processing of multibeam echosounder data. *Int. Hydrogr. Rev.* **2003**, *4*, 53–68.

23. Zimmermann, M.; Prescott, M. Bathymetry and Canyons of the Eastern Bering Sea Slope. *Geosciences* **2018**, *8*, 184. [CrossRef]

24. Bourguignon, S.; Bastos, A.; Quaresma, V.; Vieira, F.; Pinheiro, H.; Amado-Filho, G.; de Moura, R.; Teixeira, J. Seabed Morphology and Sedimentary Regimes defining Fishing Grounds along the Eastern Brazilian Shelf. *Geosciences* **2018**, *8*, 91. [CrossRef]

25. Goswami, A.; Hinnov, L.; Gnanadesikan, A.; Young, T. Realistic Paleobathymetry of the Cenomanian–Turonian (94 Ma) Boundary Global Ocean. *Geosciences* **2018**, *8*, 21. [CrossRef]

26. Porskamp, P.; Rattray, A.; Young, M.; Ierodiaconou, D. Multiscale and Hierarchical Classification for Benthic Habitat Mapping. *Geosciences* **2018**, *8*, 119. [CrossRef]

27. Ryabchuk, D.; Sergeev, A.; Krek, A.; Kapustina, M.; Tkacheva, E.; Zhamoida, V.; Budanov, L.; Moskovtsev, A.; Danchenkov, A. Geomorphology and Late Pleistocene–Holocene Sedimentary Processes of the Eastern Gulf of Finland. *Geosciences* **2018**, *8*, 102. [CrossRef]

28. Lecours, V.; Devillers, R.; Edinger, E.N.; Brown, C.J.; Lucieer, V.L. Influence of artefacts in marine digital terrain models on habitat maps and species distribution models: A multiscale assessment. *Remote Sens. Ecol. Conserv.* **2017**, *3*, 232–246. [CrossRef]

29. Lecours, V.; Devillers, R.; Lucieer, V.L.; Brown, C.J. Artefacts in marine digital terrain models: A multiscale analysis of their impact on the derivation of terrain attributes. *IEEE Trans. Geosci. Remote Sens.* **2017**, *55*, 5391–5406. [CrossRef]

30. Hughes Clarke, J. The Impact of Acoustic Imaging Geometry on the Fidelity of Seabed Bathymetric Models. *Geosciences* **2018**, *8*, 109. [CrossRef]

31. Spina, R. The pockmark stars: Radial structures in the seabed surrounding the Hawaii Islands. *J. Environ. Geol.* **2017**, *1*, 33–50. [CrossRef]

32. Diesing, M.; Thorsnes, T. Mapping of Cold-Water Coral Carbonate Mounds Based on Geomorphometric Features: An Object-Based Approach. *Geosciences* **2018**, *8*, 34. [CrossRef]

33. Masetti, G.; Mayer, L.; Ward, L. A Bathymetry- and Reflectivity-Based Approach for Seafloor Segmentation. *Geosciences* **2018**, *8*, 14. [CrossRef]

34. Jasiewicz, J.; Stepinski, T.F. Geomorphons-a pattern recognition approach to classification and mapping of landforms. *Geomorphology* **2012**, *182*, 147–156. [CrossRef]

35. Gardner, J. The Morphometry of the Deep-Water Sinuous Mendocino Channel and the Immediate Environs, Northeastern Pacific Ocean. *Geosciences* **2017**, *7*, 124. [CrossRef]

36. Gafeira, J.; Dolan, M.; Monteys, X. Geomorphometric Characterization of Pockmarks by Using a GIS-Based Semi-Automated Toolbox. *Geosciences* **2018**, *8*, 154. [CrossRef]

37. Sánchez-Guillamón, O.; Fernández-Salas, L.; Vázquez, J.-T.; Palomino, D.; Medialdea, T.; López-González, N.; Somoza, L.; León, R. Shape and Size Complexity of Deep Seafloor Mounds on the Canary Basin (West to Canary Islands, Eastern Atlantic): A DEM-Based Geomorphometric Analysis of Domes and Volcanoes. *Geosciences* **2018**, *8*, 37. [CrossRef]

38. Greene, H.; Cacchione, D.; Hampton, M. Characteristics and Dynamics of a Large Sub-Tidal Sand Wave Field—Habitat for Pacific Sand Lance (Ammodytes personatus), Salish Sea, Washington, USA. *Geosciences* **2017**, *7*, 107. [CrossRef]

39. Picard, K.; Radke, L.; Williams, D.; Nicholas, W.; Siwabessy, P.; Howard, F.; Gafeira, J.; Przeslawski, R.; Huang, Z.; Nichol, S. Origin of High Density Seabed Pockmark Fields and Their Use in Inferring Bottom Currents. *Geosciences* **2018**, *8*, 195. [CrossRef]

40. Maleika, W.; Koziarski, M.; Forczmański, P. A Multiresolution Grid Structure Applied to Seafloor Shape Modeling. *ISPRS Int. J. Geo-Inf.* **2018**, *7*, 119. [CrossRef]

41. Holland, M.; Hoggarth, A. Hydrographic processing considerations in the big data age: A focus on techonolgy trends in ocean and coastal surveys. *IOP Conf. Ser.: Earth Environ. Sci.* **2016**, *34*, 012016. [CrossRef]

geosciences

MDPI

Concept Paper

The Nippon Foundation—GEBCO Seabed 2030 Project: The Quest to See the World's Oceans Completely Mapped by 2030

Larry Mayer [1,*] , Martin Jakobsson [2], Graham Allen [3], Boris Dorschel [4] , Robin Falconer [5], Vicki Ferrini [6] , Geoffroy Lamarche [7,8] , Helen Snaith [3] and Pauline Weatherall [3]

1   Center for Coastal and Ocean Mapping, University of New Hampshire, Durham, NH 03824, USA
2   Department of Geological Sciences, Stockholm University, 106 91 Stockholm, Sweden;
    martin.jakobsson@geo.su.se
3   British Oceanographic Data Centre, National Oceanography Centre, Liverpool L3 5DA, UK;
    graham.allen@noc.ac.uk (G.A.); h.snaith@bodc.ac.uk (H.S.); paw@bodc.ac.uk (P.W.)
4   Alfred Wegener Institute, 27570 Bremerhaven, Germany; boris.dorschel@awi.de
5   GEBCO Guiding Committee, 112 Rimu Road, Paraparaumu 5032, New Zealand;
    robinfalconerassociates@gmail.com
6   Lamont-Doherty Earth Observatory, Palisades, NY 10964, USA; ferrini@ldeo.columbia.edu
7   National Institute of Water and Atmospheric Research Ltd. (NIWA), Private Bag 14-901,
    Wellington 6241, New Zealand; Geoffroy.Lamarche@niwa.co.nz
8   School of Environment, University of Auckland, Auckland 1149, New Zealand
*   Correspondence: larry@ccom.unh.edu; Tel.: +1-603-862-2615

Received: 23 January 2018; Accepted: 5 February 2018; Published: 8 February 2018

Abstract: Despite many of years of mapping effort, only a small fraction of the world ocean's seafloor has been sampled for depth, greatly limiting our ability to explore and understand critical ocean and seafloor processes. Recognizing this poor state of our knowledge of ocean depths and the critical role such knowledge plays in understanding and maintaining our planet, GEBCO and the Nippon Foundation have joined forces to establish the Nippon Foundation GEBCO Seabed 2030 Project, an international effort with the objective of facilitating the complete mapping of the world ocean by 2030. The Seabed 2030 Project will establish globally distributed regional data assembly and coordination centers (RDACCs) that will identify existing data from their assigned regions that are not currently in publicly available databases and seek to make these data available. They will develop protocols for data collection (including resolution goals) and common software and other tools to assemble and attribute appropriate metadata as they assimilate regional grids using standardized techniques. A Global Data Assembly and Coordination Center (GDACC) will integrate the regional grids into a global grid and distribute to users world-wide. The GDACC will also act as the central focal point for the coordination of common data standards and processing tools as well as the outreach coordinator for Seabed 2030 efforts. The GDACC and RDACCs will collaborate with existing data centers and bathymetric compilation efforts. Finally, the Nippon Foundation GEBCO Seabed 2030 Project will encourage and help coordinate and track new survey efforts and facilitate the development of new and innovative technologies that can increase the efficiency of seafloor mapping and thus make the ambitious goals of Seabed 2030 more likely to be achieved.

Keywords: global bathymetry; Seabed 2030; Nippon Foundation/GEBCO; seafloor mapping technologies; seafloor mapping standards and protocols

1. Introduction

The oceans cover 71% of the Earth's surface [1], are fundamental to sustaining life, controlling climate, facilitating commerce and they represent a vast source of resources and economic wealth. Our understanding of the ocean and natural processes occurring at the seafloor is quite limited due to the difficulties in operating in this environment, especially the fact that electromagnetic waves, (e.g., light and radar), are highly attenuated in ocean water. The suite of optical and electromagnetic sensors that are used to map, observe, and understand land topography cannot penetrate more than tens of meters at best in ocean waters. This has left the vast majority of our planet virtually unmapped, unobserved, and unexplored. Satellite measurements of ocean surface height have been used to provide a general view of the shape of the deep ocean floor through altimetry-derived predicted seafloor depths [2]. However, the phenomena referred to as the "upward continuation" of gravity signals from bathymetric features at depth, limits the horizontal resolution of undersea features to those greater than approximately double the regional ocean depth [3]. Considering that the recent estimate of the mean world ocean depth is ~3.9 km [1], the average achievable resolution is on the order of 8 km. Thus, while satellite altimetry-derived bathymetry is excellent for a regional tectonic studies (e.g., [4]), it does not provide enough spatial resolution or accuracy to perform the detailed geomorphometric analyses [5] required to understand the origin and significance of, for example, bottom current features [6], submarine glacial landforms [7], benthic habitats [8–10], geohazards such as shallow faults [11,12], pockmarks, or mass transport complexes [13].

Knowing seabed depths, (bathymetry) is of vital importance for a growing variety of uses fundamental to understanding the workings of our planet. Early echo-sounding profiles across the Atlantic Ocean enabled Bruce Heezen and Marie Tharp to understand the relationship between mid-ocean ridges and earthquake seismicity and played an important part in the formulation of one of the most significant paradigm shifts in science—the development of the hypothesis of seafloor spreading and plate tectonics [14]. Bathymetric data are also critical for establishing the limits of the extended continental shelf under the United Nations Convention on Law of the Sea [15], play a key role in a host of military and defense applications [16], and represent a fundamental dataset for addressing the growing challenges associated with climate change [17,18].

Recognizing the need for global bathymetric data, The General Bathymetric Chart of the Oceans (GEBCO), operating under the auspices of the International Hydrographic Organization (IHO) and the Intergovernmental Oceanographic Commission (IOC) of the United Nations Educational, Scientific and Cultural Organization (UNESCO), was initiated more than 100 years ago with the vision of providing authoritative, publicly-available bathymetry data for the world's oceans [19]. Since its initiation in 1903 through the efforts of Prince Albert I of Monaco and Professor Julien Thoulet of The University of Nancy, GEBCO, and its mostly volunteer community, has produced a series of bathymetric charts that represent the best knowledge of seafloor depths at the time of their production. While the earliest products were contour plots based on very sparse lead-line measurements [19], the most recent product of GEBCO (GEBCO_2014) is now a digital, gridded product with a cell-spacing of 30 arc sec that incorporates data derived from both single-beam and modern high-resolution multibeam echo sounders, superimposed on a base that is derived from satellite altimetry data [1] (Figure 1).

Despite the appearance of full global coverage of ocean depths, even the most recent GEBCO or other global compilations of ocean bathymetry (e.g., NOAA's National Centers for Environmental Information (NCEI, which serves as the IHO Data Center for Digital Bathymetry), or the Global Multi-Resolution Topography (GMRT)) are deceptive, the product of the power of modern interpolation and visualization techniques that produce what appears to be a complete compilation of ocean depth from fundamentally sparse data. Upon close inspection, when divided into its resolution of 30 arc-second grid cells (926 m at the equator), approximately 82 percent of the grid cells in the GEBCO_2014 grid have no depth measurements in them [1] (Figure 1). While the details of current global bathymetric coverage will be discussed below, there is no question that even from even this cursory view, most of the earth's surface that lies beneath the seas is vastly under-sampled, limiting our ability to perform the detailed quantitative analyses of the seafloor morphology, i.e., geomorphometry, needed to understand ocean and seafloor processes.

Recognizing the poor state of our knowledge of ocean depths and the critical role that such knowledge plays in understanding and maintaining our planet, the Nippon Foundation and GEBCO launched the Seabed 2030 Project at the "Forum for the Future of Ocean Floor Mapping" in Monaco in June of 2016. With planned funding from the Nippon Foundation of approximately $18.5M US over ten years, The Nippon Foundation—GEBCO Seabed 2030 Project aims at establishing an infrastructure to facilitate the complete mapping of the world ocean floor by 2030, with the view of empowering the world to make policy decisions, use the ocean sustainably, and undertake scientific research based on detailed bathymetric information of the Earth's seabed. To implement this ambitious goal, the Seabed 2030 Project Team was established and met for the first time in late October 2017 in Southampton U.K. In creating this strategy, Seabed 2030 builds on one hundred years of GEBCO's legacy of ocean mapping as well as established regional connections in all corners of the world and a global community of alumni built over 12 years through the Nippon Foundation-GEBCO training program in ocean bathymetry at the University of New Hampshire (http://www.gebco.net/training/training_programme/).

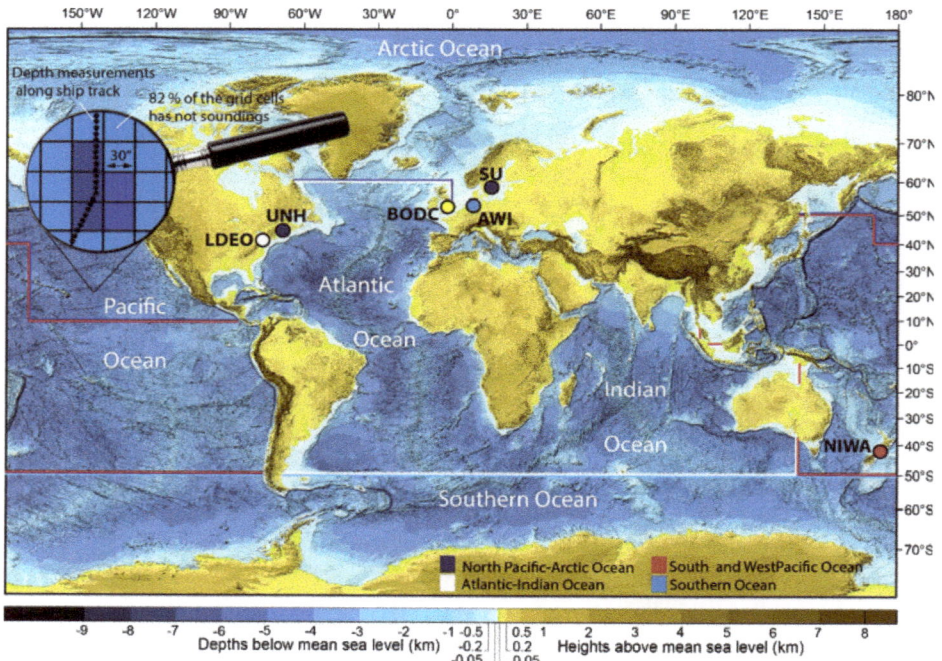

Figure 1. A shaded relief of the GEBCO_2014 grid modified from [1]. The oceans have been divided into four geographic regions each managed by a Seabed 2030 Regional Data Assembly and Coordination Center: AWI—Alfred Wegner Institute, Germany—responsible for Southern Ocean; LDEO—Lamont Doherty Earth Observatory, USA—responsible for Atlantic and Indian Oceans; NIWA—National Institute of Water and Atmospheric Research , New Zealand—responsible for South Pacific Ocean; Stockholm University, Sweden and the University of New Hampshire, USA—responsible for Arctic and North Pacific Oceans. BODC—British Oceanographic Data Center will serve as the Global Data Assembly and Coordination Center and be responsible for distributing the final Seabed 2030 products.

This paper, based upon "The Nippon-Foundation GEBCO—Seabed 2030 Roadmap for Future Ocean Floor Mapping" [20] (which served as the proposal for the project to the Nippon Foundation), presents the rationale behind the Seabed 2030 Project, describes its structure, outlines its initial implementation strategy, and discusses the benefit of developing such a product. Of particular relevance with respect to geomorphometry, we present these issues in the context of geomorphometric analyses and address the critical question "at what resolution should Seabed 2030 aim to map the world's ocean's floors?"

2. Why Seabed 2030?

2.1. How Much of the Seafloor Is Actually Mapped?

Answering the question of "how much of the seafloor is actually mapped" is far from simple. Aside from the issues of knowing how much data exists from hundreds of potential sources around the world (an issue that will be addressed later), there is the fundamental question of what does "mapped" mean. When a lead-line makes a measurement on the seafloor, the actual spot sampled is, in reality just a few tens of centimeters wide, as represented by the diameter of the weight (Figure 2). When a single-beam echo-sounder makes a depth measurement, the shallowest point in the area ensonified (with a diameter on the order of one half the water depth for a sounder with a beam width of 30°; Table 1) is reported, but with no angular information and thus the location of the sounding is reported as the downward projection of the position of the vessel on the seafloor no matter where the shallow feature really is within the ensonified area (Figure 2). For example, in 4000 m of water depth, a single-beam sounder with a 30° beam width will ensonify a diameter of approximately 2100 m on the seafloor. The shoalest feature within that circle will be the reported depth at a position that is at the center of the ensonified diameter (Figure 1). We can thus be confident that the depth of shoalest feature within the ensonified area has been measured, but we do not know where in that area the feature is and there is no other information about what else is in the 2100 m diameter ensonified area. Should we consider the entire ensonified area mapped for each single-beam sounding?

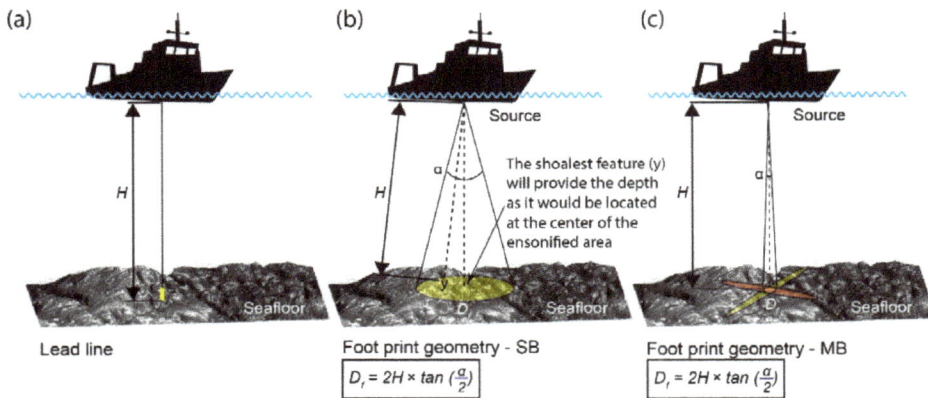

(a) **(b)** **(c)**

Source

The shoalest feature (y) will provide the depth as it would be located at the center of the ensonified area

Source

H H H

Seafloor   Seafloor   Seafloor

Lead line   Foot print geometry - SB   Foot print geometry - MB

$$D_f = 2H \times \tan\left(\frac{\alpha}{2}\right)$$ $$D_f = 2H \times \tan\left(\frac{\alpha}{2}\right)$$

Figure 2. The mapped area of the seafloor (ensonified area) using different methods. (**a**) Lead line. (**b**) Single-beam echo-sounder (SB). (**c**) Multibeam echo-sounder (MB).

The question of what is mapped or not is somewhat simpler for modern multibeam sonars that use beam forming and orthogonal transmitter and receiver arrays (Mills Cross) to form many narrow beams across a wide (athwartship) swath. The fundamental lateral resolution on the seafloor is controlled by the length of the arrays and the intersection (cross-product) of the transmit and receive beams (Figure 2). Most modern deep-water multibeam sonars form beams that are 1° × 1°, 1° × 2° or 2° × 2° producing ensonified areas on the seafloor (and thus fundamental feature resolutions) that vary with depth (Table 1).

GEBCO's latest product (GEBCO_2014 [1]) is a mixture of both single-beam, multibeam, and satellite altimetry-derived bathymetry assembled into a model gridded at a regular interval of 30 arc-seconds (Figure 3). The Arctic Ocean in GEBCO_2014 is comprised of a separate grid provided by the International Bathymetric Chart of the Arctic Ocean (IBCAO; [21]) and the Southern Ocean consists of a similar grid created by the International Bathymetric Chart of the Southern Ocean (IBCSO; [22]), two Regional Mapping Projects working within GEBCO. GEBCO_2014 also includes the GMRT compilation of multibeam bathymetry from research expeditions throughout the world (http://gmrt.marine-geo.org; [23]), which covers ~8% of the world ocean and is the largest source of multibeam-derived soundings contributing to the GEBCO_2014 grid. In addition, some regions are covered by other regional projects such as EMODnet, covering European waters (www.emodnet-bathymetry.eu), and the Baltic Sea Bathymetry Database (http://data.bshc.pro; [24]). The GEBCO_2014 grid was based on all publically available bathymetric data at the time of compilation (Figure 3). Track-line plots such as Figure 3 are particularly deceptive in terms of depicting coverage. Given the scale of these figures, each of the track-lines shown covers a width of 100s of kilometers while in reality, a multibeam swath is typically on the order of four times the water depth (~16 km in 4000 m water depth) and a single beam would ensonify a diameter of approximately 2 km in 4000 m water depth. Lines this thin would be impossible to display on global (or even regional-scale) maps and thus one gets the impression of much greater coverage than actually exists.

Table 1. Calculated foot prints of ensonified seafloor area at different water depths for single-beam (SB) and multibeam (MB) echo-sounders with various beam widths.

	SB Beam Width Foot Print Diameter in Meters	MB Beam Width (Transmit × Receive) Footprint Diameter in Meters. (For the 1° × 2° Configuration, the Semi-Minor and Semi-Major Axes of the Footprint Ellipse Is Given)		
Depth	30°	1° × 1°	1° × 2°	2° × 2°
0	0	0		0
500	268	9	9 × 19	17
1000	536	17	17 × 35	35
2000	1072	35	35 × 70	70
3000	1608	52	52 × 105	105
4000	2144	70	70 × 140	140
5000	2679	87	87 × 175	175
6000	3215	105	105 × 209	209
7000	3751	122	122 × 244	244
8000	4287	140	140 × 279	279
9000	4823	157	157 × 314	314
10000	5359	175	175 × 349	349
11000	5895	192	192 × 384	384

As previously mentioned, the vast majority of the world ocean has not been sampled by echo sounders even at a resolution of about 30 arc seconds. Considering that many of the approximately 1 km grid cells only have a single sounding in them, the percentage of the seafloor that has actually been measured by echo-sounders is considerably less than 18%.

Between track-lines, large areas of the GEBCO_2014 grid are based on interpolation guided by satellite altimetry-derived bathymetry data except in most of the two polar regions where sea ice precludes the use of satellite altimetry-derived data. GEBCO_2014 started with the base grid from the previous GEBCO_08 version, which included the altimetry-derived bathymetry model SRTM30_PLUS [25]. Satellite altimetry-derived bathymetry has represented a remarkable advancement in our mapping of oceans, providing global coverage of general estimates of depths and coarsely filling gaps between sparse ship soundings [2]. Satellite altimetry-derived bathymetry is, however, far less precise than echo sounder-derived data and has far less resolution than modern multibeam sonars, but the method is objective and, in most areas without sea ice, superior to interpolation between sparse ship tracks by mathematical algorithms or hand-contouring. Altimetry-derived data are particularly good for mapping tectonic-scale features such as spreading ridge segments and the fracture zones that offset them, but much finer-scale features and accurate depths are often difficult to derive (Figure 4).

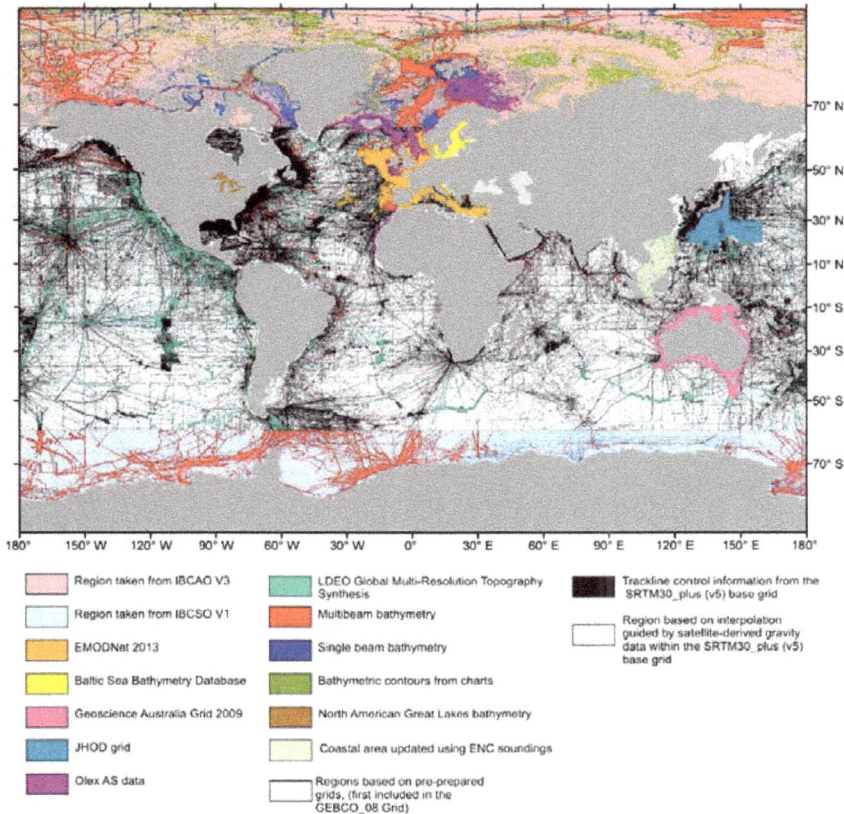

Region taken from IBCAO V3	LDEO Global Multi-Resolution Topography Synthesis	Trackline control information from the SRTM30_plus (v5) base grid
Region taken from IBCSO V1	Multibeam bathymetry	Region based on interpolation guided by satellite-derived gravity data within the SRTM30_plus (v5) base grid
EMODNet 2013	Single beam bathymetry	
Baltic Sea Bathymetry Database	Bathymetric contours from charts	
Geoscience Australia Grid 2009	North American Great Lakes bathymetry	
JHOD grid	Coastal area updated using ENC soundings	
Olex AS data	Regions based on pre-prepared grids, (first included in the GEBCO_08 Grid)	

Figure 3. GEBCO_2014 bathymetric data coverage. At this scale, the World Ocean appears much better covered with ship soundings than it is. The fact is that the available bathymetric data used to compile GEBCO_2014 provided depth control points to only 18% of all the 30 arc-second (926 m at the equator) grid cells. Figure is from [1].

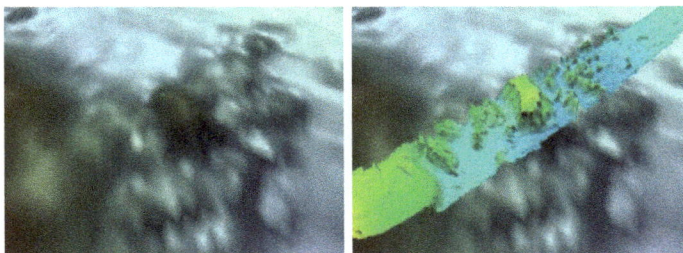

Figure 4. Satellite-derived bathymetry (including single-beam sonar data) from SRTM 30_Plus off Hawaii (**left**). Single swath of 12 kHz multibeam sonar over the same area—swath width = 20 km for scale (**right**).

 The achievable resolution of a bathymetric model (grid) is a function of the underlying data density—in the case of the GEBCO_2014 grid, the coverage of satellite tracks for depth estimates and ship soundings. Ship-borne multibeam surveys collect depth measurements at very high density and may be designed with overlapping swaths to provide full map coverage. When multibeam data are incorporated into a bathymetry model, their very high-lateral resolution (70 m × 140 m for a 1° × 2° multibeam system in 4000 m water depth) and high data density (up to 800 soundings across a 20 km wide swath in 4000 m water depth) improves the resolution of seafloor details (Figure 4). If one only displays the multibeam bathymetry data included in the GEBCO_2014 grid, the global coverage is greatly reduced as compared to the plot of all bathymetric data (Figures 3 and 5). Approximately 9% of the seafloor is covered by high-resolution multibeam sonar data, but for this 9%, the seafloor is truly covered. This implies that even if an analysis of the existing bathymetric coverage incorporated into the GEBCO-2014_grid reveals that 18% of the approximately 1 km² grid cells have at least one sounding in them, only about 9% are covered with high-resolution multibeam sonar data. This does not mean that all of the 1 km² grid cells in the 9% are filled with multibeam sonar data just that some multibeam data are present in the 1 km² grid cell. Thus, the actual global coverage of multibeam sonar data is even less than 9% of the seafloor. The harsh reality is that there is desperately little knowledge of the depths of the oceans and thus the motivation for the Nippon Foundation—GEBCO Seabed 2030 Project.

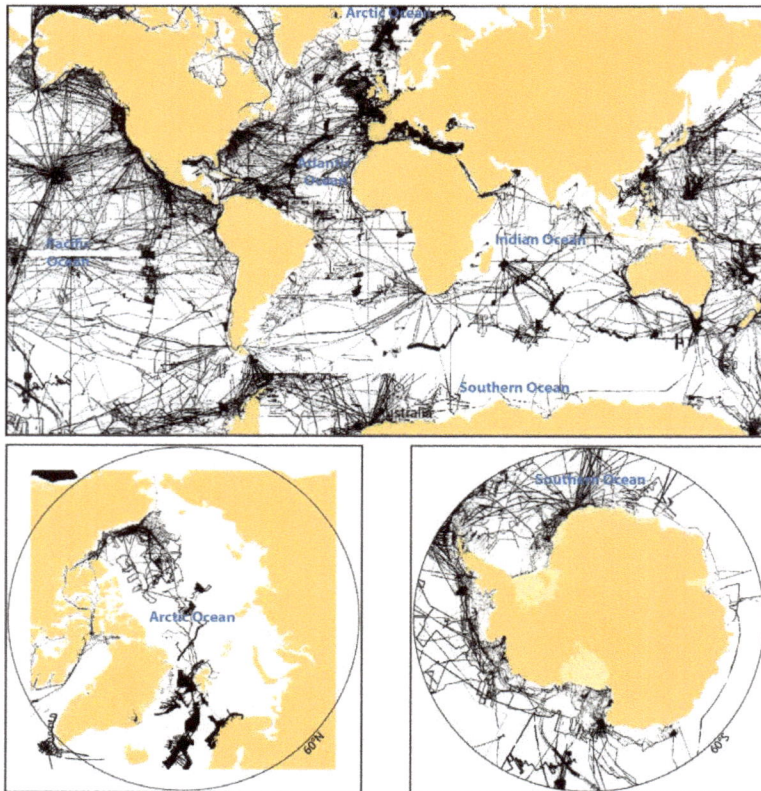

Figure 5. Tracklines of multibeam bathymetry included in the GEBCO-2014 grid. Map is from the Nippon Foundation—GEBCO—Roadmap for Future Ocean Floor Mapping [20].

2.2. At What Resolution Should We Map the Seafloor?

As we plan for the Nippon Foundation—GEBCO Seabed 2030 Project, we must address the question "at what resolution should we be mapping the seafloor?" Ideally, one would want a resolution commensurate with that at which our land surfaces are mapped (sub-meter resolution). While this would be ideal, the limited propagation of electromagnetic waves in the ocean must be considered, and thus achieving sub-meter resolution with either acoustic or optical systems requires that the mapping system be within tens of meters of the seafloor. This is technologically achievable with surface vessels operating in shallow water environments. However, more than 90% of the ocean is deeper than 200 m water depth, and this would require the delivery of the measurement system close to the seafloor by either remotely operated vehicles (ROV) or autonomous underwater vehicles (AUV). Such systems are available and are capable of collecting very high-resolution acoustic or optical data near the bottom; however, at present, given their expense and slow speed of operation, both ROVs and AUVs are not realistic options for global mapping. This may change in the future (see new technology section below); however, if we truly aspire to global coverage, we will continue to rely largely on surface vessels, which will dictate achievable spatial resolution.

Any discussion of seafloor mapping resolution must be presented in the context of changing depths, and as the effort to map the world's oceans is global, the resolution specified for a given sonar system must vary as a function of depth (Table 1). With the hope that much of the new data collected in support of the Nippon Foundation—GEBCO Seabed 2030 Project will be multibeam sonar data, we look at the achievable beam footprints for a range of water depths along with the percentage of seafloor area in each depth bin. Since few multibeam systems are $1° \times 1°$ configuration, we have adopted a conservative approach and base the achievable mapping resolution on a $2° \times 2°$ system deep water system (Table 1). Furthermore, as the beams are steered away from nadir, their footprints expand, decreasing the lateral resolution on the seafloor across the swath; at 60° from nadir (approximately 1.7 times the water depth) on each side, the beam footprint is double that found at nadir (Figure 6).

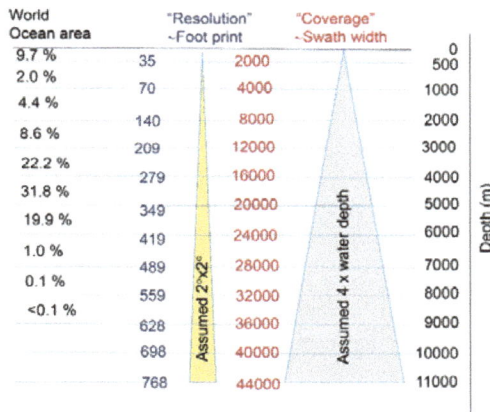

Figure 6. Footprint of outer (60° to one side) beam on seafloor (achievable lateral resolution) for $2° \times 2°$ multibeam sonar, as a function of water depth. Swath width coverage (assuming 4× water depth) is shown in red. Left column is the percentage of the global seafloor in each depth bin. Note that in deepest water depths (>10 km) swath widths can be limited by attenuation.

Considering that high-quality multibeam sonar data typically can be retrieved to a maximum swath width of about four times the water depth we will adopt a feasible resolution that is constrained by the worst-case solution of the near outermost beams (at about 60° to either side of nadir). Thus, for the Nippon Foundation GEBCO Seabed 2030 Project, we will establish a depth-variable resolution goal and consider the seafloor "mapped" if at least one sounding falls in a grid cell of the size described in Table 2. The depth regions corresponding to the calculated feasible grid resolutions are shown in Figure 7.

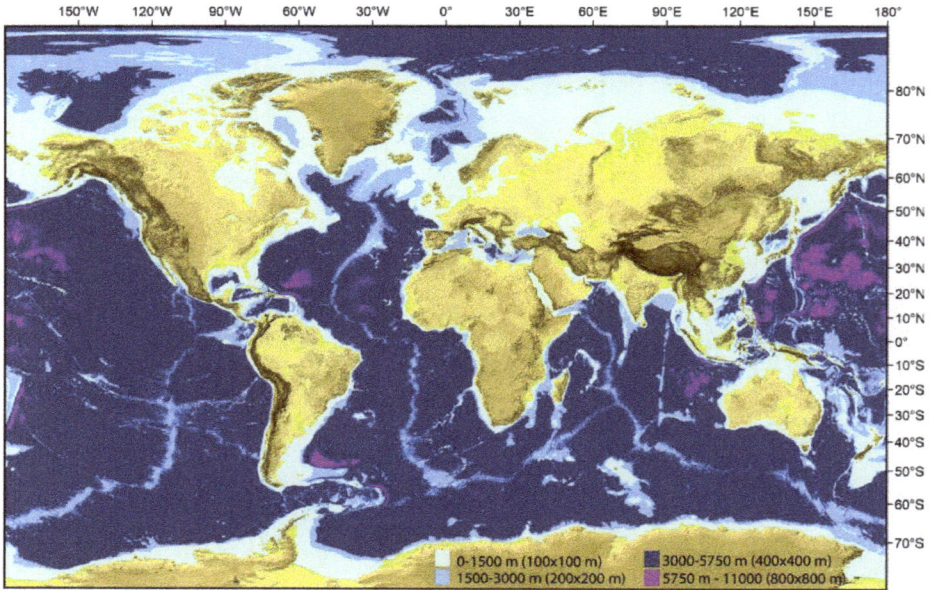

Figure 7. Depth regions based on state-of-the-art 2° × 2° deep-water multibeam installed on surface vessels, calculated at 60° from nadir shown in Table 2.

Table 2. Feasible resolution based on state-of-the-art 2° × 2° deep-water multibeam installed on surface vessel, calculated at 60° from nadir.

Depth Range	Grid-Cell Size	% of World Ocean
0–1500 m	100 × 100 m	13.7
1500–3000 m	200 × 200 m	11
3000–5750 m	400 × 400 m	72.6
5750–11,000 m	800 × 800 m	2.7

The needs of the broad community of stakeholders who use deep ocean bathymetry vary with respect to required resolution. Mapping at the resolutions described above may not serve all needs; however, baseline bathymetry from the Nippon Foundation GEBCO Seabed 2030 Project will help identify those regions where higher-resolution data may be required. It will also serve as a baseline against which additional surveys can be compared to better understand processes that can change the seafloor in some regions.

The inadequacy of our current baseline knowledge of seafloor bathymetry became particularly evident in the search for the Malaysia Airlines flight MH370, which disappeared 8 March 2014 (Figure 8). The satellite altimetry predicted bathymetry and database of available soundings in the southeastern Indian Ocean at the time of the crash proved so sparse that it was impossible to deploy the AUVs needed to find the wreckage without first conducting new surface ship deployed multibeam sonar bathymetric surveys [26,27]. The search for MH370 facilitated the collection of ~710,000 km² of new surface ship-derived multibeam sonar data representing a remarkable improvement and addition to Indian Ocean bathymetry (Figure 9) as well as many new insights into the geology of the southeastern Indian Ocean [27].

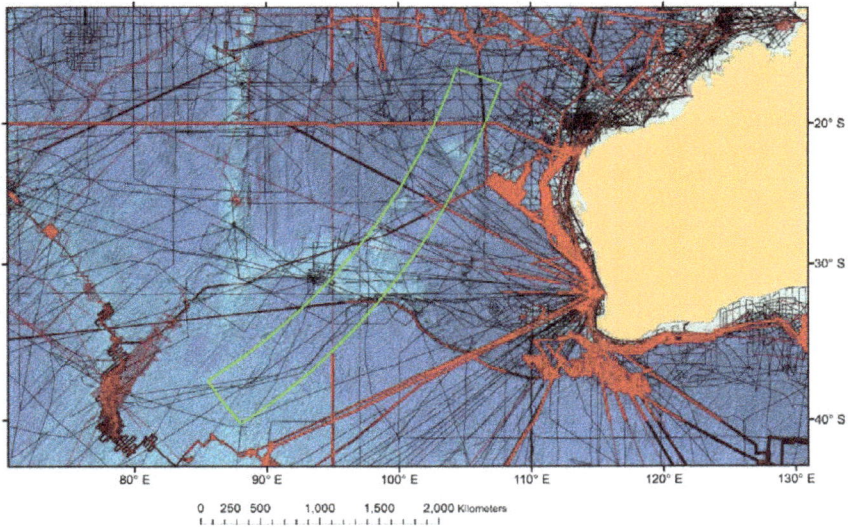

Figure 8. The need for high-resolution bathymetry in the deep ocean became strikingly evident outside of the scientific community following the disappearance of Malaysia Airlines flight MH370, en route from Kuala Lumpur to Beijing on 8 March 2014. The seafloor in the projected search area (green box) was very poorly mapped at the time the search for parts of the fuselage began [26]. Black lines are single-beam sonar ship track lines, and red lines show multibeam bathymetry available at the time the search began. The bathymetric data coverage was far from sufficient to navigate underwater vehicles for detailed inspection of the seafloor. Map is from the Nippon Foundation—GEBCO—Seabed 2030 Roadmap for Future Ocean Floor Mapping [20].

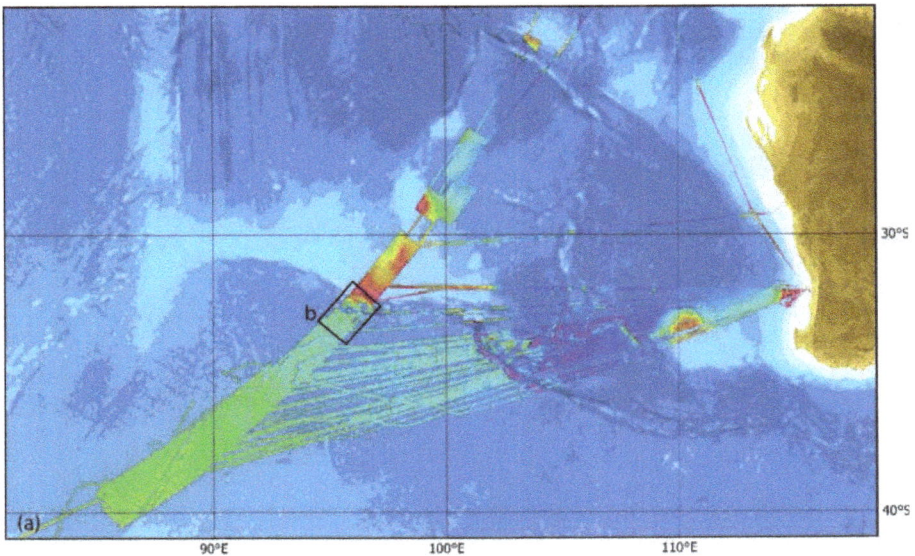

Figure 9. *Cont.*

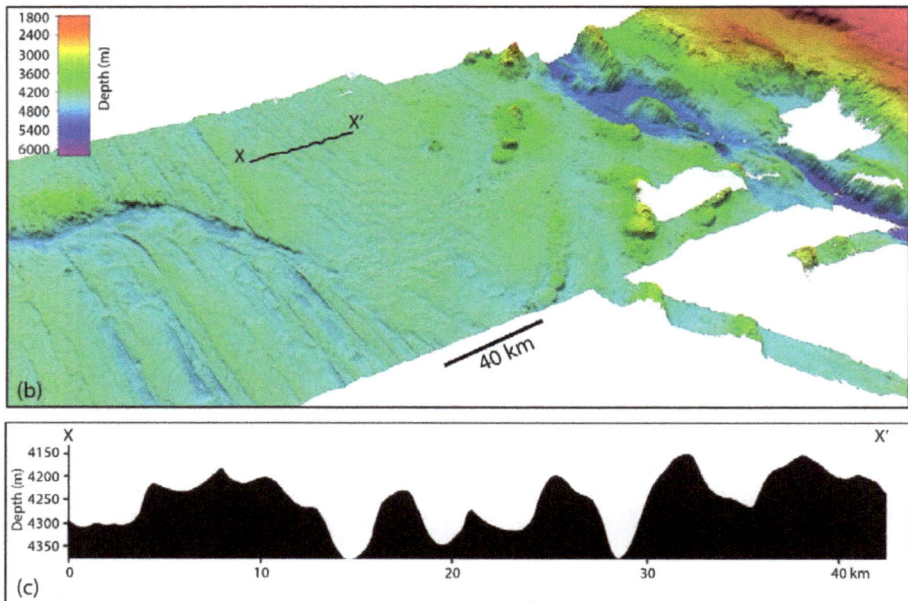

Figure 9. (**a**) Surface ship multibeam sonar data collected (~710,000 km^2) in support of the search for wreckage of Malaysian Airlines Flight MH370. (**b**) close-up revealing bathymetric detail provided by multibeam sonar including tectonic fabric and deep-sea trench (see [27] for details) VE = 6×) (**c**) cross-section of small area (black line in (**b**) demonstrating roughness of topography and why surficial mapping was necessary before deploying near-bottom mapping systems (AUVs or ROVs).

2.3. What Is the Magnitude of the Task?

Mapping the world ocean from surface vessels is a costly and time-consuming endeavor of global extent that is best achieved through international coordination. Through analyses of the present GEBCO bathymetric database [1], we can make a first-hand approximation of the mapping effort needed to obtain a continuous grid of bathymetric information at any given resolution. To assess feasible scales of resolution, GEBCO_2014 grid nodes originating from altimetry were selected using the Source Identification (SID) grid produced during the compilation of GEBCO_2014. The grid nodes were converted to surfaces and classified into water depth intervals that represent the broad geomorphological features of the continental shelf, continental slope, and deep sea area, in recognition that modern multibeam techniques and coverage are highly dependent on the water depth (Table 3). Using this depth distribution, the effort required to completely map the seafloor was calculated.

In order to compute an estimate of surveying effort given in Table 3, the following assumptions were made: (1) for each water depth interval, the average value was used to represent the distribution; (2) this average water depth is multiplied by a factor representing the projection of the swath width of a multibeam system on the seafloor using the conservative estimate of 3.5 times the water depth, and; (3) the speed of the survey vessel is considered to be 7.5 knots (~10 km/h). This speed value is also very conservative. However, it is used to compensate for the time needed for maneuvering, adverse weather conditions, deployment of auxiliary sensors (e.g., tide gauges principally in shallow waters, sound velocity profiling), and transiting, which in the high seas will be significant. The analyses have been carried out using the GEBCO_2014 grid with a resolution of 30 arc seconds as a base, and the grid cells with depth values from single-beam echo soundings have also been considered mapped. For this reason, the result may be underestimating the required survey time rather than overestimating it.

Table 3. Calculated survey effort needed to completely map the world ocean.

Water Depth Interval (Modal Water Depth)	Average Water Depth (km)	Proportion of Water Depth (%)	Proportion of Uncharted Surface (This Interval)	Proportion of Uncharted Surface (Overall)	Cumulated Surface of GEBCO_2014 Grid Nodes from Interpolated Altimetry (km^2)	Remaining Effort (Years) for One Survey Vessel
>3000	4	75.3	85	69	230,910,385	188
3000–1000	1.5	13.0	72	15	34,143,193	74
1000–200	0.4	4.4	66	7	10,654,693	86
200–0	0.1	7.3	71	9	18,995,603	619

Using the assumptions stated above implies that ~970 years would be required to survey the area of the GEBCO_2014 grid today unconstrained by any sounding with a single survey ship. Of this, however, ~620 years represent the time required to survey the shallow areas between 0 and 200 m depth, indicative of the ever lessening swath coverage as water depths decrease. Thus, 64% of the projected survey time is required to cover only 7% of the world's seafloor that is less than 200 m deep. Recognizing this, and the fact that these shallow seas lie mostly in the territorial waters of coastal states and thus much of the responsibility for mapping these shallow waters will rest with local hydrographic agencies, most of the effort of the Nippon Foundation GEBCO Seabed 2030 Project will be focused on the 93% of the world's oceans that is deeper than 200 m and often beyond the jurisdiction of local hydrographic agencies. Thus, the prime target for the Nippon Foundation GEBCO Seabed 2030 Project will be the approximately 350 ship years of effort required to map the 93% of the world ocean floor deeper than 200 m. Considering that there are currently more than 700 multibeam systems on survey vessels operated by national hydrographic offices, research institutions or private entities in the world (as estimated in 2003 by the IHO) the remaining surveying effort appears to be a reachable goal. For the other seven percent, The Nippon Foundation GEBCO Seabed 2030 Project, will take advantage of its connections with GEBCO and the IHO to work closely with local hydrographic agencies to obtain shallow water data wherever possible.

Coastal states are generally responsible for mapping the regimes over which they have sovereign rights or obligations. Legal regimes cover areas that include the Territorial Seas up to 12 NM, the Exclusive Economic Zone (EEZ) to 200 NM from the coastline, and the limits of the juridical continental shelf (often referred to as the Extended Continental Shelf (ECS)), a legal definition of the continental shelf that can extend the sovereign rights of a coastal state well beyond 200 NM in some cases as per the United Nations Convention on Law of the Sea (UNCLOS). The establishment of the limits of the juridical continental shelf is based on the presentation of detailed mapping data to the Commission on the Limits of the Continental Shelf. This has led many coastal states to undertake ambitions seafloor mapping programs. These programs, as is the case, e.g., New Zealand, have often morphed into national mapping programs and can be a potential source of new mapping data into a global compilation. The total areas of ocean that are not under state jurisdiction, however, (The Area) still represents ~50% (188,000,888 km^2) of the total Earth surface covered by seas and oceans [28]. For those states who do not have well-developed mapping programs, the products of the Nippon Foundation—GEBCO Seabed 2030 Project may help them meet their Law of the Sea requirements.

It is important also to consider, that our estimation is based on GEBCO's bathymetric database at the time of the compilation of the GEBCO_2014 grid. While we believe that this is the most complete bathymetric database available, we are fully aware that it is far from including all bathymetric data that has been collected (e.g., most ECS data collected by coastal states has not yet been made available to international data repositories). Identifying and retrieving data that has already been collected but that is not yet available for global compilations is thus another critical component of the Nippon Foundation GEBCO Seabed 2030 Project, highlighting the need for increased national and international collaboration and the coordination of regional and global bathymetric mapping initiatives.

3. Structure of the Nippon Foundation—GEBCO Seabed 2030 Project

To achieve its goal, the Seabed 2030 team will need to compile all available bathymetric data into a high-quality, high-resolution digital model. Inasmuch as bathymetric data and survey activities are dispersed across many countries and organizations, including governmental agencies, industry, academia and research organizations, the Nippon Foundation—GEBCO—Seabed 2030 Project will act as a coordinating body for aggregating existing data and prioritizing survey operations through the development of tools and products that highlight gaps in data coverage and facilitate new mapping expeditions, specifically targeting the poorest mapped areas.

The project sits within the existing and well-functioning IHO-IOC GEBCO framework making full use of existing bodies such as the GEBCO Sub-Committee on Regional Undersea Mapping (SCRUM), the Technical Sub-Committee on Ocean Mapping (TSCOM), the Sub-Committee for Regional Undersea Feature Names (SCUFN) and the newly founded Sub-Committee on External Relations and Communications. This structure will support Seabed 2030, by contributing to the governance structure and allowing it to benefit from the large networks already provided by the IHO and IOC. This is particularly important since all hydrographic offices of IHO's member states, with their mandate to map within their countries' territories, become de facto partners in the effort. Without the collaboration of the world's hydrographic offices, there is little chance of reaching the goal of portraying the world ocean bathymetry. IOC, on the other hand, is a self-standing body within UNESCO with a mandate to organize marine science within the UN system. This provides Seabed 2030 with a strong global network that reaches beyond the hydrographic agencies and includes the broader marine scientific community that also collects and utilizes ocean bathymetry.

The project structure is based on GEBCO's experience working with Regional Mapping Projects (e.g., IBCAO and IBCSO) that have contributed substantially to GEBCO by delivering regional bathymetric gridded compilations. Seabed 2030 is divided into four regions (Figure 1) for which data assembly, processing, and compilation fall under the responsibility of a dedicated team of experts based at globally distributed institutions that we call Regional Data Assembly and Coordination Centers (RDACC) each working in coordination with a Regional Mapping Committee:

- The North Pacific and Arctic Oceans Center, based at the University of Stockholm, Sweden and University of New Hampshire, USA
- The South and West Pacific Oceans Center based at the National Institute of Water and Atmospheric Research (NIWA), Wellington, New Zealand
- The Atlantic and Indian Oceans Center (Lamont Doherty Earth Observatory, Palisades, NY, USA)
- The Southern Ocean Center (Alfred Wegner Institute, Bremerhaven, Germany)

The RDACCs will be responsible for assembling a regional database of cleaned bathymetric data and the production of a regional bathymetric grid for incorporation into the ultimate product of the project, the Nippon Foundation GEBCO Seabed 2030 global grid. A Global Data Assembly and Coordination Center (GDACC), located at the British Oceanographic Data Center at the National Oceanographic Center in Southampton, U.K., will ultimately be responsible for assembling the regional grids and producing and delivering the Seabed 2030-GEBCO gridded product through the Internet. The GDACC will host and maintain the Seabed 2030-GEBCO website (https://seabed2030.gebco.net/) including sub-pages for the RDACCs as well as be responsible for handling user inquiries regarding the Seabed 2030 website and products.

Beyond the identification, assembly and processing of existing data sets, the RDACCs will also be responsible for identifying data gaps and opportunities for new data collection, including the facilitation of new mapping endeavors through coordination of ongoing activities among stakeholders in the region. Importantly the RDACC will not be a permanent archive, repository or distributor of the data they are assembling. As such, the RDACC will not be duplicating the responsibilities of existing data centers or data compilation efforts such as NCEI, GMRT or EMODNET, but rather will work with them as both a contributor and recipient of data.

Each RDACC will also establish a Regional Mapping Committee, a group of regional experts who will work with the RDACC in identifying sources and facilitating the collection of bathymetric data in the region. As the RDACCs build connections in their regions, they will have the benefit of more than 78 graduates of the Nippon Foundation-GEBCO training program in ocean bathymetry at the University of New Hampshire's Center for Coastal and Ocean Mapping. Over the past thirteen years this program has trained students from more than 34 countries, many of whom now occupy senior positions at hydrographic agencies, navies, and mapping centers around the world. Another potential source of data to the RDACCs is the private sector, which may be able to provide data collected during transit between commercial surveys (already being done by Fugro) or data sets whose commercial value has depreciated.

4. New Technologies

While the initial efforts of the Nippon Foundation GEBCO Seabed 2030 Project will focus on the identification and assembly of existing data sets, identifying key data gaps, and facilitating new data collection, it will also promote the development of new technologies that may increase the efficiency, accuracy, and resolution of seafloor mapping data. Seafloor mapping technology has been advancing at a rapid pace since the invention of the echosounder in the late 1920s. Over the past few years, multibeam sonars have steadily increased the number of beams that ensonify the seafloor and the density of their measurements (both along- and across-track), improved their ability to compensate for ship motion and invoked enhanced bottom detection techniques that have greatly improved the quality of the data being collected. In shallow water, when water clarity permits, bathymetric LIDAR (LIght Detection and Ranging) and satellite-derived bathymetry (bathymetry from satellite imagery as opposed to satellite altimetry-derived bathymetry which comes from the gravity field) have provided efficient ways to produce estimates of bathymetry in regions where multibeam sonar mapping is much less efficient.

While incremental improvements are ongoing, the challenge of mapping the world ocean will benefit greatly from new technological breakthroughs, and thus the Seabed 2030 Project will explore, initially through workshops and collaboration with industry, new approaches to seafloor mapping that would particularly enhance efficiency and coverage. Among the new technologies to be examined will be the possibility of low-powered echo sounders deployed on gliders or ARGO floats, the possibility of fleets of long-range AUVs equipped with interferometric swath sonars, and autonomous surface barges or sailing vessels capable of long-duration, ocean-wide transits (Figure 10). These autonomous platforms would be controlled from shore and transmit data back to shore via satellite communication links allowing 24/7 global operation with minimal human supervision. This concept becomes particularly important as we envision thousands of hours of data collection in remote areas of the world oceans where crewed vessels would be particularly inefficient given the need for long transits to accommodate crew changes and the potential monotony of long data collection campaigns.

In the more trafficked regions of the world ocean, crowd-sourced bathymetry offers tremendous potential. Using crowd-sourced bathymetry is not new to GEBCO. Bathymetry provided by the Norwegian company Olex comprised a significant source for the compilation of the International Bathymetric Chart of the Arctic Ocean (IBCAO) Version 3.0 grid released 2012, as well as in the latest GEBCO_2014 grid. The Olex depth measurements originate from their automatic charting system installed primarily on fishing vessels. Other companies using a crowd source approach have now also entered the market. Small and easy to install NMEA-loggers storing depths from any ship echo sounder already exist and are being further developed. Such methods could be used on a global scale, through adoption by shipping companies, cruise ship companies, and recreational boaters for example. IHO has a crowd-sourced working group with substantial GEBCO engagement. This working group is tasked to draft recommendations for the minimum metadata to be provided along with depth measurements, and discuss available technologies, post-processing, as well as online upload technologies and storage.

Figure 10. Autonomous multibeam sonar barge would be controlled and transmit data through telepresence and could become platform for many oceanographic and atmospheric measurements and maintain long-endurance at sea with relatively little cost.

The gap between the coastline and where depth measurements exist on the continental shelf is large in remote areas on Earth. Surveying of these areas using conventional methods from ships, and even with AUVs, may be enormously challenging and expensive. LIDAR is highly effective and relatively inexpensive for large-scale regional mapping but is limited to areas of relatively clear water. In such remote areas, where other means of seafloor mapping are not easily feasible, bathymetry derived from satellite imagery is very promising. Freely available imagery, such as Landsat 8, as well as commercial higher-resolution satellite images, comprise vast sources of data with global coverage. Conventional "water penetrating" satellite-derived bathymetry also requires relatively clear water, but the development of satellite-derived bathymetry methods that are not solely based on the optical spectrum may overcome turbid water issues, though their accuracy and resolution are considerably lower and the method will still be depth limited.

Whether from the identification of existing data sets or the collection of new data, the Nippon Foundation GEBCO Seabed 2030 Project will be responsible for bringing disparate depth measurements together into the compilation of a coherent bathymetric portrayal of the world ocean floor. Therefore, advances in bathymetric post-processing and analysis software, database technology, computing infrastructure, and gridding techniques are also important technological considerations. The present GEBCO central bathymetric data compilation, as well as compilations of the existing Regional Mapping Projects under GEBCO, reside on distributed servers at their respective host organizations. Moving towards establishing the RDACCs suggests that there may be potential benefits from establishing a shared cloud-based infrastructure for data storage as well as for gridding and processing routines (Virtual Research Environments and Infrastructures) that can also become part of new e-learning processes. The overall coordination of efforts to make use of emerging technologies will be the responsibility of the GDACC in coordination with the RDACCs.

5. Conclusions

For thousands of years, we have been probing the seas to measure its depths—originally with a stone or lead at the end of a line and now using modern multibeam echo-sounders capable of ensonifying wide swaths of the seafloor and producing hundreds of highly accurate depth measurements across this swath. Despite these thousands of years of effort, only a fraction of the world ocean's seafloor has been sampled for depth (far less than 15%), greatly limiting our ability to explore and understand critical ocean and seafloor processes. Remarkably, several planets in our solar system are mapped with far better coverage and at a higher resolution than our own planet. While it may be argued that mapping these distant planets is easier than mapping the seafloor because electromagnetic sensors can penetrate their atmospheres and thus high-resolution mapping is much simpler than trying to map the deep sea, the reality is that the overall cost of an extra-terrestrial mapping mission (a Mars mission typically costs on the order of $3B U.S.) is commensurate with the estimated cost of using existing sonar technology to completely map the 93% of the world's ocean deeper than 200 m. Unfortunately, we appear to be willing to spend these amounts to map other planets but not our own.

Recognizing this poor state of our knowledge of ocean depths and the critical role such knowledge plays in understanding and maintaining our planet, GEBCO and the Nippon Foundation have joined forces to begin to address this issue. Through the establishment of the Nippon Foundation GEBCO Seabed 2030 Project, an international effort has begun with the objective of facilitating the complete mapping of the world ocean by 2030. Achieving this goal will be far from simple. The Seabed 2030 Project will initially establish globally-distributed regional data assembly and coordination centers (RDACCs) that will focus on identifying existing data from their assigned regions that are not currently in publicly available databases and seek to make these data available. To do this, they will work with a Regional Mapping Committees made of regional experts from local hydrographic agencies and institutes as well as a network for alumni of the Nippon Foundation/GEBCO postgraduate training program, many of whom hold senior positions with hydrographic and other mapping agencies around the world.

The four RDACCs—North Pacific and Arctic; South and West Pacific; Atlantic and Indian; and Southern centers—will also be developing common software and other tools to assemble and attribute appropriate metadata to the data sets they have acquired and then to assimilate these into regional grids using standardized techniques. Along with the data assembly and gridding tools, the RDACCs will be developing tools to continuously update data coverage and to identify and prioritize critical data gaps. The RDACCs will work with regional and global mapping agencies to facilitate and coordinate efforts aimed at collecting new data to fill these gaps. The regional grids will be transmitted to a Global Data Assembly and Coordination Center (GDACC) whose primary task will be to integrate the regional grids into a global grid and distribute this new grid to users worldwide. The GDACC will also act as the central focal point for the coordination of common data standards and processing tools as well as the outreach coordinator for Seabed 2030 efforts. Finally, the Nippon Foundation GEBCO Seabed 2030 Project will track and encourage the development of new and innovative technologies that can increase the efficiency of seafloor mapping and thus make the ambitious goals of Seabed 2030 more likely to be achieved.

There can be no question that the goal of seeing the world ocean completely mapped by 2030 is a difficult one. So little of the ocean is currently mapped that an effort to achieve complete coverage by 2030 will require several hundred ship-years of new mapping effort using current technology. New technological advances may help to reduce this number (or the cost of achieving it) however, the magnitude of the task is such that it will only be done through international cooperation and collaboration. It is our sincere hope that through the initial efforts of the Nippon Foundation GEBCO Seabed 2030 Project, the world will come to recognize the great value of such collaboration and the goal of seeing the world ocean completely mapped by 2030 will be achieved.

Acknowledgments: All authors acknowledge the generous support of the Nippon Foundation in facilitating and funding the Nippon Foundation GEBCO Seabed 2030 Project. Larry Mayer acknowledges the support of NOAA Grant NA15NOS400002000.

Author Contributions: Larry Mayer and Martin Jakobsson were primary authors of this document; Graham Allen, Boris Dorschel, Robin Falconer, Vicki Ferrini, Geffroy Lamarche, Helen Snaith, and Pauline Weatherall made important contributions to the text and particularly to the development of the Seabed 2030 project and The Nippon Foundation—GEBCO Seabed 2030 Roadmap for Future Ocean Floor Mapping upon which this document is based.

Conflicts of Interest: The authors declare no conflicts of interest.

References

1. Weatherall, P.; Marks, K.M.; Jakobsson, M.; Schmitt, T.; Tani, S.; Arndt, J.E.; Rovere, M.; Chayes, D.; Ferrini, V.; Wigley, R. A new digital bathymetric model of the world's oceans. *Earth Space Sci.* **2015**, *2*, 331–345. [CrossRef]
2. Smith, W.H.F.; Sandwell, D.T. Global seafloor topography from satellite altimetry and ship depth soundings. *Science* **1997**, *277*, 1957–1962. [CrossRef]
3. Sandwell, D.T.; Gille, S.T.; Smith, W.H.F. *Bathymetry from Space: Oceanography, Geophysics, and Climate*; Bethesda: Rockville, MD, USA, 2002; pp. 1–24.
4. Harris, P.T.; Macmillan-Lawler, M.; Rupp, J.; Baker, E.K. Geomorphology of the oceans. *Mar. Geol.* **2014**, *352*, 4–24. [CrossRef]
5. Lecours, V.; Dolan, M.F.J.; Micallef, A.; Lucieer, V.L. A review of marine geomorphometry, the quantitative study of the seafloor. *Hydrol. Earth Syst. Sci.* **2016**, *20*, 3207–3244. [CrossRef]
6. Björk, G.; Jakobsson, M.; Assmann, K.; Andersson, L.G.; Nilsson, J.; Stranne, C.; Mayer, L. Bathymetry and oceanic flow structure at two deep passages crossing the Lomonosov Ridge. *Ocean Sci.* **2018**, *14*, 1–13. [CrossRef]
7. Jakobsson, M.; Gyllencreutz, R.; Mayer, L.A.; Dowdeswell, J.A.; Canals, M.; Todd, B.J.; Dowdeswell, E.K.; Hogan, K.A.; Larter, R.D. Mapping submarine glacial landforms using acoustic methods. *Geol. Soc. Lond. Mem.* **2016**, *46*, 17–40. [CrossRef]
8. Kostylev, V.E.; Todd, B.J.; Fader, G.B.J.; Courtney, R.C.; Cameron, G.D.M.; Pickrill, R.A. Benthic habitat mapping on the Scotian shelf based on multibeam bathymetry, surficial geology and sea floor photographs. *Mar. Ecol. Prog. Ser.* **2001**, *219*, 121–137. [CrossRef]
9. Cutter, G.R.; Rzhanov, Y.; Mayer, L.A. Automated segmentation of seafloor bathymetry from multibeam echosounder data using local Fourier histogram texture features. *J. Exp. Mar. Biol. Ecol.* **2003**, *285–286*, 355–370. [CrossRef]
10. Pickrill, R.A.; Todd, B.J. The multiple roles of acoustic mapping in integrated ocean management, Canadian Atlantic continental margin. *Ocean Coast. Manag.* **2003**, *46*, 601–614. [CrossRef]
11. Armijo, R.; Pondard, N.; Meyer, B.; Uçarkus, G.; de Lépinay, B.M.; Malavieille, J.; Dominguez, S.; Gustcher, M.-A.; Schmidt, S.; Beck, C.; et al. Submarine fault scarps in the Sea of Marmara pull-apart (North Anatolian Fault): Implications for seismic hazard in Istanbul. *Geochem. Geophys. Geosyst.* **2005**, *6*, Q06009. [CrossRef]
12. McAdoo, B.G.; Capone, M.K.; Minder, J. Seafloor geomorphology of convergent margins: Implications for Cascadia seismic hazard. *Tectonics* **2004**, *23*, TC6008. [CrossRef]
13. Bastia, R.; Radhakrishna, M.; Nayak, S. Identification and characterization of marine geohazards in the deepwater eastern offshore of India: Constraints from multibeam bathymetry, side scan sonar, and 3d high-resolution seismic data. *Nat. Hazards* **2011**, *57*, 107–120. [CrossRef]
14. Hess, H.H. History of Ocean Basins. In *Petrologic Studies: A Volume to Honor A. F. Buddington*; Engel, A.E.J., James, H.L., Leonard, B.F., Eds.; Geological Society of America: Boulder, CO, USA, 1962; pp. 599–620.
15. Jakobsson, M.; Mayer, L.; Armstrong, A. Analysis of data relevant to establishing outer limits of a continental shelf under law of the sea article 76. *Int. Hydrogr. Rev.* **2003**, *4*, 1–18.
16. Elmore, P.A.; Avera, W.E.; Harris, M.M. Use of the AN/AQS-20A tactical mine-hunting system for on-scene bathymetry data. *J. Mar. Syst.* **2009**, *78*, S425–S432. [CrossRef]
17. Fenty, I.; Willis, J.K.; Khazendar, A.; Dinardo, S.; Forsberg, R.; Fukumori, I.; Holland, D.; Jakobsson, M.; Moller, D.; Morison, J.; et al. Oceans melting Greenland: Early results from NASA's ocean-ice mission in Greenland. *Oceanography* **2016**, *29*, 72–83. [CrossRef]
18. Stocker, T.F.; Qin, D.; Plattner, G.-K.; Tignor, M.; Allen, S.K.; Boschung, J.; Nauels, A.; Xia, Y.; Bex, V.; Midgley, P.M. *Climate Change 2013: The Physical Science Basis*; Cambridge University Press: Cambridge, UK, 2013; pp. 1–1535.
19. Hall, J. GEBCO centennial special issue—Charting the secret world of the ocean floor: The GEBCO project 1903–2003. *Mar. Geophys. Res.* **2006**, *27*, 1–5. [CrossRef]
20. Jakobsson, M.; Allen, G.; Carbotte, S.M.; Falconer, R.; Ferrini, V.; Marks, K.; Mayer, L.; Rovere, M.; Schmitt, T.W.; Weatherall, P.; et al. The Nippon Foundation—GEBCO—Seabed 2030: Roadmap for Future Ocean Floor Mapping. Available online: https://seabed2030.gebco.net/documents/seabed_2030_roadmap_v10_low.pdf (accessed on 10 January 2018).

21. Jakobsson, M.; Mayer, L.; Coakley, B.; Dowdeswell, J.A.; Forbes, S.; Fridman, B.; Hodnesdal, H.; Noormets, R.; Pedersen, R.; Rebesco, M.; et al. The international bathymetric chart of the Arctic Ocean (IBCAO) version 3.0. *Geophys. Res. Lett.* **2012**, *39*. [CrossRef]

22. Arndt, J.E.; Schenke, H.W.; Jakobsson, M.; Nitsche, F.O.; Buys, G.; Goleby, B.; Rebesco, M.; Bohoyo, F.; Hong, J.; Black, J.; et al. The international bathymetric chart of the Southern Ocean (IBCSO) version 1.0—A new bathymetric compilation covering circum-Antarctic waters. *Geophys. Res. Lett.* **2013**, *40*, 3111–3117. [CrossRef]

23. Ryan, W.B.F.; Carbotte, S.M.; Coplan, J.; O'Hara, S.; Melkonian, A.; Arko, R.; Weissel, R.A.; Ferrini, V.; Goodwillie, A.; Nitsche, F.; et al. Global multi-resolution topography synthesis. *Geochem. Geophys. Geosyst.* **2009**, *10*, Q03014. [CrossRef]

24. Hell, B.; Öiås, H. A new bathymetry model for the Baltic Sea. *Int. Hydrogr. Rev.* **2014**, *12*, 21–32.

25. Becker, J.J.; Sandwell, D.T.; Smith, W.H.F.; Braud, J.; Binder, B.; Depner, J.; Fabre, D.; Factor, J.; Ingalls, S.; Kim, S.H.; et al. Global bathymetry and elevation data at 30 arc seconds resolution: SRTM30_PLUS. *Mar. Geodesy* **2009**, *32*, 355–371. [CrossRef]

26. Smith, W.H.F.; Marks, K.M. Seafloor in the Malaysian Airlines MH370 search area. *EOS Tran. Am. Geophys. Union* **2014**, *95*, 173–174. [CrossRef]

27. Picard, K.; Brooke, B.P.; Harris, P.T.; Siwabessy, P.J.W.; Coffin, M.F.; Tran, M.; Spinoccia, M.; Weales, J.; Macmillan-Lawler, M.; Sullivan, J. Malaysia Airlines flight MH370 search data reveal geomorphology and seafloor processes in the remote southeast Indian Ocean. *Mar. Geol.* **2018**, *395*, 301–319. [CrossRef]

28. Suárez-de Vivero, J.L. The extended continental shelf: A geographical perspective of the implementation of article 76 of UNCLOS. *Ocean Coast. Manag.* **2013**, *73*, 113–126. [CrossRef]

geosciences

MDPI

Article

Development of a Seamless, High-Resolution Bathymetric Model to Compare Reef Morphology around the Subtropical Island Shelves of Lord Howe Island and Balls Pyramid, Southwest Pacific Ocean

Michelle Linklater [1,*,†] [ID], **Sarah M. Hamylton** [1], **Brendan P. Brooke** [2], **Scott L. Nichol** [2], **Alan R. Jordan** [3] **and Colin D. Woodroffe** [1]

[1]   School of Earth and Environmental Sciences, University of Wollongong, Northfields Ave.,
      Gwynneville 2522, Australia; shamylto@uow.edu.au (S.M.H.); colin@uow.edu.au (C.D.W.)
[2]   Geoscience Australia, GPO Box 378, Canberra 2601, Australia; Brendan.Brooke@ga.gov.au (B.P.B.);
      Scott.Nichol@ga.gov.au (S.L.N.)
[3]   New South Wales Department of Primary Industries, Locked Bag 1, Nelson Bay 2315, Australia;
      Alan.Jordan@dpi.nsw.gov.au
*   Correspondence: michelle.linklater@environment.nsw.gov.au
†   Current address: Water, Wetlands and Coasts Science, New South Wales Office of Environment and
      Heritage, Locked Bag 1002, Dangar NSW 2309, Australia.

Received: 3 November 2017; Accepted: 23 December 2017; Published: 2 January 2018

Abstract: Lord Howe Island and Balls Pyramid are located approximately 600 km offshore of the southeastern Australian mainland, in the subtropical waters of the northern Tasman Sea. Lord Howe Island hosts the most southern coral reef in the Pacific Ocean, and the shelves surrounding both islands feature fossil coral reefs. This study creates a seamless, high-resolution (5 m cell size) bathymetry model of the two shelves to compare and contrast the extent of reef development and shelf morphology. This was produced by integrating satellite-derived depth data (derived to 35 m depth) and multibeam echosounder (MBES) data. Image partitioning and filtering improved the accuracy of the bathymetry estimates and the suitability for integration with MBES data. Diverse accretionary and erosional geomorphic features were mapped on both shelves, with fossil reefs dominating the shelves in 25–50 m depth. Similar patterns of shelf morphology were observed for the middle and outer shelves, while the inner shelf regions were most dissimilar, with reef development greater around Lord Howe Island compared to the more restricted inner shelf reefs around Balls Pyramid. Understanding the relative extent and morphology of shelf features provides insights into the geological and ecological processes that have influenced the formation of the shelves.

Keywords: bathymetry; DEM; satellite imagery; multi beam echosounder; filter; geomorphology; coral reefs

---

1. Introduction

Geomorphic characterisations of the seabed provide fundamental information for management of benthic biodiversity [1]. Geomorphology can be used as a physical surrogate for biodiversity, help identify areas of likely high habitat diversity, as well as stratify subsequent biological sampling. Geomorphic interpretations are often utilised as cost-effective baseline surveys for marine spatial planning, with broadscale, provincial mapping informing international and national policy [2–4] and mesoscale, regional mapping useful for local management applications [5–7].

Digital elevation models (DEMs) are a core dataset for geomorphic feature interpretations [8]. In the marine environment, DEMs can be produced from data acquired from active (e.g., sonar) or passive (e.g., satellite) sensors. Multibeam echosounders (MBES) are a common platform used to map seabed bathymetry; however their efficiency typically decreases in shallower waters. Marine light detection and ranging (LiDAR) data and satellite-derived bathymetry can be utilised to fill the void in this coastal zone and create a seamless surface of the land and seafloor [9]. The acquisition of airborne LiDAR can often be prohibitively expensive, and the approach of extracting depth from satellite imagery offers an accessible and effective means to generate DEMs [10].

Depth can be derived from satellite imagery using physics-based or empirical methods. Employing physical methods [11,12] require the input of a range of parameters measured in the field and, while providing robust seabed data, are complex to implement. Empirical methods [13] are more simplistic, requiring fewer parameters, although ground validation depth data are needed. Due to the increased parameterisation of physical methods, the resultant surface can be more accurate than provided by empirical methods. However, in certain settings, empirical approaches can provide comparably accurate surfaces [10].

To effectively derive depth from satellite imagery, the images are typically pre-processed to reduce or remove the artefacts of atmosphere, cloud cover or surface disturbance, such as sun glint or wind waves [14,15]. Noise and pixelation are commonly smoothed by image filtering methods which remove outliers in the data. As terrain derivatives can be used to explore spatial patterns in relation to benthic and pelagic communities [1,16], it is important that terrain measures derived from the satellite image are reflecting surface variation rather than pixelation artefacts. For the creation of seamless DEMs, it is important that surface smoothing of input datasets are at comparable levels to ensure consistency for subsequent geomorphic analysis of the integrated DEM.

The creation of a seamless DEM of the coast and shelf enables an holistic approach to terrain analysis and geomorphic interpretation, enabling landform features to be defined and described at the same scale. The acquisition of marine LiDAR and satellite-derived bathymetry over expansive areas along the Australian coastline have led to the creation of seamless coastal DEMs which inform the understanding of coastal processes and evolution, and reveal distributions of potential submerged habitats [17–19]. For example, along the Great Barrier Reef in Australia, the creation of a seamless DEM through integrated bathymetric sources [20] has led to the identification of extensive areas of potential suitable habitat for corals and increased understanding of reef morphology over broad spatial scales [21,22].

In this study, we report the development of a seamless bathymetric DEM for the shelf areas surrounding the world-heritage-listed Lord Howe Island and Balls Pyramid, which occur in subtropical waters offshore of the New South Wales (NSW) mainland, Australia. The creation of a seamless shelf DEM of this region would enable characterisation of the marine geomorphology of the region, which allows an assessment of geodiversity and potential biodiversity.

Accurately delineating the extent of fossil reefs is important for understanding capacity for reef accretion in marginal, subtropical settings. In this study, we utilise satellite imagery to infill gaps in coverage and produce a new shelf DEM. We explore methods of enhancing the processing of satellite-derived bathymetry to suit the purposes of integration with MBES for DEM generation, including filtering and image partitioning. The aims of this study are to: (1) improve application of depth extraction for input into an integrated DEM; (2) create an updated DEM for the region; (3) define and map shelf geomorphic features to compare and contrast shelf morphology.

2. Materials and Methods

2.1. Study Area

Lord Howe Island and Ball Pyramid are remote, pristine islands which are located 600 km offshore of the east coast of the Australian mainland (Figure 1). These islands are internationally valued for their high biodiversity and endemism, and for possessing the southernmost coral reef in the Pacific

Ocean, which fringes the west coast of Lord Howe Island [23,24]. The islands are protected by state and Commonwealth marine parks and reserves, and have been World Heritage listed since 1982.

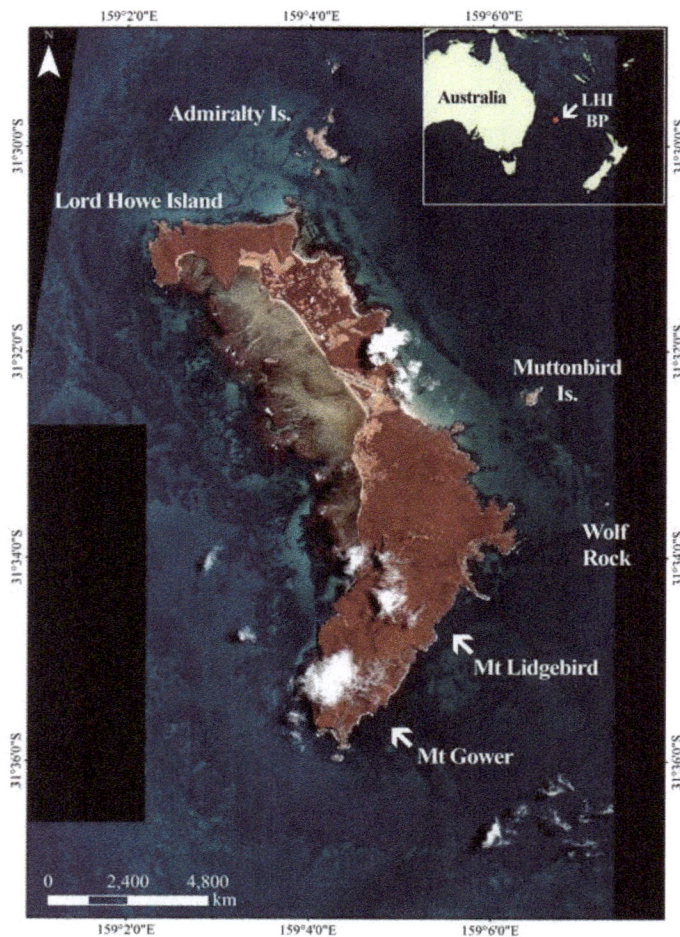

Figure 1. World View 2 (2010) image of Lord Howe Island, showing high water clarity evident with visibility to the outer shelf (~60 m water depth) in the southwest corner of the image.

The islands formed from hotspot volcanism 6–7 million years ago [25] and have eroded to a small fraction of their original size with broad shelf platforms submerged in 30–100 m water depth [26]. Eroded volcanic lavas and tuffs are overlain by Quaternary calcarenites [25,27]. A coral reef fringes the west coast of Lord Howe Island [23,28] and extensive fossil reefs have been mapped around the submerged shelves [29,30]. The fossil reef around Lord Howe Island developed during times of lower sea level and is more than 20 times larger in area than the modern fringing coral reef [29]. Material dated from the fossil reef revealed several metres of accretion from 9–2 ka [29]. A comparable submerged reef system is also present on the Balls Pyramid shelf, and inferred to similarly be a drowned fossil reef [30]. Coral reef growth is possible at this subtropical location due to the influx East Australian Current which flows east from the Australian mainland delivering warm waters to the region.

Broadscale characterisations at the provincial and biome levels have been previously undertaken for the Lord Howe region, generating datasets for international [4] and national [3,31] marine management. Meso-scale mapping at the geomorphic and primary biome classification levels have also been produced for the shelves [30,32–35] and habitat classifications have been performed for shallow and mesophotic assemblages [23,34,36,37]. Mapping of geomorphic features has been utilised in assessment reviews of the marine park zoning scheme and research planning [34,38].

High-resolution bathymetry grids were compiled for the Lord Howe region by Geoscience Australia [39], which included a land-bathymetry model with MBES data of the shelves and slopes together with depth extracted from a Quickbird image of Lord Howe Island. Since the creation of the bathymetry grids by Geoscience Australia [39], new MBES data was acquired around the shelves aboard the R.V. *Southern Surveyor* (Marine National Facility, Canberra, Australia) in 2013, presented in [30] and this study. The availability of detailed new MBES data and satellite imagery around the islands creates the opportunity to create a seamless DEM and geomorphic interpretation of the shelves from shoreline to shelf break.

2.2. Depth Estimation for the Lord Howe Island Shelf

Depth estimates were empirically derived from high-resolution World View 2 imagery 2010 (WV2, 8 spectral bands, 2 m cell size) to supplement the gaps in bathymetry data around the inner shelf of Lord Howe Island. MBES data were collected around the middle shelf in 2008 and 2013 aboard the R.V. *Southern Surveyor* using the onboard Kongsberg EM300 30 kHz system. Single beam data were collected in the lagoon in 2008 by New South Wales Roads and Maritime Services using an ODOM CVx3 echosounder. Marine LiDAR data was available around the north of the island, although the data was considered too coarse for this study (35 m cell size). Single beam and MBES data were utilised for bathymetry estimates due to their greater data density (4–5 m point spacing for multibeam; 10–15 m point spacing for single beam) which better matches the image resolution (2 m cell size).

Data were converted into the local Lord Howe Island hydro datum and tidal corrections were applied to all datasets (tide height 1.90 m above local datum at time of image acquisition). Single beam data were collected in the local datum. MBES data were collected in Australian height datum and were transformed to the local datum for analysis. A summarised workflow of satellite imagery processing is shown in Figure 2.

2.2.1. Correction for Atmospheric Interference and Sun Glint Effects

The Lord Howe Island WV2 image was calibrated to radiance units ($\mu W \cdot cm^{-2} \cdot nm^{-1} \cdot sr^{-1}$) using ENVI v4.8 (Harris Corporation, Colorado, USA). The Fast Line-of-sight Atmospheric Analysis of Spectral Hypercubes (FLAASH) algorithm was used to correct for atmospheric interference [40,41]. This algorithm adopts the MODTRAN4 (MODerate resolution atmospheric TRANsmission) radiation transfer model to remove the contributions of atmosphere from the spectral reflectance [40,41]. A mid-latitude model was used with initial visibility adjusted to 80 km.

Corrections for sun glint were undertaken on the atmospherically-corrected image using the methodology of [15]. This method achieved deglinting by using the near infra-red (NIR) band to approximate the contribution of specular reflection from the water surface, and subtracting this from the reflectance in each individual image band. To calculate the slope product and minimum NIR values required for deglinting, a subset was extracted from a region of interest (ROI) in each band (49,567 points per band). The ROI was defined from a representative area, which demonstrated a range of glint intensities over homogenous substrate. A single ROI area was extracted from the image as this was found to produce a stronger correlation. Extraction of data from multiple ROI's produced data clusters that expressed similar slopes within the regression, though the spread of data reduced the overall fit of the linear regression. Land and cloud artefacts were masked from the image using a digitised land and cloud polygon.

SATELLITE IMAGE PROCESSING

BATHYMETRY MODEL

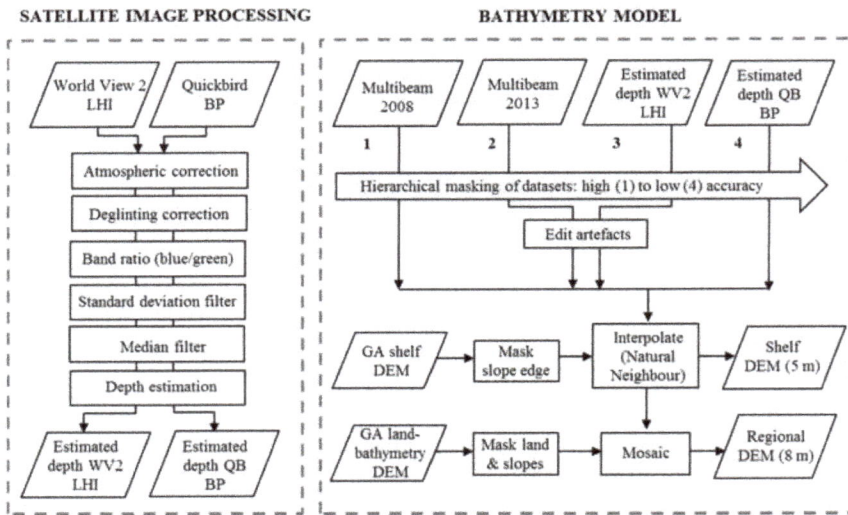

Figure 2. Process diagram of steps undertaken during processing satellite imagery and producing bathymetry model. Geoscience Australia bathymetry grids sourced from [39].

2.2.2. Estimating Bathymetry from Satellite Image Bands

Band ratio approaches are a type of empirical method which uses the relative attenuation of light between two bands to infer depth [13]. This approach has been shown to be effective in deriving depth using relatively few parameters. Known depth data is required for this approach (such as single beam data, MBES or marine LiDAR) to calibrate the ratio and validate the surface. This approach was selected due to the simplicity of the method and the availability of known depth data for the region (single beam and MBES).

Depth was estimated in the satellite image using the methodology described in [13] (Equation (9)). This approach functions independently of bottom type by using a ratio transform to measure the relative attenuation of light through the water column for individual bands. Ratios were calculated for different combinations of bands, and these ratios were plotted against known depth to assess the most suitable bands for depth estimation. The ratio of the blue (band 2) and green (band 3) bands were shown to have the best correlation and were used for subsequent analysis. Deglinted reflectance values for each substrate type for band 2 and 3 are provided in Appendix A, Figure A1.

Depth was estimated by correlating reference depth points against the ratio of reflectance for the blue and green bands. Reference depth points for the Lord Howe Island shelf were sourced from the 2013 multibeam (22,445 points) and 2008 single beam (6316 points). Coverage of calculation and validation points is shown in Appendix B, Figure A2. Band ratio values were plotted against known depths to generate a relationship function, which was then applied across the entire band ratio rasters using the raster calculator in ArcGIS to generate a continuous surface of estimated depth. Estimated surfaces were produced for the different filter types, and slope surfaces were generated for comparison.

2.2.3. Filter Comparison

Filters were applied using two approaches: the first approach used a single filter; while the second approach used a combination of two or more filters. Filters were generated for the blue/green ratio grid using ArcGIS 10.5 Spatial Analyst Toolbox, and these were compared to the ratio grid with no filters applied.

For the first approach, single filters were applied to the ratio grid including: (1) low pass filter; (2) median filters, with circle radii of three, five and 10 cells; and (3) standard deviation filter, with a three cell circle radius filter. Values exceeding a standard deviation threshold of 0.035 were reclassified and used to create a mask which excluded these outlier values from the ratio grid. The second approach applied a series of filters to the band ratio grid. These filters included: (1) low pass filter followed by a median filter with a 10 cell circle radius; and (2) standard deviation filter followed by median filter with a 10 cell circle radius. The surface created with the combination of standard deviation and median 10 (10 cell radius) filters was determined to be the most suitable filter for the purposes of this study.

2.2.4. Image Partitioning and Error Assessment

Root mean square error (RMSE) was performed using all remaining data points in the inner shelf region, excluding the depth points used to derive depth, to avoid bias of error calculation (Appendix B, Figure A2). Depth points overlapping the calculated surface were extracted from the 2013 (249,988 points) and 2008 (234,627 points) multibeam data and single beam data (3666 points) to calculate error. RMSE were calculated for the different ratio filters and depth intervals.

Qualitative and quantitative evaluation of the derived depth surface revealed that the surface to the west of the island was more accurate than the eastern side in waters greater than 25 m depth. Water surface disturbance is apparent on the eastern side of the image, which may be due to its exposure to prevailing conditions. Variation between image tiles is also apparent, and these factors may account for the differences in regression fit.

It was therefore considered to partition the image into eastern and western segments and apply separate regressions to each side of the image. RMSE errors were calculated for the partitioned segments and the results were compared.

2.3. Depth Estimation for the Balls Pyramid Shelf

Bathymetry estimates for the Balls Pyramid shelf were previously conducted by [30]. These estimates were recalculated by this study using the optimal approach determined by the aforementioned process of depth estimation for the Lord Howe Island shelf. This included the application of standard deviation and median (10 circle radius) filters which were applied to a 2010 Quickbird (QB, four spectral bands, 2.4 m cell size) image for the Balls Pyramid shelf. Image partitioning was not required for the Balls Pyramid shelf as there were no distinct variations across the surface. Reference depth points (57,269 points) and validation points (220,629) were sourced from the 2013 multibeam dataset.

2.4. Integrated Bathymetry Model

Bathymetry estimates from the WV2 and QB satellite images were converted to Australian height datum (AHD) using an offset of 1.10 m [42] for integration with the 2008 and 2013 MBES data, which were gridded to AHD. Estimated depth surfaces for the WV2 and QB images were clipped at 35 m due to deviation in the spread of unfiltered data and ideal depth for seamless integration with MBES data. The application of image partitioning for the WV2 image improved surface accuracy in deeper waters, which improved the integration with the MBES data. Error assessments were additionally performed on the 2008 and 2013 MBES survey data, with RMSE values calculated on areas of overlap.

Bathymetry estimates were integrated with MBES data from R.V. *Southern Surveyor* voyages (Marine National Facility, Canberra, Australia) in 2008 (4 m cell size) and 2013 (5 m cell size), and supplemented with data around the shelf slopes from the Geoscience Australia (8 m cell size) shelf grid [39] for a seamless transition. A summarised workflow of processing for the integrated bathymetry model is shown in Figure 2. Coverages were hierarchically masked based on the relative accuracy of the input data, whereby MBES data was considered the highest ranking layer, followed by the satellite-derived depth. Data were converted to points and interpolated to a 5 m grid using natural neighbour (ArcGIS v10.1, Environmental Systems Research Institute (ESRI), Redlands, CA, USA), which was shown to generate the most appropriate surface from a range of interpolation approaches available [43,44]. A gap of 20 m was applied when combining bathymetry estimates with MBES data to create a smooth transition.

The new shelf model was clipped to 300 m, as this captured the full extent of the shelf area. The shelf model was then mosaicked with coverage for the land and shelf slopes from the Geoscience Australia land-bathymetry grid [39] to create an updated regional seamless DEM of the land, shelf and slopes (8 m cell size). An overlap distance of 10 m was applied when combing bathymetry estimates with MBES data. Slope and ruggedness terrain derivatives were calculated for the shelf and regional DEMs using ArcGIS Spatial Analyst toolbox and Benthic Terrain Modeler [45], respectively.

2.5. Geomorphomic Feature Interpretation

Broad seafloor features were visually interpreted through digitisation in ArcGIS v10.1 using terminology consistent with international nomenclature [46,47] and national standards [48]. Shelves were classified into shelf region (inner, mid, outer) and geomorphic features, with definitions primarily sourced from [48] with feature definitions and sources presented in Table 1. The classification of geomorphic features extends upon the interpretation of Balls Pyramid shelf undertaken by [31]. These interpretations were further informed by previous characterisations of the Lord Howe Island shelf produced by [29,35–37].

Table 1. Definitions of feature terms used with this study.

Geomorphic Feature	Definition
Coral reef	A tract of corals growing on a massive, wave resistant structure and associated sediments, substantially built by skeletons of successive generations of corals and other calcareous biota [49]
Channel	A linear or sinuous depression on an otherwise flat surface [48]
Basin	A depression, in the seafloor, more or less equidimensional in plan and of variable extent [48]
Depression	A low lying area surrounded by higher ground and with no outlet or opening (i.e., closed) [48]
Pavement	Flat (or gently sloping), low-relief, solid, carbonate rock with little or no fine-scale rugosity. These areas can be covered with algae, hard coral, gorgonians, zooanthids, or other sessile vertebrates; the coverage may be dense enough to partially obscure the underlying surface. On less colonized pavement features, rock may be covered by a thin sand veneer [47]
Ridge	A long, narrow elevation, usually sharp crested with steep sides. Larger ridges can form an extended upland between valleys [48]
Step	A narrow area on the continental (or island) shelf that has a distinctive steep gradient [50]
Terrace	A relatively level or gently inclined surface defined along one edge by a steeper descending slope and along the other by a steeper ascending slope. Terraces may border a valley floor or shoreline, and they can represent the former position of a flood plain or shoreline [48]
Shelf break	The line along which there is a marked increase of slope at the seaward margin of a continental (or island) shelf [46]
Slope	The sloping region that deepens from a shelf to a point where there is a general decrease in gradient [46]

The inner shelf represents the zone within approximately 1 km of the shoreline to around 30–35 m depth. Inner shelf features around Lord Howe Island were defined at a 1:6000 map scale using WV2 imagery, supplemented with ADS40 (2012) (Land and Property Information, New South Wales, Australia) aerial imagery where cloud artefacts obscured the view. The spatial extent of the modern fringing reef was informed by the existing literature [23,28,36].

The middle and outer shelf zones have varying distances from the shoreline, and typically represent the seaward 50 and 130 m isobaths, respectively. For the remaining shelf, features were digitised at 1:10,000 using slope transparenly (50%) displayed over the bathymetry model. The large, mid-shelf fossil reef features were sub-divided into the upper (below 35 m depth), lower (beyond 35 m depth), and intra-reef depression (localised lows within reef structure) features.

The zonal statistics tool in ArcGIS was used to extract summary statistics of depth and slope for each feature, and the zonal histogram tool was used to create hypsometric curves. These allowed for the spatial extent, depth and slope characteristics of specific shelf features to be compared and contrasted. Linear features are included into the zonal histogram analysis, with area representing the cumulative area of individual cells directly beneath the lines. This allows for depth distribution patterns to be described for linear features as well as polygon features.

3. Results

3.1. Depth Estimation for the Lord Howe Island Shelf

The estimated depth surface for the Lord Howe Island shelf was enhanced through the selection of suitable bands, the application of filters and image partitioning (Table 2, Figures 3 and 4). Comparisons of input bands showed that the ratio of blue and green bands had the strongest relationship to known depth. The optimal correlation was achieved for the standard deviation and median 10 (10 cell radius) filter using a third order polynomial function ($R^2 = 0.935$).

Table 2. Regression results and error assessments for selected filters. Results shown for each depth interval.

Filter Type	R^2	Polynomial Type	RMSE for Each Depth Range (m)						
			0–10	0–15	0–20	0–25	0–30	0–35	0–40
No filter	0.807	Order 2	1.98	2.12	2.67	3.73	3.55	3.9	4.61
Low Pass (3 × 3)	0.894	Order 2	0.89	1.15	1.82	2.4	2.41	3.22	4.14
Median 10	0.93	Order 3	0.75	1.03	1.45	2.42	2.2	3.18	4.23
Low pass + Median 10	0.931	Order 3	0.74	1.05	1.55	2.46	2.21	3.14	4.2
Standard deviation (Std)	0.84	Order 2	1.24	1.33	1.8	3.1	3.23	3.85	5.72
Std + Median 10	0.935	Order 3	0.72	0.97	1.12	2.4	2.19	3.15	5.36
West			0.72	0.97	1.12	1.48	1.37	1.98	3.1
East			-	-	1.11	4.03	4.12	5.25	6.23
Image partition									
Std + Median 10			0.72	0.97	1.17	2.1	1.74	2.39	3.3
West	0.935	Order 3	0.72	0.97	1.12	1.48	1.37	1.98	3.1
East	0.875	Order 3	-	-	1.68	3.75	3.13	3.5	3.88

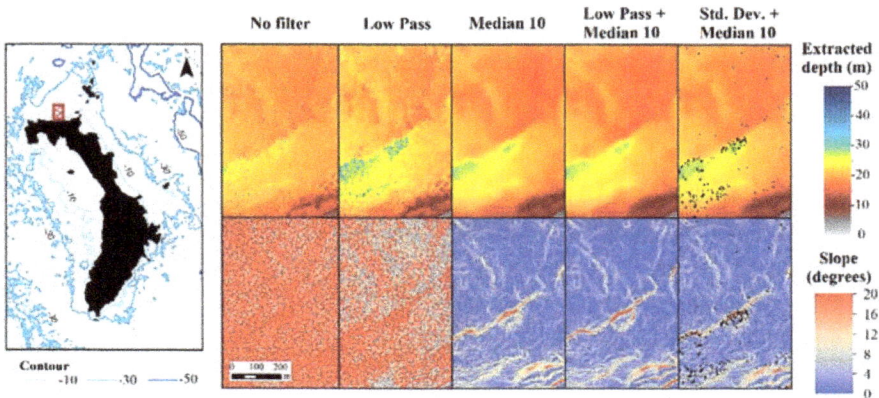

Figure 3. Comparison of bathymetry estimates and calculated slope for selected filters. Black areas for standard deviation and median 10 filter (Std. Dev. + Median 10) denotes "no data" areas.

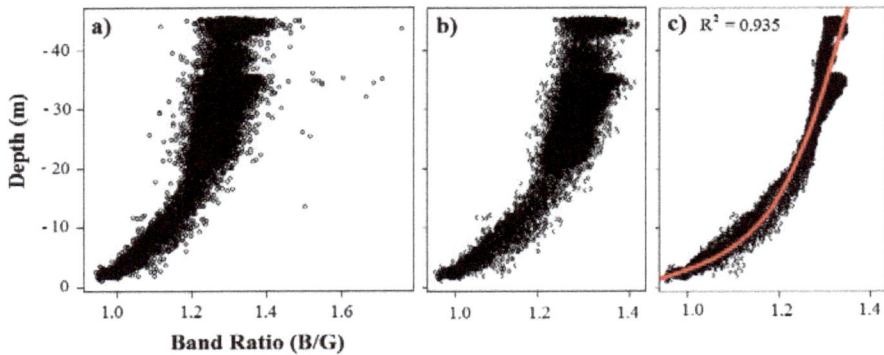

Figure 4. Polynomial regression of blue/green band ratio with: (**a**) no filters applied; (**b**) standard deviation filter applied; (**c**) standard deviation and median (10 cell circle radius) filters Fapplied.

This was further improved with image partitioning, which reduced the overall error of the surface to RMSE 2.36 in 0–35 m water depth. Applying separate regressions to the east and west portions of the image reduced the error of the eastern surface in 25–40 m water depth, although it slightly increased the error in the 0–20 depth interval. Multibeam data points were not available in water depths less than 20 m for the western side, which may have affected the accuracy in this depth interval. The overall error of the surface in the 0–35 m depth interval was reduced from an RMSE error of 3.15 prior to partitioning, to an RMSE error of 2.39 with partitioning applied (Table 2). As with the ROI selection, the reference points for depth calculation were extracted from one continuous section of the image.

In addition to RMSE calculations, residual values for each point in the subsample dataset were plotted to visually assess the areas of greater error. Errors appeared to be greatest in bathymetric depressions, where the calculation appeared to overestimate the gradient of depth. An additional source of error arises from the time difference between the acquisition of imagery and reference depth data. Coordination of data acquisition is logistically difficult, and therefore some of the observed error may originate from sediment movement between the imagery and survey dates.

The application of filters improved the correlation through removing the outliers within the surface and reducing the data spread in deeper waters (Figures 3 and 4). The singular application of a low pass filter appears to produce a strong correlation ($R^2 = 0.894$), however the resultant surface is too variable for the derivation of terrain measures, as shown by high slope values across the surface in Figure 3. The application of a median filter with a sufficiently wide radius (10 cell) provided smoothing to a level most suitable for integration with MBES data. Smaller filter radii (three and five cell) showed higher surface variability, and greater filter windows showed data loss through over-smoothing the surface.

The standard deviation filter removes artefacts of image tiling and outlier values, which can be retained with the median or low pass filters alone. If the image is not affected by surface disturbance or tiling edge effects, the median filter on its own may be sufficient for surface smoothing. The addition of the low pass filter did not significantly improve the correlation or RMSE error of the surfaces.

The pristine water clarity of this study region enabled depth estimations down to 35 m, where typically depth is not derived from satellite imagery beyond 20 m water depth [10]. For the purposes of this study, 35 m was selected as the depth limit, as this was where the deviation in the unfiltered data increased and also provided a balance between coverage and accuracy for seamless integration with MBES data. RMSE error between the overlapping coverages of the 2008 and 2013 MBES surveys was calculated to be 1.15 (3,428,021 points).

3.2. Depth Estimation for the Balls Pyramid Shelf

The optimal correlation for the Balls Pyramid Quickbird image was achieved using a linear function ($R^2 = 0.47$). The weaker correlation of the depth data for the Balls Pyramid shelf is attributable to the severe sun glint evident in the Quickbird image, which hindered the success of the correlation. Furthermore, reliable known depth points from MBES data around Balls Pyramid existed only for waters deeper than 18 m across the mapped area of the shelf. Within the area of interest around the inner shelf, the shallowest MBES data used for calculations and validations was 21 m depth.

The RMSE values for the Balls Pyramid shelf varied with depth. As no MBES data was available for the area of interest in waters shallower than 21 m, the estimated depth in this shallow-water range is considered the area of highest error. Estimated depths are presumed to be overestimated close to the island where the surface is deeper than would likely occur. RMSE values were lowest for 21–35 m, where RMSE = 1.55 (RMSE: 21–40 = 3.18; 21–35 = 1.55; 21–30 = 4.16; 21–25 = 7.76). These estimations are suitable for geomorphic interpretations although a high degree of caution is needed for other applications. This calculation could be improved with the addition of shallow water data, which would allow for the correlation to be fitted to a more representative spread of data. Data were clipped to 35 m depth for seamless integration with MBES data.

3.3. Integrated Bathymetry Model

A high-resolution bathymetry model was produced for the island shelves (5 m cell size). The new estimated depth surfaces from the satellite imagery contributed 34 km^2 of data for the Lord Howe Island shelf and 7.2 km^2 for the Balls Pyramid shelf (Figure 5a). The estimated depth from the WV2 satellite image greatly improved the bathymetric resolution of the southeast shelf of Lord Howe Island (Figure 5b). This region of the shelf is difficult to access due to high exposure to swell and winds, and the previous bathymetry data and subsequent geomorphic interpretations were heavily interpolated in this region. The satellite data were ideal for these applications, and the new bathymetry model has substantially enhanced the detail of the features inaccessible to vessel-based platforms.

The regional land and bathymetry model (8 m cell size) was also updated to include the new shelf model. The production of high-resolution bathymetry models enables the exploration of the terrain, including the calculation of metrics such as slope (Figure 5c), ruggedness (Figure 5d) and the identification of shelf features. Detailed descriptions and comparisons of geomorphic features across the seascape are presented below.

3.4. Geomorphic Interpretation and Shelf Comparison

Diverse accretionary and erosional geomorphic features were mapped on both shelves (Figure 6). The two island shelves possess broad platforms which predominantly occur in 30–60 m water depth (69% and 77% of the shelf area for the Lord Howe Island and Balls Pyramid shelves, respectively) and are dominated by extensive fossil reefs (Figure 7, Table 3). The 504 km^2 area of the Lord Howe Island shelf is almost twice the size of the 261 km^2 Balls Pyramid shelf. Similar proportions of the shelves occur in deeper waters, with 13–14% of shelf area in 60–90 m for both shelves, and 6–9% >90 m. The greatest difference occurs in the shallow waters, where <1% occurs in <30 m depth around Balls Pyramid, while 12% of the Lord Howe Island shelf comprises shallow reefs and depressions, including the modern fringing reef and shallow lagoon.

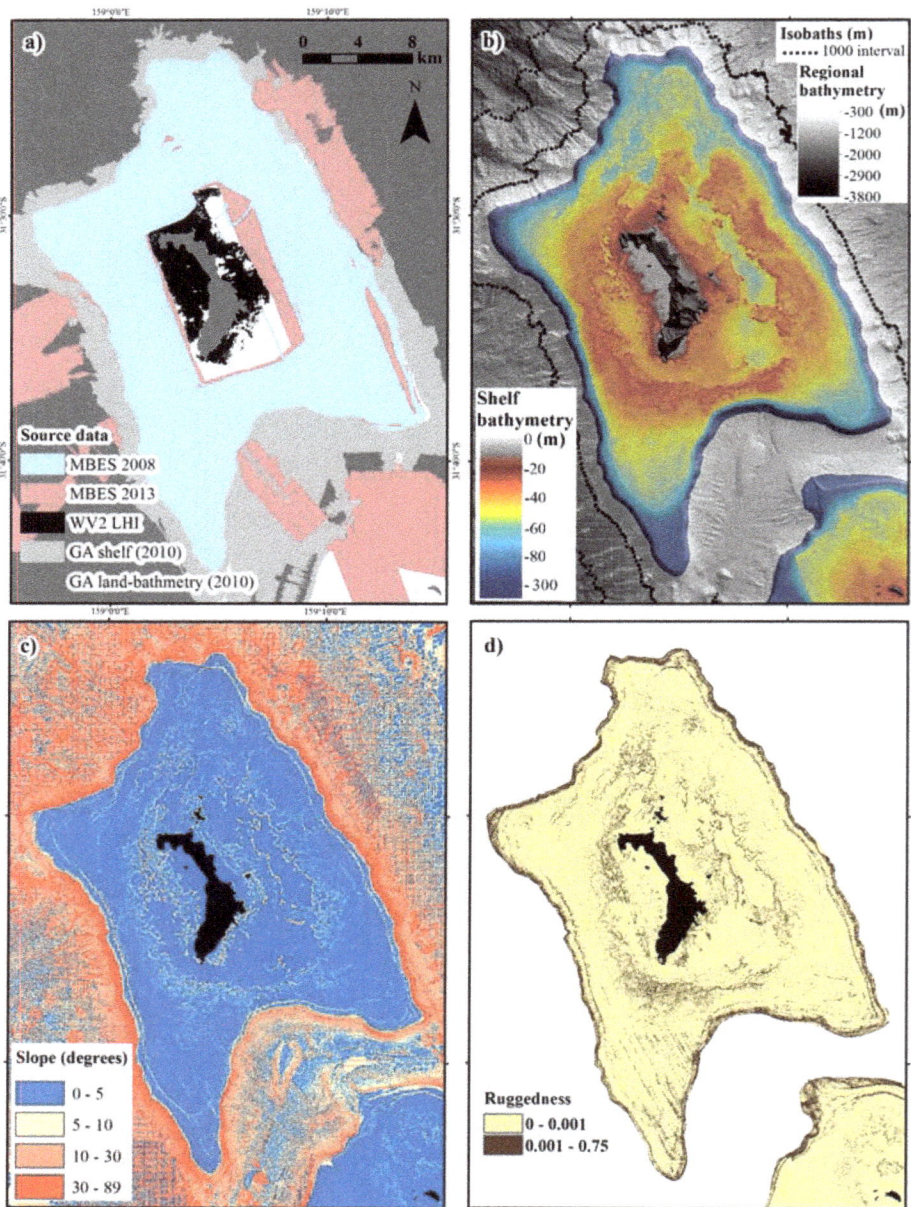

Figure 5. Lord Howe Island shelf: (**a**) Source data coverages for bathymetry model; (**b**) integrated high-resolution bathymetry model (regional grid shown, 8 m cell size). Colour scheme stretched to 100 m to emphasise shelf features. Isobaths displayed at 1000 m intervals (white dashed line) together with the 300 m isobath (shown in red) which represents the limit of the shelf bathymetry model; (**c**) slope calculated for the regional bathymetry model; (**d**) ruggedness (rugosity) calculated at a three cell window for shelf bathymetry model.

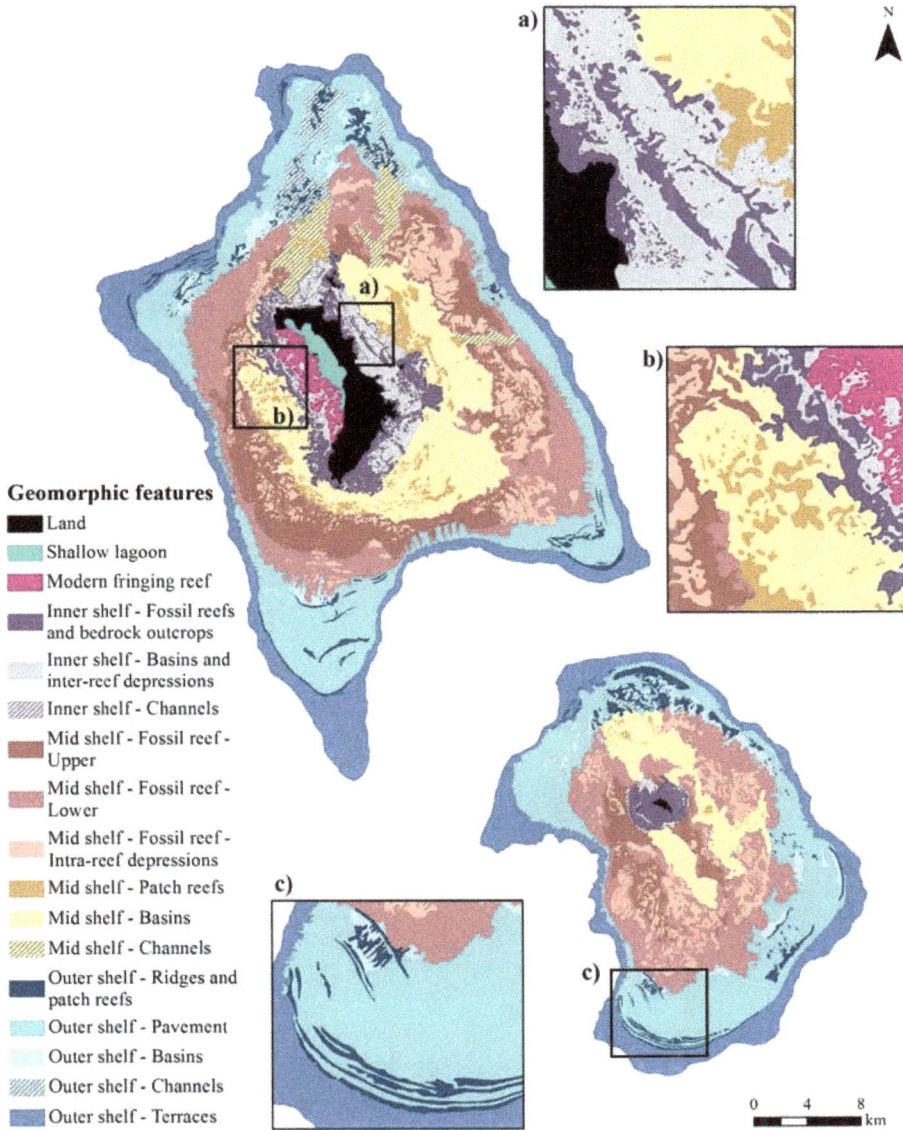

Figure 6. Geomorphic feature interpretation of the Lord Howe Island and Balls Pyramid shelves. Inset locations show: (**a**) linear reefs on the eastern inner shelf of Lord Howe Island; (**b**) patch reefs on the western mid shelf of Lord Howe Island; and (**c**) sub-parallel, linear ridges along the southern outer shelf of Balls Pyramid.

Table 3. Zonal statistics for the geomorphic features of Lord Howe Island and Balls Pyramid shelves. Depth range (R), average (Av), standard deviation (Std) and area percentage (%) of the shelf. Statistics for depth and slope calculated from the 5 m cell size shelf bathymetry grid.

Shelf Region	Geomorphic Feature	Lord Howe Island Shelf					Balls Pyramid Shelf				
		Area (km²)	% of Shelf	Depth (R)	Depth (Av ± Std)	Slope (Av ± Std)	Area (km²)	% of Shelf	Depth (R)	Depth (Av ± Std)	Slope (Av ± Std)
Inner shelf	Modern fringing reef	6.5	1.3	2 to 35	10 ± 7.4	2.9 ± 3.4	-	-	-	-	-
	Shallow lagoon	3.9	0.8	2 to 9	3 ± 0.3	0.3 ± 0.3	-	-	-	-	-
	Fossil reefs and bedrock outcrops	22.8	4.5	1 to 48	25 ± 8.9	4.5 ± 4.4	5.6	2.2	21 to 41	32 ± 2.3	1.6 ± 1.2
	Basins and intra-reef depressions	18.4	3.7	2 to 40	26 ± 8.3	1.9 ± 1.9	0.8	0.3	32 to 40	35 ± 0.7	0.9 ± 1.4
	Channels	0.3	0.1	3 to 31	10 ± 5.1	3.7 ± 4.3	0.6	0.2	31 to 38	35 ± 1.2	1.3 ± 1.1
Mid shelf	Patch reefs	13.6	2.7	19 to 63	38 ± 5.1	4.3 ± 5.3	1.1	0.4	36 to 57	46 ± 2.8	4.5 ± 4.0
	Fossil reefs: –Upper	56.0	11.1	14 to 35	31 ± 2.4	3.2 ± 2.9	13.9	5.3	18 to35	33 ± 1.4	2.7 ± 2.5
	–Lower	70.9	14.1	35 to 76	40 ± 4.3	2.3 ± 2.5	56.1	21.5	35 to 55	42 ± 4.0	2.3 ± 2.2
	–Intra-reef depressions	28.5	5.7	25 to 55	37 ± 4.6	1.7 ± 1.5	16.9	6.5	28 to 50	41 ± 4.1	1.6 ± 1.6
	Channels	13.9	2.8	22 to 67	48 ± 9.1	1.8 ± 2.3	-	-	-	-	-
	Basins	60.0	11.9	23 to 61	44 ± 6.7	1.3 ± 1.5	24.0	9.2	30 to 57	46 ± 4.3	1.0 ± 1.3
Outer shelf	Ridges and patch reefs	11.5	2.3	43 to 80	57 ± 6.5	3.4 ± 3.1	10.8	4.1	41 to 74	53 ± 5.5	2.1 ± 2.0
	Pavement	99.7	19.8	30 to 80	54 ± 5.0	1.9 ± 2.0	72.7	27.9	36 to 104	54 ± 6.4	1.5 ± 2.4
	Basins	8.1	1.6	42 to 63	55 ± 3.9	1.7 ± 1.9	7.1	2.7	41 to 67	52 ± 3.6	1.2 ± 1.2
	Channels	9.9	2.0	47 to 67	59 ± 3.5	2.0 ± 2.2	0.6	0.2	50 to 58	55 ± 1.2	1.4 ± 1.5
	Terraces	79.6	15.8	45 to 215	88 ± 18.4	5.5 ± 6.0	50.4	19.3	47 to 217	94 ± 23.6	4.9 ± 5.8
	Terrace steps (line)	-	-	45 to 213	79 ± 19.9	8.8 ± 7.4	-	-	47 to 136	80 ± 17.5	7.7 ± 6.7
Shelf break	Shelf break (line)	-	-	86 to 206	130 ± 20.6	26.8 ± 8.3	-	-	83 to 217	130 ± 19.9	26.9 ± 12.2
Total	Entire shelf	503.8	100	0 to 215	49 ± 21.6	2.8 ± 3.6	260.7	100	18 to 217	56 ± 22.8	2.4 ± 3.5

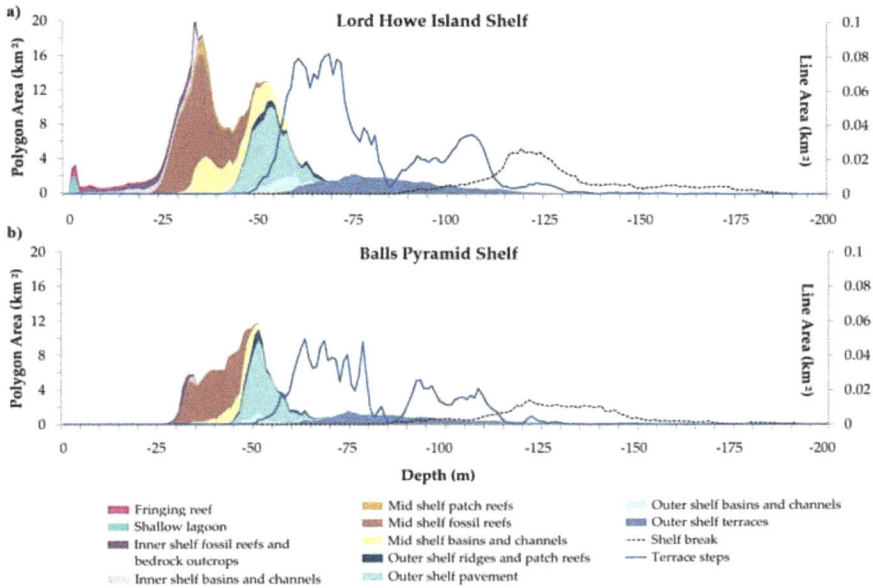

Figure 7. Zonal histogram for geomorphic features of the: (**a**) Lord Howe Island shelf; and (**b**) Balls Pyramid shelf. Line area represents the cumulative area of individual cells beneath the shelf break and terrace step lines, to provide an indication of depth distribution patterns for these linear features.

The hypsometric curve highlights the similarities in the depth distribution of features, as well as the differences in the cumulative area of the Lord Howe Island features (Table 3). A pronounced modal depth occurs on the Lord Howe Island shelf at 35–40 m from the collective contributions of the mid-shelf reefs and inner shelf features. At 30–35 m, a minor mode from the mid-shelf reefs is seen on the Balls Pyramid shelf, though the distribution spreads more broadly across 30–50 m and there is minimal contribution from inner shelf features. Both shelves exhibit a distinct mode at 50–55 m and reach a similar areal extent from the contributions of outer shelf features, although the additional contribution of the mid-shelf basins around Lord Howe Island exceed those of Balls Pyramid. Terrace-step patterns are similar, with multiple modes occurring from 65–80 m and 95–110 m. The shelf break is more distinct around Lord Howe Island at 125 m depth, whereas it is more variable around Balls Pyramid and occurs at 115–150 m.

The bathymetry estimates for the Balls Pyramid shelf generated deeper values around the inner shelf and therefore it is inferred that a greater proportion of shallower substrate <20 m depth likely occurs on the inner shelf. While the hypsometry of depth in shallow waters around Balls Pyramid must be interpreted with caution, area calculations provide for quantification and comparison of the size of the inner-shelf reefs (Table 3). The areal extent of inner-shelf reefs around Balls Pyramid was shown to be reduced by area in relation to the Lord Howe Island shelf; totalling 5.6 km^2 for the Balls Pyramid shelf (2.2% of the shelf) and 22.8 km^2 for the Lord Howe Island shelf (4.5% of the shelf).

3.4.1. Inner Shelf

The Lord Howe Island shelf possesses a more extensive inner-shelf reef system, which uniquely includes a modern fringing reef. The inner-shelf reefs around Lord Howe Island are substantially larger in area (22.8 km^2) and occur across a wider distribution of depths (average 25 ± 8.9 m) compared to the Balls Pyramid inner-shelf reefs (5.6 km^2 area, average depth 32 ± 2.3 m). Complex reef systems occur around the inner shelf of Lord Howe Island, in contrast to the less extensive inner-shelf reef features which encircle the Balls Pyramid sea stack. The previously unmapped inner-shelf around the east and south of Lord Howe Island reveal contiguous reefs extending from the island and patch reefs interspersed within the basin.

To the east of Lord Howe Island, a shore-parallel discontinuous linear reef ridge, 5 km in length, up to 420 m in width and 4 m in vertical relief, extends south from the Admiralty Islands to Muttonbird Island in 14–30 m depth (Figure 6a). Extensive patch reefs extend east of the shore-parallel fringing reef, surrounding Muttonbird Island and encroaching into the eastern mid-shelf basin. Southeast of Muttonbird Island, Wolf Rock rises to the surface at a slope of 10–15° (up to 26°) from a base around 35 m depth. On the northern shelf, smaller linear reef structures up to 2.4 km long, 140 m wide and up to 5 m in relief, occur in a sub-parallel formation in 18–25 m depth. Along the southern coast, a large, contiguous reef adjoins the coastline in 0–30 m depth beneath Mt Gower and Mt Lidgebird.

3.4.2. Middle Shelf

Mid-shelf fossil reef features, including the upper reef, lower reef and intra-reef depressions, dominate both shelves. The 155.4 km^2 fossil reef around Lord Howe Island is almost twice the area (180% larger) of the 86.9 km^2 Balls Pyramid fossil reef, although the reefs comprise a similar proportion of shelf area at 31% for Lord Howe Island and 33% for Balls Pyramid (Table 3). The Lord Howe Island reef forms a barrier-type reef morphology that encircles the island with pronounced, large basins distinctly separating the mid-shelf reefs from the inner-shelf reefs. The mid-shelf reef has a typical relief of 20 m. It is widest in the southeast (5.9 km) and southwest (4.8 km) and extends closest to the shelf break (<400 m) along the western rim. The mid-shelf reef around Balls Pyramid instead forms a platform-type morphology with basins that only partially intersect the reef structure. The reef has a typical relief of 15 m, and is similarly widest on the southwest shelf (5.2 km) where it extends to within 500 m of the shelf break.

Patch reef features are interspersed between the inner-shelf reefs and the fossil reef. On the western middle shelf, a dense network of patch reefs occur in 24–34 m, rising 10–20 m in relief from the basin floor at 42–47 m depth (Figure 6b). To the east, shore-parallel patch reefs form an 8 km chain which adjoins the margin of inner-shelf reefs on the east shelf. Along this chain, reefs rise to 20 m depth from the surrounding inner- and mid-shelf basin floor in 30–58 m depth. Around the southern inner rim of the fossil reef, low-lying reef patches appear to be inundated by sand, visible from satellite and aerial imagery.

Large forereef buttresses occur on the southern seaward rim of the Lord Howe Island shelf, reaching 5–6 m in height, 50–430 m in width and 470–800 m in length. The magnitude of these buttresses is substantially larger than the 1–4 m high forereef buttresses observed elsewhere along the rim of the Lord Howe Island and Balls Pyramid mid-shelf fossil reefs (Figure 8a–c).

Basins are prominent on the northern, eastern and southern mid shelves around both islands, with the western shelf basin reduced in size around Lord Howe Island and absent on the Balls Pyramid shelf. Basin and channel networks dissect the eastern and northern mid-shelf reefs around Lord Howe Island and Balls Pyramid, and the channels connect to outer-shelf channel systems. Steep margins are commonly observed on the inner-reef rim adjoining the basins on both shelves. The basin rims around Lord Howe Island have gradients up to 30° on the eastern basin and up to 22° observed on the eastern basin rim of Balls Pyramid (Figure 8d,e).

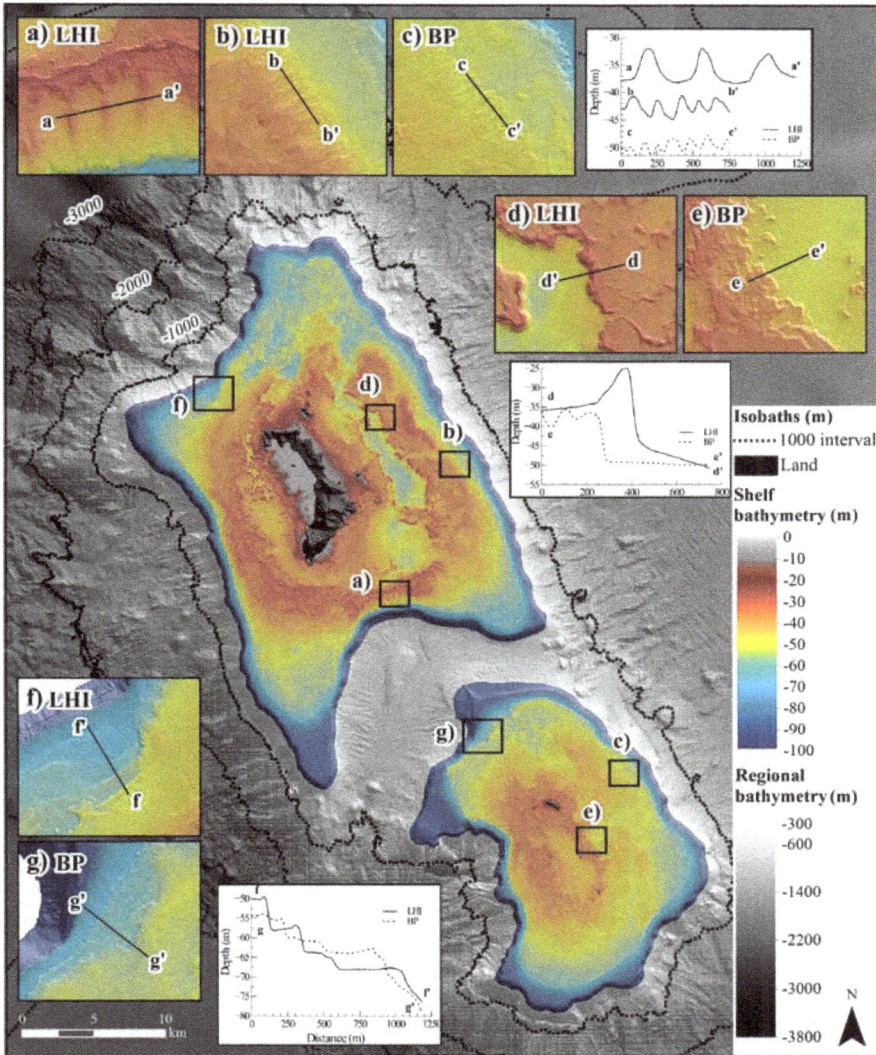

Figure 8. Hillshaded integrated bathymetry grid (8 m cell size) and profiles of: (**a**) southern forereef buttresses of Lord Howe Island (LHI); (**b**) eastern forereef buttresses of LHI; (**c**) eastern forereef buttresses of Balls Pyramid (BP); (**d**) steep rim of eastern basin of LHI; (**e**) steep rim of eastern basin of BP; (**f**) terraces on northeast outer shelf of LHI; (**g**) terraces on northeast outer shelf of BP.

3.4.3. Outer Shelf

The outer-shelf pavement encompasses a large proportion of shelf area for both the Lord Howe Island (99.7 km^2; 20%) and Balls Pyramid (72.7 km^2; 28%) shelves. Underwater imagery of the outer shelf of Balls Pyramid reveals the pavement is commonly encrusted with coralline algae with a carbonate sand veneer [37]. It is widest on the southwestern (11.3 km) and northeastern (7.8 km) section of the Lord Howe Island shelf and the southern section of the Balls Pyramid shelf (3.6 km). It is narrowest (<50 m) on the western side of both shelves where the mid-shelf reefs extend close to the shelf break. The northeast shelves are characterised by basin and channel networks. Patch reefs are more prominent on the northeast shelves whereas sub-parallel, linear ridges are more typical of the southern outer-shelf platforms (Figure 6c).

Terraces are evident along the outer shelf rim of both shelves, predominantly occurring at 65–110 m depth, with a similar average depth (88 ± 18 m for Lord Howe Island; 94 ± 24 m for Balls Pyramid) and average terrace-step depth (79 ± 20 m for Lord Howe Island; 80 ± 18 m for Balls Pyramid). Terraces appear most separated on the gentler-gradient northern and southern shelves, and conjoin along the steeper-gradient eastern and western shelves. The most distinct terrace-step sequences are observed on the northwest shelf region (Figure 8f,g). These appear more clearly defined on the Lord Howe Island shelf, occurring at 50, 57, 63 and 69 m with a raised rim of 0.5–1 m. On Balls Pyramid, steps occur at 55, 60 and 63 m with a raised rim of <0.5 m. The shelf break occurs at the same average of 130 m around both shelves (± 21 m for Lord Howe Island and ± 20 m for Balls Pyramid).

4. Discussion

The estimated depth surfaces for the inner shelves around Lord Howe Island and Balls Pyramid have substantially enhanced the detail of the features that are very difficult to access using vessel-based platforms. The methods presented here provide a more accurate and detailed seabed model derived from satellite imagery and provide data that are more amenable for integration with multibeam echosounder (MBES) data. The application of multiple filters and the use of larger filter windows (10 cell size) provide a smoothing level similar to that produced with 4–5 m cell size MBES grids. Standard deviation filters were shown to remove outliers and artefacts which may occur from tiling edge effects or surface disturbance, and median 10 (10 cell size) filters were shown to produce a level of smoothing comparable with MBES data.

The selection of filters for individual studies depends on the level of pixel-to-pixel variation within the satellite image and the resolution of bathymetry datasets used for integration. The combination of the standard deviation and median 10 filters were selected for this study due to the surface disturbance observed on the eastern inner shelf and improved RMSE performance of the filtered surface. Image partitioning further improved surface accuracy through tailoring the regression to the east–west variation observed across the image. Quantitative (e.g., RMSE calculations) and qualitative (e.g., plotting residuals) error assessments indicate the reliability of the surface and its fitness for purpose.

The empirical, band ratio method for depth estimation from satellite imagery provides a relatively simple method of shallow depth estimation [10,13]. Importantly, in this study the WV2 image for Lord Howe Island reveals high water clarity (Figure 1), and this high clarity together with the high-resolution of the WV2 image enables a reasonably accurate product to be generated for the inner shelf. For the Balls Pyramid shelf, severe sun glint reduced the suitability of the method. Physics-based approaches may offer more accurate surfaces, however such approaches would also be compromised by sun glint and require increased model parameterisation and complex data processing [10]. Therefore, the empirical, ratio-based approach employed here provides an efficient and relatively accurate method suitable for seabed geomorphic analysis.

4.1. Comparison of Shelf Morphology

The creation of a seamless bathymetry model for the entire shelf region of the two island shelves enabled geomorphic features to be mapped from the shoreline to shelf break at the same resolution and scale. The high-resolution shelf (5 m cell size) and regional (land, and shelf and slope, 8 m cell size) DEMs provide detailed information on shelf morphology, which allow for comparisons of the extent and distribution of fossil reefs. This, in turn, informs interpretations on the formation and driving processes of shelf features. The two oceanic shelves possess a diverse range of accretionary and erosional geomorphic features which have been defined and described. Submerged fossil coral reefs, basins, channels, pavements and terraces are identified on both shelves, with the expression and extent of features typically more pronounced on the larger shelf surrounding Lord Howe Island.

Mid-shelf fossil reefs dominate both shelves in 25–50 m, comprising a similar proportion of shelf area (approximately one third). The reefs form concentric patterns encircling large basins, which are inferred to be paleolagoons. The morphological similarities and comparable depth distributions suggest the mid-shelf reefs developed concurrently, with the mid-shelf reef around Balls Pyramid appearing to have drowned with postglacial sea-level rise [30] while the Lord Howe Island shelf backstepped to form the modern reef [28,29]. The Lord Howe Island mid-shelf fossil reef accreted several metres during the Holocene (9–2 ka) [29], and it is presumed the Last Interglacial (125 ka, Marine Isotope Stage, MIS 5) reef material forms a significant component of the reef foundations, and possibly deposits from preceding interglacials (MIS 7, 9, 11).

The lateral and vertical extent of reef development is greatest on the southwestern shelves, interpreted as the more exposed, windward setting. Forereef buttresses border the eastern, western, and southern rims of the mid-shelf reefs and outer-shelf pavements, indicating variable exposure gradients which are typical of the mid-ocean setting [26]. The development of larger buttresses (5–6 m height) on the southern rim of the Lord Howe Island fossil reef suggest the southern reef was exposed to significantly higher prevailing energy conditions from due south than occurred around the surrounding shelf where buttresses were reduced in size (1–2 m height).

On the outer shelf, where there was ample substrate available for colonisation on the outer-shelf pavement, the reefs cover similar relative areas on both shelves. These reefs formed as ridges and patch reefs, with ridges most developed on the southern outer shelves. Similar paleoshoreline features have been described around the Australian continental shelf [51] and the linear, sub-parallel configuration of these features suggest beach barrier or coral reef origins. Occurring at 40–80 m depth, these features may have formed during postglacial sea-level rise of the Early Holocene or during earlier interstadials (e.g., MIS 3).

The dense network of patch reefs on the mid shelf and the linear reef systems on the eastern and northern inner shelves of Lord Howe Island are interpreted as transitional fossil patch and fringing reefs that developed as the reef retreated landward with postglacial sea-level rise. As the linear reefs have a maximum relief of 4 m, the associated lagoons are likely to be shallow and therefore more typical of fringing reef than barrier systems [52]. While the less-exposed west coast is dominated by reef accretion, the more exposed eastern, northern and southern coasts have limited coral accretion and the substratum comprises volcanic and calcarenite outcrops [23,36]. Along the southern coast, the nearshore waters adjoining the steep basalt cliffs are characterised by boulder stacks and plunging cliffs, which likely extend to form the contiguous reef mapped along the southern inner shelf.

Unlike the mosaic of different reef morphologies observed on the Lord Howe Island inner shelf, the Balls Pyramid inner shelf possesses a more limited inner shelf fossil reef. The Balls Pyramid pinnacle comprises steep cliffs which plunge into shallow waters, and the contiguous inner shelf reef surrounding the island is likely dominated by volcanic bedrock. The concentric formation of the outer edge of the inner shelf reefs, intersected with narrow channels, suggests constructional fossil reef origins which may have accreted, in part, during the postglacial rise in sea-level.

In addition to accretionary geomorphic features, the shelves exhibit diverse erosional features and morphologies. Complex networks of basins and channel systems characterise the northeast shelves, interpreted as the more sheltered, leeward setting. Basin features are interconnected to the channels, which are interpreted as inter-reef passages which would have functioned to transport water from the paleolagoons when the sea level was at or near the fossil reef surface. Three prominent channels dissect the Lord Howe Island mid-shelf reef, whereas distinct channels are not apparent on the mid shelf around Balls Pyramid. However, the leeward setting is apparent on the Balls Pyramid shelf through developed channels on the northeast outer shelf and the extension of a large northern mid-shelf basin.

During periods of lower sea level when the shelf was exposed, the mid- and outer-shelf channels appear to have fed sediment off the shelf edge, as suggested by the sub-bottom profiles presented for the Balls Pyramid northeast shelf by [30]. These processes are similarly inferred for the Lord Howe Island shelf, where distinct channels are evident on the northeast middle shelf of Lord Howe Island, with a complex network of channels extending across the northeast shelf. Sediment samples collected from the slope areas by [53] indicate the transport and deposition of sediments off the shelf into slope areas during periods of lower sea level.

The karstification of limestone shelf features likely occurred during times of lower sea level, as suggested by onshore deposits of calcarenites around Lord Howe Island [27]. Karst features including dolines, caves and subaerially exposed speleothems were documented within calcarenite sequences around the island, which experienced dissolution and weathering following deposition during the MIS 7 [27]. Morphological characteristics of the mid-shelf basins, including steep basin rims and a sand-inundated low-profile reef, suggest the basin morphology may reflect karstification processes (e.g., [54]). The steeper basin rims and greater extent of the mid-shelf basins around Lord Howe Island (60 km^2 around Lord Howe Island; 24 km^2 around Balls Pyramid) likely reflect the greater volume of water drainage from the larger shelf system during periods when the sea-level was at or near the fossil reef surface and during lowstands when the shelves were exposed.

Terrace and step features are associated with lowstand sea levels during the last glacial period (Last Glacial Maximum ~21 ka) and the preceding interstadial and glacial periods of lower sea level. The depth range of these features are distributed across a wide spread of depths (45–217 m), corresponding to a range of lower sea levels. Similar mean depths of step features occur on both shelves (79–80 m), which may be associated with MIS 3. Morphologically, terrace step feature patterns are remarkably similar for the two shelves, particularly in the northwest where several distinct terraces form with rimmed margins (Figure 8f,g). The shelf break similarly varies around the island shelves (83–217 m, mean depth 130 m). Shelf planation is proposed to have occurred rapidly after the formation of the shield volcanoes (6–7 million years ago), with marine abrasion accounting for the majority (90%) of erosion [26]. Following shelf planation, carbonate sequences were deposited over the basalt platform [29,30,32], and accretionary and erosional processes during sea level lowstands shaped the variable nature of the shelf break.

The availability of substrate for coral colonisation, leading to reef formation, is a key factor differentiating the morphology of the two shelves. The larger size of the shelf and thus the original formative volcano of Lord Howe Island, translates to larger island remnants that remained after shelf planation. The greater extent of reefs around the Lord Howe Island inner shelf was likely facilitated through the availability of shallow substrates and the larger island size which presumably provided greater shelter from exposure. Possibly, slightly warmer sea temperatures and/or more favourable currents (e.g., upwelling) or levels of exposure may have enabled coral to grow faster on the Lord Howe Shelf. In contrast to Lord Howe Island, the Balls Pyramid shelf possesses minimal shallow inner-shelf substrates and the steep pinnacle provides little shelter from high wind and wave energies. Although the areal extent of shelf features are reduced in comparison to Lord Howe Island, substantial past reef development is evident on the Balls Pyramid.

4.2. Applications for Management

Previous studies of the distribution of benthic assemblages around the island shelves and broader Lord Howe region have shown strong correlations to geomorphic features and shelf regions [31,32,37,55–57]. Abundant hard corals were recently discovered growing on the mid-shelf reef of Balls Pyramid, showing increased abundance associated with the mid- and outer-shelf reef features [37]. It is likely that similar distributions of hard corals occur around the Lord Howe Island mid-shelf reef, particularly given the development of a modern fringing reef and the more extensive fossil reef. The outer shelf pavement has been characterised as an area of sand veneers and rhodolith beds [32,37,55], with gorgonian whips and fans observed on the outer shelf and shelf break [37,55]. Investigations of benthic invertebrates around the Lord Howe Island shelf have shown that the infaunal benthic community structure was significantly different between geomorphic zones (fossil reef, basins and outer shelf, [56]).

While geomorphology appears to be a useful surrogate for benthic assemblages for the mid- to outer-shelf features, benthic communities around the inner shelf appear to be strongly structured by the hydrodynamic regime [23,58,59]. Geomorphology is considered to be less useful as a surrogate in this zone, however terrain variables derived from the DEM, such as seafloor ruggedness, may provide useful proxies for explaining distribution of benthic assemblages.

Seafloor habitat mapping is an important component for marine spatial planning and fisheries management [7,60]. The high-resolution bathymetry model and geomorphic characterisation produced in this study feed directly into the management needs identified by marine park managers [34]. These macro-scale classifications of geomorphic features fit within the hierarchical framework of biome and provincial characterisations of the seafloor and biogeography for the broader Lord Howe region [3,31]. The datasets produced by this study reveal detailed bathymetric information and characterise the geodiversity of the shelf landscape. The continuous depth information and stratification of the shelf into distinct features can be utilised in the ongoing planning and management of the shelf environment. An understanding of geodiversity around the shelf can assist in the experimental design of future data collection, and can identify areas of potential biodiversity, which can be targeted for further exploration.

Future research will focus on resolving the timing of reef accretion around the Balls Pyramid shelf and the evolution of the shelf features. The distribution of benthic assemblages around the Lord Howe Island shelf will also be explored to further examine relationships between biota to underlying geomorphology and terrain variables.

5. Conclusions

The principal conclusions that arise from the present study are:

1. Filtering of the band ratio grid with standard deviation and median filters improved the surface for integration with multibeam echosounder data.
2. Image partitioning further improved the surface, accounting for east-to-west variation across the image.
3. The integration of depth derived from satellite imagery together with multibeam echosounder data produced a seamless 5 m cell size bathymetry model of the shelf from the shoreline to shelf break.
4. Geomorphic interpretations of the shelves defined diverse accretionary and erosional features, with mid-shelf fossil reefs shown to dominate both shelves in 25–50 m.
5. The mid-shelf fossil reef around Balls Pyramid is approximately half the area of the Lord Howe Island mid-shelf reef, although it represents a similar proportion of shelf area (approximately one third).
6. The morphology, size, configuration and depth distribution of outer shelf features are most similar between the two shelves, and are most dissimilar for inner shelf features.

Acknowledgments: Thank you to the Marine National Facility for the use of the R.V. *Southern Surveyor* and to all the staff and crew aboard the voyages in 2008 (SS06-2008) and 2013 (SS2013_v02). Thank you to New South Wales (NSW) Roads and Maritime Services for provision of single beam data. We express our thanks to the Lord Howe Island Marine Park Authority (LHIMPA, Permit LHIMP/R/2012/013), NSW Department of Primary Industries (DPI, Permit P12/0030-1.0) and the Department of the Environment (Permit 003-RRR-120918-02). Funding for this research was received by the Australian Government's National Environmental Research Program Marine Biodiversity Hub. Funding was also provided by NSW DPI, LHIMPA and University of Wollongong (UOW) Research Partnerships Grant. M.L. gratefully acknowledges support from the Australian Postgraduate Award. B.P.B. and S.L.N. publish with the permission of the Chief Executive Officer of Geoscience Australia.

Author Contributions: This work is based on the doctoral thesis of M.L., who conducted the data processing and analysis and wrote the paper. C.D.W., S.M.H., B.P.B., S.L.N., A.R.J. reviewed the analyses and revised the paper.

Conflicts of Interest: The authors declare no conflict of interest.

Appendix A

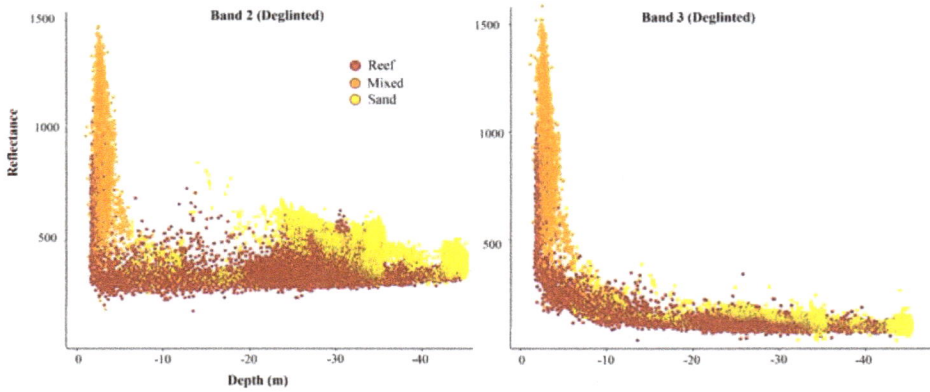

Figure A1. Deglinted band 2 and band 3 reflectance values plotted against depth and coloured by substrate type (sand, mixed sand and reef, and reef).

Appendix B

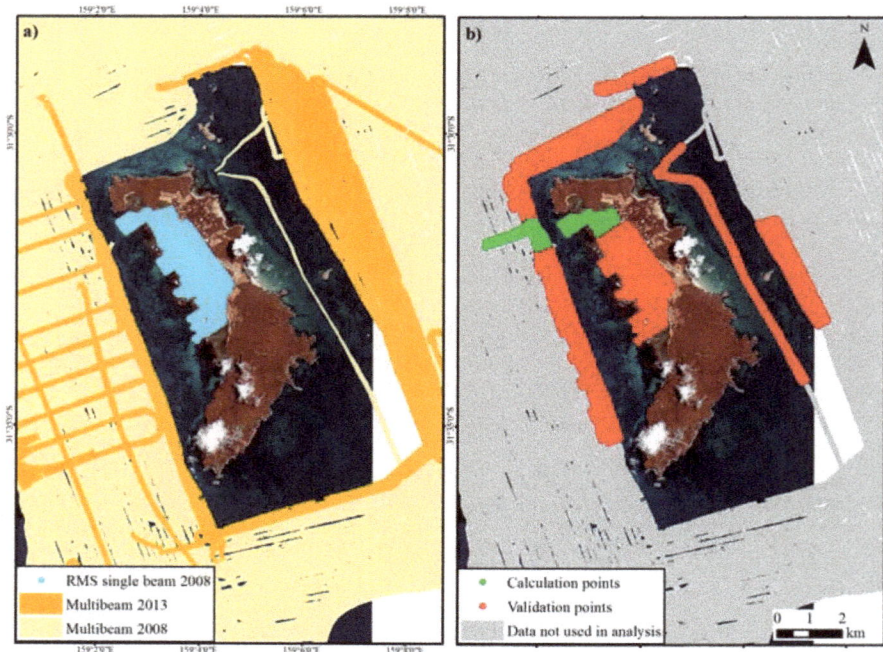

Figure A2. (a) Coverage of 2008 and 2013 multibeam data, and 2008 Roads and Maritime (RMS) single beam data; (b) Calculation points used to estimate bathymetry, which extend from the western lagoon, and validation points which include all remaining known depth points in shallower areas (approximately <40 m depth).

References

1. Harris, P.T.; Baker, E.K. *Seafloor Geomorphology as Benthic Habitat: Geohab Atlas of Seafloor Geomorphic Features and Benthic Habitats*; Elsevier Science: Burlington, ON, Canada, 2012; p. 947.

2. Directive 2008/56/EC of the European Parliament and of the Council of 17 June 2008: Establishing a Framework for Community Action in the Field of Marine Environmental Policy (Marine Strategy Framework Directive). Available online: http://eur-lex.europa.eu/eli/dir/2008/56/oj (accessed on 25 December 2017).

3. Harris, P.T.; Heap, A.D.; Whiteway, T.; Post, A. Application of biophysical information to support Australia's representative marine protected area program. *Ocean Coast. Manag.* **2008**, *51*, 701–711. [CrossRef]

4. Harris, P.T.; Macmillan-Lawler, M.; Rupp, J.; Baker, E.K. Geomorphology of the oceans. *Mar. Geol.* **2014**, *352*, 4–24. [CrossRef]

5. Stevens, T.; Connolly, R.M. Local-scale mapping of benthic habitats to assess representation in a marine protected area. *Mar. Freshw. Res.* **2005**, *56*, 111–123. [CrossRef]

6. Last, P.R.; Lyne, V.D.; Williams, A.; Davies, C.R.; Butler, A.J.; Yearsley, G.K. A hierarchical framework for classifying seabed biodiversity with application to planning and managing Australia's marine biological resources. *Biol. Conserv.* **2010**, *143*, 1675–1686. [CrossRef]

7. Jordan, A.R.; Lawler, M.; Halley, V.; Barrett, N. Seabed habitat mapping in the Kent Group of islands and its role in marine protected area planning. *Aquat. Conserv. Mar. Freshw. Ecosyst.* **2005**, *15*, 51–70. [CrossRef]

8. Evans, I.S. Geomorphometry and landform mapping: What is a landform? *Geomorphology* **2012**, *137*, 94–106. [CrossRef]

9. Leon, J.X.; Phinn, S.R.; Hamylton, S.M.; Saunders, M.I. Filling the 'white ribbon'—A multisource seamless digital elevation model for Lizard Island, Northern Great Barrier Reef. *Int. J. Remote Sens.* **2013**, *34*, 6337–6354. [CrossRef]

10. Gao, J. Bathymetric mapping by means of remote sensing: Methods, accuracy and limitations. *Prog. Phys. Geogr.* **2009**, *33*, 103–116. [CrossRef]

11. Brando, V.E.; Anstee, J.M.; Wettle, M.; Dekker, A.G.; Phinn, S.R.; Roelfsema, C. A physics based retrieval and quality assessment of bathymetry from suboptimal hyperspectral data. *Remote Sens. Environ.* **2009**, *113*, 755–770. [CrossRef]

12. Botha, E.; Brando, V.; Dekker, A. Effects of per-pixel variability on uncertainties in bathymetric retrievals from high-resolution satellite images. *Remote Sens.* **2016**, *8*, 459. [CrossRef]

13. Stumpf, R.P.; Holderied, K.; Sinclair, M. Determination of water depth with high-resolution satellite imagery over variable bottom types. *Limnol. Oceanogr.* **2003**, *48*, 547–556. [CrossRef]

14. Hernandez, W.; Armstrong, R. Deriving bathymetry from multispectral remote sensing data. *J. Mar. Sci. Eng.* **2016**, *4*, 8. [CrossRef]

15. Hedley, J.D.; Harborne, A.R.; Mumby, P.J. Technical note: Simple and robust removal of sun glint for mapping shallow-water benthos. *Int. J. Remote Sens.* **2005**, *26*, 2107–2112. [CrossRef]

16. McArthur, M.A.; Brooke, B.P.; Przeslawski, R.; Ryan, D.A.; Lucieer, V.L.; Nichol, S.L.; McCallum, A.W.; Mellin, C.; Cresswell, I.D.; Radke, L.C. On the use of abiotic surrogates to describe marine benthic biodiversity. *Estuar. Coast. Shelf Sci.* **2010**, *88*, 21–32. [CrossRef]

17. Sagar, S.; Roberts, D.; Bala, B.; Lymburner, L. Extracting the intertidal extent and topography of the Australian coastline from a 28 year time series of Landsat observations. *Remote Sens. Environ.* **2017**, *195*, 153–169. [CrossRef]

18. Kennedy, D.M.; Ierodiaconou, D.; Schimel, A. Granitic coastal geomorphology: Applying integrated terrestrial and bathymetric LiDAR with multibeam sonar to examine coastal landscape evolution. *Earth Surf. Process. Landf.* **2014**, *39*, 1663–1674. [CrossRef]

19. Quadros, N.; Rigby, J. *Construction of a High Accuracy Seamless, State-Wide Coastal DEM*; FIG Coastal Zone Special Publication: Sydney, Australia, 2010.

20. Beaman, R. *3DGBR: A High-Resolution Depth Model for the Great Barrier Reef and Coral Sea*; Marine and Tropical Sciences Research Facility (MTSRF) Project 2.5i.1a Final Report; Reef and Rainforest Research Centre: Cairns, Australia, 2010.

21. Bridge, T.; Beaman, R.; Done, T.; Webster, J. Predicting the location and spatial extent of submerged coral reef habitat in the Great Barrier Reef World Heritage Area, Australia. *PLoS ONE* **2012**, *7*, e48203. [CrossRef] [PubMed]

22. Harris, P.T.; Bridge, T.C.; Beaman, R.J.; Webster, J.M.; Nichol, S.L.; Brooke, B.P. Submerged banks in the Great Barrier Reef, Australia, greatly increase available coral reef habitat. *ICES J. Mar. Sci. J. Cons.* **2013**, *70*, 284–293. [CrossRef]

23. Veron, J.; Done, T. Corals and coral communities of Lord Howe Island. *Aust. J. Mar. Freshw. Res.* **1979**, *30*, 203–236. [CrossRef]

24. Roberts, C.M.; McClean, C.J.; Veron, J.E.; Hawkins, J.P.; Allen, G.R.; McAllister, D.E.; Mittermeier, C.G.; Schueler, F.W.; Spalding, M.; Wells, F.; et al. Marine biodiversity hotspots and conservation priorities for tropical reefs. *Science* **2002**, *295*, 1280–1284. [CrossRef] [PubMed]

25. McDougall, I.; Embleton, B.J.J.; Stone, D.B. Origin and evolution of Lord Howe Island, southwest Pacific Ocean. *J. Geol. Soc. Aust.* **1981**, *28*, 155–176. [CrossRef]

26. Dickson, M. The development of talus slopes around Lord Howe Island and implications for the history of island planation. *Aust. Geogr.* **2004**, *35*, 223–238. [CrossRef]

27. Brooke, B.P.; Woodroffe, C.D.; Murray-Wallace, C.V.; Heijnis, H.; Jones, B.G. Quaternary calcarenite stratigraphy on Lord Howe Island, southwestern Pacific Ocean and the record of coastal carbonate deposition. *Quat. Sci. Rev.* **2003**, *22*, 859–880. [CrossRef]

28. Kennedy, D.M.; Woodroffe, C.D. Holocene lagoonal sedimentation at the latitudinal limits of reef growth, Lord Howe Island, Tasman sea. *Mar. Geol.* **2000**, *169*, 287–304. [CrossRef]

29. Woodroffe, C.D.; Brooke, B.P.; Linklater, M.; Kennedy, D.M.; Jones, B.G.; Buchanan, C.; Mleczko, R.; Hua, Q.; Zhao, J. Response of coral reefs to climate change: Expansion and demise of the southernmost pacific coral reef. *Geophys. Res. Lett.* **2010**, *37*, 15. [CrossRef]

30. Linklater, M.; Brooke, B.P.; Hamylton, S.M.; Nichol, S.L.; Woodroffe, C.D. Submerged fossil reefs discovered beyond the limit of modern reef growth in the pacific ocean. *Geomorphology* **2015**, *246*, 579–588. [CrossRef]

31. Przeslawski, R.; Williams, A.; Nichol, S.L.; Hughes, M.G.; Anderson, T.J.; Althaus, F. Biogeography of the Lord Howe Rise region, Tasman sea. *Deep Sea Res. Part II Top. Stud. Oceanogr.* **2011**, *58*, 959–969. [CrossRef]

32. Kennedy, D.M.; Woodroffe, C.D.; Jones, B.G.; Dickson, M.E.; Phipps, C.V.G. Carbonate sedimentation on subtropical shelves around Lord Howe Island and Balls Pyramid, southwest Pacific. *Mar. Geol.* **2002**, *188*, 333–349. [CrossRef]

33. Brooke, B.P.; Woodroffe, C.D.; Linklater, M.; McArthur, M.A.; Nichol, S.L.; Jones, B.G.; Kennedy, D.M.; Buchanan, C.; Spinoccia, M.; Mleczko, R.; et al. *Geomorphology of the Lord Howe Island shelf and Submarine Volcano: SS06-2008 Post-Survey Report Record 2010/26*; Geoscience Australia: Canberra, Australia, 2010; p. 125.

34. Lord Howe Island Marine Park: Zoning Plan Review Report. Available online: https://www.yumpu.com/en/document/view/49288746/lord-howe-island-marine-park-zoning-plan-review-report (accessed on 25 December 2017).

35. Linklater, M. An Assessment of the Geomorphology and Benthic Environments of the Lord Howe Island Shelf, Southwest Pacific Ocean, and Implications for Quaternary Sea Level. Ph.D. Thesis, University of Wollongong, Wollongong, Australia, 2009.

36. Environment Australia and Marine Parks Authority. *Lord Howe Island Marine Park Issues Paper: A Planning Issues Paper for the Lord Howe Island Marine Park (State and Commonwealth Waters)*; Environment Australia, Canberra NSW Marine Parks Authority: Canberra, Australia, 2001; p. 116.

37. Linklater, M.; Carroll, A.G.; Hamylton, S.M.; Jordan, A.R.; Brooke, B.P.; Nichol, S.L.; Woodroffe, C.D. High coral cover on a mesophotic, subtropical island platform at the limits of coral reef growth. *Cont. Shelf Res.* **2016**, *130*, 34–46. [CrossRef]

38. New South Wales (NSW) Marine Parks Authority. *Marine Parks: Strategic Research Framework 2010–2015*; NSW Marine Parks Authority: Sydney, Australia, 2010.

39. Mleczko, R.; Sagar, S.; Spinoccia, M.; Brooke, B.P. *The Creation of High Resolution Bathymetry Grids for the Lord Howe Island Region*; Geoscience Australia: Canberra, Australia, 2010; p. 58.

40. Berk, A.; Bernstein, L.; Robertson, D. *MODTRAN: A Moderate Resolution Model for LOWTRAN7*; GL-TR-89-0122; AFG Lab., Hanscom Air Force Base: Bedford, MA, USA, 1989; p. 38.

41. Matthew, M.W.; Adler-Golden, S.M.; Berk, A.; Richtsmeier, S.C.; Levine, R.Y.; Bernstein, L.S.; Acharya, P.K.; Anderson, G.P.; Felde, G.W.; Hoke, M.P.; et al. Status of atmospheric correction using a MODTRAN4-based algorithm. In Proceedings of the SPIE Proceedings, Algorithms for Multispectral, Hyperspectral, and Ultraspectral Imagery VI, Orlando, FL, USA, 24–26 April 2000.

42. Jacobs, R. *NSW Ocean and River Entrance Tidal Levels Annual Summary 2015–2016*; Manly Hydraulics Laboratory: Sidney, Australia, 2016; p. 122.

43. Li, J.; Heap, A.D. *A Review of Spatial Interpolation Methods for Environmental Scientists*; Geoscience Australia: Canberra, Australia, 2008; p. 137.

44. Arun, P.V. A comparative analysis of different DEM interpolation methods. *Egypt. J. Remote Sens. Space Sci.* **2013**, *16*, 133–139.

45. *ArcGIS Benthic Terrain Modeler (BTM)*; version 3.0; Environmental Systems Research Institute, NOAA Coastal Services Center: Boston, MA, USA, 2012.

46. International Hydrographic Organisation (IHO). *Standardization of Undersea Feature Names: Guidelines, Proposal Form, Terminology, Bathymetric Publication no.6*, 4th ed.; International Hydrographic Bureau: Monaco, 2013; p. 38.

47. Madden, C.; Goodin, K.; Allee, R.; Cicchetti, G.; Moses, C.; Finkbeiner, M.; Bamford, D. *Coastal and Marine Ecological Classification Standard*; National Oceanic and Atmospheric Administration and NatureServe: Washington, DC, USA, 2009.

48. Nichol, S.L.; Huang, Z.; Howard, F.; Porter-Smith, R.; Lucieer, V.L.; Barrett, N. *Geomorphological Classification of Reefs-Draft Framework for an Australian Standard; Geoscience Australia: Report to the National Environmental Science Program*; Marine Biodiversity Hub: Hobart, Australia, 2016; p. 27.

49. Done, T. Coral reef, definition. In *Encyclopedia of Modern Coral Reefs: Structure, Form and Process*; Hopley, D., Ed.; Springer: Dordrecht, The Nederland, 2011; pp. 261–267.

50. Beaman, R.J.; Webster, J.M.; Wust, R.A.J. New evidence for drowned shelf edge reefs in the Great Barrier Reef, Australia. *Mar. Geol.* **2008**, *247*, 17–34. [CrossRef]

51. Brooke, B.P.; Nichol, S.L.; Huang, Z.; Beaman, R.J. Palaeoshorelines on the Australian continental shelf: Morphology, sea-level relationship and applications to environmental management and archaeology. *Cont. Shelf Res.* **2017**, *134*, 26–38. [CrossRef]

52. Kennedy, D.; Woodroffe, C.D. Fringing reef growth and morphology: A review. *Earth-Sci. Rev.* **2002**, *57*, 255–277. [CrossRef]

53. Kennedy, D.; Brooke, B.P.; Woodroffe, C.D.; Jones, B.; Waikari, C.; Nichol, S.L. The geomorphology of the flanks of the Lord Howe Island volcano, Tasman Sea, Australia. Deep Sea Research Part II: Topical Studies. *Oceanography* **2011**, *58*, 899–908.

54. Stoddart, D.R.; Spencer, T.; Woodroffe, C.D. *Mauke, Mitiaro and Atiu: Geomorphology of Makatea Islands in the southern Cooks*; National Museum of Natural History; Smithsonian Institution: Washington, DC, USA, 1990; p. 65.

55. Speare, P.P.; Cappo, M.M.; Rees, M.M.; Brownlie, J.J.; Oxley, W.W. *Deeper Water Fish and Benthic Surveys in the Lord Howe Island Marine Park (Commonwealth Waters): February 2004*; Australian Institute of Marine Science: Townsville, Australia, 2004; p. 36.

56. Anderson, T.; McArthur, M.; Syms, C.; Nichol, S.L.; Brooke, B.P. Infaunal biodiversity and ecological function on a remote oceanic island: The role of biogeography and bio-physical surrogates. *Estuar. Coast. Shelf Sci.* **2013**, *117*, 227–237. [CrossRef]

57. Brooke, B.P.; McArthur, M.A.; Woodroffe, C.D.; Linklater, M.; Nichol, S.L.; Anderson, T.J.; Mleczko, R.; Sagar, S. Geomorphic features and infauna diversity of a subtropical mid-ocean carbonate shelf: Lord Howe Island, southwest Pacific Ocean. In *Seafloor Geomorphology as Benthic Habitat: Geohab Atlas of Seafloor Geomorphic Features and Benthic Habitats*; Harris, P.T., Baker, E.K., Eds.; Elsevier Science: Burlington, ON, Canada, 2012; pp. 375–386.

58. Harriott, V.J.; Harrison, P.L.; Banks, S.A. The coral communities of Lord Howe Island. *Mar. Freshw. Res.* **1995**, *46*, 457–465. [CrossRef]

59. Edgar, G.J.; Davey, A.; Kelly, G.; Mawbey, R.B.; Parsons, K. Biogeographical and ecological context for managing threats to coral and rocky reef communities in the Lord Howe Island Marine Park, south-western Pacific. *Aquat. Conserv. Mar. Freshw. Ecosyst.* **2010**, *20*, 378–396. [CrossRef]
60. Shumchenia, E.J.; King, J.W. Comparison of methods for integrating biological and physical data for marine habitat mapping and classification. *Cont. Shelf Res.* **2010**, *30*, 1717–1729. [CrossRef]

geosciences

MDPI

Article

Bathymetry and Canyons of the Eastern Bering Sea Slope

Mark Zimmermann [1,*] and **Megan M. Prescott** [2]

[1] National Marine Fisheries Service, NOAA, Alaska Fisheries Science Center, 7600 Sand Point Way NE, Bldg. 4, Seattle, WA 98115-6349, USA

[2] Lynker Technologies, Under contract to Alaska Fisheries Science Center, 7600 Sand Point Way NE, Bldg. 4, Seattle, WA 98115-6349, USA; megan.prescott@noaa.gov

* Correspondence: mark.zimmermann@noaa.gov; Tel.: +1-206-526-4119

Received: 28 February 2018; Accepted: 15 May 2018; Published: 21 May 2018

Abstract: We created a new, 100 m horizontal resolution bathymetry raster and used it to define 29 canyons of the eastern Bering Sea (EBS) slope area off of Alaska, USA. To create this bathymetry surface we proofed, edited, and digitized 18 million soundings from over 200 individual sources. Despite the vast size (~1250 km long by ~3000 m high) and ecological significance of the EBS slope, there have been few hydrographic-quality charting cruises conducted in this area, so we relied mostly on uncalibrated underway files from cruises of convenience. The lack of hydrographic quality surveys, anecdotal reports of features such as pinnacles, and reliance on satellite altimetry data has created confusion in previous bathymetric compilations about the details along the slope, such as the shape and location of canyons along the edge of the slope, and hills and valleys on the adjacent shelf area. A better model of the EBS slope will be useful for geologists, oceanographers, and biologists studying the seafloor geomorphology and the unusually high productivity along this poorly understood seafloor feature.

Keywords: bathymetry; thalwegs; canyons; Alaska; Bering Sea

1. Introduction

The eastern Bering Sea (EBS) slope is an abrupt, sinuous seafloor feature, ranging approximately 1250 km in length from Bering Canyon in US waters to the Vityaz Sea Valley in Russian waters (Figure 1). Depths to the east of the slope on the Bering Sea shelf are shallow (~130 m), depths to the west of the slope in the Aleutian Basin are deep (>3000 m), and numerous canyons, some of which are the largest in the world [1], scallop the edge of the shelf. This vast vertical and horizontal expanse of seafloor has been only partially explored and charted. The only land along the EBS slope is the Pribilof Islands, so navigating with visual fixes to triangulation shore stations, the traditional and primary method used by the National Ocean Service (NOS, Silver Spring, MD, USA) for producing smooth sheets [2] was not well-suited to this area. The absence of high quality smooth sheets, which are the detailed records of the original charting surveys, is a major detriment to understanding the bathymetry in this area. While the Aleutian Islands (AI) portion of this area was well-charted with smooth sheets mostly in the late 1930's [3], the waters surrounding islands in the Bering Sea such as St. Matthew Island (1951), the Pribilof Islands (1951–53), and St. Lawrence Island (1951–54, 1968–70) were charted much later, incompletely, and often at small scales. The Aleutian Basin (1952) and portions of the EBS slope (1951–53) were only charted at a coarse scale of 1:500,000.

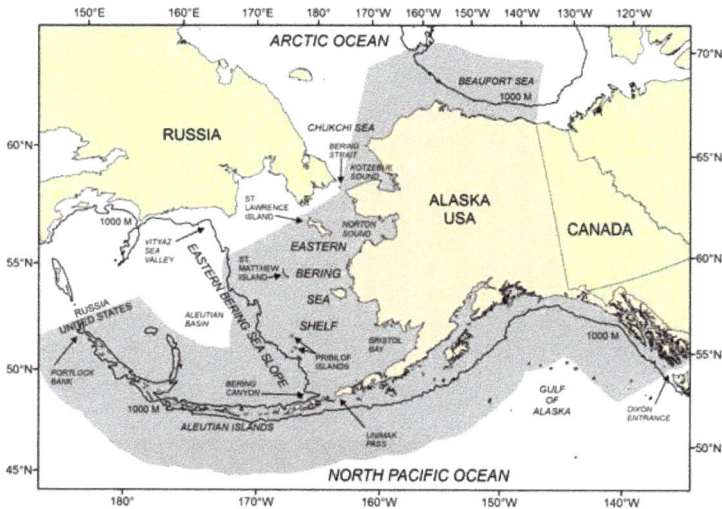

Figure 1. Overview map of the study area, with the US Exclusive Economic Zone in gray.

1.1. Early Charting of the Eastern Bering Sea Slope and Canyons

Prior to 1950, the EBS slope and a few canyons were only partially depicted with 100 and 1000 fathom contours on the most detailed US chart (9302) of this area (Figure 2). While Bering Canyon was clearly placed at the intersection between the slope and the Aleutian Islands, and portions of adjoining Umnak Canyon were depicted, Pribilof Canyon was only vaguely outlined, and none of the northern canyons (Zhemchug, Middle, St. Matthew, Pervenets, and Navarin) were known. Many of the soundings depicted on Chart 9302 are from *Albatross* cruises conducted in the late 1800s.

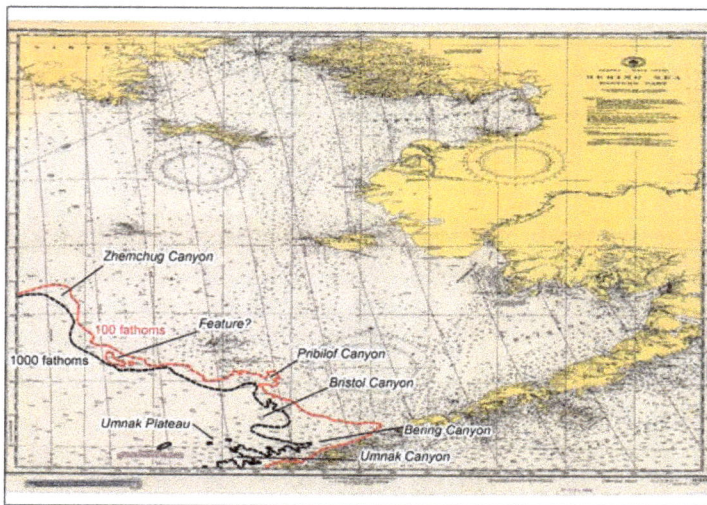

Figure 2. NOS navigational Chart 9302 from 1948. Eastern Bering Sea slope canyons are poorly defined by dashed lines for the 100 (traced in red) and 1000 (traced in black) fathom depth contours. This chart was replaced by Chart 16006 in 1975.

In the early 1950s, the NOS coarsely charted the US portion of the slope from the Aleutian Islands up to 59°30′ N. Pervenets and Zhemchug Canyons were clearly depicted on smooth sheet H08103 (1953, Scale 1:500,000) but only assigned the generic name of "Marine Canyon" in the descriptive report (https://data.ngdc.noaa.gov/platforms/ocean/nos/coast/H08001-H10000/H08103/DR/H08103.pdf). Zhemchug Spur (945 fathoms or 1728 m) and two ridges on St. Paul Spur (northern ridge as shallow as 47 fathoms or 86 m, southern ridge as shallow as 51 fathoms or 93 m) were also charted by H08103 but only identified as notable shallow soundings. Smooth sheet H07949 (1951–53, Scale 1:500,000) covered the southern half of the US portion of the slope, but a digital version of the smooth sheet is not available for analysis, and no seafloor features were named in the descriptive report (https://data.ngdc.noaa.gov/platforms/ocean/nos/coast/H06001-H08000/H07949/DR/H07949.pdf).

The Russian fishery research vessel *Zhemchug* explored and named four large canyons in the mid- to late-1950s: Navarin Canyon (1955), Pervenets Canyon (1958, along with the *R/V Pervenets*), Pribilof Canyon (1958), and Zhemchug Spur and Canyon (1959) (https://www.ngdc.noaa.gov/gazetteer/). The Russians were probably unaware of the earlier US mapping effort, as smooth sheets H07949 and H08103 were routinely classified as confidential by the US government. Following the new knowledge gained in the 1950s, the 1967 version of NOS chart 8802 (Scale 1:1,023,188) shows a much more definitive Pribilof Canyon and the 1968 version of chart 9302 (Scale 1:1,534,076) shows Zhemchug Canyon, a few soundings from Zhemchug Spur, and a single, shallow sounding representing the two St. Paul Spur ridges.

St. Matthew and Middle Canyons [4] were discovered in the northern slope area by the USGS (US Geological Survey) in the early 1980s [5–7]. Some northern canyon metrics, such as areas and volumes, were described [7]. In 1986, an early version of sidescan sonar (GLORIA: Geological LOng-Range Inclined Asdic) was used to map the backscatter of the Aleutian Basin and eastern Bering Sea shelf edge, including the major canyons (see https://pubs.usgs.gov/of/2010/1332/), and the towing vessel *Farnella* provided a path of singlebeam depths, which was a significant mapping contribution to the Aleutian Basin.

Multibeam surveys are rare in the EBS slope. In 2003, a 210 km section of the Beringian Margin near Pervenets Canyon was mapped from depths of ~1000 to 3700 m (http://ccom.unh.edu/theme/law-sea/beringian-margin-bering-sea). The NOS has been multibeam mapping in the Unimak Pass area since the mid-2000s (https://maps.ngdc.noaa.gov/viewers/bathymetry/). Pribilof Canyon was mapped with multibeam in 2009, making it the only EBS slope canyon with accurate bathymetry (https://data.ngdc.noaa.gov/platforms/ocean/nos/coast/H12001-H14000/H12115/DR/H12115.pdf).

1.2. Chart 16006 and the Zhemchug Canyon Pinnacles

There are four pinnacles reported to be in the Zhemchug Canyon area (Figure 3). Discoveries of previously unknown and dangerous pinnacles, such as the 650 ft (198.1 m) tall "Washington Monument" pinnacle in South East Alaska, rising to a depth of just 2.75 fathoms (5.0 m), have been a very real reminder of the danger for navigation in Alaska's uncharted waters for over a century [8]. The Zhemchug Canyon pinnacles are legendary and members of the research, environmental, and commercial fishing communities are all familiar with their vague and inconvenient location in relatively deep water along the edge of the Bering Sea self.

The Zhemchug Canyon pinnacles originated with the US submarine *Bergall*, which surfaced during its first training patrol in the northern portion of Zhemchug Canyon and reported a day time sounding of 8 fathoms (14.6 m) on 19 April, 1947 and a nearby sounding of 13 fathoms (23.8 m) on the following day (https://data.ngdc.noaa.gov/platforms/ocean/nos/coast/H06001-H08000/H07951/DR/H07951.pdf). In response to this US Navy report of pinnacles, survey H07951 (smooth sheet not available) was conducted in 1951; the *USS Bergall* pinnacles were disproved, and no pinnacles were reported on the most detailed NOS chart of this area (NOS Chart 9302, Scale 1:1,600,000) through the 1971 edition (22nd edition). The third pinnacle (3.75 fathoms or 6.9 m) was reported, potentially by

Russia, in 1971 through the Notice to Mariners (http://msi.nga.mil/NGAPortal/MSI.portal?_nfpb= true&_pageLabel=msi_portal_page_61), and the next edition of NOS Chart 9302 (1973) had the first depiction on a US navigational chart of this new pinnacle, as well as the two previously disproven *USS Bergall* pinnacles. Chart 9302 was replaced by Chart 16006 in 1975, and that edition adds the fourth and final pinnacle, this one—of unknown depth, near the southern edge of Zhemchug Canyon—was reported to the Notice to Mariners by the *US Coast Guard Cutter Jarvis* on 10 May, 1974.

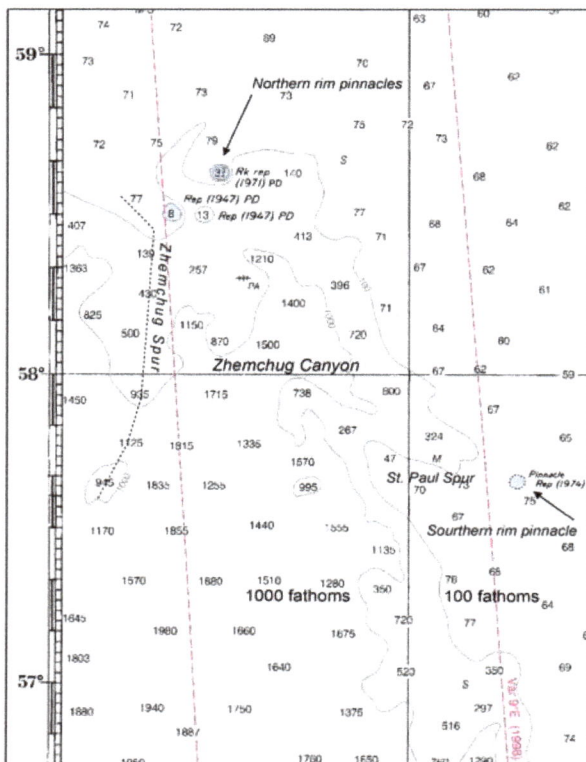

Figure 3. Detail of NOS navigational chart 16006 (Edition 37, 2015, Scale 1:1,534,076), showing reported pinnacles in the Zhemchug Canyon area.

1.3. Bathymetry Compilations, Oceanographic and Biological Research

Global and regional bathymetry compilations of various bathymetric sources have advanced our knowledge of the Alaska seafloor. Global bathymetry surfaces of 2 arc-minute resolution (~3704 m) [9], 1 arc-minute resolution (~1852 m) (ETOPO1: [10]), and 30 arc-second resolution (~926 m) (General Bathymetric Chart of the Oceans, GEBCO: [11]) were published using a mixture of soundings and calibrated remote sensing data. Regional bathymetry compilations include those of the AKRO (National Marine Fisheries Service, Alaska Regional Office (Anchorage, AK, USA): https://alaskafisheries.noaa.gov/, variable resolution, 2005; updated to 40 m resolution in 2017, https://inport.nmfs.noaa.gov/inport/item/27377), the International Bathymetric Chart of the Arctic Ocean (IBCAO 3.0, 500 m resolution, [12]), and the Alaska Regional Digital Elevation Model, which is a compilation that includes US and Russian chart soundings (ARDEM, ~1 km resolution, [13]).

A global analysis of canyons was conducted using the ETOPO1 data set [14], producing 26 thalwegs within our study area. Additional seafloor features in the EBS slope area were defined from Shuttle Radar Topography Mapping (SRTM30_PLUS) in another global analysis [15]. Locations and names of seafloor features—mostly canyons—were available from GEBCO (https://www.gebco.net/data_and_products/undersea_feature_names/) and the National Geospatial Intelligence Agency (NGA, Springfield, VA, USA: https://www.ngdc.noaa.gov/gazetteer/).

Marine researchers have benefitted greatly from these bathymetry compilations and subsequent geomorphological analyses, but still there has been an unmet need for more detailed and more accurate depth data for fisheries habitat studies at the Alaska Fisheries Science Center (AFSC, Seattle, WA, USA). To meet this need, we have been publishing 100 m resolution bathymetry compilations in recent years: AI [3], Cook Inlet (50 m resolution: [16]), Norton Sound [17], and central Gulf of Alaska (GOA, AK, USA) [18]. These bathymetry compilations have been utilized for a variety of fishery research purposes including fish vertical migration [19]; coral and sponge distribution modeling in the AI [20] and GOA [21]; quantifying inshore study sites in the central GOA [22], eastern GOA [23], and bathymetry groundtruthing [24]; bathymetric steering of seafloor current flow [25]; inshore habitat loss [26]; Essential Fish Habitat modeling in the EBS [27], GOA [28,29], and AI [30]; juvenile groundfish habitat suitability models [31]; and capelin (*Mallotus villosus*) distribution modeling [32].

The EBS slope area is important for several commercial fisheries, and the impact of these fisheries has been the focus of significant research. The majority of the commercial fishery for walleye pollock (*Gadus chalcogrammus*)—the largest fishery in the world [33]—takes place on the slope and adjoining shelf. The biological importance of the canyons versus the inter-canyon areas was examined (https://www.npfmc.org/bering-sea-canyons/; [34–36]), and there has been interest in documenting the ecosystem function of pinnacles in the Zhemchug canyon area [pers. comm. John Hocevar 2017]. Skate egg case nurseries for three species of *Bathyraja* are found in the canyons [37], and biannual stock assessment bottom trawl surveys for several commercially important species are conducted at depths of 200 to 1200 m along the slope [38].

The EBS slope is an area of particular oceanographic and biological importance beyond just fish, corals, and sponges. For example, the slope plays an important role in limiting the extent of winter ice formation on the EBS shelf [39] (see Figure 3). Researchers coined the term "Bering Sea Green Belt" to describe the productivity of the EBS slope (see Figure 2) [40], summarized the flow of ocean currents; and plotted the peak abundances of phytoplankton, zooplankton, squids, fishes, sea birds, and mammals along this shelf edge. Wong et al. [41] demonstrated that it was an important habitat for marine birds such as red-legged (*Rissa brevirostris*) and black-legged kittiwakes (*R. tridactyla*); northern fulmars (*Fulmarus glacialis*); sooty, great, and short-tailed shearwaters (*Puffinus griseus, P. gravis, P. tenuirostris*, respectively); fork-tailed, Leach's, and Wilson's storm-petrels (*Oceanodroma furcata, O. leucorhoa*, and *Oceanites oceanicus*, respectively); surface and diving piscivores; and surface and diving planktivores. The North Pacific Pelagic Seabird Database also shows that this area is important for the endangered short-tailed albatross (*Phoebastria albatrus*) (https://alaska.usgs.gov/science/biology/nppsd/index.php), one of the rarest marine birds in the world. A better depiction of the slope and canyons should improve our understanding of how this undersea feature affects so many physical and biological aspects of the EBS.

2. Materials and Methods

2.1. Bathymetry Data Sources and Typical Errors

We utilized 18 million soundings from over 200 individual sources that can be grouped into ten general categories, each of which came with its own advantages and disadvantages (Figure 4). Initially, we edited each file by searching for outliers (e.g., depths of zero) or incorrect positions. When possible, we corrected or deleted errors by comparing depth values to original sources (e.g., smooth sheets and echosounder files). We also rejected soundings that overlapped with a data set we judged to be superior. Reducing overlaps of data sets from the 10 different sources helped to minimize disagreements and avoid conflicts such as differing vertical datums, as nearshore surveys were generally corrected to a vertical tidal datum, while offshore surveys generally were not corrected. NOS smooth sheets and multibeam surveys covered only a small amount of the study area and, therefore, we had to rely heavily on other bathymetric sources. Unpublished underway files, either collected from navigational software or extracted from the raw echosounder files, covered the vast majority of this project area.

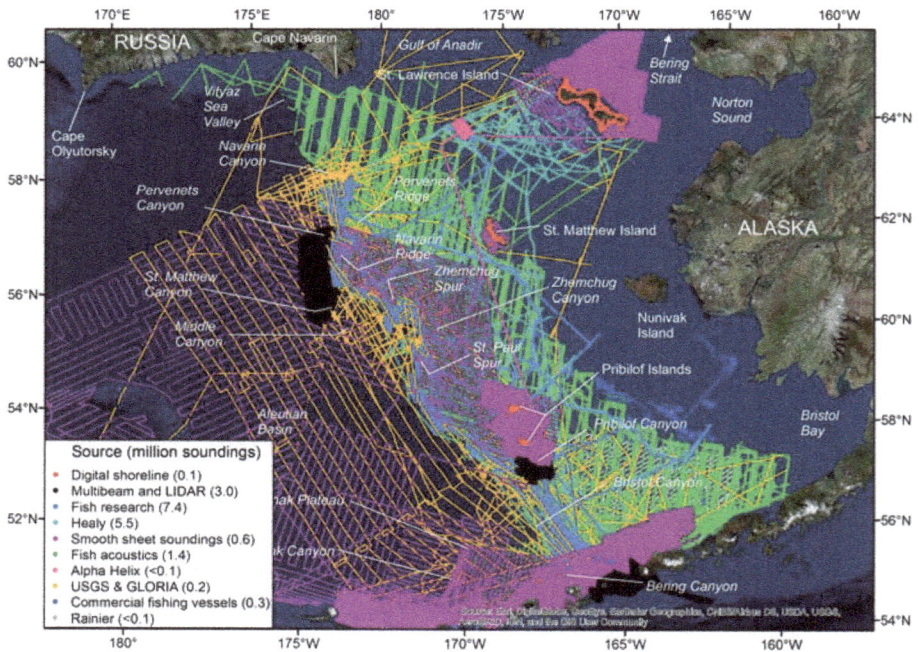

Figure 4. Sources of bathymetry data used in this EBS slope compilation.

2.1.1. Smooth Sheets and Multibeam

Most smooth sheet and all multibeam data were collected at a vertical datum of Mean Lower Low Water (MLLW). In general, offshore smooth sheets, such as those soundings depicted northwest of the Pribilof Islands and those covering the Aleutian Basin, are not corrected to any vertical datum and thus are considered to approximate Mean Sea Level (MSL). For example, the Aleutian Basin soundings were collected at depth intervals of 5 to 10 fathoms (9.1 to 18.3 m) in recognition of the fact that inaccuracies in speed of sound estimates, poor navigation (due to vast distance from shore stations), heave, and errors in distinguishing the exact start of the seafloor reflection far exceeded tidal differences. Smooth sheet soundings (digital files available at https://www.ngdc.noaa.gov/) [42] for this compilation had errors that were familiar to us [43] from previous compilations, including improper horizontal datum shifting, random digitization errors, and undigitized soundings—for example, smooth sheets H08103 and H08001B needed to be digitized entirely.

We downloaded the National Geodetic Survey shorelines (NOAA Shoreline Explorer web site: http://www.ngs.noaa.gov/NSDE/) for the Pribilof Islands and St. Lawrence Island, and annotated them with MHW (mean high water) from corresponding NOS smooth sheets. The MLLW and MHW vertical datums are compatible with each other and utilized on every navigational chart of Alaska.

This useful resource provides shoreline very similar to that of the smooth sheets, but derived from a different survey product, sometimes contradicting findings from the smooth sheet inshore work. We also digitized the shoreline from the smooth sheets for Bogoslof Island and St. Matthew Island, again annotating them with MHW from the corresponding smooth sheets.

All multibeam data sets were subsampled to a resolution of 100 m. The UNCLOS (United Nations Convention of the Law of the Sea) multibeam from the Bering Margin (http://ccom.unh.edu/theme/law-sea/law-of-the-sea-data/bering-sea) reported positions in decimal latitudes and longitudes, rather than in projected meters, resulting in off-center raster cells.

The 1970 cruise of the *Rainier*, a NOS hydrographic vessel, provided an underway file that was similar in resolution to that of the smaller scale NOS smooth sheets, although it was a USGS research cruise (https://walrus.wr.usgs.gov/infobank/r/r170bs/html/r-1-70-bs.meta.html). This cruise was collected without tidal correction and thus is at a vertical datum of MSL.

2.1.2. Underway Files

Underway files, collected without tidal correction and without heave correction, are similar to the 1970 *Rainier* cruise. Thus, the underway files approximate a vertical datum of MSL. These underway files typically had random depth errors but were most significantly plagued with the echosounder losing track of the seafloor and either repeating a depth for long distances or creating false depths by incorrectly recording reflections from a mid-water scattering layer. Both of these "lost seafloor" sources of error were difficult to detect, as repeated depths on the EBS shelf were common because the shelf is extremely flat, while depths along the slope were often unknown in areas of rapid depth change.

The *Alpha Helix* mooring recovery cruise of 2004 (http://www.rvdata.us/catalog/HX291) provided an unusually thorough, gridded search pattern of underway soundings between Navarin Canyon and St. Lawrence Island due to the unfortunate loss of a mooring [Pers. comm. Cal Mordy, 2015].

AFSC EBS slope bottom trawl biennial cruises (e.g., [38]) provided underway files from Seaplot software, Globe software, or bottom picks from EchoView analysis of raw echosounder files. Random and "lost seafloor" errors were also common in this source, but each cruise was different and had to be proofed and edited carefully. For example, one cruise collected soundings in fathoms for over one hundred km before switching to meters—the standard for all other cruises. Additional AFSC fish research cruises on the Zhemchug ridges (SCS *Miller Freeman* 2007 and Globe 2008 *F/V Vesteraalen*) provided detailed parallel transects [44]. These appear to be the only AFSC cruises conducted for seafloor mapping purposes on the EBS slope.

AFSC walleye pollock acoustic cruises covering the Bogoslof Island area (e.g., [45]) or the outer EBS shelf and slope area (e.g., [46]) occur on an almost annual basis. Older *Miller Freeman* data were in the format of depth averaged across distance along transects; no raw acoustic files or individual soundings are available (1991–1996). Underway files of raw soundings, typically collected at 30 second intervals, were recorded in more recent years with Scientific Collection System software (https://www.unols.org/sites/default/files/200110rvtap16.pdf) on the *Miller Freeman* (1997–2006) and the *Oscar Dyson* (2007–2016). These more recent pollock surveys accounted for hull depth but had the same types of errors as in other underway sources (random and "lost seafloor" errors), but much less frequently. We attempted to incorporate data from the *F/T Continuity* cruise of 1991 and the *Miller Freeman* cruise of 1992, but depths from these cruises disagreed with all other cruises—we also had to delete data from legs 1 and 2 of the 1994 *Miller Freeman* cruise but were able to utilize data from leg 3.

BEST (Bering Sea Ecosystem Study: https://www.eol.ucar.edu/projects/best/) cruises on the *US Coast Guard Cutter Healy* (2006–2009) seemed to have higher incidences of random errors, and we screened sections of transects by examining for continuous distribution of depths, deleting soundings outside of a central range of values. Northern portions of transects, which appear to have been conducted in the sea ice, had too many random depth errors to use in this project.

Commercial fishery vessel depth data were all from Globe files, and these typically only had random errors. A benefit of this source was that vessels tended to conduct new tracks adjacent to previous tracks, which provided depth data over new areas, rather than consistently on top of old tracks, as in research cruises.

USGS explorations (https://cmgds.marine.usgs.gov/) utilized various vessels (*Discoverer* 1980–81; *Sam Phillips Lee* 1975–1983; *Sea Sounder* 1976–77; and *Maurice Ewing* 1994). These cruises seemed to suffer from navigational errors, similar to the smooth sheets, but were able to produce reliable depths in deep water. GLORIA cruises conducted on the *Farnella* (1986) produced depths in the Aleutian Basin (>3000 m), which were generally deeper than overlapping smooth sheet depths. With two-way travel time sometimes exceeding five seconds for a sounding, we suspected that the difference in depth between the two data sets is due to differences in speed-of-sound utilized for estimating depth. However, the GLORIA data utilized speed-of-sound estimates [pers. comm. Jim Gardner, 2017] derived from Carter [47] or field observations, and these were generally slower than used by the smooth sheets (1500 m/s). Thus, attempts to correct one data set to the other by correcting for speed-of-sound differences resulted in greater differences; therefore, we did not perform this correction.

2.2. Bathymetry Raster Creation

Smooth sheet bathymetry formed the foundation of this compilation, as in all of our previous compilations, but smooth sheets covered a smaller area than in our previous compilations. We reused smooth sheet bathymetry from our AI compilation [3] for the southern portion of this compilation to provide a more complete spatial extent of the EBS slope. Most of the effort on this project was devoted to proofing and editing the underway files, which, along with covering the majority of the area also seemed to have the bulk of the errors. For the first time in our bathymetry compilations, we encountered the problem that the edited underway files from vessels with GPS navigation, but with completely uncalibrated echosounders, often exceeded the quality of the smooth sheets, which were small scale (1:500,000). Therefore we eventually had to delete large areas of smooth sheet soundings to construct this bathymetric surface. Multibeam data only covered about two percent of our study area. All data sets (smooth sheets, multibeam, and underway files) were combined to create TINs (Triangulated Irregular Networks). In an iterative process, TINs were plotted with color ramps for slope (see Figure 30 in reference [43]) and depth to reveal sounding outliers to target for editing. After numerous rounds of editing, the TIN was converted to a raster of 100 m horizontal resolution by using local area weighting (termed natural neighbors in ArcMap), a method and resolution that has worked well in our previous compilations [2], refs. [16–18] for depicting seafloor features as detailed as earthquake faults with low resolution data [18]. Our final raster surface is a mix of onshore MLLW and offshore MSL data sets, as NOAA's vertical datum tidal correction software is not available for Alaskan waters (https://vdatum.noaa.gov/about.html).

2.3. Thalweg Creation

We determined the location, size, and number of canyons within our compilation by utilizing the Hydrology toolbox in ArcMap (v.10.2.2, ESRI: Environmental Systems Research Institute, Redlands, CA, USA). In this method, raster cells are examined to determine how water would flow downhill based on depth, slope, and aspect; sinks are filled; and then the few cells that receive the most calculated runoff (we set our lower limit as >1000 cells) are labeled as rivers. Runoff may only travel in the eight cardinal or inter-cardinal compass directions, so straight river sections are common, and parallel river sections are allowed. In our submerged environment we treat these "rivers" as thalwegs to define the center-lines of canyons. We conducted this analysis with both positive and negative depth surfaces to create thalwegs and ridgelines, both of which helped identify cruises that were slightly (~1 m) shallower or deeper than other cruises. Underway cruises with the largest depth errors were compared to multibeam depths by using the Ordinary Least Squares function in ArcMap:

$$y = m \times x + b, \tag{1}$$

in which y are the underway soundings and x are the multibeam depths. To make the depth correction, we subtracted the y-intercept (b) from the underway depths and divided by the slope (m):

$$(y - b)/m = x, \tag{2}$$

As thalweg creation was still highly sensitive to cruises with depth errors that were too small to correct with multibeam comparisons, we smoothed the bathymetry raster with a low pass filter using a rectangular neighborhood of 20 cells. Thalwegs were then created from the smoothed surface, but canyon metrics were derived by comparing the thalweg paths to the unsmoothed 100 m resolution bathymetry and slope surfaces. A shape file of the thalwegs is provided as Supplementary Material with this manuscript (Thalwegs.zip).

2.4. Zhemchug Canyon Pinnacles

We digitized the warning circles drawn on NOS Chart 16006 around each of these pinnacles and extracted the corresponding soundings from our compilation. We examined these soundings for any indication of the pinnacles.

3. Results

3.1. Eastern Bering Sea Slope Bathymetry

Our bathymetry compilation (Figure 5) covered almost one million km^2, spanning 1400 km from west to east and 1300 km from north to south, joining our Norton Sound compilation [17] in the north and overlapping with our AI compilation in the south [3]. Imagery of the bathymetry is provided as Supplementary Material with this manuscript (EBS Slope Bathymetry.zip). We compared our bathymetry surface to previously published cartographic information of the area: seafloor gazetteers from GEBCO and NGA, NOS navigation Chart 16006 (as the source for the Zhemchug Canyon pinnacles), published geomorphic features, and other bathymetry compilations. We attempted to reconcile differences between our bathymetry compilation results and the seafloor interpretations of others in order to minimize cartographic confusion. NOS Chart 16006 is still the most detailed US navigational chart of the area, and still only uses the 100 and 1000 fathom depth contours to describe the EBS slope. Previously published canyon shape parameters and thalwegs [14], and geomorphology polygons [15], were also used for comparisons. Global and regional bathymetry compilations utilize a wide variety of input data sets and publish at several different resolutions.

3.2. Eastern Bering Sea Slope Canyon Thalwegs

Umnak Canyon occupied the southern portion of our compilation, running roughly parallel to the Aleutian Islands, with several smaller canyons running perpendicular to the main Umnak Trunk. The southeastern area was dominated by Bering Canyon and several other connected or nearby canyons, including Bristol and Pribilof Canyons. Navarin and Zhemchug Canyons occupied most of the northern slope. The thalwegs north of Navarin Canyon and around St. Lawrence Island did not connect with the EBS slope canyons, indicating that the bathymetry compilation extended sufficiently far north in extent, but the Bering, Pribilof, Zhemchug, and Navarin Canyon thalwegs reached near the eastern edge of our bathymetry compilation, suggesting that this portion of the compilation could be expanded. The interior section of Bristol Bay and adjoining shallow areas were not included in this project but may be the subject of a future compilation.

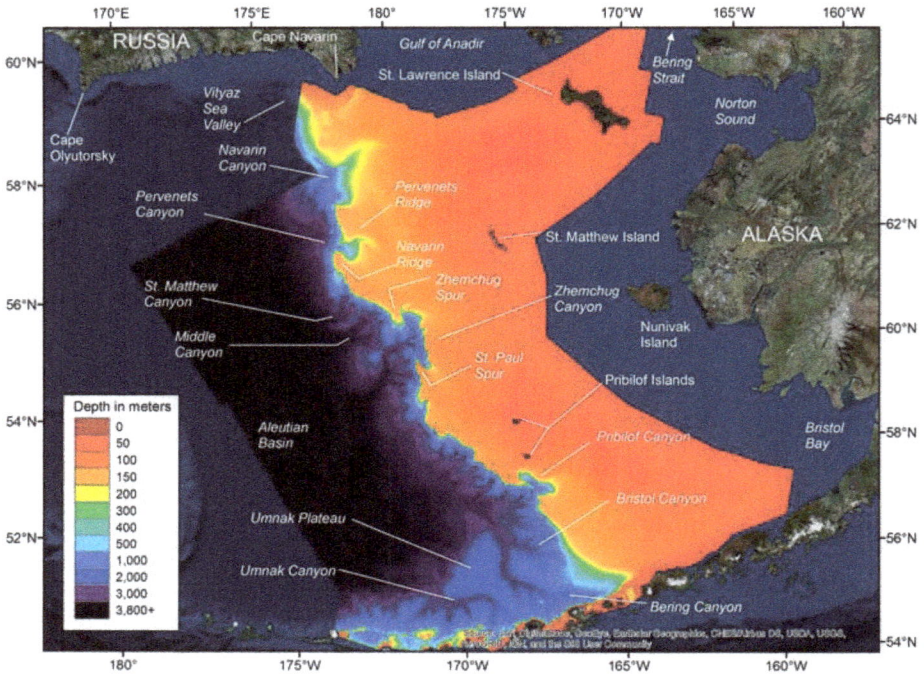

Figure 5. Bathymetry of the eastern Bering Sea slope.

3.2.1. Umnak Canyon Thalwegs

The GEBCO gazetteer only names 5 canyons (plus a canyon basin) in the Umnak area, while the NGA gazetteer repeats these same features and adds 7 more canyons, all extending northward from the Aleutian Islands into deeper water, to a large and deep canyon (Figure 6, Table 1). Starting on the western side, our thalwegs agree with NGA's placement of Korovin Canyon. To the east of Korovin is Atka Canyon, and the NGA label is on a western thalweg (1015–3303 m) that joins with a similar, unnamed eastern thalweg (935–3304 m) at a depth of 3305 m. Both Atka thalwegs form a trunk that extends down to 3512 m, nearly joining with Korovin Canyon at the western edge of our compilation. Therefore, we suggest changing NGA's Atka Canyon to include West, East, and Trunk thalwegs. Both GEBCO and NGA recognize Amlia Canyon and Amlia Basin, and our results concur with a long, single thalweg. To the east of Amlia is an unnamed canyon with a single thalweg extending 156 km—it is blocked on the shallow end by the Amlia Basin enclosure, so it begins at a relatively deep depth of 842 m and extends down to 3698 m. NGA's Seguam Canyon joins a similar, unnamed thalweg on the east at a depth of 2462 m, forming a trunk that extends 56 km to the deep canyon to the north. The Seguam East thalweg extends south into a partially enclosed basin, similar to that of Amlia Basin. Therefore, we suggest changing the designation of Seguam Canyon to include West, East, Basin, and Trunk thalwegs. Our results agree with the next 5 canyon designations of the NGA gazetteer: Amukta, Chagulak, Yunaska, Herbert, and Carlisle (Chagulak and Herbert are not recognized by GEBCO). Amukta and Chagulak merge into a trunk and Yunaska and Herbert also merge into a trunk before joining the large canyon to the north. At the eastern end of the Umnak area, the GEBCO and NGA gazetteers disagree with each other—GEBCO shows Umnak as a canyon curving from deep water south toward Samalga Pass, while NGA shows only the deep portion of the canyon as Umnak, with shallower thalwegs ending in Uliaga, and possibly Okmuk Canyons.

Nothing in our analysis corresponds with NGA's Okmuk Canyon label (it falls on Umnak Plateau), and therefore we suggest deleting this canyon name. Our results show that the next to last canyon of the Umnak area drains an area more than twice as large as the easternmost canyon (4810 vs. 1944 km^2), and therefore we suggest that this extension from the deep Umnak Trunk continue the name of Umnak. We suggest naming the easternmost canyon after Inanudak Bay of Umnak Island, where the canyon ends, and, where Inanudak Canyon merges with Umnak, we suggest naming the large canyon Umnak Trunk. Altogether, Umnak Trunk and the canyons directly linked to it in our analysis drains an area of 56,385 km^2.

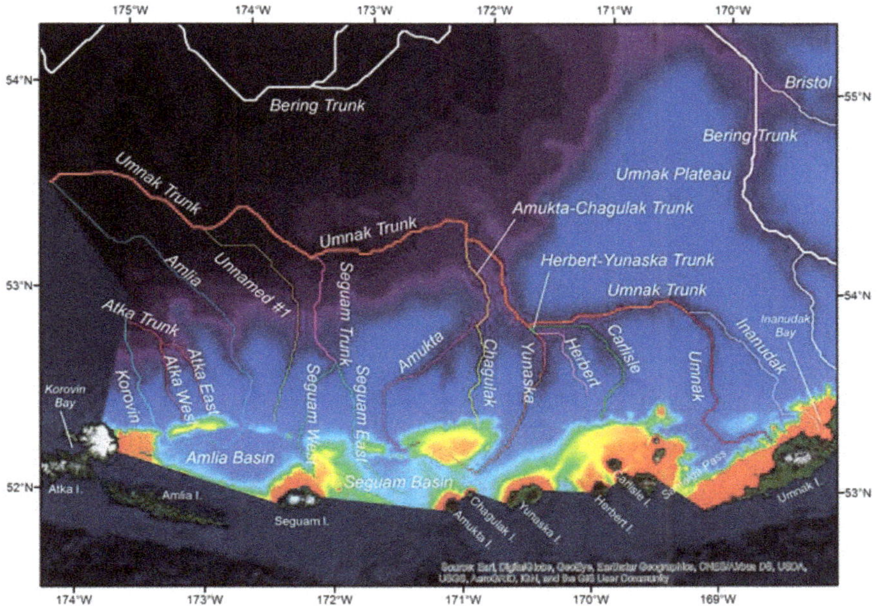

Figure 6. Thalwegs of the Umnak Canyon area of the eastern Bering Sea slope.

3.2.2. Bering Canyon Thalwegs

The Bering complex of canyons (Figure 7, Table 2) is to the northeast of the Umnak complex, and both NGA and GEBCO gazetteers place the Bering Canyon name label in a relatively deep area. There are 4 Bering complex canyons arrayed along 74 km of a deeper trunk, near the Bering label, extending south past Bogoslof Island and toward the Aleutian Islands. The NGA gazetteer names the westernmost canyon as Inanudak, and we suggest renaming this as we have shown that a canyon in the Umnak complex extends into Inanudak Bay of Umnak Island. The two middle canyons are unnamed but both end near Okmok volcano on Umnak Island. The fourth and easternmost of this group is labeled as Bogoslof Canyon by NGA at a location where our results show that West, East, and Trunk branches meet. Also, there is a disconnected Basin component in Makushin Bay, Unalaska Island. Therefore, we suggest that the Bogoslof Canyon name recognize these West, East, Trunk, and Basin divisions. The main thalweg of Bering extends for a distance of 771 km east of Bogoslof Canyon, ending just outside of Port Moller at a depth of 40 m. Bristol Canyon, named in both gazetteers, is much shorter (319 km) than Bering, extending onto the shelf only up to a depth 134 m and joining the Bering Trunk at a depth of 2920 m. The tributaries of both Bering Canyon and Pribilof Canyon cover most of the shelf, blocking the eastern reach of Bristol Canyon.

Table 1. Umnak Canyon names and metrics described from west to east.

Canyon	Area Drained km²	Start Depth m	End Depth m	Mean Slope Degrees	Thalweg Length km
Korovin	611	686	3511	3.8	71
Atka Complex	2418	935	3512	3.4	132
Trunk	543	3305	3512	0.7	26
West	737	1015	3303	4.2	51
East	1138	935	3304	3.8	55
Amlia Complex	7102	112	3714	1.9	281
Thalweg	4549	1064	3714	2.1	210
Basin Thalweg	2553	112	1034	1.6	72
Unnamed#1	3700	842	3698	1.6	156
Seguam Complex	5597	170	3555	2.6	235
Trunk	1563	2462	3555	1.6	56
West	1496	170	2460	3.4	79
East	674	558	2462	3.0	56
Basin Thalweg	1864	565	558	1.2	43
Amukta	3570	270	2945	3.6	119
Chagulak	1222	376	2936	3.8	67
Amuk.-Chag. Trunk (includes drainage of both canyons)	5871	2942	3295	1.8	60
Yunaska	4124	163	2789	2.4	121
Herbert	374	1563	2789	4.6	57
Herb.-Yun. Trunk (includes drainage of both canyons)	4821	2789	2712	0.9	6
Carlisle	1944	892	2713	4.1	101
Umnak	4810	422	2444	4.4	126
Umnak Trunk (includes all drainage except Korovin and Atka)	56,385	2445	3713	1.8	453
Inanudak	2352	71	2453	4.2	121

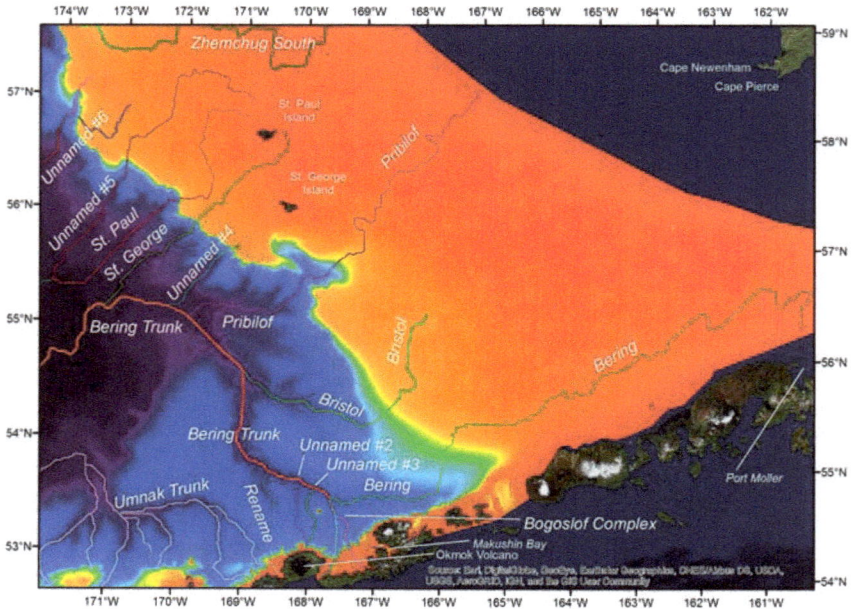

Figure 7. Thalwegs of the Bering Canyon area of the eastern Bering Sea slope.

Table 2. Bering Canyon names and metrics described from south to north.

Canyon	Area Drained km²	Start Depth m	End Depth m	Mean Slope Degrees	Thalweg Length km
Rename	2536	560	2601	2.7	98
Unnamed#2	1117	558	2460	2.4	77
Unnamed#3	806	893	2312	2.0	67
Bogoslof Complex	3192	34	2194	3.0	183
Trunk	669	1899	2194	1.6	21
West	1053	106	1897	3.2	73
East	1471	56	1897	2.8	71
Basin Thalweg	208	34	215	4.3	18
Bering Thalweg	97,547	40	2193	0.4	771
Bering Trunk (includes all drainage)	329,961	2192	3694	0.9	696
Bristol	20,019	134	2920	1.0	319
Pribilof	74,422	61	3334	0.7	529
Unnamed#4	2375	128	3352	2.4	111
St George	9971	46	3468	1.0	311
St Paul	15,025	70	3522	0.8	360
Unnamed#5	11,634	73	3522	0.9	404
St P.-Un.#5 Trunk (includes drainage from both canyons)	33,485	3522	3629	0.3	150
Unnamed#6 Complex	9712	115	3694	1.5	511
Trunk	6325	1533	3694	0.8	347
North	1454	124	1524	3.7	58
South	1932	115	1541	1.5	106

We expected that Pribilof Canyon would resolve into two main thalwegs due to the kidney shape of the canyon, but the southern thalweg is so long and drains such as large area, while the northern thalweg is so short and drains such a small area between the canyon and St. George Island, that we identify only the main thalweg. Several other relatively small canyons, some extending onto the shelf, join the Pribilof thalweg in deeper water (not shown). Since these join the Pribilof thalweg before joining the Bristol Canyon trunk, and none of them were named by GEBCO or NGA, we consider them all to be tributaries of Pribilof.

Just to the north of Pribilof Canyon is an unnamed, relatively short (111 km) canyon that joins the Bristol Trunk separately and is therefore potentially deserving of its own name (Unnamed#4). The next 3 canyons are all about equal in length (311–404 km) and depth range (<100 to >3500 m), also joining the Bering Trunk in deep water. The southern of the three drains the area between the Pribilof Islands, while the two others drain the area north of St Paul Island. The central canyon of the three was named St. George by the NGA gazetteer, and therefore we suggest changing it to St. Paul, and also suggest naming the southern unnamed canyon as St. George. The northern of the three canyons is Unnamed#5, and it merges with St. Paul to form a 150 km long trunk that extends to the Bering Trunk (not shown). Just to the north of Unnamed#5 is another canyon without a name (Unnamed#6), having two thalwegs that extend onto the shelf, and connecting with a trunk of 347 km to the Bering Trunk (not shown). Altogether, Bering Trunk and the canyons directly linked to it in our analysis drain an area of 329,961 km².

3.2.3. Navarin Canyon Thalwegs

The northern canyons (Figure 8, Table 3), which include Navarin, Pervents, St. Matthew, Middle, and Zhemchug, all have at least two main thalwegs. These canyons all eventually connect together in the deep water of the Aleutian Basin (not shown), but this is very poorly resolved in our compilation due to the disagreement between the GLORIA and smooth sheet depths. We derived three thalwegs from Navarin. The northern (551 km) and central (935 km) thalwegs are the longest, merging at a depth of 1698 m, well within the main body of the canyon, but the short southern (227 km) thalweg merges with the Pervenets Canyon trunk at a deep place (3356 m) between the canyons. This Navarin South-Pervenets Trunk then merges with the Navarin North-Central Trunk at a depth of 3679 m to form the Navarin Trunk. This was the only instance in our analysis where a thalweg from one canyon merged with a thalweg from another canyon rather than first merging with one of its own canyon thalwegs. Thus, it might be possible to consider South Navarin as a separate canyon from the North and Central Navarin thalwegs. Both the GEBCO (179°15′ E) and NGA labels (179°45′ E) for Navarin Canyon are too far to the west, and we suggest shifting them to 179°15′ W.

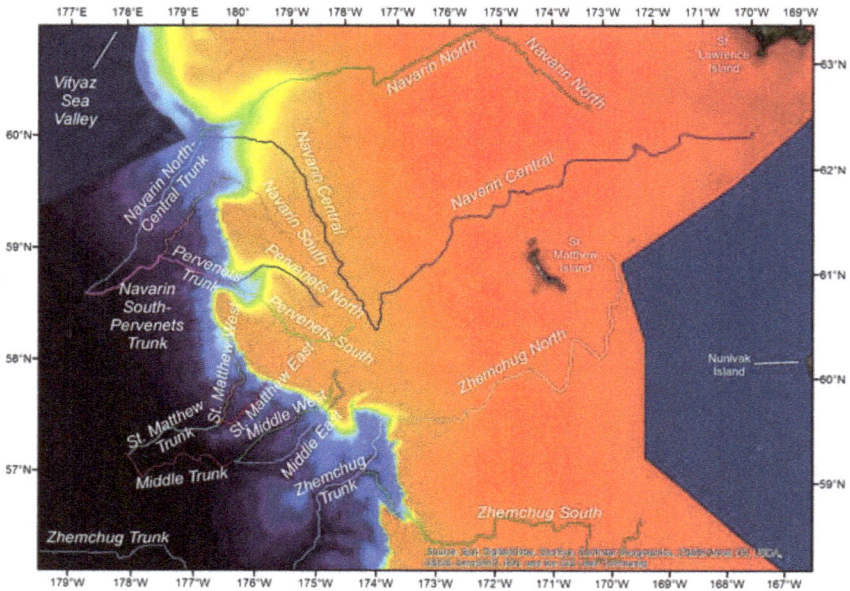

Figure 8. Thalwegs of the Navarin Canyon area of the eastern Bering Sea slope.

Pervenets Canyon has two relatively short thalwegs (North = 127 km and South = 172 km), as their reach onto the shelf is blocked by the south thalweg of Navarin Canyon. These Pervenets thalwegs join together at a depth of 821 m, near the edge of the canyon. Middle and St. Matthew Canyons do not have large, distinct incisions onto the shelf due to the South Pervenets and Central Navarin thalwegs. The East and West thalwegs of Middle and St. Matthew Canyons are both relatively short (<200 km), and they both join into trunks in deep water (3400–3600 m), far away from the shelf edge. Zhemchug Canyon has long northern (554 km) and southern (494 km) thalwegs, which join together near the edge of the canyon mouth, at a depth of 3021 m. Altogether, the Navarin area canyons drain an area of 632,670 km^2; this value would increase if our deeper bathymetry were of sufficient resolution to connect them all.

Table 3. Navarin Canyon names and metrics described from south to north.

Canyon	Area Drained km²	Start Depth m	End Depth m	Mean Slope Degrees	Thalweg Length km
Zhemchug Complex	428,642	64	3810	0.9	1548
Trunk	345,451	3021	3810	0.6	500
North	50,833	64	3013	0.9	554
South	32,359	68	3022	1.1	494
Middle Complex	23,242	131	3743	1.3	498
Trunk	12,028	3589	3743	0.2	162
West	5544	133	3589	2.1	146
East	5670	131	3589	1.6	190
St Matthew Complex	10,340	138	3743	2.3	346
Trunk	5232	3465	3743	0.4	124
West	3414	138	3464	3.0	142
East	1693	645	3464	3.7	81
Pervenets Complex	17,507	136	3365	0.7	387
Trunk	5945	821	3365	2.2	88
North	4442	136	819	0.4	127
South	7121	137	819	0.3	172
Navarin Complex	152,940	38	3680	0.4	2054
North	37,638	56	1691	0.3	551
Central	97,765	38	1695	0.2	935
Nav. N.-Cent. Trunk	5652	1698	3680	0.9	227
South	7483	138	3363	1.5	241
Nav. S.-Perv. Trunk (not including Pervenets drainage)	4401	3356	3679	0.2	99

3.3. Chart 16006 and the Zhemchug Canyon Pinnacles

There was no evidence in our bathymetry compilation of the Zhemchug pinnacles depicted on NOS Chart 16006. The warning circles (size range 21.0–32.2 km2) on Chart 16006 around each of these pinnacles overlapped nearly 17,000 corresponding soundings from our compilation (Table 4). The 8 fathom (14.6 m) pinnacle had 33 soundings from two cruises with a minimum depth of 147.9 m. The 13 fathom (23.8 m) and the 3.75 fathom (6.9 m) pinnacles occurred in relatively steep and deep places, with 3154 soundings from six cruises having a minimum depth of 463.7 m for the former, and 10,712 soundings from 11 cruises having a minimum depth of 224.0 m for the latter. The fourth pinnacle, of unknown depth, had 2756 soundings from 13 cruises with a minimum depth of 137.2 m.

Table 4. We summarized the soundings from numerous different cruises used for our compilation at four sites near Zhemchug Canyon where pinnacles are reported as potential navigational hazards on NOS navigational Chart 16006.

Pinnacle	Reported by	Year Reported	Depth in Meters	Number of Cruises	Depth Mean (m)	Depth Range (m)	Soundings (n)
8 fathoms	*USS Bergall*	1947	14.6	2	169.0	147.9 to 197.0	33
13 fathoms	*USS Bergall*	1947	23.8	6	940.4	463.7 to 1106.1	3154
3.75 fathoms	Russian source	1971	6.9	11	409.1	224.0 to 570.8	10,712
Unknown	*USCGC Jarvis*	1974	N/A	13	139.9	137.2 to 149.3	2756

3.4. Canyon Thalwegs and Other Seafloor Features

Our canyon designations generally agreed with those of [14] but did have some differences worth noting. The main difference was in thalweg length, with those of [14] shorter at both the shallow and deep ends, owing to differences in methods. Harris and Whiteway [14] defined their thalwegs based on significant deflections in 100 m depth contours, while our thalwegs were based on the ArcMap Hydrology function (minimum lower limit needed to drain 1000 upstream cells), and canyon designations were mostly based on those canyons named in the NGA and GEBCO gazetteers. Thus, we have fewer canyons composed of numerous connected thalwegs that extend farther onto the shelf and farther into the Aleutian Basin, while [14] has more numerous, shorter canyons generally restricted just to the slope area. We did identify some canyons that [14] did not, such as Korovin, Unnamed#1, and Herbert in the Umnak complex and Unnamed#3 in the Bering complex.

We did not divide our bathymetry surface into geomorphic provinces, but our surface differed significantly from the geomorphic features of the EBS shelf [15]. Most of the EBS shelf was interrupted by three classes of relief (low, medium, and high) [15], while our compilation did not have much relief on the shelf at all. Nor do we have the shelf valleys or basins perched on the shelf. We hypothesize that these features are all most likely to have come from bathymetry errors. The linear "medium" shelf relief features that extend north from Unimak Pass to Nome are most likely vessel paths. The "basin perched on shelf" and "Moderate size shelf valley" just to the north of Navarin Canyon are most likely derived from errant, shallow depths. Our bathymetry has some discontinuous features in this area and could use more soundings for clarification.

3.5. Other Compilations

The global [9–11] and AKRO 2005 compilations were very similar to each other. These were also generally similar to our compilation in the area around the Pribilof Islands and southern EBS shelf, corresponding to an area covered by smooth sheets and the 1970 *Rainier* cruise, and they were generally similar to our compilation in the Aleutian Basin and EBS slope canyons. Beyond these areas of agreement, these compilations had numerous hills and pits, especially along the shelf edge near the canyons, where our compilation is relatively flat. For example, in the GEBCO compilation [11], pits just south of Zhemchug Canyon exceeded 400 m in depth (Figure 9A) in areas we characterize as about 120 m deep (Figure 9B). ETOPO1 [10] had a pit almost 700 m deep and about 100 km wide just inshore of Navarin Canyon, an area we estimate to be about 100 to 200 m deep. All also had some smaller pits along the rim of Pribilof Canyon; one in the GEBCO compilation [11] exceeded 800 m and was surrounded by depths equivalent to the sea surface (Figure 9A), all in an area we characterize as about 140 to 160 m deep (Figure 9B). These compilations also had straight lines, presumably from underway files from vessel transects, both shallower and deeper than surrounding areas; these created significant ridges and narrow canyons (Figure 9A). As mentioned previously, the AKRO compilation of 2005 also had several shallow spots, regarded as pinnacles, near the Zhemchug Canyon area.

The AKRO 2017 and ARDEM [13] compilations are the most recent published for this area. The AKRO 2017 compilation has much in common with the previous compilations, such as numerous pinnacles and pits along with numerous visible ship tracks. Within Zhemchug Canyon there is a pit >3000 m in depth in an area we show is about half of that depth. Between Zhemchug Canyon and the Pribilofs there are some large pits extending down to about 300 and 400 m deep—all areas we have as being ~100 to 120 m deep. South of Zhemchug there are two groups of some very small pits; the northern group is as deep as ~3200 m in areas of the shelf that we have as ~200 to 600 m, while the southern group is as deep as ~3400 m in an area we have as ~140 to 150 m deep. The ARDEM compilation greatly reduced the appearance of deep hills and valleys on the EBS shelf that were so visible in the other compilations and in the geomorphic provinces of [15]. We presume that the improvement comes from avoidance of satellite-derived bathymetry. This depiction of the area, specifically the EBS shelf, corresponds to our AFSC cruises in this area: no pinnacles, no pits or hills, and no slender ridges or canyons. Our mid-water acoustic surveys for pollock provide a continuous record of the seafloor and show kilometer after kilometer of flat seafloor on the EBS shelf. Our bottom trawl surveys test the slope and rugosity of the seafloor by dragging nets on it. The EBS shelf bottom trawl survey is our only trawl survey in which the fishing captain of our chartered survey vessels can simply drop the net to the seafloor and drag it in any direction for 30 min without risk of tears or hangs, because the seafloor is so smooth and so well-covered with unconsolidated sediments. The AFSC's other current bottom trawl surveys, such as in the GOA, AI, the EBS slope, and our historical surveys in the US west coast shelf and slope, all require significant searching and planning for obtaining successful bottom tows without incurring damage to the net. Still, ARDEM [13] has several shallow outliers that appear as pinnacles in Navarin Canyon, and in other places just above and just below the shelf break. We trace some of these features to soundings from very small scale charts that probably could be excluded from the ARDEM compilation [13] without the loss of any information.

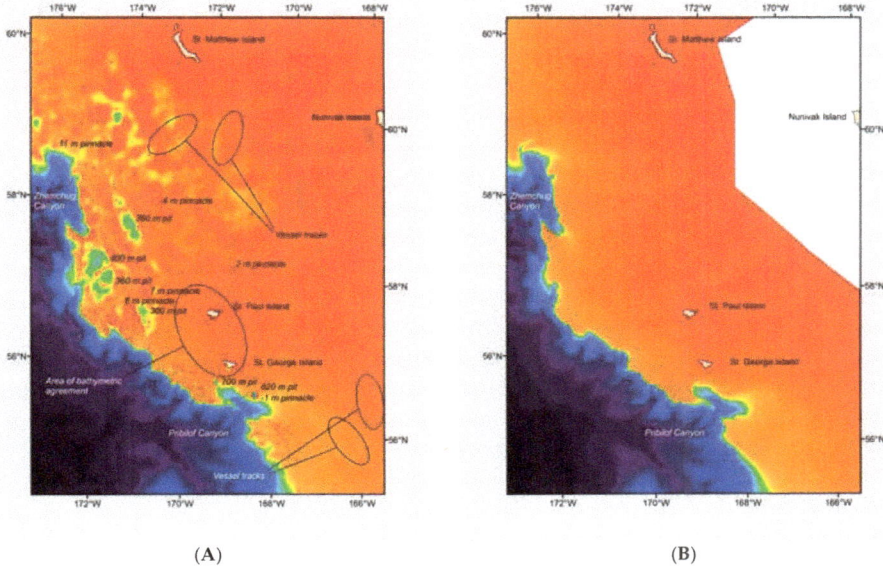

(A)   (B)

Figure 9. Comparison of eastern Bering Sea bathymetry between (**A**) the global GEBCO compilation [11] with numerous artifacts in the area between Zhemchug and Pribilof canyons and (**B**) our bathymetry which depicts a much smoother surface.

4. Discussion

This EBS slope 100 m resolution bathymetry compilation is our latest regional seafloor map of Alaska. Soon, we expect to publish similar compilations for the western Gulf of Alaska and the eastern Gulf of Alaska. Together, these new data sets, along with our previously published data sets, will create a continuous 100 m resolution surface of the Alaska seafloor ranging from Dixon Entrance in the southwest to Stalemate Bank in the west and to the Bering Strait in the north (Figure 1). These data are already being incorporated into the global lower resolution GEBCO map [11] and may also serve as an extension of a new, higher resolution version of IBCAO [12], providing a bridge from the Arctic into the North Pacific for one of Regional Data Assembly and Coordination Centers (RDACCs) of the Nippon Foundation—GEBCO Seabed 2030 Project [48]. Due to the general agreement of the ARDEM [13] regional bathymetry and our EBS slope compilation, we propose that combining data sources, especially with the Russian navigational charts, would result in an improved coverage across the entire Bering Sea. The Interferometric Synthetic Aperture Radar (ifsar) 5 m Digital Elevation Model of Alaska is in progress (http://ifsar.gina.alaska.edu/) and would serve as an excellent source for those areas above sea level.

4.1. Data Quality

With our heavy reliance on non-hydrographic underway files from numerous sources, on older data sets, and on a paucity of multibeam data, there is substantial room for improvement in our compilation. The large area of the Aleutian Basin in our map could simply and easily be improved by developing an appropriate correction between the smooth sheet and GLORIA underway files that are the only information for that area. A Gulf of Mexico bathymetry compilation covering a similar depth range was constructed entirely out of high-quality 3-D seismic data, allowing a 40 ft (12.2 m) pixel size [49], which represents a 67-fold increase in resolution over our 100 m pixel size. We did not have any high-quality 3-D seismic data for our project area, and it was nearly three times the size of the Gulf of Mexico compilation. Any dedicated bathymetry cruises in the EBS slope area, especially those that range across the steepest portion of the slope, would be a valuable addition to future versions. Lack of bathymetry data, or reliance on inaccurate bathymetry data, produces depth maps with large errors or uncertainties. If utilized, these low quality maps perpetuate their errors into plans for research cruises, survey strata boundaries, station placement, physical and biological analyses, and conclusions about management of resources in the area. Many of the bathymetry errors we noted in the regional and global compilations even occur in the global background map provided by our ArcMap software (ESRI), which we used in our figures. Better bathymetry can be a useful guide for better ocean science.

4.2. Data Editing

Visualizing and editing the data were the most important and extensive parts of creating this bathymetry surface and these canyon thalwegs. We rely on extended color ramps and utilizing very narrow color bands for distinguishing differences between neighboring depths. Slope surfaces of TINs emphasize depth differences in the raw point data and lead to most of the outliers that we investigate for correction or deletion. Since we were interested in describing the canyons in this EBS slope regional bathymetry compilation, the processing required additional steps of creating aspect surfaces and thalwegs. This, in turn, led to additional rounds of editing, as we created vertical depth corrections for individual cruises in an attempt to keep the thalwegs from following individual cruise tracks. We have used these methods as a means of working with lower quality data sets and as a substitute for automated outlier detection algorithms.

4.3. Gazetteer Names

We tried to align our canyon thalwegs with previously published canyon names, but our results indicated numerous differences. Six of our canyons are unnamed, and we suggested several name changes due to differences between our thalweg paths and the locations of source names, such as islands. For example, Okmuk Canyon of the NGA gazetteer did not appear to exist in our analysis, but this name might be re-used for one of the other unnamed canyons in the area. The Inanudak Canyon label of the NGA gazetteer was used for a canyon that extends to Okmok Volcano, not Inanudak Bay. We hope to work with GEBCO's Sub-Committee on Undersea Feature Names (SCUFN: https://www.gebco.net/about_us/committees_and_groups/) to update the canyon names in our study area.

4.4. Pinnacles

Our lack of finding the pinnacles was worrisome, as they pose a significant navigational danger, but not completely unexpected. The four Zhemchug pinnacles do not show up on global compilations such as [9], ETOPO1 [10], and GEBCO [11], nor on regional bathymetry compilations such as the AKRO of 2005, ARDEM [13], and the AKRO of 2017. The 1986 GLORIA cruise also did not reveal any pinnacles within Zhemchug canyon. Despite not confirming the four Zhemchug pinnacles from NOS Chart 16006, the 2005 AKRO bathymetry compilation depicted three new pinnacles to the east of Zhemchug Canyon and five new pinnacles south of the canyon and west of St. Paul Island. While exploring Pribilof and Zhemchug Canyons with a Deep-Worker mini-submersible and an ROV in 2007 [34], researchers on the *MV Esperanza* led an unsuccessful effort to find and document the four NOS Chart 16006 pinnacles and the eight 2005 AKRO pinnacles [pers. comm. John Hocevar 2017].

A potential source of the pinnacles might be the nearby seafloor feature of St. Paul Spur, which was mapped partially with singlebeam on two AFSC cruises [44]. These AFSC cruises clearly defined two ridges, both of which rise only to a depth of about 85 m [44], which is not much different than the depths reported on smooth sheet H08103, and are too shallow to be candidates for the reports of pinnacles. However, a primary finding of [44] acoustic analysis on the Zhemchug Canyon ridges was that extremely dense schools of juvenile Pacific ocean perch (*Sebastes alutus*) and northern rockfish (*S. polyspinus*) rise off the seafloor during daylight hours [44] (see Figure 8). We propose that these fish, which are good backscatter targets, are the source of the reports of pinnacles in this area. An additional cruise on the *MV Esperanza* exploring Zhemchug and Pribilof canyons was conducted in 2012 [35]: once again, no pinnacles were found, but no time was devoted to the search [pers. comm. John Hocevar, 2017].

Since dozens of AFSC cruises have passed through this Zhemchug area and none of them have encountered the pinnacles, we suggest that the status of the pinnacles be changed from legendary to mythical.

Supplementary Materials: The following are available online at http://www.mdpi.com/2076-3263/8/5/184/s1, Thalwegs.zip: a zipped ArcMap shape file of the thalweg polylines, and EBS Slope Bathymetry.zip, a zipped file of bathymetry image and world file for plotting as a backdrop to the thalwegs.

Author Contributions: M.Z. was the primary author of this document; M.Z. and M.M.P. both extracted, formatted, and edited bathmetry data, M.M.P. mostly extracted the comparative data and did the digitizing, and both participated in the thalweg creation.

Acknowledgments: Thanks to Wayne Palsson, Jodi Pirtle, Jennifer Jencks, Michael Martin, and H. Gary Greene for helpful reviews of the manuscript. The staff at NGDC (National Geophysical Data Center (Boulder, CO, USA): http://www.ngdc.noaa.gov) provided frequent assistance with smooth sheets and multibeam data sets. Doug Graham and Maryellen Sault helped with National Geodetic Survey shoreline products (NOAA Shoreline Explorer web site: http://www.ngs.noaa.gov/NSDE/). The staff at Pacific Hydrographic Branch, Office of Coast Survey, National Ocean Service, assisted with multibeam survey products. Abigail McCarthy and Scott R. Furnish assisted with accessing and understanding the MACE acoustic data (https://data.noaa.gov/dataset). Frank Parker and Tara Wallace helped with Notice to Mariner reports about the Zhemchug pinnacles and Robert Akro (Lamont-Doherty Earth Observatory, Columbia University, New York, NY, USA) supplied *Alpha Helix* data. Jerry Hoff helped with EBS slope bottom trawl cruises and *Miller Freeman* 200712. David Doyle (Base 9 Geodetic Consulting Services, formerly NGS) and David Grosh (NOS) provided key assistance with St. Matthew Island smooth sheet georegistration. Underway data from the BEST cruises were provided by NCAR/EOL (http://data.eol.ucar.edu/) under sponsorship of the National Science Foundation. The GEBCO bathymetry data were supplied by data request through the British Oceanographic Data Center (Liverpool, UK) (https://www.bodc.ac.uk). Funding for much of the work was provided by NOAA's Essential Fish Habitat (EFH), Habitat and Ecological Processes Research (HEPR) through the NMFS Alaska Regional Office. The findings and conclusions in the paper are those of the authors and do not necessarily represent the views of the National Marine Fisheries Service, NOAA. Reference to trade names does not imply endorsement by the National Marine Fisheries Service, NOAA. Otherwise we received no funds for covering the costs to publish in open access.

Conflicts of Interest: The authors declare no conflict of interest. The founding sponsors had no role in the design of the study; in the collection, analyses, or interpretation of data; in the writing of the manuscript; or in the decision to publish the results.

References

1. Normark, W.R.; Carlson, P.R. Giant submarine canyons: Is size any clue to their importance in the rock record. In *Extreme Depositional Environments*; Chan, M.A., Archer, A.W., Eds.; Geological Society of America: Boulder, CO, USA, 2003.

2. Hawley, J.H. *Hydrographic Manual*; U.S. Department of Commerce, U.S. Coast and Geodetic Survey, Special Publication No. 143; U.S. Government Printing Office: Washington, DC, USA, 1931.

3. Zimmermann, M.; Prescott, M.M.; Rooper, C.N. *Smooth Sheet Bathymetry of the Aleutian Islands*; U.S. Department Commerce: Washington, DC, USA, 2013.

4. Carlson, P.R.; Karl, H.A. Discovery of two new large submarine canyons in the Bering Sea. *Mar. Geol.* **1984**, *56*, 159–179. [CrossRef]

5. Carlson, P.R.; Karl, H.A. *High-Resolution Seismic Reflection Profiles: Navarin Basin Province, Northern Bering Sea*; U.S. Geological Survey: Washington, DC, USA, 1981.

6. Carlson, P.R.; Karl, H.A. *High-Resolution Seismic Reflection Profiles Collected in 1981 in Navarin Basin Province, Bering Sea*; U.S. Geological Survey: Washington, DC, USA, 1982.

7. Fischer, J.M.; Carlson, P.R.; Karl, H.A. *Bathymetric Map of Navarin Basin Province, Northern Bering Sea*; U.S. Geological Survey: Washington, DC, USA, 1982.

8. Jones, E.L. *Safeguard the Gateways of Alaska: Her Waterways*; U.S. Department of Commerce, U.S. Coast and Geodetic Survey, Special Publication No. 50; U.S. Government Printing Office: Washington, DC, USA, 1918.

9. Smith, W.H.F.; Sandwell, D.T. Global seafloor topography from satellite altimetry and ship depth soundings. *Science* **1997**, *277*, 1957–1962. [CrossRef]

10. Amante, C.; Eakins, B.W. ETOPO1 1 Arc-Minute Global Relief Model: Procedures, Data Sources and Analysis. In *NOAA Technical Memorandum ESDIS NGDC-24*; NOAA: Silver Spring, MD, USA, 2009.

11. Weatherall, P.; Marks, K.M.; Jakobsson, M.; Schmitt, T.; Tani, S.; Arndt, J.E.; Rovere, M.; Chayes, D.; Ferrini, V.; Wigley, R. A new digital bathymetric model of the world's oceans. *Earth Space Sci.* **2015**, *2*, 331–345. [CrossRef]

12. Jakobsson, M.; Mayer, L.; Coakley, B.; Dowdeswell, J.A.; Forbes, S.; Fridman, B.; Hodnesdal, H.; Noormets, R.; Pedersen, R.; Rebesco, M.; et al. *The International Bathymetric Chart of the Arctic Ocean (IBCAO)*; Version 3.0; John Wiley & Sons, Inc.: Hoboken, NJ, USA, 2012.

13. Danielson, S.L.; Dobbins, E.L.; Jakobsson, M.; Johnson, M.A.; Weingartner, T.J.; Williams, W.J.; Zarayskaya, Y. Sounding the northern seas. *Eos* **2015**, *96*. [CrossRef]

14. Harris, P.T.; Whiteway, T. Global distribution of large submarine canyons: Geomorphic differences between active and passive continental margins. *Mar. Geol.* **2011**, *285*, 69–86. [CrossRef]

15. Harris, P.T.; Macmillan-Lawler, M.; Rupp, J.; Baker, E.K. Geomorphology of the oceans. *Mar. Geol.* **2014**, *352*, 4–24. [CrossRef]

16. Zimmermann, M.; Prescott, M.M. *Smooth Sheet Bathymetry of Cook Inlet, Alaska*; U.S. Department Commerce: Washington, DC, USA, 2014.

17. Prescott, M.M.; Zimmermann, M. *Smooth Sheet Bathymetry of Norton Sound*; U.S. Department Commerce: Washington, DC, USA, 2015.

18. Zimmermann, M.; Prescott, M.M. *Smooth Sheet Bathymetry of the Central Gulf of Alaska*; U.S. Department Commerce: Washington, DC, USA, 2015.

19. Nichol, D.G.; Kotwicki, S.; Zimmermann, M. Diel vertical migration of adult Pacific cod Gadus macrocephalus in Alaska. *J. Fish Biol.* **2013**, *83*, 170–189. [CrossRef] [PubMed]

20. Rooper, C.N.; Zimmermann, M.; Prescott, M.M.; Hermann, A.J. Predictive models of coral and sponge distribution, abundance and diversity in bottom trawl surveys of the Aleutian Islands, Alaska. *Mar. Ecol. Prog. Ser.* **2014**, *503*, 157–176. [CrossRef]

21. Rooper, C.N.; Zimmermann, M.; Prescott, M.M. Comparison of modeling methods to predict the spatial distribution of deep-sea coral and sponge in the Gulf of Alaska. *Deep Sea Res. Part I Oceanogr. Res. Pap.* **2017**, *126*, 148–161. [CrossRef]

22. Zimmermann, M.; Reid, J.A.; Golden, N. Using smooth sheets to describe groundfish habitat in Alaskan waters. *Deep Sea Res. II Top. Stud. Oceanogr.* **2016**, *132*, 210–226. [CrossRef]

23. Zimmermann, M. Comparison of the physical attributes of the central and eastern Gulf of Alaska IERP inshore study sites. *Deep Sea Res. II Top. Stud. Oceanogr.* **2018**, in press.

24. Zimmermann, M.; De Robertis, A.; Ormseth, O. Verification of historical smooth sheet bathymetry. *Deep Sea Res. II Top. Stud. Oceanogr.* **2018**. submitted for publication.

25. Mordy, C.W.; Stabeno, P.J.; Kachel, N.B.; Kachel, D.; Ladd, C.; Zimmermann, M.; Doyle, M. Importance of canyons to the northern gulf of Alaska ecosystem. *Deep Sea Res. II Top. Stud. Oceanogr.* **2018**, submitted for publication.

26. Zimmermann, M.; Ruggerone, G.T.; Freymueller, J.T.; Kinsman, N.; Ward, D.H.; Hogrefe, K. Volcanic ash deposition, eelgrass beds, and inshore habitat loss from the 1920s to the 1990s at Chignik, Alaska. *Estuar. Coast. Shelf. Sci.* **2018**, *202*, 69–86. [CrossRef]

27. Laman, E.A.; Rooper, C.N.; Rooney, S.C.; Turner, K.A.; Cooper, D.W.; Zimmermann, M. *Model-Based Essential Fish Habitat Definitions for Bering Sea Groundfish Species*; U.S. Department Commerce: Washington, DC, USA, 2017.

28. Rooney, S.; Rooper, C.N.; Laman, E.A.; Turner, K.; Cooper, D.; Zimmermann, M. *Model-Based Essential Fish Habitat Definitions for Gulf of Alaska Groundfish Species*; U.S. Department Commerce: Washington, DC, USA, 2018.

29. Laman, E.A.; Rooper, C.N.; Turner, K.; Rooney, S.; Cooper, D.; Zimmermann, M. Using species distribution models to define essential fish habitat in Alaska. *Can. J. Fish. Aquat. Sci.* **2017**. [CrossRef]

30. Turner, K.; Rooper, C.N.; Laman, E.A.; Rooney, S.C.; Cooper, D.W.; Zimmermann, M. *Model-Based Essential Fish Habitat Definitions for Aleutian Island Groundfish Species*; U.S. Department Commerce: Washington, DC, USA, 2017.

31. Pirtle, J.; Shotwell, S.K.; Zimmermann, M.; Reid, J.A.; Golden, N. Habitat Suitability Models for Groundfish in the Gulf of Alaska. *Deep Sea Res. II Top. Stud. Oceanogr.* **2017**, in press. [CrossRef]

32. McGowan, D.W.; Horne, J.K.; Thorson, J.T.; Zimmermann, M. Influence of environmental factors on capelin distributions in the Gulf of Alaska. *Deep Sea Res. II Top. Stud Oceanogr.* **2017**, in press. [CrossRef]

33. Food and Agriculture Organization. The state of world fisheries and aquaculture 2016. In *Contributing to Food Security and Nutrition for All*; FAO: Rome, Italy, 2016.

34. Miller, R.J.; Hocevar, J.; Stone, R.P.; Fedorov, D.V. Structure-forming corals and sponges and their use as fish habitat in Bering Sea submarine canyons. *PLoS ONE* **2012**, *7*. [CrossRef] [PubMed]

35. Miller, R.J.; Juska, C.; Hocevar, J. Submarine canyons as coral and sponge habitat on the eastern Bering Sea. *Glob. Ecol. Conserv.* **2015**, *4*, 85–90. [CrossRef]

36. Sigler, M.F.; Rooper, C.N.; Hoff, G.R.; Stone, R.P.; McConnaughey, R.A.; Wilderbuer, T.K. Faunal features of submarine canyons on the eastern Bering Sea slope. *Mar. Ecol. Prog. Ser.* **2015**, *526*, 21–40. [CrossRef]

37. Hoff, G.R. Identification of skate nursery habitat in the eastern Bering Sea. *Mar. Ecol. Prog. Ser.* **2010**, *403*, 243–254. [CrossRef]

38. Hoff, G.R. *Results of the 2012 Eastern Bering Sea Upper Continental Slope Survey of Groundfish and Invertebrate Resources*; U.S. Department of Commerce: Washington, DC, USA, 2013.

39. Nghiem, S.V.; Clemete-Colon, P.; Rigor, I.G.; Hall, D.K.; Neumann, G. Seafloor control on sea ice. *Deep Sea Res. Part II Top. Stud. Oceanogr.* **2012**, *77–80*, 52–61. [CrossRef]

40. Springer, A.M.; McRoy, C.P.; Flint, M.V. The Bering Sea green belt: Shelf-edge processes and ecosystem production. *Fish. Oceanogr.* **1996**, *5*, 205–223. [CrossRef]

41. Wong, S.N.P.; Gjerdrum, C.; Morgan, K.H.; Mallory, M.L. Hotspots in cold seas: The composition, distribution, and abundance of marine birds in the North American Arctic. *J. Geophys. Res. Oceans* **2014**, *119*, 1691–1705. [CrossRef]

42. Wong, A.M.; Campagnoli, J.G.; Cole, M.A. Assessing 155 years of hydrographic survey data for high resolution bathymetry grids. In Proceedings of the Oceans 2007, Vancouver, BC, Canada, 29 September–4 October 2007.

43. Zimmermann, M.; Benson, J. *Smooth Sheets: How to Work with Them in a GIS to Derive Bathymetry, Features and Substrates*; U.S. Department Commerce: Washington, DC, USA, 2013.

44. Rooper, C.N.; Hoff, G.R.; De Robertis, A. Assessing habitat utilization and rockfish (Sebastes spp.) biomass on an isolated rocky ridge using acoustics and stereo image analysis. *Can. J. Fish. Aquat. Sci.* **2010**, *67*, 1658–1670. [CrossRef]

45. McKelvey, D.; Steinessen, S. *Results of the March 2014 Acoustic-Trawl Survey of Walleye Pollock (Gadus chalcogrammus) Conducted in the Southeastern Aleutian Basin Near Bogoslof Island, Cruise DY2014–02*; NOAA: Silver Spring, MD, USA, 2015.

46. Honkalehto, T.; McCarthy, A. *Results of the Acoustic-Trawl Survey of Walleye Pollock (Gadus Chalcogrammus) on the U.S. and Russian Bering Sea Shelf in June–August 2014 (DY1407)*; NOAA: Silver Spring, MD, USA, 2015.

47. Carter, D.J.T. *Echo-Sounding Correction Tables*, 3rd ed.; Hydrographic Department Ministry of Defence: Taunton, UK, 1980.

48. Mayer, L.; Jakobsson, M.; Allen, G.; Dorschel, B.; Falconer, R.; Ferrini, V.; Lamarche, G.; Snaith, H.; Weatherall, P. The Nippon Foundation—GEBCO seabed 2030 project: The quest to see the world's oceans completely mapped by 2030. *Geosciences* **2018**, *8*. [CrossRef]

49. Kramer, K.V.; Shedd, W.W. A 1.4-billion-pixel map of the Gulf of Mexico seafloor. *Eos* **2017**, *98*. [CrossRef]

geosciences

MDPI

Article

Realistic Paleobathymetry of the Cenomanian–Turonian (94 Ma) Boundary Global Ocean

Arghya Goswami [1,*] [iD]**, Linda Hinnov** [2]**, Anand Gnanadesikan** [3] **and Taylor Young** [1]

1   Department of Natural Sciences, Northwest Missouri State University, Maryville, MO 64468, USA;
    tayloryoung031497@gmail.com
2   Department of Atmospheric, Oceanic, and Earth Sciences, George Mason University, Fairfax, VA 22030, USA;
    lhinnov@gmu.edu
3   Earth & Planetary Sciences, Johns Hopkins University; Baltimore, MD 21218, USA; gnanades@jhu.edu
*   Correspondence: goswami@nwmissouri.edu; Tel.: +1-660-562-1719

Received: 6 November 2017; Accepted: 10 January 2018; Published: 15 January 2018

Abstract: At present, global paleoclimate simulations are prepared with bathtub-like, flat, featureless and steep walled ocean bathymetry, which is neither realistic nor suitable. In this article, we present the first enhanced version of a reconstructed paleobathymetry for Cenomanian–Turonian (94 Ma) time in a $0.1° \times 0.1°$ resolution, that is both realistic and suitable for use in paleo-climate studies. This reconstruction is an extrapolation of a parameterized modern ocean bathymetry that combines simple geophysical models (standard plate cooling model for the oceanic lithosphere) based on ocean crustal age, global modern oceanic sediment thicknesses, and generalized shelf-slope-rise structures calibrated from a published global relief model of the modern world (ETOPO1) at active and passive continental margins. The base version of this Cenomanian–Turonian paleobathymetry reconstruction is then updated with known submarine large igneous provinces, plateaus, and seamounts to minimize the difference between the reconstructed paleobathymetry and the real bathymetry that once existed.

Keywords: Cretaceous; Cenomanian–Turonian; paleobathymetry; paleoclimate; paleoceanography; reconstruction; simulation; shelf-slope-rise

1. Introduction

Understanding past climate is critical for the scientific study of ancient, modern, and future climate change. Computer-generated models, also known as general circulation models (GCMs) are often used to study the plausible dynamic effects of diverse boundary conditions on climate. Paleoclimate data provide a useful framework from which these models can be used to predict past, present, and future climate change scenarios. The use of numerical models to investigate Earth's climate history has increased significantly over the past decade [1–3]. These models have grown in sophistication, with increased space–time resolution, coupled ocean-atmosphere GCMs that incorporate vegetation, soil (weathering), ice, and chemistry modules.

Accordingly, more robust and detailed boundary conditions are needed. One of the most difficult and often overwhelming tasks in paleoclimate modeling is the assembly of surface boundary conditions for past time periods [2]. These boundary conditions include but are not limited to continental configurations, topography, ocean bathymetry, ocean chemistry, terrestrial vegetation distribution, atmospheric trace gas concentrations, insolation, etc. Many of these boundary conditions have been the subjects of extensive research, and abundant data are available. A notable exception is paleobathymetry, which is often modeled as a bathtub with steep walls or oversimplified bathymetry to such a great extent that the role of bathymetry on climate change is greatly undermined. A simple scholarly search yields dozens of publications with such simplified ocean/paleo-ocean bathymetries and/or slab oceans that are being used to produce paleoclimate simulations with GCMs [4–25]. Among them there are a few global paleobathymetries [1,2,15–17] available in the literature. Bice paleobathymetry [18] is not for Cenomanian–Turonian (C–T) time. The data from Müller [15,16] and Sewall [2] are available and used in this article for comparison, while some data are not available [1]. In addition to the global paleobathymetries, there are several regional/local paleobathymetries for the C–T time (for example, [19–25]) are available, which are either impossible to reverse engineer or will be extremely time consuming to recover in an usable format [19,20]. Moreover, these data are not used in this study keeping in mind the global nature of the current work.

The great importance of bathymetry on global climate is illustrated by the following studies. Krupitsky [26] discussed how present-day Southern Ocean bathymetry blocks flow through the Drake Passage and that this affects the magnitude of the circumpolar current. In the Northern Hemisphere, Wright and Miller [27] have hypothesized that variations in the depth of the Greenland–Iceland–Scotland Ridge modulate North Atlantic Deep Water formation. At the global scale, tidal dissipation is concentrated in shallow marine environments; Simmons [28] proposed that the generation of tides in the presence of rough bathymetry plays a major role in driving deep ocean mixing. Sijp and England [29] have demonstrated that the stability of the thermohaline circulation is related to the present-day Southern Ocean bathymetry. Thus, detailed knowledge of paleobathymetry is critical for quantifying analogous processes in the geologic past.

The geometric rules of plate tectonics and seafloor spreading provide an objective method for paleobathymetric reconstruction; consequently, much progress has been made in reconstructing paleobathymetry younger than 200 Ma. In particular, the relationship discovered between ocean crustal age and depth to basement [30] was quickly exploited to estimate paleobathymetry of the Atlantic and Indian oceans [19,20]. Pacific Ocean paleobathymetry has proven to be more challenging with its multiple spreading centers, plates of various sizes, ages and orientations, and active subduction zones [31], as well as the now lost Tethys Ocean [32].

By contrast, the dynamics of ocean currents in the proximity of continental margins is controlled by the regional morphology of the conjugate continental shelves, slopes and rises [33]. An important element missing from these margin reconstructions is the shelf-slope-rise region between oceanic crust and continental shoreline. Despite these difficulties, today a convincing case has been made for the general validity of paleobathymetric reconstructions for oceans that overlie oceanic crust of known age [15,16,32,34]. For near-present-day reconstructions, this region can be adapted from modern bathymetry. However, further back in geologic time the structure of the continent-ocean transition becomes increasingly less certain or unknown. Yet this region represents a critical zone for many biological, sedimentary, and oceanographic processes that influence the Earth system.

This paper presents a deep-time Cenomanian–Turonian (C–T; 94 Ma) bathymetric reconstruction based on the geometric rules of plate tectonics, seafloor spreading, relation of ocean crustal age to depth to basement, knowledge of global modern ocean sediment cover and modeled shelf-slope-rise structure that is suitable at the global scale, takes into account the heterogeneity of these compound structures and based on modern-day geometric relationships between ocean crust and shoreline. Goswami [35] describe the modern-day version of this bathymetric reconstruction methodology as a test case. This methodology, called "OESbathy" (OES: Open Earth Systems; www.openearthsystems.org), is used here for the first time to produce C–T global ocean bathymetry. This C–T bathymetry also includes known Large Igneous Provinces (LIPs) and seamounts. In this work OES reconstructed C–T bathymetry is compared with other available or equivalent reconstructions.

The C–T transition intersects with the following important geological events:

1. The occurrence of a major oceanic anoxic event (OAE2) associated with large accumulations of organic rich sediments in the oceans and leading to the largest petroleum resources being utilized by society today [36,37];

2. Major carbon cycle acceleration and biogeochemical perturbation in the form of a major global marine carbon isotope excursion [38];

3. Dramatic global warming from mantle-plume volcanism (main phase Caribbean and Madagascar, late phase High Arctic and Ongtong Java LIPs) [39];

4. Extensive mid-ocean ridge formation, high crustal heat flow and rapid ocean basin expansion [15];

5. Major change in ocean circulation from eddy-dominant with widespread cosmopolitan biogeography in the Cenomanian to thermohaline with restricted marine biogeographic zones in the Turonian [40];

6. Highest sea levels of the Mesozoic with development of major, multiple epicontinental seas, some serving as gateways (e.g., the Western Interior Seaway connecting the tropical Atlantic with polar Arctic) [16,41] and

7. Completion of opening of the Central Atlantic Seaway connecting the North and South Atlantic sectors [42,43].

8. Onset of the Alpine orogeny as indicated by the relative plate motions and paleogeographic evidence from Africa, Iberia and Europe (especially, the Carpatho-Balkan and Sicily regions) [44–47].

All of these events involved the oceans and climate change, and so it is clear that a detailed C–T paleobathymetry model will be an important contribution for future paleoclimatic investigations.

2. Materials and Methods

2.1. Materials

To reconstruct the C–T bathymetry for the global ocean with generalized continental shelf-slope-rise structures using the OESbathy methodology (version 1.0, [35]) five major datasets were used. A brief description of these individual datasets is provided below.

Ocean crust age: Müller [16] published global reconstructions of ocean crust age in 1 Myr intervals for the past 140 million years. For each reconstructed age in [16], ocean crust age, depth to basement, and bathymetry are given. This reconstructed bathymetry obtained as part of the dataset provided by Müller [16] is referred to hereafter as EB08 (EB: EarthByte). These data are in 0.1° × 0.1° resolution (3601 longitude × 1801 latitude points). For this project, (modern) crustal age reconstruction data are used (Figure 1).

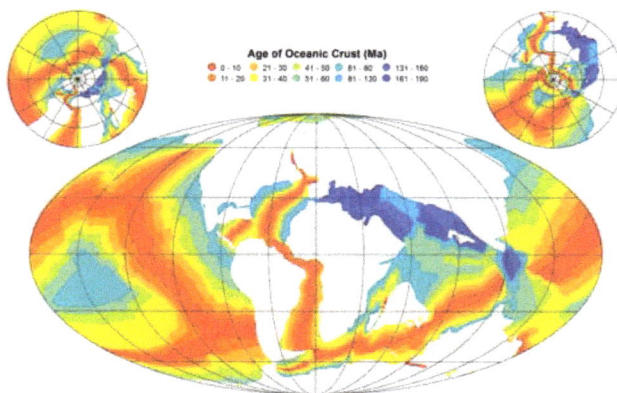

Figure 1. Age of the ocean crust for Cenomanian–Turonian (C–T) time (94 Ma), derived from Müller et al. [15].

Three geologic time scales that concern the age of the Cenomanian–Turonian (C–T) boundary are summarized in Table 1. The geologic time scale used by Müller et al. (2008) assigns a 36.9 m.y. duration from the top of C-Sequence chron C34 to the base of M-Sequence chron M0. The Cretaceous time scale in GTS2012 assigns a 42.66 m.y. duration and GTS2016 a 42.1 m.y. duration for the same interval, based on revised ages for the top of chron C34 and base of M0.

The C–T boundary age is now extremely well constrained at 93.9 Ma [48]. The duration of the interval from the C–T boundary to the top of C34 is estimated to be from 8.875 m.y. (GTS2012) to 8.5 m.y. (GTS2016), i.e., 23% to 24% of the duration of the C34 to M0 interval. This scales the C–T boundary age in the Müller GTS to between 92.375 Ma and 92.0 Ma. This differs slightly from the 93.5 Ma C–T boundary age of Gradstein [49], whose time scale was used to calibrate the digital age grid for the oceanic crust by Müller [31] and later used in EB08 [15].

Table 1. Geologic time scales, durations and ages of key boundaries. GTS2012 = The Geologic Time Scale for the Cretaceous Period [50]; GTS2016 = Concise Geologic Time Scale 2016 [48]; Müller GTS according to [15].

	GTS2012	GTS2016	Müller GTS
Top C34 age	83.64 Ma	84.2 Ma	83.5 Ma
Base M0 age	126.3 Ma	126.3 Ma	120.4 Ma
Top C34 to base M0 duration	42.66 m.y.	42.1 m.y.	36.9 m.y.
Top C34 to C–T boundary duration	8.875 m.y.	8.5 m.y.	8.875 [a] m.y. or 8.5 [b] m.y.
C–T boundary age	93.9 Ma	93.9 Ma	92.375 [a] Ma or 92.0 [b] Ma

[a] Based on GTS2012 C34 to C–T boundary duration as 24% of the top C34 to base M0 duration. [b] Based on GTS2016 C34 to C–T boundary duration as 23% of the top C34 to base M0 duration.

Our study assumed a "traditional" age of 90 Ma for the C–T boundary for the ocean crustal component (construction of EB08). However, our reconstruction is actually between 2.375 m.y. and 2 m.y. younger than the C–T boundary age according to the Müller GTS (see Table 1), the latter presumably equivalent to the 93.9 Ma age of the C–T boundary assigned by GTS2012 and GTS2016.

Modern ocean sediment thickness: Modern ocean sediment thickness data from Divins [51] and Whittaker [52] are used. These data are derived from seismic profiling of the world's ocean basins and other sources. The reported thicknesses are calculated using seismic velocity profiles that yield minimum thicknesses. Data values represent the distance between the seafloor and "acoustic basement". The data are given in 5s × 5s resolution and have been regridded to 0.1° × 0.1° resolution values to match the EB08 grid [35].

ETOPO1: To reconstruct shelf-slope-rise structures, large igneous provinces and seamounts, modern topography and bathymetry from ETOPO1 [53] is used. The "Bedrock" version of ETOPO1 is utilized for this work, which is available in a 1s × 1s resolution (earthmodels.org), and regridded to 0.1° × 0.1° resolution [35] in order to match the EB08 grid. This version of ETOPO1 includes relief of Earth's surface depicting the bedrock underneath the ice sheets. However, only the oceanic points in this dataset are used, so that there is no impact on the reconstructed bathymetry.

Large igneous provinces: In this reconstruction, the large igneous provinces (LIPs) dataset [54] is used. This dataset includes a geographic description of each LIP (Figure 2), plate identification number, and times of appearance and disappearance (in Ma). These LIPs are distinguished from mid-ocean ridge and subduction-related rocks based on petrology, geochemistry, geochronology, physical volcanology, and geophysical data. For this study only oceanic LIPs are considered. The modern distribution of the oceanic LIPs (Figure 2a) were rotated back to 94 Ma (Figure 2b).

Global Seamount Database: In this study, the seamounts from the Global Seamount Database [55] are used, which is based on satellite-derived vertical gravity gradient (VGG) data with globally 24,643 potential seamounts that are higher than or equal to 0.1 km, and located away from continental margins. The modern distributions of these seamounts (Figure 3a) were rotated back to 94 Ma (Figure 3b).

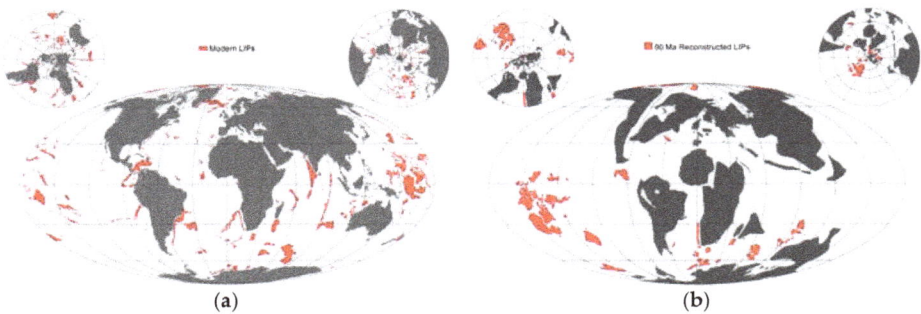

Figure 2. Distribution of Large Igneous Provinces (LIPs) in the ocean basins. (**a**) Modern; (**b**) 94 Ma.

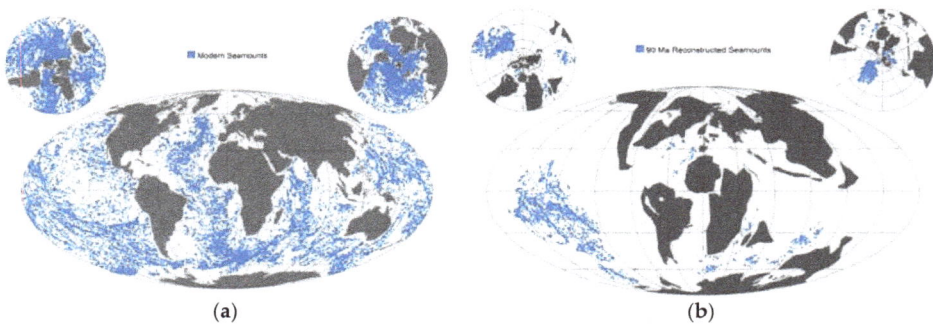

Figure 3. Distribution of seamounts in the ocean basins. (**a**) Modern; (**b**) 94 Ma.

2.2. Methods

For this reconstruction, a standard plate cooling model for the oceanic lithosphere is combined with the age of the oceanic crust, global oceanic sediment thicknesses, plus generalized shelf-slope-rise structures calibrated at modern active and passive continental margins. As with the modern ocean basins, there are different types of crust: oceanic crust, submerged continental crust, and a transitional zone between the two (e.g., [35]). In this reconstruction, the regions underlain by oceanic crust to which an age has been assigned are termed "open ocean" regions. Here, a negative sign is assigned to signify depths below mean sea level (MSL).

The reconstruction of open ocean bathymetry starts with ocean crust age (Figure 1). This information is available only at locations where oceanic crust is preserved or has been reconstructed. The ocean depth-to-basement (Figure 4) is calculated based on a cooling plate with an average depth of -2639.8 m for the depth of mid-oceanic ridge [35].

The addition of sediment and an isostatic correction from sediment loading of the depth-to-basement oceanic crust (e.g., [56]) is critical to complete the bathymetric reconstruction of the open ocean region. A parameterized multilayer sediment cover, called "OES sediment thickness", based on a third-degree polynomial fit between area corrected global sediment thickness data of modern oceans and marginal seas [51,52] and the age of the underlying oceanic crust [35], was isostatically added on top of the depth-to-basement. The sediment loading approach used here is similar to procedures used by Crough [57] and Sykes [58], was adopted using a multicomponent sediment layer with varying sediment densities calculated from a linear extrapolation of published sediment densities [35]. Once the sediment loading was added, the reconstruction of the open ocean regions was considered complete (Figure 5), and ready for merging with shelf-slope-rise structures, and adding LIPs, plateaus and seamounts, as will be described next.

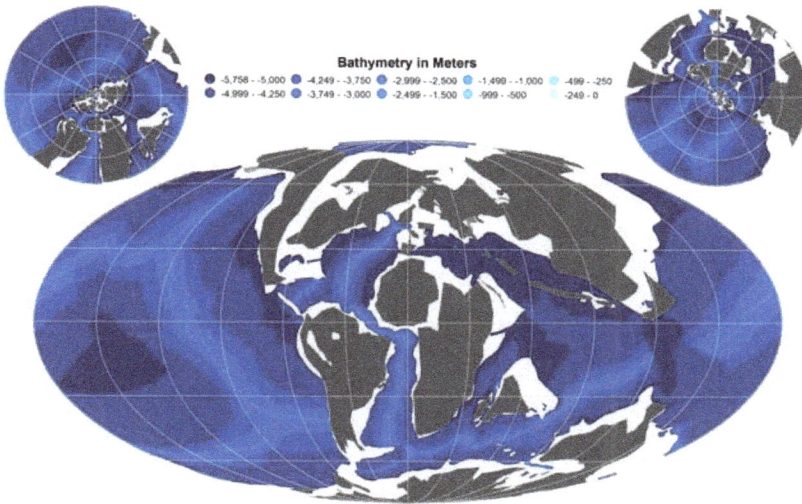

Figure 4. Depth-to-basement calculated from the age of oceanic crust at 94 Ma.

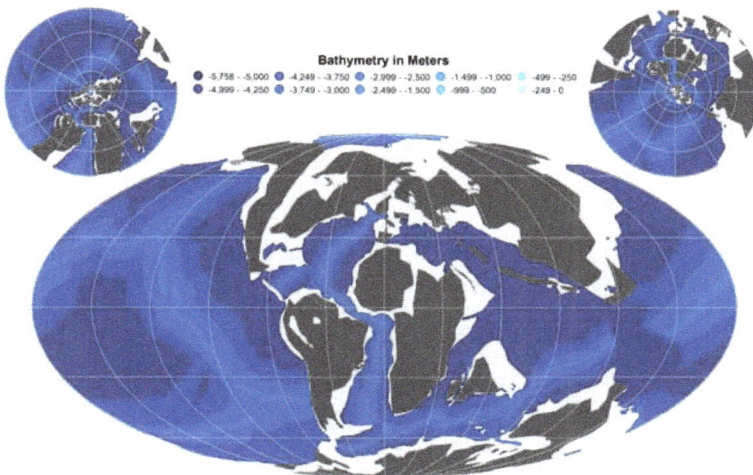

Figure 5. Depth-to-basement, with isostatically adjusted, multicomponent sediment layer at 94 Ma.

The parts of the ocean basins that occupy the transitional zone between oceanic crust and the emergent continental crust are termed here as "shelf-slope-rise" regions. These regions typically extend from the boundary of open ocean regions to the coastline. The three-parameter "shelf-slope-rise" model from Goswami [35], based on analysis of various modern shelf-slope-rise structures at active and passive margin regions from ETOPO1, along with their corresponding sediment thicknesses taken from [51], was used to reconstruct the C–T "shelf-slope-rise" regions, extending from the maximum extent ocean crust to the coastlines (Figure 6).

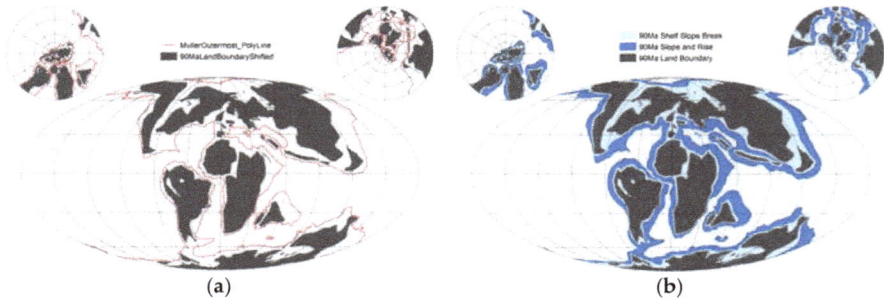

Figure 6. The extent of different 94 Ma ocean boundaries. (**a**) The maximum extent of the ocean crust in the ocean basins; (**b**) the extent of 94 Ma land, shelf, slope, and rise.

The paleo-shoreline model used in this reconstruction was obtained from the C–T paleo-DEM obtained from the PALEOMAP Project [41]. In this model, a zero meter coastline was extracted from the paleo-DEM to be used along with the EB08 ocean age grid map. When the extracted continental shorelines from the PALEOMAP Project reconstructions were incorporated into the EB08 CT age grid, there was a very small difference in the rotation of the continental South America block. This was manually adjusted to match with the EB08 grid.

The construction of the shelf-slope model is a combination of geographic information system and hand drawing activity. The process starts with drawing perpendicular lines to the paleo-coastline. Once the normal distance between the coastline and the shoreline has been defined, the nearest ocean crust is determined, and the shelf, slope and rise [35] dimensions are marked.

Accordingly, as detailed in [35] the C–T open-ocean regions (Figure 5) are then merged with C–T shelf-slope-rise regions (Figure 6). To accomplish this merging, map-based operations such as computing distances between locations were carried out using ArcGIS, and local calculations with interpolation were carried out in Matlab. The merged C–T base reconstructed bathymetry is shown in Figure 7.

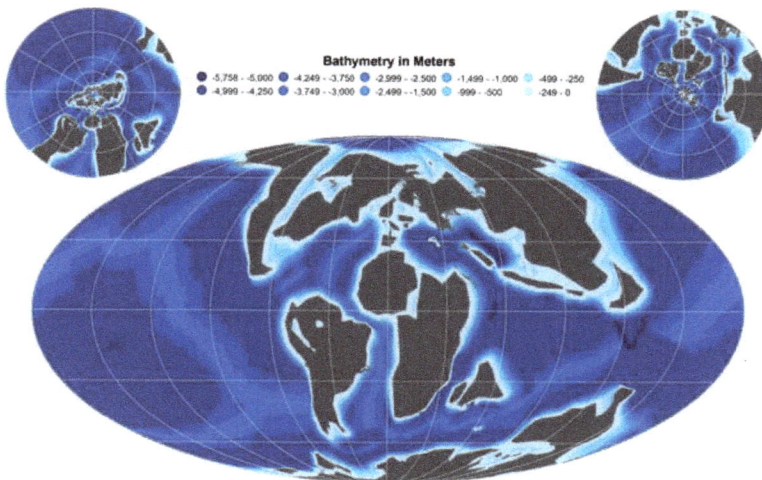

Figure 7. The 94 Ma reconstructed base bathymetry with isostatically adjusted, multicomponent sediment layer and shelf-slope-rise structure.

This reconstructed C–T bathymetry was then updated with known oceanic LIPs and seamounts (Figures 2b and 3b), for which GPlates (www.gplates.org) was used to rotate the modern distribution of the LIPs and seamounts back to C–T time. This updated reconstructed bathymetry was then smoothed and cleaned in Matlab using kriging interpolation. The final map of C–T reconstructed bathymetry is presented in Figure 8.

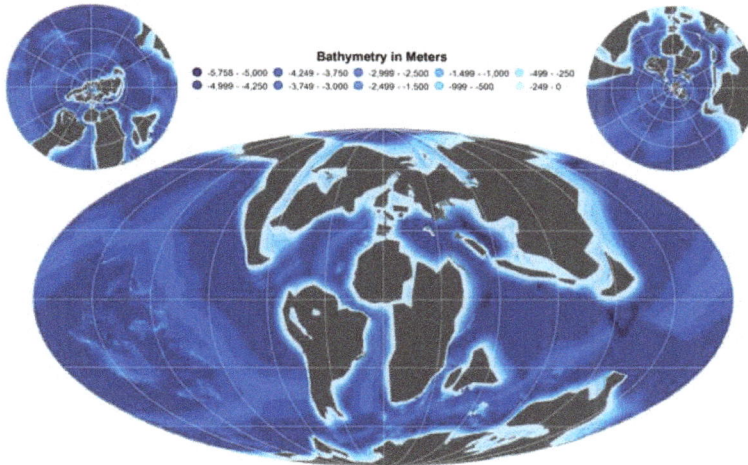

Figure 8. Interpolated 94 Ma reconstructed bathymetry with isostatically adjusted, multicomponent sediment layer, continental shelf-slope-rise structures, and known LIPs and seamounts.

In sum, the final products are global maps of C–T depth-to-basement at 0.1° × 0.1° resolution (Figure 4), C–T paleobathymetry for the open oceans with an isostatically adjusted multicomponent sediment layer at 0.1° × 0.1° resolution (Figure 5), C–T paleobathymetry with reconstructed continental shelf-slope-rise structures (Figure 7) and a realistic combined C–T paleobathymetry with known LIPs and seamounts, continental shelf-slope-rise structures, and isostatically adjusted multicomponent sediment layer at 0.1° × 0.1° resolution (Figure 8).

3. Results

Here, the general nature of the C–T reconstructed bathymetry is discussed. One important point is that this bathymetric reconstruction was updated only with known LIPs and seamounts. As is clear from Figures 2b and 3b, only certain areas in the central Pacific, Indian, proto-Caribbean and the North Atlantic Ocean therefore could be updated with seamounts and LIPs.

3.1. Reconstructed C–T Shelf-Slope-Rise Regions

The reconstructed C–T bathymetry (Figure 8) indicates that the global distribution of the continental blocks was more clustered than the modern world. North America and Eurasia were still connected, a young north Atlantic was spreading, and the central and southern Atlantic was in a juvenile state. In the south, Indian Ocean was also in its infancy while in between Indian peninsula and Eurasia was huge and mature Tethys Ocean. This unique paleogeography was quite different from the modern world and consequently influenced the C–T shelf-slope-rise regions.

Table 2 presents a comparison between the percentages of the global area for continents, oceans, continental shelf, slope and rise. Of the first order, during the C–T time, the area of the land was about 4% less of the global total area than the present world. Here, land refers to the part of the continental plate that is not submerged. Comparing with modern land areas, it is almost 14% less.

All of the continental blocks were separated at this point of time, as the rifting of the supercontinent Pangea continued, though they were not far from each other. Accordingly, the ocean area was 4% larger in terms of total global area, and 5.66% greater the present-day world. These differences may be attributed to the growth in the size of the continents due to collisional tectonics [59].

Oceanic eustasy coupled with terrigenous sediment supply from newly separated continental interiors contributed to the development of mature shelf-slope-rise regions. As discussed in [35], typically the extent of these mature shelf-slope-rise regions extends beyond the maximum extent of ocean crust towards the shoreline. During C–T time, besides the major continental blocks including Africa, America, Eurasia, Antarctica, and India separating from the main Pangean aggregated supercontinent, many small-scale tectonic events took place that contributed to further expansion of shelf-slope-rise regions. For example, Madagascar separated from India, and the Kohistan-Ladakh Arc was rapidly approaching the northern boundary of the Indian peninsula [60]. In the northern Tethys between the North Atlantic and NW Tethys there were several smaller continental blocks. Surrounding both big and small continental blocks, development of shelf-slope rise structures was critical. The presence of an equatorial-tropical seaway connecting the Tethys to the Atlantic oceans, and the shallow but significantly wide Western Interior Seaway connecting the Gulf of Mexico to the Arctic Ocean, and a shallow African Seaway all added to the larger fraction of the shelf-slope-rise area during C–T time.

Table 2. Comparison of area between modern and C–T oceans.

	Modern World [1]	C–T World [2]	C–T/Modern
Continents	29.00%	24.98%	−13.86%
Ocean	71.00%	75.02%	5.66%
Continental shelf *	5.68%	8.60%	51.47%
Continental slope and rise *	11.72%	13.70%	16.93%
Combined shelf, slope and rise *	17.40%	22.31%	28.20%
Open Ocean	53.60%	52.71%	−1.66%

[1] Data for modern world is taken from Encyclopedia of Britannica [61]; [2] Data for C–T world are according to this project; * Percent of total global area.

As summarized in Table 2, the total area occupied by all of the shelf-slope-rise regions was about 22.3% of the global area. Just shelf alone was 8.60% of the global area. Compared to modern combined shelf-slope-rise area, C–T shelf-slope-rise area was larger by more than 30%. The C–T shallow shelf was 8.60% of global area compared to 5.68% for the modern world. In other words, this was more than 60% of modern shelf area. Considering the continental slope and rise together the trend is not so similar, and was about 16% larger during C–T time. In sum, the most significant difference between modern and C–T oceans was in the area of the shelf region. Consequently, from a bathymetric point of view, this additional shallow shelf area and intermediate depth slope-rise area gave the C–T oceanic world a unique dimension.

3.2. Reconstructed C–T Open Ocean Region

Beyond the shelf-slope-rise region is the open ocean region. According to Table 2, the area of the C–T open ocean region is most comparable to that of the modern open ocean regions, only 2.61% less than today. The majority of the C–T open ocean region was part of the Pacific Ocean (Figure 1), extending around the newly rifted and separated continental landmasses. It was a giant young ocean basin with well-defined mid-oceanic ridge system creating new ocean floor. The North Atlantic had very limited area as open ocean. The South and Central Atlantic had negligible open ocean regions as these ocean basins had just started rifting South America from Africa. Also, the proto Caribbean Sea was rifting the South American block from North America. To the east, the other major ocean was the Tethys, which later would be completely subsumed between the Eurasian block and the fast-moving Indian block. The northern and northwestern part of the Tethys Ocean floor was the oldest among the global C–T ocean floors. Among all the other C–T oceans, the Tethys was unique as it did not have a well-developed, active ridge system. The Indian Ocean began to rift open between the Indian peninsula and Antarctica-Australia block. This long and linear rift extended to the eastern part of the Tethys. The Indian mid-oceanic ridge was separated from the Pacific ridge system by a smaller old oceanic block, which may have originally been part of eastern Tethys.

The maximum depth of the C–T open ocean region was about 5758 m, with a mean depth of 3262 m and median depth of 4080 m. C–T open oceans were shallower compared with modern open oceans. But looking at the median depth of the C–T open ocean was slightly deeper than that of the modern open oceans. In this regard, one of the final processing steps involved kriging interpolation to eliminate outlier calculation points. The total number of such points (19,285) was very small compared to the total number of points that define the entire C–T open ocean (4,498,703), i.e., less than 0.29%. Positive bathymetric values were changed to 0 using kriging type interpolation methods. Figure 9 shows the distribution of such points on top of the Interpolated 94 Ma reconstructed bathymetry (Figure 8).

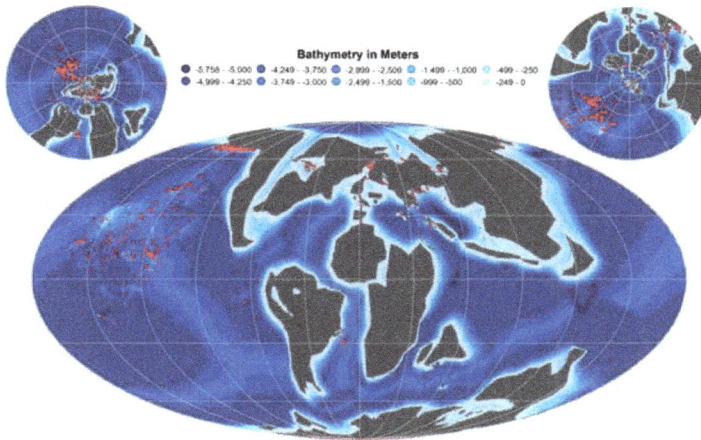

Figure 9. Distribution of the interpolated points (in red) on top of the interpolated 94 Ma reconstructed bathymetry with isostatically adjusted, multicomponent sediment layer, continental shelf-slope-rise structures, and known LIPs and seamounts.

Table 3 shows the statistics for modern and C–T ocean floor age, and the various steps of reconstructed C–T bathymetry in contrast to modern bathymetry (obtained from ETOPO1). The mean age of C–T ocean crust was about 40 million years (Myr) whereas the modern ocean crust has a mean age of about 63 Myr. Thus, on average, C–T oceanic crust was younger than the modern equivalent. The median ages corroborate the same view with C–T being about 36 Myr while the modern ocean is about 55 Myr. Looking at the mode value for C–T oceanic crust a drastic difference is revealed. While the age of modern oceanic crust has most common value of 0 Myr, for C–T oceanic crust it was 190 Myr. Thus C–T oceanic crust had overall older crustal age than the modern ocean. The overall statistics for the C–T reconstructed bathymetry are also provided in Table 3 and Supplementary Figure S3. The output of the plate model, listed as C–T depth-to-basement has maximum and minimum depths of 5555 m and 2663 m, respectively; the average depth is 4465 m. With loading from modeled sediment layers, these values change to 4920, 1385, and 4025 m, respectively. When the shelf-slope-rise wedge model is added to the open ocean reconstruction to obtain the C–T base bathymetry the signature changes to 5760 m, 677 m (above sea level) and 3305 m respectively for maximum, minimum and average depths (Supplementary Figure S4).

Table 3. Comparative statistics of Open Earth Systems (OES), Modern and EB08 bathymetries.

Variable	Max	Min	Mean	Mean %[9]	Median	Median %[9]	Mode	Std.Dev.	Std.Dev. %[9]	Variance	Variance %[9]
C–T OC[1] Age[2]	190.00	0.00	40.29	63.56%	36.20	65.94%	190.00	32.51	72.66%	1057.17	52.81%
Modern OC[1] Age[2]	*280.00*	*0.00*	*63.39*	*100.00%*	*54.90*	*100.00%*	*0.00*	*44.74*	*100.00%*	*2001.77*	*100.00%*
OES D2B[3]	−2662.69	−5554.11	−4464.67	131.79%	−4592.02	118.87%	−5554.11	604.00	34.69%	364,819.64	12.03%
OES D2B[3] + Sediment	−1385.96	−4920.04	−4024.87	118.81%	−4084.91	105.74%	−4671.99	527.11	30.27%	277,844.40	9.17%
OES Base[4] Bathymetry	677.31	−5758.30	−3303.11	97.51%	−4128.30	106.87%	−6.52	1769.18	101.61%	3,129,987.14	103.25%
OES Bathymetry[5]	5087.88	−5758.30	−3229.23	95.32%	−4005.90	103.70%	−6.52	1764.88	101.36%	3,114,784.82	102.75%
OES Difference[6]	2283.91	−5551.14	−385.27	–	−48.74	–	−676.01	1018.68	–	1,037,718.60	–
OES Interpolated Bathymetry	**0.00**	**−5758.30**	**−3233.20**	**95.44%**	**−4005.90**	**103.70%**	**−6.52**	**1756.50**	**100.88%**	**3,085,309.63**	**101.77%**
Modern Ocean Bathymetry[7]	*−1.00*	*−10,714.00*	*−3387.63*	*100.00%*	*−3863.00*	*100.00%*	*−1.00*	*1741.13*	*100.00%*	*3,031,539.27*	*100.00%*
OES Interpolated EB08 Extent[8]	0.00	−5753.00	−4070.70	120.16%	−4394.60	113.76%	−4878.10	995.00	57.15%	990,018.77	32.66%
EB08 D2B[3]	2290.43	−5638.43	−4520.08	133.43%	−4724.14	122.29%	−5638.43	803.68	46.16%	645,909.44	21.31%
EB08 Bathymetry	2402.45	−5266.97	−4269.14	126.02%	−4423.02	114.50%	−5150.16	753.15	43.26%	567,234.14	18.71%
EB08 Difference[6]	−96.22	−733.19	−250.94	–	−204.79	–	−488.27	140.55	–	19,753.06	–
SW07 C–T Ocean	0.00	−5755.90	−3442.90	101.63%	−4508.39	116.71%	−5040.00	1970.61	113.18%	3,883,318.54	128.10%

[1] OC = Ocean Crust; [2] Age in million annum (Ma). All other numbers are in meters; [3] D2B = Depth-to-basement; [4] Base = with sediment and shelf-slope-rise structure; [5] This maximum value and correlated positive values represent some points that become positive when different equations are applied. The number of such points with positive values (19,285) are very small compared to the total number of points in the entire ocean basin (4,498,703), approximately 0.29%. The calculated positive bathymetric values of these points were changed to 0 using kriging type interpolation methods; [6] difference = (depth-to-basement—final bathymetry); [7] Data source ETOPO1; [8] OES bathymetry trimmed to EB08 extent; [9] With respect to modern value.

The minimum depth of 677 m above sea level is the result of outliers generated by the mathematical calculations. These outliers were later corrected using kriging type interpolation. The number of these outliers were very small compared to the scope of the entire dataset, (19,285 out of 4,498,703, which is less than 0.29%). The addition of oceanic LIPs and seamounts (both before and after interpolation) did not change the result of fixing the values programmatically for these 19,285 points. These erroneous points (Supplemental Figure S1) are mostly clustered, distributed in low to high latitudes of the Northern Hemisphere, and cover 0.031% of global C–T area (0.04% of global C–T ocean area).

Modern ocean bathymetric statistics derived from ETOPO1 are comparable to the C–T statistics. Although the modern ocean has much deeper trenches, the C–T ocean has a maximum depth of 5758 m, with an average depth of 3233 m, which is surprising similar to the modern (95.44%), which is 3388 m. The median depths of the modern and C–T oceans are also significantly similar, 3863 and 4005 m (103.7%), respectively. The difference may be attributed to the much older overall crustal age of the C–T ocean (see above). The most striking similarity is that the standard deviation of the OES C–T dataset is 100.88% of modern dataset. Another surprising similarity is in variance of these two datasets is within 2% range (Table 3).

4. Discussion

The reconstructed global C–T bathymetry presented here is the first of its kind based on the observations from modern ocean bathymetry. The main goal of OES is to reconstruct global paleobathymetry, provided age of the oceanic crust, onto which a generalized shelf-slope-rise structure model (based on present-day geometry) is anchored. The EB08 dataset was adopted for oceanic crustal ages. Along with the age of the oceanic crust for the last 140 Myr, EB08 also provides a version of depth-to-basement calculation and thereby a paleobathymetry (global in nature but without shelf-slope-rise) for each of the last 140 Myr in one Myr increments. The comparison of different statistical parameters between models are shown in Supplementary Figure S5.

Accordingly, the discussion starts with the comparison between EB08 (Supplementary Figure S1) and OES. The depth-to-basement reconstructions of OES and EB08 are similar in terms of minimum, mean, median, and mode depths. The difference is observed in the standard deviation, as the EB08 depth-to-basement has a maximum value which is positive. This is due to the fact that EB08 includes LIPs in the depth-to-basement whereas OES reconstruction does not. In Table 3, C–T OES final interpolated reconstructed bathymetry and its various components are compared with EB08, and SW07 bathymetries, which are very different. While OES bathymetry has a mean depth of 3233 meters, EB08 bathymetry has a mean depth of 4269 m. While compared to OES bathymetry trimmed to EB08 extent the mean value for OES bathymetry is 4070 meters. So overall, EB08 reconstructed bathymetry is on an average 200 m deeper in the open ocean regions. The OES full extent bathymetry is but much shallower than EB08 because of the shelf-slope-rise regions. The major difference in the EB08 bathymetry and the OES bathymetry EB08 extent is clear in variance, the measure of how far each value in the data set is from the mean. The variance for the OES bathymetry dataset (full bathymetry) is extraordinarily analogous to the variance of modern world bathymetry. The OES bathymetry dataset (EB08 extent), compared to EB08, has much smaller variance signifying gradual changes in bathymetry than that in EB08. This is confirmed by the modal and median values which shows that OES bathymetry has shallower median values as well as very shallow depths as the most common values. In EB08, the mode is very high at 5150 m as opposed to 6.5 m in OES. When compared to the OES bathymetry trimmed to EB08 extent, high modal value 4878 m (though much less than EB08, 5150 m) is observed. This exactly spells out the difference between the two bathymetry dataset and highlights the importance of reconstructing the shelf-slope-wedge. This has special significance especially for the C–T time, as for decades the paleoclimate community has argued for extensive shallow epicontinental seas [36,62–76] for example, to explain oceanic anoxic events. The OES C–T Ocean compared to Modern Ocean, has 4.75% more area (ocean area) less than 1000 m deep (maximum extent of combined euphotic and disphotic zone depth). For areas shallower than 500 m and 200 m, the OES C–T Ocean has 3.69% and 1.21% more compared to the Modern Ocean (Supplementary Table S1). The corresponding maps showing the extent of the reconstructed OES C–T ocean with these depths are shown in Supplementary Figure S6. These increased areas of shallower depths (compared to total ocean areas) are encouraging because OES reconstructed bathymetry at least in the case of C–T time, reconstructed larger areas of shallower depths in agreement with previously postulated arguments.

The difference between the depth-to-basement and final bathymetry also sheds some light regarding how much sediment was used in these bathymetric reconstructions, although caution is needed as direct measurement of sediment amount was not involved. Isostatic loading of sediment is complicated owing to density and thickness of sediment layers. The OES model shows a significant difference in the depth-to-basement to final bathymetric reconstruction (Table 3, OES Difference). Addition of sediment, modeled shelf-slope-rise structures and known seamounts and oceanic LIPs significantly change the final bathymetry with a much larger standard deviation. This is also seconded by the median values of the two models: OES median value is 48 m and EB08 median value is 205 m. This implies that the changes in depth-to-basement to final bathymetry in the OES reconstruction does not involve many abrupt changes (or less abrupt changes) in depth, and rather, follows a natural gradient, which may be more realistic. A high median value, on the other hand, implies that in EB08, there are unexpected abrupt variations in the bathymetric values. The most common value (mode) in the OES difference dataset is 675 m, while the mean is 385 m. In EB08, these values are considerably lower, 490 m and 250 m, respectively.

When compared with the modern ocean bathymetry the OES C–T reconstructed bathymetry reflects interesting similarity. Though it is created using the very different ocean crustal age (see Table 3) than the modern one, the average and median depth, mode, standard deviation and the variance of the modern ocean and C–T ocean are highly comparable (Table 3). Surprisingly, these two datasets (modern and OES C–T) have less agreement with the EB08 reconstructed bathymetry.

Other than statistical comparison, a difference map between the EB08 and OES C–T reconstructions is given in Figure 10. The red tone signifies that EB08 reconstruction is deeper while blue tone signifies the OES reconstruction is deeper. Statistics for this difference map are provided in Table 3, but also computed EB08 extent (mainly open oceans regions) as EB08 does not have shelf-slope-rise structures. Both maximum and minimum values for the difference plots are very large. The maximum value of 5640 m are in the regions where there are underwater LIPs. In EB08, these areas have positive values signifying they are above the water and sometimes even up to 2000+ m. In the OES reconstruction, all known LIPs are added, but they all have an average depth of at least 1500 m. Similarly, the high minimum number is near the outer (oldest) edge of the oceanic crust, where in the OES reconstruction, the shelf-slope-rise model is added to estimate natural coastal landforms. Much of the open ocean regions in the two models are in agreement and their mean difference is 190 m with a high standard deviation.

Another interesting comparison can be made with the C–T bathymetric model of Sewall [2] (SW07). The SW07 model is available in a 64 × 128 grid size, and here is resized to 1801 × 3601 (Supplementary Figure S2) in order to compare with the OES model. The SW07 model also contains topography; only values less than or equal to zero were considered for comparison (Figure 11). The statistical comparison between the OES and SW07 models indicates some fascinating similarity. The minimum, mean and median depths are within 10% of each other. The modal depth, however, is very different, in OES bathymetry being very shallow (6.5 m), while that in SW07 is very deep (5040 m). The standard deviations of these two models are 10% (OES) and 20% (SW07). In the SW07-OES difference map (Figure 11), regions with positive values indicate deeper depths in OES. Most of the disagreements are in the continental shelf-slope-rise regions and may be attributed to the shelf-slope-rise structure of the OES model. There are also some differences in the depth of mid-oceanic ridges: in OES, the mid-oceanic ridges are much shallower than in SW07. Both models include a larger and more open

North Atlantic Ocean; extensive Cretaceous interior seaways in North and South America, North Africa, and Eastern Europe, and a narrowly open South Atlantic Ocean.

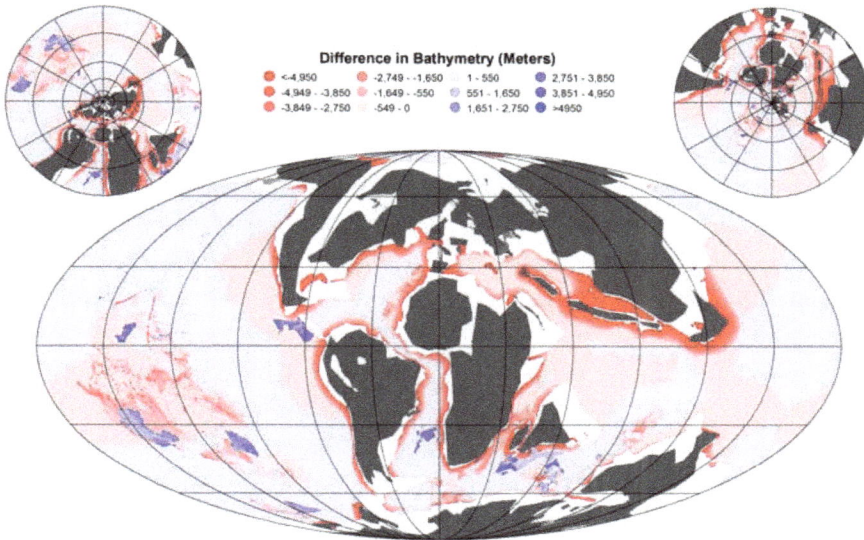

Figure 10. Difference map between the EB08 and OES C–T reconstructions.

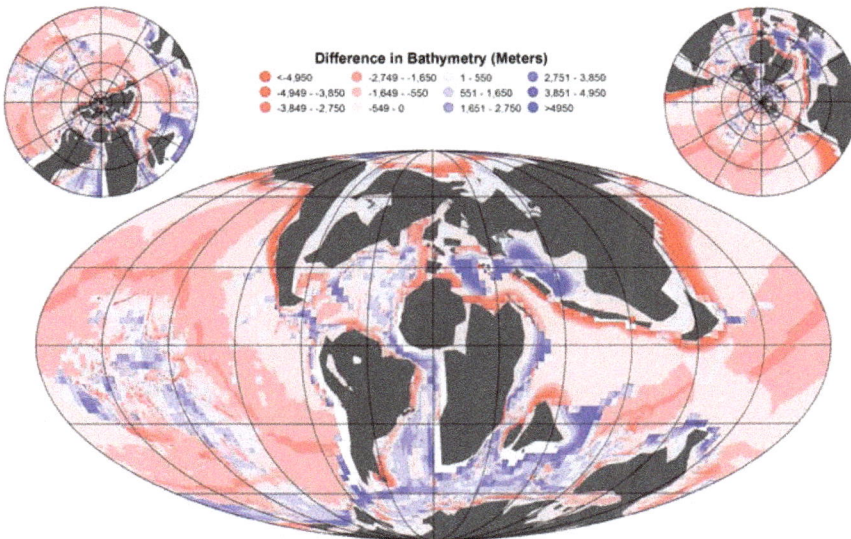

Figure 11. Difference map between the SW07 and OES C–T reconstructions.

In addition to the Cretaceous paleobathymetric reconstruction presented here, there are other important related developments that are very encouraging. Hochmuth and Gohl [33] are working on improving geodynamic and paleoceanographic aspects related to ocean crustal age with a high-resolution plate kinematic model. The alternative seafloor age grid for the South Atlantic Ocean based on a recent high-resolution plate kinematic model [43] is of particular interest due to its connection to the Cretaceous Normal Superchron. Enhanced studies like this will open the door for better understanding of long-offset fracture zones, currently mostly overlooked in paleobathymetric reconstructions.

The C–T transition occurred during the wettest, hottest climates of the Cretaceous greenhouse with assists from greenhouse gases emitted to the atmosphere by unusually large mid-ocean ridge systems and mantle plume magmatism [77]. The opening of the critical N-S Equatorial Atlantic Seaway coincided with a change from an "eddy ocean" to a "thermohaline ocean" [78]. Evidence for this change is provided by North Atlantic nannofossils which evolved from homogeneous Cenomanian distributions with cosmopolitan ranges to heterogeneous Turonian distributions with restricted ranges [40].

The Earth's equator-to-pole temperature gradient is a fundamental climatological property related to equator-to-pole heat transport efficiency, ocean circulation, hydrological cycling and atmospheric chemistry [79]. The meriodional thermal gradient may have been at its lowest in the Cretaceous during the C–T transition. Unfortunately, Cretaceous paleoclimate modeling with GCMs has consistently underestimated the warm polar temperatures that are indicated by fossil and geochemical data (e.g., [74,80,81]).

Realistic paleobathymetries are among the improvements that can be tested in GCM paleoclimate simulations. The goal of this project has been to address this by producing a more realistic ocean floor with continental shelf-slope-rise, underwater ridges, trenches, seamounts, LIPs and a corresponding ocean bottom roughness. In particular, seafloor roughness and topography exert critical controls on vertical ocean mixing and circulation [82,83]. Will our paleobathymetry provide clues for understanding the eddy-dominated Cenomanian ocean? What would cause ocean reorganization in the Turonian—development of a less rough seafloor due to more rapid seafloor spreading [42,83].

5. Conclusions

The ongoing goal of OESbathy reconstruction is to provide a means for reconstructing realistic paleobathymetries for geologic times that can be tied directly to ocean crustal age. The innovation of OESbathy is the generalized shelf-slope-rise model to provide bathymetric interpolation across the transitional zone between open ocean and coastlines; this model is described in detail for the test case modern day OESbathy 1.0 reconstruction [35]. Here OESbathy has been extended to Cenomanian–Turonian (C–T) boundary time (94 Ma), and updated with known oceanic LIPs and seamounts. The main results are as follows:

- Parameterized shelf-slope-rise structures result in a much shallower mean ocean depth than other available C–T bathymetry reconstructions.
- C–T ocean crust was on average 40.29 million years old whereas modern ocean crust has a mean age of 63.39 million years.
- The OES C–T ocean has an average depth of 3262 m, while the modern ocean average depth is 3388 m. Despite that the OES C–T Ocean is about 125 m shallower, the overall ocean bathymetric pattern matches with the modern bathymetric values as both have similar standard deviations (within 1%) despite significantly different maximum and average depths.
- Seafloor roughness is provided from mid-ocean ridges, and oceanic LIPS and sea mounts rotated to their C–T positions (using known features).

The following improvements are being planned: (1) The OES sediment model will be validated and corrected using data from DSDP/ODP/IODP. (2) The OES shelf-slope-rise model will be revisited with a more detailed parameterization of modern active and passive margins. (3) We are working on the reconstruction of ocean bottom surface fractures, especially those near the mid-ocean ridge areas, and experimenting with statistical models to simulate seafloor roughness. (4) An OES reconstruction of Paleocene–Eocene Thermal Maximum (55 Myr) paleobathymetry is now underway.

Supplementary Materials: The following are available online at www.mdpi.com/2076-3263/8/1/21/s1; Figure S1: EB08 C–T bathymetry; Figure S2: SW07 [2] C–T bathymetry (from the total land ocean topography bathymetry, only the values zero or below have been plotted here); Figure S3: Comparative statistics of modern and C–T ages of the ocean crust (in Ma). The minimum is not shown as both have zero as minimum ocean crust age; Figure S4: Comparison of the three models. OES: Interpolated OES CT Bathymetry, OES*: Interpolated OES CT Bathymetry trimmed to EB08 extent, EB08: EB08 CT Bathymetry, and SW07: SW07 CT Bathymetry; Figure S5: Comparison between different of OES C–T bathymetry development. Step 1: OES Depth-to-Basement. Step 2: OES Depth-to-Basement isostatically loaded with multilayer sediments. Step 3: OES C–T Base Bathymetry with shelf-slope-rise. Step 4: OES Bathymetry with known LIPs and Seamounts, and Step 5: OES Interpolated Bathymetry; Figure S6: OES C-T Ocean showing (A) areas of 200, 500 and 1000 meters depths; (B) area of 200 meters depth; (C) area of 500 meters depth; and (D) area of 1000 meters depth. Table S1: Comparison of shallow ocean areas between Modern and C–T oceans.

Acknowledgments: A significant portion of this work was supported by Frontiers in Earth System Dynamics grant EAR-1135382 from the National Science Foundation. Also, the authors are thankful to C. R. Scotese for sharing the SW07 datasets used in this study for comparison. The authors highly appreciate the invitation from the MDPI publication group for the invitation to this special marine geomorphometry issue.

Author Contributions: Data for the open ocean reconstruction and sediment wedge model were acquired and analyzed by Arghya Goswami. The reconstruction algorithm was developed by Arghya Goswami under the guidance of Ananad Gnanadesikan) and Linda Hinnov. Taylor Young acquired the data related to LIPs and seamounts, and after preparation and analysis, updated the C–T paleobathymetry. Arghya Ggoswami did the main writing while Linda Hhinnov did the editing.

Conflicts of Interest: The authors declare no conflict of interest.

References

1. Markwick, P.J.; Valdes, P.J. Palaeo-digital elevation models for use as boundary conditions in coupled ocean–atmosphere GCM experiments: A Maastrichtian (late Cretaceous) example. *Palaeogeogr. Palaeoclimatol. Palaeoecol.* **2004**, *213*, 37–63. [CrossRef]

2. Sewall, J.O.; Van De Wal, R.S.W.; Van Der Zwan, K.; Van Oosterhout, C.; Dijkstra, H.A.; Scotese, C.R. Climate model boundary conditions for four Cretaceous time slices. *Clim. Past* **2007**, *3*, 647–657. [CrossRef]

3. Lunt, D.J.; Huber, M.; Anagnostou, E.; Baatsen, M.L.J.; Caballero, R.; DeConto, R.; Dijkstra, H.A.; Donnadieu, Y.; Evans, D.; Feng, R.; et al. The DeepMIP contribution to PMIP4: Experimental design for model simulations of the EECO, PETM, and pre-PETM (version 1.0). *Geosci. Model Dev.* **2017**, *10*, 889–901. [CrossRef]

4. Barron, E.J.; Washington, W.M. The role of geographic variables in explaining paleoclimates: Results from Cretaceous climate model sensitivity studies. *J. Geophys. Res. Atmos.* **1984**, *89*, 1267–1279. [CrossRef]

5. Barron, E.J. Explanations of the Tertiary global cooling trend. *Palaeogeogr. Palaeoclimatol. Palaeoecol.* **1985**, *50*, 45–61. [CrossRef]

6. Chandler, M.A.; Rind, D.; Ruedy, R. Pangaean climate during the Early Jurassic: GCM simulations and the sedimentary record of paleoclimate. *Geol. Soc. Am. Bull.* **1992**, *104*, 543–559. [CrossRef]

7. Barron, E.J.; Fawcett, P.J.; Pollard, D.; Thompson, S. Model simulations of Cretaceous climates: The role of geography and carbon dioxide. *Philos. Trans. R. Soc. Lond. B Biol. Sci.* **1993**, *341*, 307–316. [CrossRef]

8. Barron, E.J.; Moore, G.T. *Climate Model Application in Paleoenvironmental Analysis*; Society for Sedimentary Geology (SEPM): Tulsa, OK, USA, 1994.

9. Price, G.D.; Sellwood, B.W.; Valdes, P.J. Sedimentological evaluation of general circulation model simulations for the "greenhouse" Earth: Cretaceous and Jurassic case studies. *Sediment. Geol.* **1995**, *100*, 159–180. [CrossRef]

10. Poulsen, C.J.; Barron, E.J.; Peterson, W.H.; Wilson, P.A. A reinterpretation of Mid-Cretaceous shallow marine temperatures through model-data comparison. *Paleoceanography* **1999**, *14*, 679–697. [CrossRef]

11. Upchurch, G.R.; Otto-Bliesner, B.L.; Scotese, C.R. *Terrestrial Vegetation and Its Effects on Climate during the Latest Cretaceous*; Special Papers; Geological Society of America: Boulder, CO, USA, 1999; pp. 407–426.

12. DeConto, R.M.; Brady, E.C.; Bergengren, J.; Hay, W.W. Late Cretaceous climate, vegetation, and ocean interactions. In *Warm Climates in Earth History*; Cambridge University Press: Cambridge, UK, 2000; pp. 275–296.

13. Poulsen, C.J.; Gendaszek, A.S.; Jacob, R.L. Did the rifting of the Atlantic Ocean cause the Cretaceous thermal maximum? *Geology* **2003**, *31*, 115–118. [CrossRef]

14. Hay, W.W.; Floegel, S. New thoughts about the Cretaceous climate and oceans. *Earth-Sci. Rev.* **2012**, *115*, 262–272. [CrossRef]

15. Müller, R.D.; Sdrolias, M.; Gaina, C.; Roest, W.R. Age, spreading rates, and spreading asymmetry of the world's ocean crust. *Geochem. Geophys. Geosyst.* **2008**, *9*, 1–19. [CrossRef]

16. Müller, R.D.; Sdrolias, M.; Gaina, C.; Steinberger, B.; Heine, C. Long-term sea-level fluctuations driven by ocean basin dynamics. *Science* **2008**, *319*, 1357–1362. [CrossRef] [PubMed]

17. Wells, M.R.; Allison, P.A.; Piggott, M.D.; Hampson, G.J.; Pain, C.C.; Gorman, G.J. Tidal modeling of an ancient tide-dominated seaway, part 1: Model validation and application to global early Cretaceous (Aptian) tides. *J. Sediment. Res.* **2010**, *80*, 393–410. [CrossRef]

18. Bice, K.L.; Barron, E.J.; Peterson, W.H. Reconstruction of realistic early Eocene paleobathymetry and ocean GCM sensitivity to specified basin configuration. *Oxf. Monogr. Geol. Geophys.* **1998**, *39*, 227–250.

19. Sclater, J.G.; Abbott, D.; Thiede, J. Paleobathymetry and sediments of the Indian Ocean. In *Indian Ocean Geology and Biostratigraphy*; Wiley: New York, NY, USA, 1977; pp. 25–59.

20. Sclater, J.G.; Hellinger, S.; Tapscott, C. The paleobathymetry of the Atlantic Ocean from the Jurassic to the present. *J. Geol.* **1977**, *85*, 509–552. [CrossRef]

21. Herkat, M. Application of correspondence analysis to palaeobathymetric reconstruction of Cenomanian and Turonian (Cretaceous) rocks of Eastern Algeria. *Palaeogeogr. Palaeoclimatol. Palaeoecol.* **2007**, *254*, 583–605. [CrossRef]

22. Sageman, B.B.; Arthur, M.A. Early Turonian paleogeographic/paleobathymetric map, western interior, US. In *Mesozoic Systems of the Rocky Mountain Region*; Society for Sedimentary Geology (SEPM): Tulsa, OK, USA, 1994.

23. Kyrkjebø, R.; Kjennerud, T.; Gillmore, G.K.; Faleide, J.I.; Gabrielsen, R.H. Cretaceous-Tertiary palaeo-bathymetry in the northern North Sea; integration of palaeo-water depth estimates obtained by structural restoration and micropalaeontological analysis. *Nor. Pet. Soc. Spec. Publ.* **2001**, *10*, 321–345.

24. Butt, A. Micropaleontological bathymetry of the Cretaceous of Western Morocco. *Palaeogeogr. Palaeoclimatol. Palaeoecol.* **1982**, *37*, 235245265–242261275. [CrossRef]

25. Topper, R.P.M.; Trabucho Alexandre, J.; Tuenter, E.; Meijer, P.T. A regional ocean circulation model for the mid-Cretaceous North Atlantic Basin: Implications for black shale formation. *Clim. Past* **2011**, *7*, 277–297. [CrossRef]

26. Krupitsky, A.; Kamenkovich, V.M.; Naik, N.; Cane, M.A. A linear equivalent barotropic model of the Antarctic Circumpolar Current with realistic coastlines and bottom topography. *J. Phys. Oceanogr.* **1996**, *26*, 1803–1824. [CrossRef]

27. Wright, J.D.; Miller, K.G. Control of North Atlantic deep water circulation by the Greenland-Scotland Ridge. *Paleoceanography* **1996**, *11*, 157–170. [CrossRef]

28. Simmons, H.L.; Jayne, S.R.; Laurent, L.C.S.; Weaver, A.J. Tidally driven mixing in a numerical model of the ocean general circulation. *Ocean Model.* **2004**, *6*, 245–263. [CrossRef]

29. Sijp, W.P.; England, M.H. Role of the Drake Passage in controlling the stability of the ocean's thermohaline circulation. *J. Clim.* **2005**, *18*, 1957–1966. [CrossRef]

30. Parsons, B.; Sclater, J.G. An analysis of the variation of ocean floor bathymetry and heat flow with age. *J. Geophys. Res.* **1977**, *82*, 803–827. [CrossRef]

31. Müller, R.D.; Roest, W.R.; Royer, J.-Y.; Gahagan, L.M.; Sclater, J.G. Digital isochrons of the world's ocean floor. *J. Geophys. Res. Solid Earth* **1997**, *102*, 3211–3214. [CrossRef]

32. Xu, X.; Lithgow-Bertelloni, C.; Conrad, C.P. Global reconstructions of Cenozoic seafloor ages: Implications for bathymetry and sea level. *Earth Planet. Sci. Lett.* **2006**, *243*, 552–564. [CrossRef]

33. Hochmuth, K.; Gohl, K. Paleobathymetry of the Southern Ocean and its role in paleoclimate and paleo-ice sheet variations–A call for a sequence of community paleobathymetric grids. In Proceedings of the SCAR Open Science Conference, Kuala Lumpur, Malaysia, 10–20 August 2016.

34. Hayes, D.E.; Zhang, C.; Weissel, R.A. Modeling paleobathymetry in the Southern Ocean. *Eos Trans. Am. Geophys. Union* **2009**, *90*, 165–166. [CrossRef]

35. Goswami, A.; Olson, P.L.; Hinnov, L.A.; Gnanadesikan, A. OESbathy version 1.0: A method for reconstructing ocean bathymetry with generalized continental shelf-slope-rise structures. *Geosci. Model Dev.* **2015**, *8*, 2735–2748. [CrossRef]

36. Jenkyns, H.C. Geochemistry of oceanic anoxic events. *Geochem. Geophys. Geosyst.* **2010**, *11*. [CrossRef]

37. Klemme, H.D.; Ulmishek, G.F. Effective petroleum source rocks of the world: Stratigraphic distribution and controlling depositional factors. *AAPG Bull.* **1991**, *75*, 1809–1851.

38. Jarvis, I.; Lignum, J.S.; Gröcke, D.R.; Jenkyns, H.C.; Pearce, M.A. Black shale deposition, atmospheric CO_2 drawdown, and cooling during the Cenomanian-Turonian Oceanic Anoxic Event. *Paleoceanography* **2011**, *26*. [CrossRef]

39. Bryan, S.E.; Ferrari, L. Large igneous provinces and silicic large igneous provinces: Progress in our understanding over the last 25 years. *Geol. Soc. Am. Bull.* **2013**, *125*, 1053–1078. [CrossRef]

40. Linnert, C.; Mutterlose, J.; Herrle, J.O. Late Cretaceous (Cenomanian–Maastrichtian) calcareous nannofossils from Goban Spur (DSDP Sites 549, 551): Implications for the palaeoceanography of the proto North Atlantic. *Palaeogeogr. Palaeoclimatol. Palaeoecol.* **2011**, *299*, 507–528. [CrossRef]

41. Scotese, C.R. *Atlas of Late Cretaceous Paleogeographic Maps, PALEOMAP Atlas for ArcGIS, Volume 2, The Cretaceous, Maps 16–22, Mollweide Projection, PALEOMAP Project, Evanston, IL*; Technical Report for PALEOMAP Project: Evanston, IL, USA, 2014.

42. Granot, R.; Dyment, J. The Cretaceous opening of the South Atlantic Ocean. *Earth Planet. Sci. Lett.* **2015**, *414*, 156–163. [CrossRef]

43. Pérez-Díaz, L.; Eagles, G. A new high-resolution seafloor age grid for the South Atlantic. *Geochem. Geophys. Geosyst.* **2017**, *18*, 457–470. [CrossRef]

44. Channell, J.E.T.; Horvath, F. The African/Adriatic promontory as a palaeogeographical premise for Alpine orogeny and plate movements in the Carpatho-Balkan region. *Tectonophysics* **1976**, *35*, 71–101. [CrossRef]

45. Rosenbaum, G.; Lister, G.S.; Duboz, C. Relative motions of Africa, Iberia and Europe during Alpine orogeny. *Tectonophysics* **2002**, *359*, 117–129. [CrossRef]

46. Basilone, L.; Perri, F.; Sulli, A.; Critelli, S. Paleoclimate and extensional tectonics of short-lived lacustrine environments. Lower Cretaceous of the Panormide Southern Tethyan carbonate platform (NW Sicily). *Mar. Pet. Geol.* **2017**, *88*, 428–439. [CrossRef]

47. Basilone, L.; Sulli, A. Basin analysis in the Southern Tethyan margin: Facies sequences, stratal pattern and subsidence history highlight extension-to-inversion processes in the Cretaceous Panormide carbonate platform (NW Sicily). *Sediment. Geol.* **2018**, *363*, 235–251. [CrossRef]

48. Ogg, J.G.; Ogg, G.; Gradstein, F.M. *A Concise Geologic Time Scale: 2016*; Elsevier: Amsterdam, The Netherlands, 2016.

49. Gradstein, F.M.; Agterberg, F.P.; Ogg, J.G.; Hardenbol, J.; Veen, P.; Thierry, J.; Huang, Z. A Mesozoic time scale. *J. Geophys. Res. Solid Earth* **1994**, *99*, 24051–24074. [CrossRef]

50. Divins, D.L. *Total Sediment Thickness of the World's Oceans and Marginal Seas*; NOAA National Geophysical Data Center: Boulder, CO, USA, 2003.

51. Whittaker, J.M.; Goncharov, A.; Williams, S.E.; Müller, R.D.; Leitchenkov, G. Global sediment thickness data set updated for the Australian-Antarctic Southern Ocean. *Geochem. Geophys. Geosyst.* **2013**, *14*, 3297–3305. [CrossRef]

52. Ogg, J.G.; Hinnov, L.A.; Huang, C. Cretaceous. In *The Geologic Time Scale*; Chapter 27; Elsevier: Amsterdam, The Netherlands, 2012; pp. 793–853.

53. Amante, C.; Eakins, B.W. *ETOPO1 1 Arc-Minute Global Relief Model: Procedures, Data Sources and Analysis*; Marine Geology and Geophysics Division: Boulder, CO, USA, 2009.

54. Coffin, M.F.; Duncan, R.A.; Eldholm, O.; Fitton, J.G.; Frey, F.A.; Larsen, H.C.; Mahoney, J.J.; Saunders, A.D.; Schlich, R.; Wallace, P.J. Large igneous provinces and scientific ocean drilling: Status quo and a look ahead. *Oceanography* **2006**, *19*, 150–160. [CrossRef]

55. Kim, S.-S.; Wessel, P. New global seamount census from altimetry-derived gravity data. *Geophys. J. Int.* **2011**, *186*, 615–631. [CrossRef]

56. Celerier, B. Paleobathymetry and geodynamic models for subsidence. *Palaios* **1988**, *3*, 454–463. [CrossRef]

57. Crough, S.T. The correction for sediment loading on the seafloor. *J. Geophys. Res. Solid Earth* **1983**, *88*, 6449–6454. [CrossRef]

58. Sykes, T.J. A correction for sediment load upon the ocean floor: Uniform versus varying sediment density estimations—Implications for isostatic correction. *Mar. Geol.* **1996**, *133*, 35–49. [CrossRef]

59. Niu, Y.; Zhao, Z.; Zhu, D.-C.; Mo, X. Continental collision zones are primary sites for net continental crust growth—A testable hypothesis. *Earth-Sci. Rev.* **2013**, *127*, 96–110. [CrossRef]

60. Encyclopedia Britannica | Britannica.com. Available online: https://www.britannica.com/ (accessed on 12 December 2017).

61. Chatterjee, S.; Goswami, A.; Scotese, C.R. The longest voyage: Tectonic, magmatic, and paleoclimatic evolution of the Indian plate during its northward flight from Gondwana to Asia. *Gondwana Res.* **2013**, *23*, 238–267. [CrossRef]

62. Schlanger, S.O.; Jenkyns, H.C. Cretaceous oceanic anoxic events: Causes and consequences. *Geol. Mijnb.* **1976**, *55*, 179–184.

63. Schlanger, S.O.; Arthur, M.A.; Jenkyns, H.C.; Scholle, P.A. The Cenomanian-Turonian Oceanic Anoxic Event, I. Stratigraphy and distribution of organic carbon-rich beds and the marine δ13C excursion. *Geol. Soc. Lond. Spec. Publ.* **1987**, *26*, 371–399. [CrossRef]

64. Arthur, M.A.; Schlanger, S.T.; Jenkyns, H.C. The Cenomanian-Turonian Oceanic Anoxic Event, II. Palaeoceanographic controls on organic-matter production and preservation. *Geol. Soc. Lond. Spec. Publ.* **1987**, *26*, 401–420. [CrossRef]

65. Thurow, J.; Brumsack, H.-J.; Rullkötter, J.; Littke, R.; Meyers, P. The Cenomanian/Turonian boundary event in the Indian Ocean: A key to understand the global picture. In *Synthesis of Results from Scientific Drilling in the Indian Ocean*; Geophysical Monograph Series; Wiley: New York, NY, USA, 1992; Volume 70, pp. 253–273.

66. Damsté, J.S.S.; Köster, J. A euxinic southern North Atlantic Ocean during the Cenomanian/Turonian oceanic anoxic event. *Earth Planet. Sci. Lett.* **1998**, *158*, 165–173. [CrossRef]

67. Tsikos, H.; Jenkyns, H.C.; Walsworth-Bell, B.; Petrizzo, M.R.; Forster, A.; Kolonic, S.; Erba, E.; Silva, I.P.; Baas, M.; Wagner, T. Carbon-isotope stratigraphy recorded by the Cenomanian–Turonian Oceanic Anoxic Event: Correlation and implications based on three key localities. *J. Geol. Soc.* **2004**, *161*, 711–719. [CrossRef]

68. Parente, M.; Frijia, G.; Di Lucia, M. Carbon-isotope stratigraphy of Cenomanian–Turonian platform carbonates from the southern Apennines (Italy): A chemostratigraphic approach to the problem of correlation between shallow-water and deep-water successions. *J. Geol. Soc.* **2007**, *164*, 609–620. [CrossRef]

69. Parente, M.; Frijia, G.; Di Lucia, M.; Jenkyns, H.C.; Woodfine, R.G.; Baroncini, F. Stepwise extinction of larger foraminifers at the Cenomanian-Turonian boundary: A shallow-water perspective on nutrient fluctuations during Oceanic Anoxic Event 2 (Bonarelli Event). *Geology* **2008**, *36*, 715–718. [CrossRef]

70. Gebhardt, H.; Friedrich, O.; Schenk, B.; Fox, L.; Hart, M.; Wagreich, M. Paleoceanographic changes at the northern Tethyan margin during the Cenomanian–Turonian Oceanic Anoxic Event (OAE-2). *Mar. Micropaleontol.* **2010**, *77*, 25–45. [CrossRef]

71. Owens, J.D.; Lyons, T.W.; Li, X.; Macleod, K.G.; Gordon, G.; Kuypers, M.M.M.; Anbar, A.; Kuhnt, W.; Severmann, S. Iron isotope and trace metal records of iron cycling in the proto-North Atlantic during the Cenomanian-Turonian oceanic anoxic event (OAE-2). *Paleoceanography* **2012**, *27*. [CrossRef]

72. Zheng, X.-Y.; Jenkyns, H.C.; Gale, A.S.; Ward, D.J.; Henderson, G.M. Changing ocean circulation and hydrothermal inputs during Ocean Anoxic Event 2 (Cenomanian–Turonian): Evidence from Nd-isotopes in the European shelf sea. *Earth Planet. Sci. Lett.* **2013**, *375*, 338–348. [CrossRef]

73. Zhou, X.; Jenkyns, H.C.; Owens, J.D.; Junium, C.K.; Zheng, X.-Y.; Sageman, B.B.; Hardisty, D.S.; Lyons, T.W.; Ridgwell, A.; Lu, Z. Upper ocean oxygenation dynamics from I/Ca ratios during the Cenomanian-Turonian OAE2. *Paleoceanography* **2015**, *30*, 510–526. [CrossRef]

74. Friedrich, O.; Erbacher, J.; Mutterlose, J. Paleoenvironmental changes across the Cenomanian/Turonian boundary event (oceanic anoxic event 2) as indicated by benthic foraminifera from the Demerara Rise (ODP Leg 207). *Rev. Micropaléontol.* **2006**, *49*, 121–139. [CrossRef]

75. Gertsch, B.; Keller, G.; Adatte, T.; Berner, Z.; Kassab, A.S.; Tantawy, A.A.A.; El-Sabbagh, A.M.; Stueben, D. Cenomanian–Turonian transition in a shallow water sequence of the Sinai. *Egypt. Int. J. Earth Sci.* **2008**, *99*, 165–182. [CrossRef]

76. Gertsch, B.; Adatte, T.; Keller, G.; Tantawy, A.A.A.M.; Berner, Z.; Mort, H.P.; Fleitmann, D. Middle and late Cenomanian oceanic anoxic events in shallow and deeper shelf environments of western Morocco. *Sedimentology* **2010**, *57*, 1430–1462. [CrossRef]
77. Hay, W.W. Toward understanding Cretaceous climate—An updated review. *Sci. China Earth Sci.* **2017**, *60*, 5–19. [CrossRef]
78. Hay, W.W. Evolving ideas about the Cretaceous climate and ocean circulation. *Cretac. Res.* **2008**, *29*, 725–753. [CrossRef]
79. Cronin, T.M. *Paleoclimates: Understanding Climate Change Past and Present*; Columbia University Press: New York, NY, USA, 2009.
80. DeConto, R.M.; Hay, W.W.; Thompson, S.L.; Bergengren, J. *Late Cretaceous Climate and Vegetation Interactions: Cold Continental Interior Paradox*; Special Papers; Geological Society of America: Boulder, CO, USA, 1999; pp. 391–406.
81. Sellwood, B.W.; Valdes, P.J. Mesozoic climates: General circulation models and the rock record. *Sediment. Geol.* **2006**, *190*, 269–287. [CrossRef]
82. Jayne, S.R.; St Laurent, L.C.; Gille, S.T. Connections between ocean bottom topography and Earth's climate. *Oceanography* **2004**, *17*, 65–74. [CrossRef]
83. Sandwell, D.T.; Smith, W.H.; Gille, S.; Kappel, E.; Jayne, S.; Soofi, K.; Coakley, B.; Géli, L. Bathymetry from space: Rationale and requirements for a new, high-resolution altimetric mission. *C. R. Geosci.* **2006**, *338*, 1049–1062. [CrossRef]

geosciences

MDPI

Article

The Impact of Acoustic Imaging Geometry on the Fidelity of Seabed Bathymetric Models

John E. Hughes Clarke [ID]

Center for Coastal and Ocean Mapping/National Oceanic and Atmospheric Administration (NOAA) Joint Hydrographic Center, University of New Hampshire, Durham, NH 03824, USA; jhc@ccom.unh.edu; Tel.: +1-603-862-5505

Received: 4 March 2018; Accepted: 22 March 2018; Published: 24 March 2018

Abstract: Attributes derived from digital bathymetric models (DBM) are a powerful means of analyzing seabed characteristics. Those models however are inherently constrained by the method of seabed sampling. Most bathymetric models are derived by collating a number of discrete corridors of multibeam sonar data. Within each corridor the data are collected over a wide range of distances, azimuths and elevation angles and thus the quality varies significantly. That variability therefore becomes imprinted into the DBM. Subsequent users of the DBM, unfamiliar with the original acquisition geometry, may potentially misinterpret such variability as attributes of the seabed. This paper examines the impact on accuracy and resolution of the resultant derived model as a function of the imaging geometry. This can be broken down into the range, angle, azimuth, density and overlap attributes. These attributes in turn are impacted by the sonar configuration including beam widths, beam spacing, bottom detection algorithms, stabilization strategies, platform speed and stability. Superimposed over the imaging geometry are residual effects due to imperfect integration of ancillary sensors. As the platform (normally a surface vessel), is moving with characteristic motions resulting from the ocean wave spectrum, periodic residuals in the seafloor can become imprinted that may again be misinterpreted as geomorphological information.

Keywords: swath geometry; multibeam spatial resolution; integration artefacts

1. Introduction

Geomorphometry is a mature field in terrestrial sciences but is less well known in the marine realm. While the concepts were applied to multibeam data several decades ago [1–4], only now, with the easier availability of GIS tools, are they routinely being applied toward seabed topographic applications (a comprehensive review can be found in [5]). As with the terrestrial terrain measurements, those parameters extracted from a digital bathymetric model (DBM) are ultimately limited by the quality of the original observations. This paper focuses on the achievable spatial resolution of the most widespread seafloor bathymetric survey instrument used: Integrated multibeam sonar systems.

This paper is part of a special edition of *Geosciences* devoted to marine geomorphometry. The other papers [6–10] present specific applications, predominately utilizing multibeam sonar observations. The scales of features investigated vary by two orders of magnitudes including deep-water (2000–5200 m) observations [6,7], outer shelf (135–376 m) [8], and shallow (10–50 m) water [9–11]. For all cases, the features of interest range from 0.2 to 10% of the total water depth in the vertical and 2–20% of the elevation in the horizontal. The lower end of these dimensions approaches the limits commonly achievable by these sonar systems. As such, artefacts specific to the acoustic imaging geometry, can be developed in the data at the same scale as the morphologic features of interest.

This paper is arranged to first introduce the multibeam sonar geometry and then explain the controls on achievable resolution as a function of a wide variety of parameters. A particular focus is provided for those system artefacts that are developed close to the limit of the achievable resolution. This is because, as noted by several researches [5,12], the seafloor geomorphic features of interest are often at, or close to, the limit of achievable resolution and are thus prone to potential distortion.

Each aspect discussed is illustrated by specific example data so that the reader may recognize the net result in a derived terrain model. Specific formulas to attempt to calculate solution density are omitted as they are strongly dependent on sonar configurations and motion history which are continuously variable and rarely stored with derived gridded products.

2. Multibeam Imaging Geometry

A multibeam echosounder is an acoustic scanner that delivers sequential topographic profiles aligned approximately orthogonal to the platform (a surface or submerged vehicle) trajectory. It does this by taking advantage of the Mills Cross array geometry to ensonify a corridor and then receiving the backscattered energy through a number of discrete beamformed channels at a variety of elevation angles [13,14]. As the vehicle advances, successive profile solutions build up a swath corridor. Different sonar models utilize different beam dimensions, spacings and density. As these sonar systems operate predominantly from underway platforms, they vary in their approach to stabilization in order to compensate for motions primarily within the ocean wave spectrum. It is those resultant data that are the underlying input to digital bathymetric models (DBM). Two primary, but quite separate, factors concerning the input data will become apparent in this discussion: The solution density, and the achievable resolution based on the footprint within which a single solution in derived.

Were those swaths of data made up of evenly spaced solutions, each of which represented the depth from an identically sized footprint, the data might be expected to be equivalent. In reality, however, as the vehicle does not follow a straight line path and rotates on three axes (roll, pitch and yaw), the solution density is uneven. Furthermore, the ensonified area, defined by the product of the transmission and receiver patterns (and/or pulse duration and processing), is not of uniform size and the incidence angle of acoustic energy is highly variable. As a result, both the solution density and the solution resolution vary strongly with both elevation and across a single swath. These factors will impact the resolved terrain roughness thereby affecting any classification scheme based on surface characteristics.

The closest analogous instrument used for terrestrial surveying is an airborne laser scanner (ALS). To contrast a multibeam sonar to an ALS, the geometric differences need to be appreciated (Figure 1).

The laser scanner sector elevation angle is typically only varying from vertical to ~20–25° incidence angle. For a typical ALS, the projected laser beam divergence is on the order of ~1 millradian (0.0573°) [15]. At a flying height of ~550 m this has a footprint then of 0.55 m at nadir growing to only 0.64 m at the edge of the swath. As the flying height is normally much larger than the variation in elevation of the feature of interest (e.g., examining rock ridges of ±20 m scale), the beam footprint only varies by ±3–4% as a result of within swath surface elevation fluctuations. For larger scale terrain fluctuations that occur over longer wavelengths, the aircraft is free to adjust its vertical trajectory to maintain a similar altitude.

In contrast, a good multibeam has a beam width of typically 1° (qualifications on this will be discussed later), but projected over an angular sector of up to ±65°. For a continental shelf depth of 50 m, this results in footprints ranging from 0.87 m at nadir to 2.1 m (along) or 4.9 m (across-track) at the edge of the swath. Additionally, for the case of within swath topography that varies within the same range (±20 m), those dimensions fluctuate by up to 40%.

The preceding simplified geometric scaling calculations demonstrate that variations in the effective resolution are going to be much more pronounced across a multibeam swath than an ALS swath. Even with the reduced changes, the variation in the ALS footprint is known to have an impact in the effective data quality [16]. Such effects are thus magnified for the multibeam geometry which, in turn, will impact the achievable terrain discrimination using geomorphometric techniques.

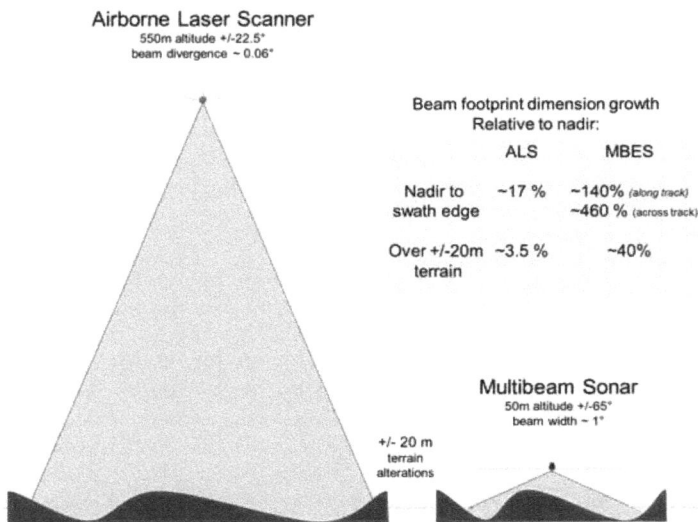

Figure 1. Comparing the imaging geometry of an airborne laser scanner and an equivalent resolution inner shelf multibeam sonar. ALS: Airborne (terrestrial) lidar scanner. MBES: Multibeam echo sounder.

3. Component Contributors to Resolution Limits

Given the grossly different scale of imaging geometry discussed, the components that influence the multibeam imaging geometry are now broken down and described separately. Specific real examples are provided where that change in the component influences the appearance of a morphologic feature that might be of interest.

3.1. Projected Beam Width Aspects

For a singular beam bottom detection, the range solution is constrained to be from within an area on the seafloor that lies within the projected beam dimension. That in turn reflects the beam widths of the transmitter and receiver respectively. Those are controlled by the array dimensions in wavelengths and the shading applied. Typical quoted beams widths range from 0.5 to 2.0 degrees and are often not the same for transmit and receive (thus the footprint for normal incidence is already elliptical). Figure 2a–d illustrate the loss in feature definition for the same target with differing Tx and Rx beam widths. If a critical geomorphic parameter is the presence of short wavelength targets such as boulder fields (a common habitat indicator), their visibility will strongly depend on the specific sonar beam width, and vary with elevation and obliquity.

3.1.1. Steering, Range, Obliquity and Seafloor Slope

For any quoted beam width the nominal value reported in the specifications is only true when unsteered. A typical multibeam system (level receiver) will steer on receive more than $60°$ to achieve the across-track coverage. The steered receiver beam width is correspondingly expanded ($\times 2$ for $60°$ $\times 4$ for ~$75°$). That beam is then projected over the slant range onto an increasingly oblique seafloor incidence. The net result, for a level receiver and a flat seafloor, is a ~$1/\cos^3$ (steering angle) dependence on the across-track dimension and a $1/\cos$ dependence for the along-track dimension. This of course is modulated by the local seafloor slope, with poorer definition on seafloors tilted away from the beam vector.

This elongation would rapidly become detrimental if the range estimation were purely defined from center-of-mass type algorithms (as is the norm for ALS). To ameliorate that effect, almost all multibeam systems use a split aperture approach [17] to estimate the geometric center of the beam in the across-track direction (phase detection, see Section 3.2.1). Unlike fisheries sonars however, the split aperture is only applied across-track and thus the along-track dimension (the projection of the transmit beam) remains a constraining limit on resolution (thus only a 1/cosine dependence) [18].

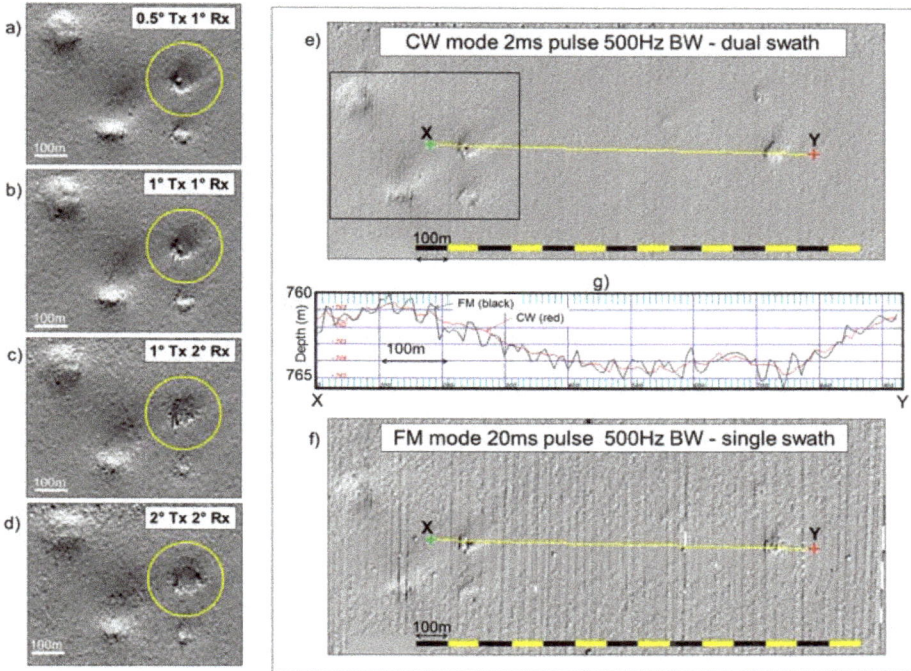

Figure 2. (left), Effect of changing beam width on the resolution of a specific small target at near normal incidence. EM710 system, 760m depth. (**a**) 0.5° Tx × 1.0° Rx; (**b**) 1.0° Tx × 1.0° Rx; (**c**) 1.0° Tx × 2.0° Rx; (**d**) 2.0° Tx × 2.0° Rx. (right) illustrating the impact of switching from CW to FM pulse over the same target (again located at near normal incidence). (**e**) Using a CW pulse; (**f**) using an FM pulse; (**g**) resultant apparent topographic relief across a featureless part of the seafloor as observed with the two pulse types (red CW, black, FM).

3.1.2. Elevation

Unlike airborne scanners in which the vehicle is free to alter its altitude above the terrain to optimize the solution density, a surface-mounted multibeam is forced to accept the altitude change as the topography rises and falls. If the roughness characteristic of interest is close to the resolution limit, then the feature may be distorted or even disappear with increasing depth.

Unlike surface mounted systems, however, the increasing use of autonomous underwater vehicles (AUVs) allows the decoupling of the resolution from the depth. Figure 3 illustrates the impact of altitude on feature resolution.

An additional benefit of AUV operations is that the altitude can be adjusted to trade-off required resolution versus coverage. Figure 4 illustrates the achieved definition of short wavelength roughness targets (flutes) as a function of elevation. Notably, even before the real relief is lost, the impact of noise due to the bottom detection method (see Section 3.2), starts to appear and can be confused with shorter wavelength relief. Again this will impact the morphometric parameters extracted from the terrain leading potentially to a false classification.

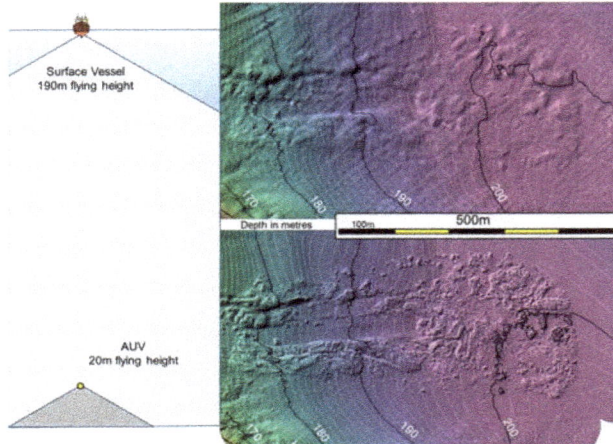

Figure 3. Contrasting the resulting derived digital bathymetric models (DBM) from a surface mounted system in 190 m of water compared with that derived from an AUV flying 20 m off the same seabed. As can be seen the detail on this landslide deposit is markedly clearer. The available scales of derived morphometric parameters are thus quite different.

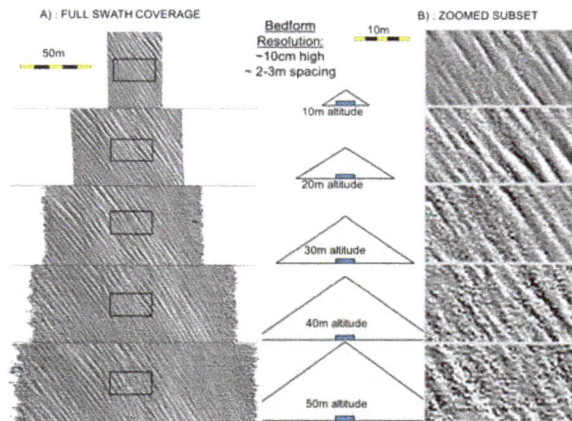

Figure 4. Illustrating the trade-off between AUV coverage and AUV resolution resulting from changing vehicle flying height. EM2040 $0.7° \times 0.7°$ system utilizing a $\pm60°$ sector. (**A**) Full swath coverage; (**B**) zoom of an identically sized (15 × 30 m) area showing the loss in feature definition and relative rise in bottom detection noise with vehicle elevation (the feature alignment is not identical due to limitations in the AUV navigation).

3.2. Bottom Detection Effects

For a single receiver channel, the intensity time series received from the location at which the main lobe of the beam strikes the bottom represents a finite time interval. That duration is strongly dependent on the beam dimension and the obliquity as discussed in Section 3.1.1. Depending on the elongation of the echo envelope two main bottom detection algorithms (amplitude and phase) are employed [17,19]. Both vary in their resulting quality of range estimation as a function of incidence angle. As a single swath sweeps through the full range of incidence angles and systematically switches from one to the other bottom detection method, the noise characteristics of each bottom detection algorithm become embedded in the swath sounding corridor in a systematic manner. Unless this is recognized, the resultant varying seafloor rugosity can be confused with a real seafloor morphometric change.

3.2.1. Amplitude vs. Phase Detection

The two common approaches employed to identify the location of the center of the beam within the received echo envelope are amplitude and phase detection. Details on the algorithmic approach can be found in [17,19]. The amplitude method relies heavily on the length of the echo envelope, and is thus optimal for the near normal incidence methods. As the angle of incidence (AOI) becomes more oblique, the range estimation can be increasingly biased towards hotspots in backscatter strength within the beam footprint. This manifests as apparent random noise in the range estimation that grows rapidly with obliquity.

In contrast, the phase estimation uses a split aperture approach and is thus relative insensitive to variations in the seabed backscatter strength within the beam footprint. As a result, for an AOI at which both bottom detection methods can be employed, the phase detection will generally be lower noise. As a result when the system switches between the two range estimation methods, there is often a characteristic change in the apparent roughness of the seafloor at that location (Figure 5 edge of white arrows).

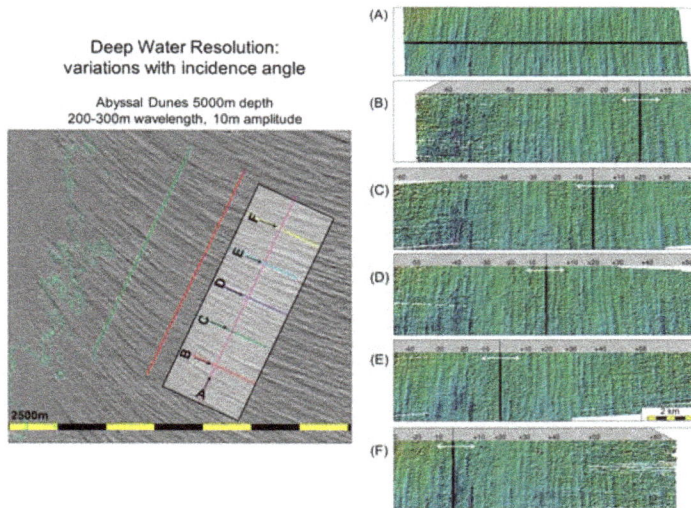

Figure 5. Illustrating the identical seafloor as imaged by the same multibeam sonar with different offsets. The seafloor consists of a field of abyssal dunes in 5000 m of water. The dune spacing (~5% of depth) and amplitude (0.2% of depth) is close to the limit of the system resolution (an EM122 1° × 1°). Left: Regional map showing location and orientation of acquired survey lines. (**A**) Shows an optimal bathymetric rendition acquired with a reduced swath width of ±25° with the ship-track oriented orthogonal to the dune crests; (**B–F**) shows the same area with 5 survey lines using ±65°, each of which is offset 2 km from the next. The same specific dunes are thus observed at 5 different grazing angles while progressing from amplitude to phase detects (and switching to FM pulses at 35°). The white arrowed section indicated the amplitude detection zone.

Unlike the amplitude detection approach, the phase detection need not utilize all the information across the wide across-track beam footprint. As long as the signal to noise ratio is high enough, the phase detection zero crossing may be based on fewer phase samples and thus relief across-track at scales significantly smaller than the projected beam footprint are possible [18]. Thus, the apparent scaling by projected beam footprint, presented in Figure 1 is not so pronounced.

At the outer limits of the swath, as the signal to noise ratio reduces, the phase detection algorithm gradually gets less accurate and there is a corresponding increase in the sounding noise again (Figure 5, outer edges of swaths in B and F).

Figure 5 illustrates this phenomena, showing abyssal dunes that have a distinct morphometric signature. As can be seen, the appearance of the same dune at differing grazing angles is inconsistent in its apparent rugosity and thus might be classed as slightly different geomorphic character. The only way to remove this differential bottom detection noise would be to grid the data as a scale that suppresses this beam-to-beam uncorrelated noise, although this could then partially obscure the feature of interest.

3.2.2. FM vs. CW Pulses

For a subset of the available multibeam systems on the market, the benefits of pulse compression are being employed to increase the range performance. Most commonly, linear frequency modulated (LFM) pulses are being employed that increase the signal to noise. While the range performance improves there are three complications that impact the appearance of low relief seafloors that could be confused with real rugosity.

1. The first is a result of the Doppler heave distortion of the pulse [20].
2. The second is the loss of the dual swath capability due to duty-cycle limitations.
3. The third is a result of the imperfection of the matched filter process [21].

All three effects will alter the short wavelength attributes of a derived DBM and thus could confound geomorphic analysis.

Figure 2e–g illustrates the combined first two effects. The extracted seafloor profiles (Figure 2g, X-Y) of the same terrain are only corrugated when the FM pulse is used. This is a result of the Doppler heave effect where the array motions at transmit and receive distort both the outgoing and incoming LFM pulse and result in a small time shift in the matched filter process [20]. The example shown has not correctly had the shift accounted for (this correction is now routinely applied). Note the scale of this artefact, 0.7 m in 700+ m of water, a ~0.1% of depth anomaly. Even though this is a small effect it is of the same order as the small targets (see Figure 2a–d) that the sonar is being used to resolve.

The second LFM-related effect manifests as a result of the fact that the sonar system involved (EM710) normally utilizes a dual swath strategy (see Section 3.3.3) using 3 sectors per swath. A total of 6×3 ms pulse (18 ms total transmission time) are required for continuous wave (CW) pulse operation in the depths considered. For the LFM configuration, a single swath now utilizes 2×40 ms and one 20 ms pulse (a total of 100 ms). A second transmission cannot be employed as the duty cycle of the transmitter would be exceeded. The net result is the along-track data density is now half that of the CW mode. This accentuates the visibility of the Doppler heave artefact.

For the case of multi-sector systems (see Section 3.3.2) which are able to employ differing pulse types in each sector, often the central sectors utilize CW pulses whereas the outer sectors switch to LFM pulses to maintain signal to noise. The net result is that the bottom detection noise in notably different. Figure 6 illustrates the higher noise evident when the equivalent bandwidth FM pulse is substituted instead of a CW pulse for the outer sectors. With these mixed pulse type modes, dual swath is maintained and the FM pulse has the same bandwidth so the noise would be expected to be similar. Comparing Figure 6b,d, however, this is clearly not the case as an abrupt jump in the surface roughness is apparent at the CW-FM sector boundary. The sounding noise statistics jump up by a factor of 2 (Figure 6c). This effect is believed to be due to matched filter sidelobes [21]. It should be noted that this noise is at the 0.2% of depth (Z) level (one standard deviation) and thus is well within standard hydrographic performance specifications (~±1.3% Z). Although this is a small effect, it is at the same scale as genuine low relief sedimentary structures such as bedforms or depressions (such as developed on the left side of the swath in Figure 6b) and thus potentially can confuse geomorphic analysis. This particular sonar system switches automatically between these two modes at about 500 m, so this spatially coherent noise contribution would abruptly appear and disappear within a DBM, potentially producing spatially varying rugosity unrelated to real seafloor characteristics.

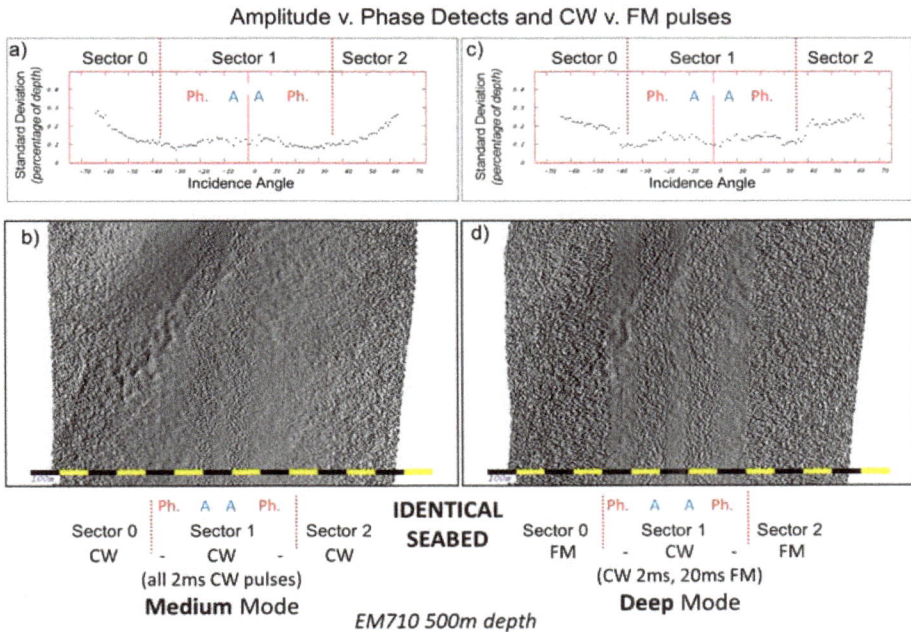

Figure 6. Illustrates the impact on changing from a CW to FM pulse on the identical seafloor. A terrain in ~500 m water depth is imaged by the same sonar sequentially with FM pulses turned on (**c,d**) and off (**a,b**). Frames (**b,d**) show a sun-illuminated representation of the data gridded at ~1% of the water depth. Frames (**a,c**) show derived sounding noise statistics (one standard deviation scaled by water depth) for the corresponding examples. The location of the sector boundaries is indicated. Ph: phase detection, A: Amplitude detection.

3.2.3. Pulse Length Changes

In order for a surface mounted multibeam to be able to cope with range dependent attenuation that changes continuously with water depth, it must alter its configuration to maintain adequate signal to noise. Almost all systems are operating at full power so usually the only means of adapting is to alter the outgoing pulse. A lengthening of the pulse provides both larger ensonified area (thereby increasing the bottom target strength) and, for the case of a CW pulse, a narrower bandwidth thus allowing better noise suppression. Such a switch, however, compromises range resolution which in turn impacts feature definition. Figure 7 illustrates two simultaneous views of the identical seafloor using two sonars with identical beam widths but different pulse lengths. The data were collected from co-mounted sonars on the same vessel so they share exactly the identical motion history and sound speed environment.

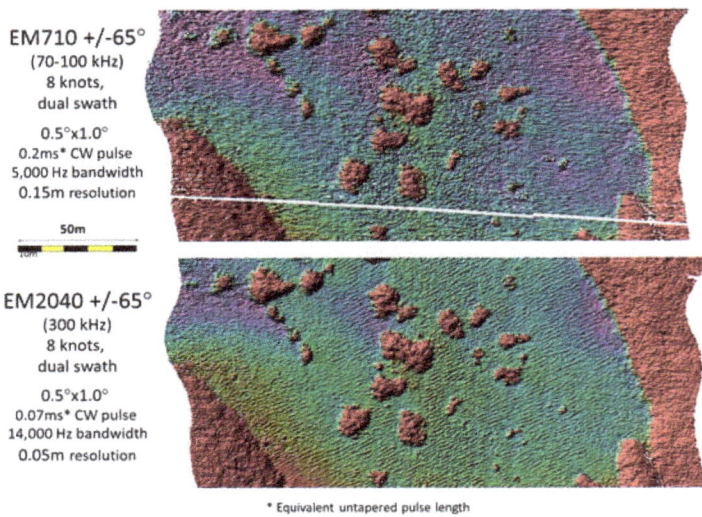

EM710 +/-65°
(70-100 kHz)
8 knots,
dual swath

0.5°x1.0°
0.2ms* CW pulse
5,000 Hz bandwidth
0.15m resolution

50m

EM2040 +/-65°
(300 kHz)
8 knots,
dual swath

0.5°x1.0°
0.07ms* CW pulse
14,000 Hz bandwidth
0.05m resolution

* Equivalent untapered pulse length

Figure 7. Two ±65° swaths of multibeam data collected simultaneously from the same vessel over a seabed in ~35 m water depth. Both sonars have the same transmit and receive beam widths (0.5° × 1.0°). The higher frequency system, however, employs a shorter (higher bandwidth) pulse length (CW: continuous wave). The impact of this is clear in the definition of the short wavelength ripples which are not resolved using the longer pulse.

Pulse length changes are usually implemented as an abrupt and discrete shift in the sonar configuration (often referred to as modes, consisting of a new set of pulse lengths, receiver bandwidths and potentially beam spacing and angular sector). If the system is operating in automatic mode, the location and timing of such shifts are driven primarily by depth. Thus, within a derived DBM, abrupt changes in the achievable resolution may be visible, implemented as the survey tracks go up and down through a critical depth threshold.

The operator may disable such automated changes. Doing so, however, risks other data quality issues (such as poor signal to noise in deeper water). If, however, the depth ranges are small, then the operator may take advantage of better range resolution (potentially thereby accepting a narrow swath) if it is important to maintain definition of a critical seabed roughness element that can be used in geomorphometric analysis (see Section 3.2.2).

This issue of altitude-dependent resolution is of course of less concern for AUVs, which usually operate in terrain-following mode. The flying height can be pre-selected to maintain the required resolution (as illustrated in Figure 4).

3.2.4. Impact of Operator Selection of Pulse Lengths

Given the trade-offs between range performance and bottom detection noise, if the operator is interested in seafloor roughness at the limit of the system resolution, the choice of operating setting can influence the visibility of the target of interest. Figure 8 illustrates the appearance of smaller dunes, developed on the back of larger ones as a function of settings for the same sonar system.

Several aspects, including those previously discussed, are visible as the five different modes are utilized in turn. For the FM pulse (Figure 8a), the impact of loss of dual swath is evident. For the longest CW pulse (Figure 8b) the amplitude phase boundary is evident and only switches at about 20° incidence angle. As the pulse is shortened (Figure 8c), the switchover from amplitude to phase moves in and the noise associated with the amplitude detection is reduced. At the same time however, the loss of signal to environmental noise in evident in the outer part of the swath. Additionally, part of the apparent outer swath noise in Figure 5C,D is manifested as apparent across-track ribbing related to integration and will be explained in Sections 3.5 and 3.6.

As the pulse is again shortened (Figure 8d), the visibility of a population of a smaller dunes superimposed on the back of the larger dunes increases. Finally, by switching to a slightly higher frequency (Figure 8e), with the corresponding narrowing of the beams, the best view is obtained, although at the expense of swath width.

If common geomorphometric parameters were extracted from each of these seafloor realizations in turn, the same seafloor could potentially be classified differently.

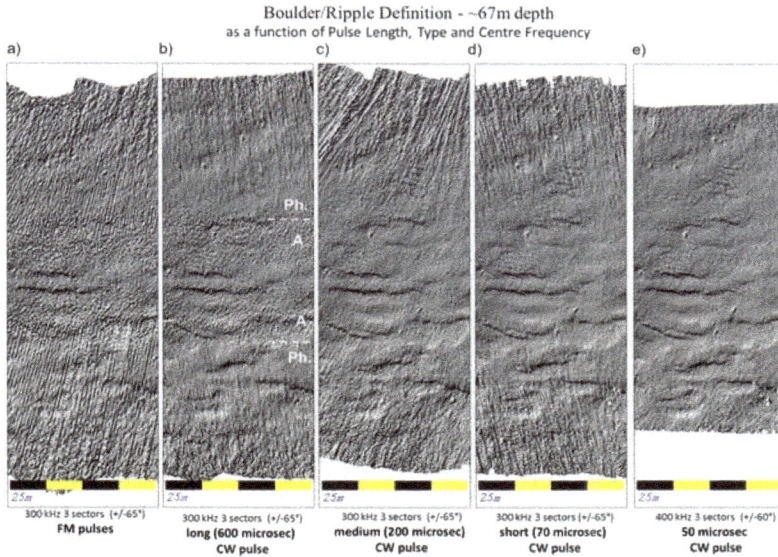

Figure 8. Identical seafloor (~67 m depth) viewed using different available sonar settings. Vessel steaming from left to right. From left to right they are (**a**) FM pulse—note drop to single swath apparent as across-track ribbing; (**b**) 0.6 ms CW, with dual-swath so ribbing removed, but note high noise at amplitude (A)-phase (Ph.) transition; (**c**) 0.2 ms CW, improved and narrower amplitude to phase transition, first resolution of the secondary dune population; (**d**) 0.07 ms CW, better view of the secondary dunes, but sensitive to noise in outer swath; (**e**) switching center frequency upward to reduce beamwidth, best view of geomorphology, but with reduced swath width.

3.3. Beam Density Considerations

For a given pulse length and bottom detection strategy, the full benefit of the achievable resolution is only gained if the solution density is significantly tighter than the projected beam footprint. The beam density is not a constant however. Vessel speed, angular sector selected, stabilization strategy and availability of multi-swath are the main factors that influence this.

In the absence of sufficient solution density, the visibility of the shortest potentially-resolvable features can be compromised.

3.3.1. Beam Spacing

In the 1990's it was common for multibeam systems to have an across-track beam spacing that was only as tight as the 3 dB beam width limit. Since then, increases in the available number of beam forming channels, together with the ability to dynamically adjust their spacing as the total angular sector is changed, has allowed tighter beam density (commonly to at least less than $\frac{1}{2}$ of the beam width). These aspects manifest in both the across and along-track density.

With heavily overlapping beam footprints in the across-track dimension, the bottom detection solution from beam to beam shares much of the same signal and is thus no longer fully independent. Nevertheless, uncorrelated noise contributions between individual bottom detections will be suppressed in an oversampled situation. Thus reducing the angular sector, thereby producing higher across-track solution density, will suppress sounding noise and promote the visibility of true short wavelength seabed relief.

As a byproduct of reducing the angular sector, the two-way travel time (twtt) to the outermost beam is reduced. As a result, the ping rate of the system is normally allowed to increase (typical ping rates are about 1.2 × this twtt). Thus reducing the angular sector also benefits the along-track solution spacing. For common operating speeds of ~8 knots (4 m/s), a ±65° swath will result in the along-track solution density being ~1.5% of depth. This is only about the 3 dB dimension (for a 1° Tx) and thus there is minimal along-track oversampling. If, however, the sector is compressed to ~±25° (e.g., Figure 5A) the swath width is reduced by a factor of 4.6 and the twtt by 2.1, resulting in ~10 × the sounding density resulting in better definition than the wider sector (Figure 5B–F).

For most surface mounted multibeam systems, the along-track density is generally the more critical limiting factor (e.g., Figure 9d). Furthermore this simplified calculation ignores the additional complication for the along-track data density of azimuth and pitch rotations of the survey platform from ping to ping.

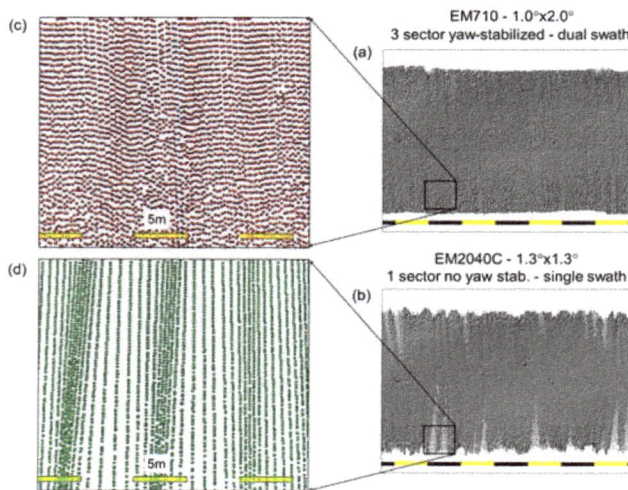

Figure 9. Impact of multisector yaw and pitch stabilization and dual swath. Two simultaneously acquired swaths of multibeam data (systems mounted on the same vessel): (**a**) Yaw stabilized coverage (and dual swath); (**b**) coverage without yaw stabilization; (**c**) sounding density showing even coverage; (**d**) sounding density showing irregular coverage.

3.3.2. Stabilization, Pitch and Heading

The preceding along-track data density calculations ignored both pitch and yaw perturbations. Pitch perturbations displace successive swaths along-track and yaw perturbations can dramatically degrade the along-track density on the outer side of a turn.

To address this, multi-sector yaw stabilization is now available that breaks the swath into sub sectors across-track and drastically reduces the impact of yaw and pitch on the along-track solution density [22].

Figure 9 illustrates the impact of having multi-sector stabilization. The two examples were acquired simultaneously from two systems mounted on the same vessel operating in open water. Both systems were operating with the same angular sector ($\pm65°$) and the same ping rate (synchronized). The scale of the yaw motion was $\sim\pm2°$ over ~10 s and the pitching was up to $\pm2°$ in 4 s. As can be seen, the sounding density is very uneven for the single sector system (Figure 9d) and thus compromised with potential data gaps larger than 1m. In this water depth, both those systems are capable of resolving 1 m targets if sufficient sounding solutions actually cover the target. To reliably distinguish a real target from a spurious sounding at least a 3×3 sounding density matrix over the feature is normally required [23]. A 0.25 m grid could be utilized for the yaw stabilized system, but that would not be justified for the single sector system.

From the point of view of geomorphic analysis, a coarser grid would be required for the single-sector single-swath system even though the across-track data density supports shorter wavelength discrimination at that orientation. Again critical rugosity (in this case boulder distribution) would be obscured.

3.3.3. Along-Track Density—Multi-Swath

Even with pitch and yaw stabilization, the along-track density is generally still the limiting factor. A recent innovation is the use of multiple swaths within a single transmission cycle. This is merely an extension of the multi-sector concept, with additional sectors now separated along-track rather than across-track. The net result is the ability to double the along-track data density without compromising swath width.

Figure 10 illustrates the impact of not having dual swath. Two simultaneous passes over the same seafloor have very different definitions of the shorter wavelength population of dunes that sit on the back of the larger dunes. Although both passes have the same across-track beam density, the smaller dunes would be missed without dual swath. Thus, the geomorphometric parameterization could again be compromised.

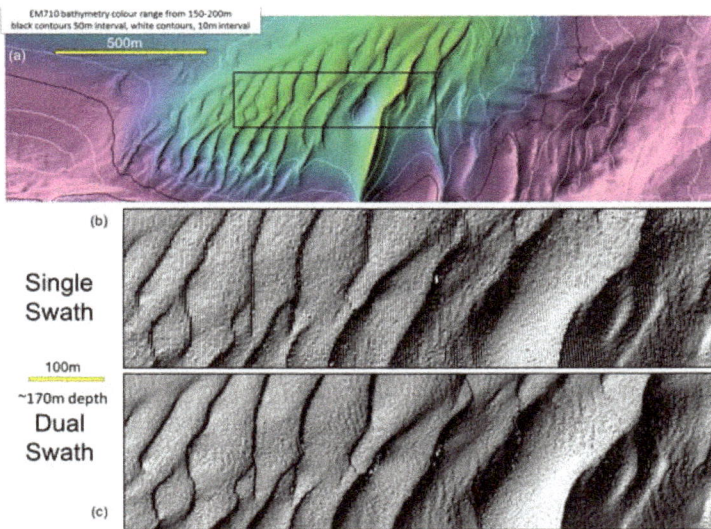

Figure 10. Impact of multi-swath. Two successive passes of the same multibeam system (EM710 0.5° × 1.0°) over a field of multi-scale dunes in 170 m of water. Speeds from 7–9 knots, both passes using a ±65° sector. (**a**) Regional location; (**b**) resolved relief using single swath mode; (**c**) resolved relief using dual swath mode.

3.3.4. Sector Width—Within Swath Beam Spacing

Almost all multibeam systems today allow the operator a choice of angular sector over which the finite number of beam forming channels can be redistributed. Thus, at the expense of swath width, the sounding density can be enhanced. Figure 5A shows the significant improvement in the definition of the abyssal dunes when a reduced angular sector is utilized (the data were collected at the same speed). Such a sector width ($\pm 25°$) is not generally practical, and thus an optimal intermediate setting is usually chosen.

Figure 11 illustrates the impact on resolution over a range of different angular sectors. The target considered is a solitary coral head that lies within a field of sand ripples. Both the coral head and the ripples are typical morphologic features that are of great interest in defining habitat. Yet both are close to the resolution limit of the sonar system. For the reasons discussed in Section 3.3.1, both the along and across-track sounding density increase with reduced swath. Four progressively wider sector widths are utilized. The ripple definition (oriented across-track and thus most sensitive to the along-track density), clearly declines with expanding sector width.

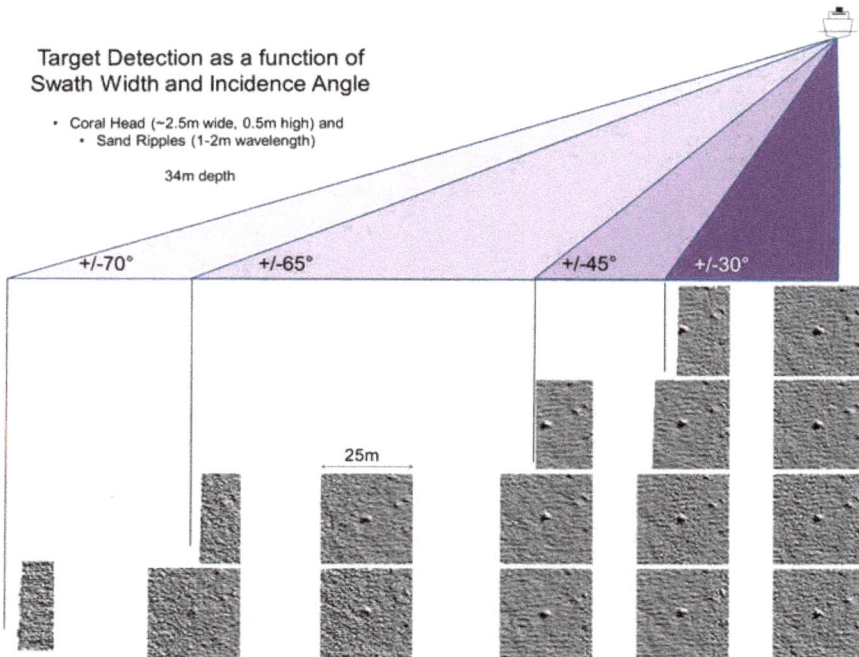

Figure 11. Impact of sector width choice on achievable resolution. A singular coral head in ~34 m of water, lying within a field of sand ripples, is viewed using the same sonar system (EM710 0.5° × 1.0°) using a range of elevation angles and swath widths. Each subarea view is a 25 m square gridded at 0.25 m resolution which is centered on the known position of the coral head. Each view is at a specific incidence angle (right to left: nadir, 30, 45, 60, 65 and 70 degrees) and with a specific sector width (top to bottom: ±30, 45, 65 and 70 degrees).

Superimposed on the effects of sounding density, the impact of bottom detection type and signal to noise (discussed in Section 3.2.1) are also visible in Figure 11. The amplitude detection noise (generally within $\pm 15°$) is comparable to the vertical scale of the ripple relief and thus gridding coarser, to smooth out the bottom detection noise would also result in removing the real morphology of interest.

At grazing angles beyond about 60° where signal to noise issues start to plague the phase detection, the coral head itself becomes poorly resolved. Thus, if boulder definition is of interest, only a subset of a swath corridor can properly characterize their presence and scale (dimensions, spacing etc.).

3.4. Corridor Overlap Considerations

For many of the artefacts presented, their manifestation is most apparent in the outer part of the swath corridor (e.g., Figures 5, 6, 8, 9 and 11). Thus, depending on the level of swath to swath overlap, their visibility may or may not be preserved in the final grid product. As long as the gridding process weights the outermost beams lower than the inner beams, higher levels of overlap will progressively remove these features. Figure 12 illustrates progressively higher levels of overlap on the same seafloor.

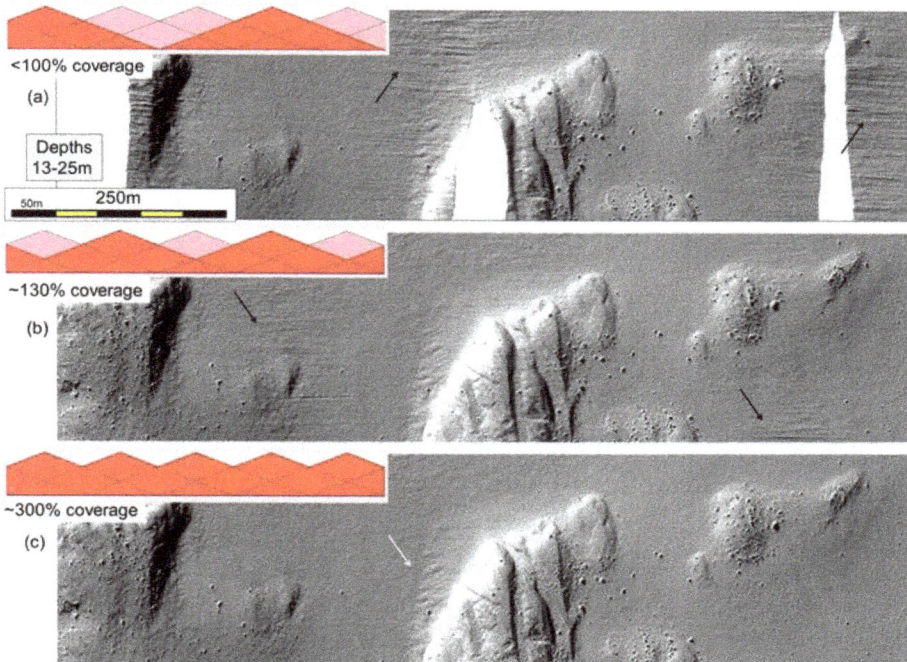

Figure 12. Impact of swath to swath overlap. Showing the manifestation of oceanographically driven instability in the outer part of a multibeam swath. (**a**) Full swath extent, including outer beam noise, revealed when <100% coverage is utilized; (**b**) partial suppression of the outer beam noise when ~30% overlap is used; (**c**) full suppression over artifact when ~300% overlap is used.

The data in question, were acquired using a ±72° sector in an area with an undulating near surface thermocline. The net result is a faint residual in the outer swath data (black arrows in Figure 12a). When minimal overlap is utilized (Figure 12b) those features (black arrows in Figure 12b) are maintained in the final grid. If 300% coverage is used (equivalent to utilizing a ±40° sector without overlap), the false relief is absent. Only with this level or overlap does one then adequately resolve the real bedform morphology (white arrow, Figure 12c) that is present whose relief is actually smaller than the artifact.

An alternate approach to increasing overlap seen in Figure 11, could have been to reduce the angular sector so that the noisier outer swath data were not collected at all. This would have the benefit of higher along-track density too. Such an approach works well where there is little topographic variation. In the example here, however, the rock highs are ~50% shallower than the surrounding sedimented areas and thus wider angular sectors allow the minimum coverage requirements to be maintained even if local knolls exist. The geometry utilized in this example was deliberate in order to provide optimal unidirectional low grazing angle backscatter imagery of this terrain.

To best represent a surface, the approaches of Sections 3.3.4 and 3.4 can be combined by having both narrower sectors and higher overlap density. Figure 13 illustrates the same seafloor alternately surveyed using 200% coverage using a ±65° sector, versus, higher overlap using a ±40° sector and slowing down.

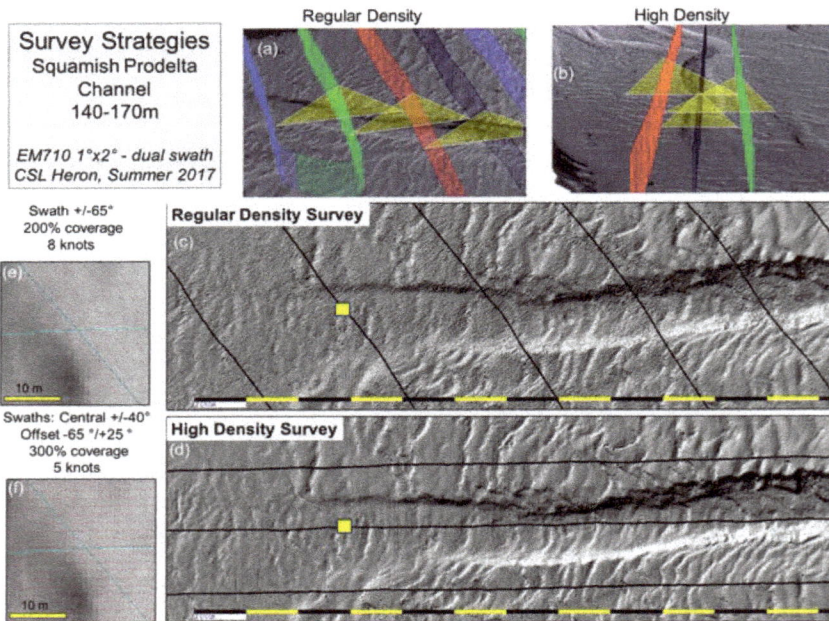

Figure 13. Impact of simultaneously changing angular sector, overlap and survey speed. Seabed channel in ~140–170 m of water surveyed twice using the same sensor (EM710 1° × 2°). Regular 200% coverage using ±65° at 8 knots, and ±40° with 300% at 5 knots. (**a**,**b**) Illustrate the two survey geometries; (**c**,**d**) show the resulting bedform definition (black lines show ship-tracks) and (**e**,**f**) reveal the resulting sounding density (zoom of yellow box in (**c**,**d**)).

A third party user of a DBM is unlikely to be aware of the underlying source data density utilized. Standard metadata for distributed grids do not usually specify such imaging geometry aspects.

3.5. Motion Residuals

All the previous discussions have assumed that the sonar-relative range and angle solutions are located without compromise due to imperfect position and orientation. With the latest technology, 3D positioning is almost always better than resolution and achievable orientation measurement (<0.02°) is again usually better than beam widths or equivalent angular discrimination through differential phase.

Such benefits, however, are only achieved if the position and orientation is applied correctly as part of the system integration. This requires that the corresponding position/orientation sensor to sonar offset, alignments and timing be at least as good as the source measurements. In reality, this is one of the largest single sources of additional error in multibeam sonar data [24]. The full discussion of the integration imperfections and their impact on the sounding solutions is beyond the scope of this paper, but the reader is referred to [25].

For the purposes of geomorphometric analysis, the main impact is that ship-track orthogonal ribbing in data becomes apparent (Figure 14A). The horizontal spacing of such ribbing reflects the main driving periodic motions of the vessel (the ocean wave spectrum in the 5–15 s period range). With ship speeds of 4–5 m/s this translates to false roughness in the 20 to 75 m range.

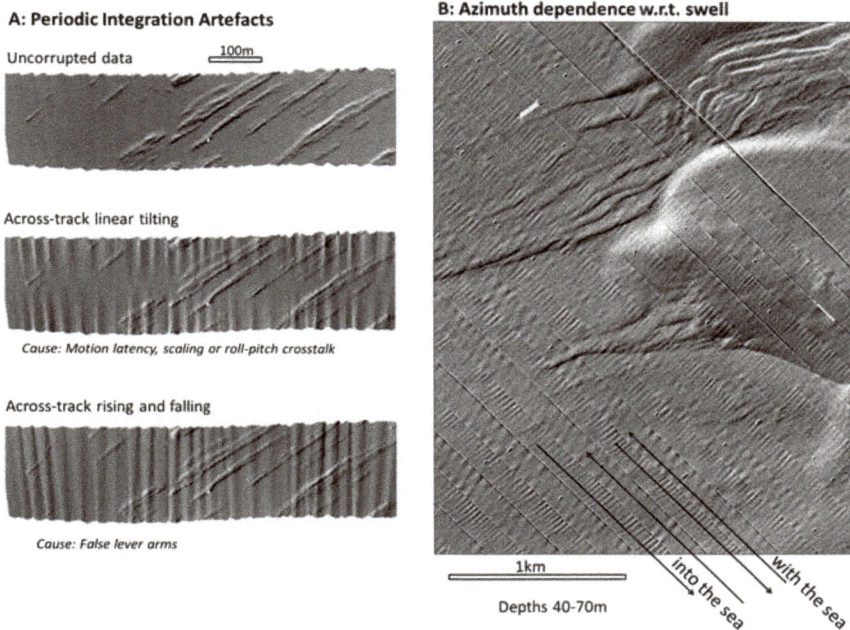

Figure 14. (**A**) The two main characteristics of periodic bathymetric residuals due to imperfect integration; (**B**) the expression of such residuals over real morphology and the Doppler shifting of the roughness as a function of vessel azimuth w.r.t. swell.

Their visibility will vary with seastate and their wavelength will also be significantly affected by the survey line orientation as the apparent wave period is shifted depending on whether the vessel is steaming into or away from the wave direction (Figure 14B). When present, these artifacts are usually immediately apparent to the interpreter as they are clearly orthogonal to the survey direction. They may however, be of a similar amplitude and wavelength to real geomorphic features of interest (as is the case in both Figure 14A,B) and thus would confound automated detection through terrain analysis.

3.6. Sound Speed Residuals

All multibeam beam bottom detection solutions have to be adjusted for the distortion of the initial beam vector by refraction through a layered, heterogeneous ocean. Sound speed profiles are routinely acquired to perform this calculation. A precursor to the beam vector calculation is utilizing the transducer depth sound speed estimate to correctly apply beam steering. If either the beam steering and/or the ray trace calculation is erroneous, false geomorphology may be overprinted on the seabed model.

The simplest artifact is an across-track symmetric, but non-linear distortion (commonly referred to as a smile or frown). This only changes over time constant of 10's of minutes and will thus only impact swath to swath boundaries (providing false relief parallel to the ship-track at a wavelength of the swath width). While degrading the DBM, this is usually immediately obvious and thus the associated false terrain attribute can be quickly discounted.

What is more concerning for geomorphometric analysis is those sound speed related residuals that result in false short wavelength roughness within a single swath that might be confused with real geomorphology. There are three common types of artifact that come under that description:

1. Heaving towards/away from a strong sound speed gradient.
2. Rolling with an erroneous surface sound speed error.
3. Internal waves or turbulence on a sound speed gradient.

The first two artefacts are clearly motion correlated and are developed strictly ship-track orthogonal and thus appear very similar to the motion artefacts shown in Figure 14. They are distinguished, however, by the fact that the magnitude of the error grown nonlinearly away from nadir. Figure 15A shows a terrain in which across-track ribbing is developed due to motion relative to near surface sound speed fluctuations. Full details of the origin of these can be found in [25].

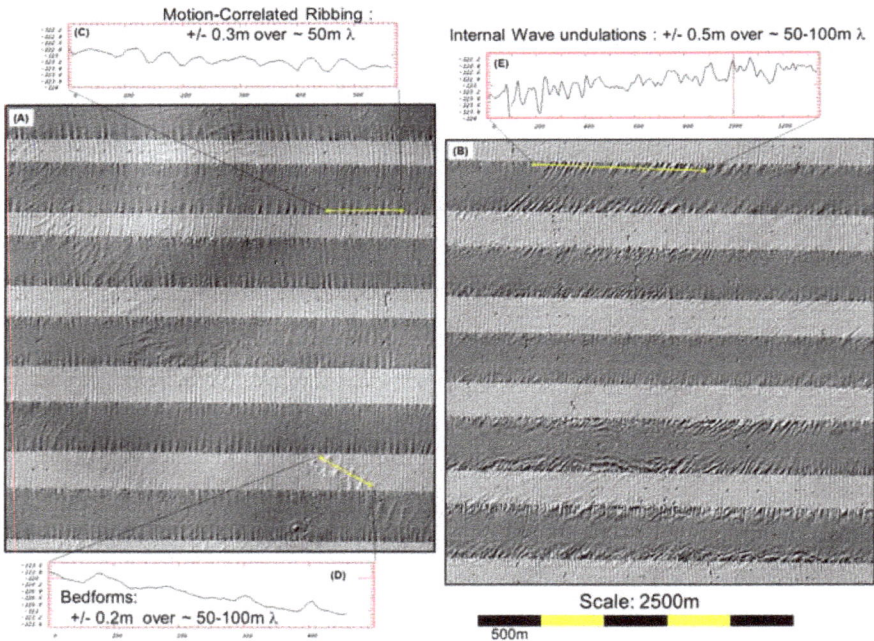

Figure 15. (**A,B**) Showing two different examples of sound speed related artifacts overprinted on a very smooth seafloor in ~120 m of water. The data were collected using a ±61° sector with just 20% overlap. The data are deliberately presented as a sun-illuminated terrain with eastbound lines stenciled separately on top of the west bound coverage. In that manner, the mismatch in the outer swath roughness of the eastbound lines (presented in a darker greyscale) can be compared with corresponding adjacent relief on the westbound lines (a lighter greyscale). Section (**C**) shows the relief due to motion-correlated errors due to near surface sound speed effects. Section (**D**) shows the strongest true geomorphic relief (faint sand ridges). Section (**E**) shows the false topography generated by the passage of an internal wave packet. Further sinuous morphology, indicative of internal wave activity, can be seen on the outer edges of the southernmost two lines.

The third artefact is perhaps the most concerning for analysts looking for short wavelength roughness elements. Applied ray tracing models assume that the sound speed gradients in the ocean occur only in the vertical. If in fact the veloclines are tilted then the ray path will be diverted from that simple assumption. The end result is that if there are wave-like undulations of a sound speed gradient (such as internal waves or turbulence resulting from baroclinic shear), a periodic bathymetric artifact (e.g., Figure 15B) will be projected from the interface onto the seafloor [26].

Figure 15B illustrates the impacts of intense internal wave activity. The data were collected in ~120 m of water in the Celtic Sea during the summertime when a pronounced thermocline is developed and perturbed by tidal activity. A false apparent seafloor roughness is developed preferentially on the edge of the swaths which shows absolutely no correlation from line to line. The artefacts grow nonlinearly toward the edge of the swath. The scale of these artifacts can exceed 1% of depth [26] with wavelengths comparable to or larger than typical shelf bedforms (e.g., Figure 15D). Their orientation is unrelated to the survey line geometry and is often sinuous (see southernmost two lines in Figure 15B), thereby being easy to confuse with real relief. Comparable oceanographically-driven artifacts are common when multibeam data are acquired across abrupt tidal fronts.

4. Discussion

Each of the component limitations of a swath bathymetric system impact the minimum resolvable scale of topographic features. Herein it has been illustrated that the resolution thus varies systematically with depth, with obliquity, with sonar design characteristics and with operator choice of options such as mode, speed and angular sector. The net result is that, within a region of investigation, the achievable resolution will be continuously variable.

Most DBMs utilize fixed grid spacing from which the geomorphometric parameters are derived. Unless the terrain model grid size is chosen to be significantly larger than the poorest achievable resolution, morphologically significant rugosity may appear and disappear within a terrain reflecting the changing imaging geometry rather than the real spatial distribution of natural morphology.

Often, in order to preserve some of the fine scale relief seen in shallower locations, deeper portions of the DBM are gridded finer than is locally justifiable. Under these conditions, uncorrelated sounding noise may appear in those locations and be misinterpreted as natural geomorphic character. That noise is likely to vary systematically across the swath and abruptly with changing sonar settings.

In addition to system resolution limitations, imperfect integration of any of position, orientation or sound speed information can generate false seafloor roughness elements that overprint true geomorphology. Horizontal dimensions for these are typically significantly larger than the sonar resolution and thus can appear similar to natural resolvable features. The vertical scale of these artefacts are usually below the standard hydrography accuracy standards (~±1.3% of depth). Nevertheless, sedimented continental shelves are routinely covered with low relief, long wavelength periodic bedforms whose vertical scale is as little as 0.2 to 0.5% of the depth (e.g., Figure 15D). Thus the presence of these integration artefacts potentially compromises automated geomorphometric analysis.

5. Conclusions

With the growing availability of continuous seafloor terrain models, routine application of geomorphometric classification techniques will become increasingly common. As the interpreter becomes increasingly removed from the field acquisition, however, there is a danger that residual systematic or environmental artefacts in swath bathymetric data may be mistaken for real seabed morphology.

While morphology at scales large with respect to the limit of sonar resolution will usually remain unambiguous, relief close to either the resolution limit or the scale of integration artefacts can be misinterpreted. Unfortunately the resolution limit cannot be easily approximated by a single dimension as this scales with the imaging geometry. Similarly the scale of integration errors is linked to the vessel motion periodicity and scale and the nature of the local oceanography, both of which normally change over the duration of a typical survey period.

Acknowledgments: The author is supported by NOAA grant NA15NOS400002000 and funding from Kongsberg Maritime. Data examples were compiled, with permission, from a variety of sources including the U.S. Naval Oceanographic Office, Kongsberg Maritime (MUNIN AUV, Craig Wallace), the Irish Marine Institute and the Canadian Hydrographic Service.

Conflicts of Interest: The author declares no conflict of interest. The funding sponsors had no role in the design of the study; in the collection, analyses, or interpretation of data; in the writing of the manuscript, and in the decision to publish the results.

References

1. Fox, C.G.; Hayes, D.E. Quantitative methods for analyzing the roughness of the seafloor. *Rev. Geophys.* **1985**, *23*, 1–48. [CrossRef]
2. Shaw, P.R.; Smith, D.K. Statistical methods for describing seafloor topography. *Geophys. Res. Lett.* **1987**, *14*, 1061–1064. [CrossRef]
3. Goff, J.A.; Jordan, T.H. Stochastic modeling of seafloor morphology: Inversion of Sea Beam data for second-order statistics. *J. Geophys. Res.* **1988**, *93*, 13589–13609. [CrossRef]
4. Mitchell, N.C.; Hughes Clarke, J.E. Classification of seafloor geology using multibeam sonar data from the Scotian Shelf. *Mar. Geol.* **1994**, *121*, 143–160. [CrossRef]
5. Lecours, V.; Dolan, M.F.J.; Micallef, A.; Lucieer, V.L. A review of marine geomorphometry, the quantitative study of the seafloor. *Hydrol. Earth Syst. Sci.* **2016**, *20*, 3207–3244. [CrossRef]
6. Sánchez-Guillamón, O.; Fernández-Salas, L.M.; Vázquez, J.-T.; Palomino, D.; Medialdea, T.; López-González, N.; Somoza, L.; León, R. Shape and Size Complexity of Deep Seafloor Mounds on the Canary Basin (West to Canary Islands, Eastern Atlantic): A DEM-Based Geomorphometric Analysis of Domes and Volcanoes. *Geosciences* **2018**, *8*, 37. [CrossRef]
7. Gardner, J.V. The Morphometry of the Deep-Water Sinuous Mendocino Channel and the Immediate Environs, Northeastern Pacific Ocean. *Geosciences* **2017**, *7*, 124. [CrossRef]
8. Diesing, M.; Thorsnes, T. Mapping of Cold-Water Coral Carbonate Mounds Based on Geomorphometric Features: An Object-Based Approach. *Geosciences* **2018**, *8*, 34. [CrossRef]
9. Di Stefano, M.; Mayer, L.A. An Automatic Procedure for the Quantitative Characterization of Submarine Bedforms. *Geosciences* **2018**, *8*, 28. [CrossRef]
10. Masetti, G.; Mayer, L.A.; Ward, L.G. A Bathymetry- and Reflectivity-Based Approach for Seafloor Segmentation. *Geosciences* **2018**, *8*, 14. [CrossRef]
11. Greene, H.G.; Cacchione, D.A.; Hampton, M.A. Characteristics and Dynamics of a Large Sub-Tidal Sand Wave Field—Habitat for Pacific Sand Lance (*Ammodytes personatus*), Salish Sea, Washington, USA. *Geosciences* **2017**, *7*, 107. [CrossRef]
12. Wilson, M.F.J.; O'Connell, B.; Brown, C.; Guinan, J.C.; Grehan, A.J. Multiscale Terrain Analysis of Multibeam Bathymetry Data for Habitat Mapping on the Continental Slope. *Mar. Geod.* **2007**, *30*, 3–35. [CrossRef]
13. De Moustier, C. State of the art in swath bathymetry survey systems. *Int. Hydrogr. Rev.* **1988**, *65*, 25–54.
14. Lurton, X. *An introduction to Underwater Acoustics: Principles and Applications*, 2nd ed.; Springer: Berlin, Germany, 2010; Chapter 8.
15. Baltsavias, E.P. Airborne laser scanning: Basic relations and formulas. *ISPRS J. Photogramm. Remote Sens.* **1999**, *54*, 199–214. [CrossRef]
16. Ussyshkin, R.V.; Ravi, R. Mitigating the impact of the laser footprint size on airborne lidar data accuracy. In Proceedings of the ASPRS 2009 Annual Conference, Baltimore, MD, USA, 9–13 March 2009.
17. De Moustier, C. Signal processing for swath bathymetry and concurrent seafloor acoustic imaging. In *Acoustic Signal Processing for Ocean Exploration*; Moura, J.M.F., Louttie, I.M.G., Eds.; Springer: Dordrecht, The Netherlands, 1993; pp. 329–354.
18. Hughes Clarke, J.E.; Gardner, J.V.; Torresan, M.; Mayer, L.A. The limits of spatial resolution achievable using a 30 kHz multibeam sonar: Model predictions and field results. In Proceedings of the OCEANS 98 Conference Proceedings, Nice, France, 28 September–1 October 1998; Volume 3, pp. 1823–1827. [CrossRef]

19. Lurton, X.; Augustin, J.-M. A measurement quality factor for swath bathymetry. *IEEE J. Ocean. Eng.* **2010**, *35*, 852–862. [CrossRef]

20. Vincent, P.; Sintes, C.; Maussang, F.; Lurton, X.; Garello, R. Doppler effect on bathymetry using frequency modulated multibeam echo sounders. In Proceedings of the 2011 Oceans, Santander, Spain, 6–9 June 2011; pp. 1–5. [CrossRef]

21. Vincent, P.; Maussang, F.; Lurton, X.; Sintes, C.; Garello, R. Bathymetry degradation causes for frequency modulated multibeam echo sounders. In Proceedings of the 2012 Oceans, Hampton Roads, VA, USA, 14–19 October 2012; pp. 1–5. [CrossRef]

22. Hughes Clarke, J.E. Optimal use of multibeam technology in the study of shelf morphodynamics. In *Sediments, Morphology and Sedimentary Processes on Continental Shelves: Advances in Technologies, Research, and Applications*; Wiley Online Library: Hoboken, NJ, USA, 2012; pp. 1–28. [CrossRef]

23. Contract Specifications for Hydrographic Surveys, Version 1.2, New Zealand Hydrographic Authority. Available online: https://www.linz.govt.nz/docs/hydro/stds-and-specs/hyspec-v1-2-aug2010.pdf (accessed on 10 February 2018).

24. Hughes Clarke, J.E.; Mayer, L.A.; Wells, D.E. Shallow-water imaging multibeam sonars: A new tool for investigating seafloor processes in the coastal zone and on the continental shelf. *Mar. Geophys. Res.* **1996**, *18*, 607–629. [CrossRef]

25. Hughes Clarke, J.E. Dynamic motion residuals in swath sonar data: Ironing out the creases. *Int. Hydrogr. Rev.* **2003**, *4*, 6–23.

26. Hughes Clarke, J.E. Coherent refraction "noise" in multibeam data due to oceanographic turbulence. In Proceedings of the US Hydrographic Conference 2017, Galveston, TX, USA, 20–23 March 2017.

geoscciences

MDPI

Article

Unified Geomorphological Analysis Workflows with Benthic Terrain Modeler

Shaun Walbridge *, Noah Slocum, Marjean Pobuda and Dawn J. Wright

Esri (Software Company); 380 New York St., Redlands, CA 92373, USA; nslocum@esri.com (N.S.); mpobuda@esri.com (M.P.); dwright@esri.com (D.J.W.)
* Correspondence: swalbridge@esri.com

Received: 18 January 2018; Accepted: 8 March 2018; Published: 11 March 2018

Abstract: High resolution remotely sensed bathymetric data is rapidly increasing in volume, but analyzing this data requires a mastery of a complex toolchain of disparate software, including computing derived measurements of the environment. Bathymetric gradients play a fundamental role in energy transport through the seascape. Benthic Terrain Modeler (BTM) uses bathymetric data to enable simple characterization of benthic biotic communities and geologic types, and produces a collection of key geomorphological variables known to affect marine ecosystems and processes. BTM has received continual improvements since its 2008 release; here we describe the tools and morphometrics BTM can produce, the research context which this enables, and we conclude with an example application using data from a protected reef in St. Croix, US Virgin Islands.

Data Set: https://doi.org/10.6084/m9.figshare.5946463

Data Set License: CC-BY

Keywords: geomorphometry; terrain analysis; bathymetry; surface roughness; benthic habitat mapping; python

1. Introduction

The marine environment is under heavy pressure from a host of human activities, with a large portion of the system strongly impacted by anthropogenic threats [1]. At the same time, our understanding of the ocean suffers from a paucity of data detailing its inner workings, with little mapped in detail [2]. While satellite derived gravity data has increased the nominal resolution of global bathymetry data to 30 arc-seconds [3], and recent efforts have mapped global marine units [4], these data are unable to resolve finer-scale oceanic processes. Applications in seascape ecology, marine spatial planning, morphological evolution, and understanding human ocean uses require higher resolution data [5]. Where terrestrial applications can rely on global-scale coverage from low cost remote sensing platforms, ocean observations have relied on local-scale ship borne or in situ sensor based platforms, with high data acquisition costs and limited swath width. Recent technological advances have lowered costs and increased the extent and quality of bathymetry and a small but growing portion of the ocean is becoming well mapped. Habitat mapping is a first step to further our understanding of the ocean environment, and is typically informed by bathymetric gradients, which influence species abundance and distribution [6,7]. Geomorphic derivatives can be used to infer the effects of bathymetric gradients, increasing the value of limited data. The geomorphic derivatives are often used directly as a measure of the environment, or as covariates in predictive models. Increased availability of bathymetric data has led to new methodologies to infer bathymetric gradients from geomorphic derivatives and map seafloor habitat [8]. Because geomorphometry lacks a formalized unifying theory [9], a single approach is not yet possible. With consistent methodologies and better understanding of the algorithms in use, the field will lend itself more readily to expanding understanding of our oceans.

Today, modern seafloor mapping technologies such as light detection and ranging (LiDAR) and multibeam bathymetry have done much to lower the costs, and provide consistent bathymetric products which incorporate in-sensor error models, provide high point density, and accurate georeferencing making them suitable for scientific analysis [10,11]. Bathymetric data are often captured, processed to remove artifacts and errors, and integrated into a consistent geomorphological model. Creation and interpretation of such a model has been aided by geomorphometry analysis tools in packages such as Benthic Terrain Modeler (BTM) [12,13]. BTM is built on top of the popular ArcGIS platform, is open source, has an intuitive interface, and has received continual development. When BTM was first introduced in 2006, transforming collected data to research results required mastery of many different pieces of software.

Recent methodological developments have reduced the complexity of doing such analysis, but this has coincided with the emergence of even more sophisticated tools and the ability to collect much higher resolution data. Here, we describe the tools provided with the current release of BTM (v3.0), and highlight how it can be used in powerful analytical workflows for seafloor classification, by combining it with the capabilities of ArcGIS, the Python scientific stack, and the R statistical programming language. As the sphere of scientific software continues to grow, it will be important to provide easy-to- use tools to aid the broader community in performing rigorous analysis of marine geomorphometry.

2. Materials and Methods

Bathymetric digital terrain models (DTMs) are commonly used in three morphometric applications: deriving terrain attributes, seafloor classification, and object detection [11]. BTM is used for both terrain attribute creation and seafloor classification. Researchers have adopted BTM to create terrain attributes (e.g., [14–16]) for use as covariates to predictive modeling applications. BTM is also used for performing semi-automated classified seafloor mapping. By combining the bathymetric position index (*BPI*; see Section 2.1.2) with slope and depth, BTM provides a simple tool which maps ranges of these terrain attributes to a classification dictionary. This can be used to identify benthic zones (e.g., crests, depressions, flats and slopes) [17], but also to provide more detailed habitat maps, with the additional step of empirical evaluation of the relationship between the terrain attributes in use and the species of interest.

The following section describes the algorithms implemented by BTM (see Table 1), which cover a broad swath of use cases, and span five classes of terrain attributes: (1) Surface gradients, computed from cell neighborhoods, form the foundation of raster analysis. BTM includes basic gradients implemented by the ArcGIS extension Spatial Analyst, and complements these with specialized gradients useful for geomorphometry; (2) Measures of relative position are closely tied to many physical and biological processes, and provide useful information on both habitat and suitability; (3) Measures of surface roughness, or rugosity, quantify local heterogeneity and are important for understanding seafloor composition; (4) Distributional moments capture statistical measures of the terrain and include simple statistics like mean and variance, but also more complex measures like kurtosis; (5) Multidimensional tools assist with the important step of determining the appropriate scale of analysis.

Table 1. Terrain attributes computed by Benthic Terrain Modeler.

Name	Algorithm	Notes	References
Slope	$\arctan\sqrt{\left(\frac{dz}{dx}\right)^2 + \left(\frac{dz}{dy}\right)^2}$	Computed over a 3×3 neighborhood	[18]
Statistical Aspect	$57.29578 \times \arctan 2\left(\frac{dz}{dy} + \frac{dz}{dx}\right)$	Generates *Easterness* and *Northerness*	[19,20]
Mean Depth	$\frac{\sum_{i=x-(n+1)/2}^{x+(n+1)/2} \sum_{j=y-(n+1)/2}^{y+(n+1)/2} z_{ij}}{n^2}$	n is the size of a square neighborhood of cells	
Standard Deviation	$\sqrt{\frac{\sum_{i=x-(n+1)/2}^{x+(n+1)/2} \sum_{j=y-(n+1)/2}^{y+(n+1)/2} (z_{ij}-\bar{z})^2}{n^2}}$	$\bar{z}$ is the mean elevation of cells within the analysis neighborhood	
Variance	σ^2	σ is the standard deviation of a neighborhood of cells	
Interquartile Range	$CDF^{-1}(0.75) - CDF^{-1}(0.25)$	CDF is the cumulative distribution function of all cells in the analysis neighborhood	[21]
Kurtosis	$\frac{\mu_4}{\sigma^4}$	μ_4 is the fourth central moment and σ is the standard deviation of all cells in the analysis neighborhood	[21]
BPI	$z_{xy} - \bar{z}_{annulus}$	$\bar{z}_{annulus}$ is the mean elevation value of all cells within an annulus-shaped neighborhood	[12]
VRM	see Section 2.1.3		[22]
SAPA	see Section 2.1.3		[23]
SAPA (Slope-corrected)	see Section 2.1.3	Decoupled from slope as per ACR	[24]
ACR	see Section 2.1.3		[24]

2.1. Marine Geomorphometry Algorithms

2.1.1. Surface Gradients

The surface gradient provides valuable information for environmental applications of DTMs. *Slope* and *aspect* form a gradient representing the first order derivative of the surface, and are implemented in ArcGIS and BTM using the method of Horn [18], which uses a 3×3 neighborhood (also known as the Moore neighborhood [25]) for its analysis, operating on a planar surface. Instead of using the Moore neighborhood, a quadratic can also be fitted to the surface at the point of interest. The Wood-Evans method [26,27] implements this approach to slope, and can be computed at arbitrary window sizes. The algorithm is implemented in Jo Wood's Landserf [27], and is also available in ArcGIS with the ArcGeomorphometry extension [28]. Dolan et al. [29] compare slope algorithms for benthic geomorphometry and the effects of scale on the results, and recommend both the Horn and Wood-Evans methods for capturing slope in planar bathymetric data over other available methods.

Geodesic slope: In 2017, ArcGIS added support for computing *slope* on a geodesic. Instead of computing the results in a Cartesian plane, the geodesic method computes directly on the ellipsoid [30]. It retains the use of a 3×3 neighborhood, but improves on the planar method by measuring the angle between the surface and the geodetic datum for each of the 8 adjacent cells, and is fitted with least squares described in Figure 1. This produces more accurate results, particularly when used in conjunction with new high resolution bathymetry sensors with low positional uncertainty [10]. For typical sonar and lidar applications, the positional uncertainty of the observations and the effects of creating a DEM surface will be primary determinants in the accuracy of the slope calculation.

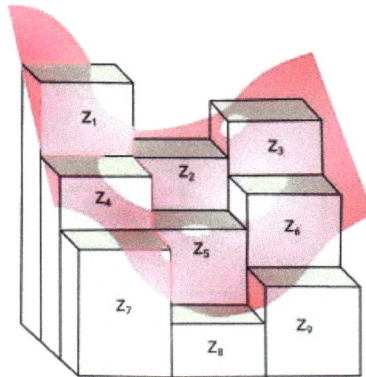

Figure 1. Least squares fitting of the curvature calculations, including that used by the *geodesic slope* computation. The pink curve is a polynomial fit against the cells in each 3×3 neighborhood, here labeled Z_1 to Z_9 where Z_5 is the origin cell.

Aspect: *Aspect* measures surface direction. In ArcGIS, it ranges from 0 to 359.9 degrees, measured clockwise from north, and -1 for locations of no slope, based on the same Horn [18] method. BTM augments this with an additional aspect calculation, which converts aspect into two variables: *Northerness* A_N, due south to due north, and *easterness* A_E, due west to due east [19,20].

$$A_N = cos(A) : -1 \leq A_N \leq 1 \tag{1}$$

$$A_E = sin(A) : -1 \leq A_E \leq 1 \tag{2}$$

This transformation is beneficial in many modeling contexts where circular variables violate the model constraints, and these two linear scaled derivatives are commonly used in models such as in Lecours et al. [31] and Davies et al. [32]. Results of this transformation are shown in the center and bottom panels of Figure 2.

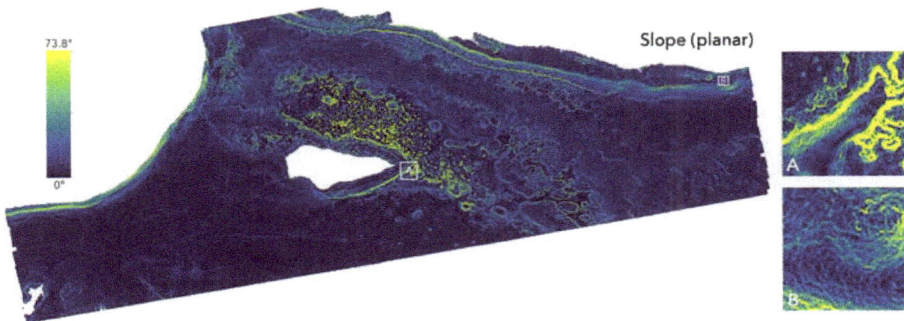

Figure 2. *Cont.*

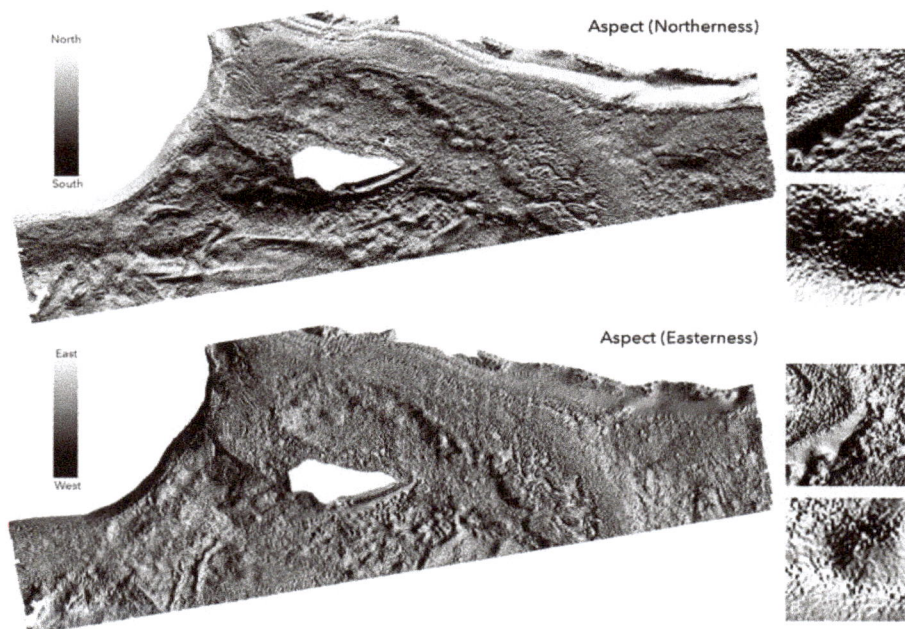

Figure 2. Buck Island Reef National Monument LiDAR depth data 2.3, and the surface gradients of *planar slope*, *northerness* and *easterness*. Upper left: *planar slope*, displayed with the perceptually uniform color map *viridis* [33], in degrees, values normalized by the standard deviation ($\sigma = 4.05$). Center left: *Northerness of aspect*, from due north to due south (1 to -1), linear ramp. Lower left: *Easterness of aspect*, from due east to due west (1 to -1), linear ramp. Inset maps: (**A**) is east of island, with large depth variation along a reef edge; (**B**) is along the bank in the northeast corner of the region.

2.1.2. Curvature/Relative Position

Analyzing the rate of change of the slope is linked to both physical and biological processes [12,34,35]. Physically these attributes can affect marine flow, internal waves and current channeling. Biologically, these attributes largely determine exposure, and are often a useful proxy of habitat suitability [36]. BTM uses the existing ArcGIS implementation for *curvature*, and implements two methods of computing relative position: *bathymetric position index (BPI)* and the *relative difference to the mean*.

Curvature: *Curvature* is the second derivative of the bathymetric surface, or the first derivative of slope, computed in ArcGIS using the method of Zevenbergen et al. [37]. Curvature is evaluated by first calculating the second derivative for each cell in the surface using a moving 3×3 window, and then fitting a fourth order polynomial to the values within the window (see Figure 1). Curvature evaluated only parallel to the slope (*profile curvature*) can describe the acceleration or deceleration of benthic flow, while curvature evaluated perpendicular to slope (*planiform curvature*) can help describe convergence or divergence of flow.

Bathymetric Position Index: *BPI* is a marine focused version of the *Topographic position index (TPI)* [38] which classifies landscape structure (e.g., valleys, plains, hill tops) based on the change in slope position over two scales. BPI quantifies where a location on a bathymetric surface is relative to the overall seascape [12]. For each cell in a surface, BPI is evaluated by finding the difference between the elevation value of the cell and the mean elevation of all cells in an annulus (a ring shape bounded by two concentric circles) surrounding the location. The use of an annulus allows for the exclusion of immediately adjacent cells when measuring mean surrounding elevation. The resulting values are positive near crests and ridges, and negative near cliff bases and valley bottoms.

BPI is used in the semi-automated classification provided with BTM, and has been linked to habitat preferences [39]. For a typical application, two scales of BPI are computed, the *broad* and *fine scale* BPIs. Analyzing BPI at two scales helps capture scale dependent phenomena within the data, and when matched to dominant scales, can be a simple way to incorporate basic multiscale analysis. For historical reasons BTM provides separate tools for *broad* and *fine scale* BPI, but these are the same algorithm internally, and future releases may provide automated multiscale analysis of this parameter. As pointed out in Lecours et al. [11], further improvements may be obtained by computing this measure against a quadratic surface, and this is an area for potential improvement for BTM.

Relative difference to the mean: *Relative difference to the mean* measures relative position of a location using the mean of a continuous neighborhood of cells rather than an annulus. Similar to *BPI*, it is evaluated by finding the difference between the mean elevation of all cells in the neighborhood and the elevation of the focal cell. This difference is divided by the range of the focal neighborhood, where range is the difference between the maximum and minimum elevations in the neighborhood.

Relative difference to the mean can be an important descriptor of bathymetric structure [40]. While both relative difference to mean and BPI are measures of relative location on a surface, the two metrics showed low correlation (<0.1) at the scales used in our analysis (see Table A2). A generalized evaluation of their correlation is out of the scope of this paper, but the results suggest that the two metrics may be independently valuable as descriptors of benthic terrain.

2.1.3. Rugosity/Surface Roughness

Terrain heterogeneity is an important predictor of habitat location and species density [41]. Rugosity, also known as surface roughness or complexity, is a common descriptor of terrain heterogeneity in both terrestrial and marine applications. Rugosity is traditionally evaluated in-situ across a two dimensional terrain profile by draping a chain over the surface and comparing the length of the chain with the linear length of the profile (see Figure 3).

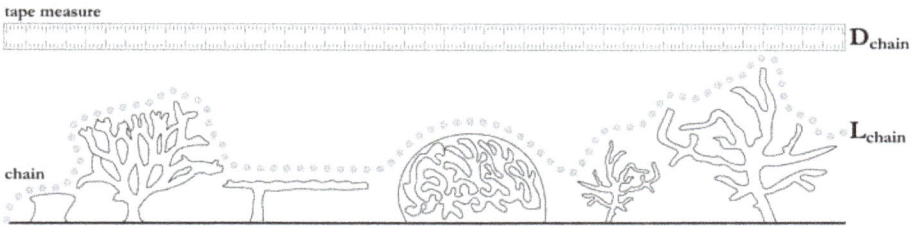

Figure 3. The tape-chain rugosity measurement is an in-situ method of evaluating terrain heterogeneity. The ratio of the chain length (L_{chain}) to profile length (D_{chain}) describes the rugosity of the two dimensional profile. Used under Creative Commons Attribution license (CC BY) from Friedman et al. [42].

With the widespread availability of terrestrial elevation data and the increasing availability of bathymetric surfaces, many methods of evaluating rugosity ex-situ on a three dimensional surface have been proposed. These methods measure a ratio of areas rather than lengths, as shown in Equation (3). BTM supports four methods of measuring rugosity across three tools: **Surface Area to Planar Area**, **Vector Ruggedness Measure**, and **Arc-Chord Ratio**.

$$R = \frac{A_{\text{contoured area}}}{A_{\text{planar area}}} \tag{3}$$

Surface Area to Planar Area: Following the work of Jeff Jenness [23], BTM has implemented the **Surface Area to Planar Area (SAPA)** tool since its release. **SAPA** evaluates rugosity using a 3 × 3 neighborhood, by drawing a line from the center of each cell in the window to the center of the central cell in three dimensions. The result is a network of eight triangles in the central cell which approximates the contoured surface at the cell location. The sum area of these triangles is divided by the two dimensional cell area to obtain a measure of rugosity. Because of its calculation method, **SAPA** is tightly coupled with slope, and is surpassed by the other methods in BTM for computing rugosity.

Vector Ruggedness Measure: Sappington at al. [22] describe a measure of surface roughness titled *Vector Ruggedness Measure (VRM)* that is also calculated using a moving 3 × 3 window, based on an earlier method proposed by Hobson [43]. For each cell in the window, a unit vector orthogonal to the cell is decomposed using the three dimensional location of the cell center, the local slope, and the local aspect. A resultant vector for the window is evaluated and divided by the number of cells in the moving window. The magnitude of this standardized resultant vector is subtracted from 1 to obtain a dimensionless value that ranges from 0 (no variation) to 1 (complete variation). Typical values are small (≤ 0.4) in natural data.

Arc-Chord Ratio: The *Arc-Chord Ratio (ACR)* was introduced in DuPreez [24]. Similar to Friedman et al. [42], ACR evaluates surface ruggedness using a ratio of contoured area (surface area) to the area of a plane of best fit (POBF), where the POBF is a function of the boundary data. By using a POBF rather than a horizontal plane to determine planar area, rugosity is decoupled from slope at the scale of the surface, and is being adopted as an improved measure of surface roughness [44]. DuPreez [44] provides two three dimensional methods of deriving ACR that are independent of data dimensionality and scale; both are supported by BTM.

The first method of calculating three dimensional ACR is exposed in BTM through the **Arc-Chord Ratio** tool, and calculates a single ACR value for an area of interest (AOI) rather than creating a surface of local ACR values. This AOI is chosen by the user based on the context and scale of the analysis, as well as the resolution of the data. The depth surface is converted to a triangulated irregular network (TIN) and clipped to the AOI. Contoured area is determined by summing the triangle areas within the TIN. Planar area is determined by fitting a POBF to the elevation values along the boundary of the AOI, using the Global Polynomial Interpolation tool in ArcGIS to obtain the fitted plane, and as shown in Figure 4. The result is a single ratio representing global ACR rugosity for the area of interest.

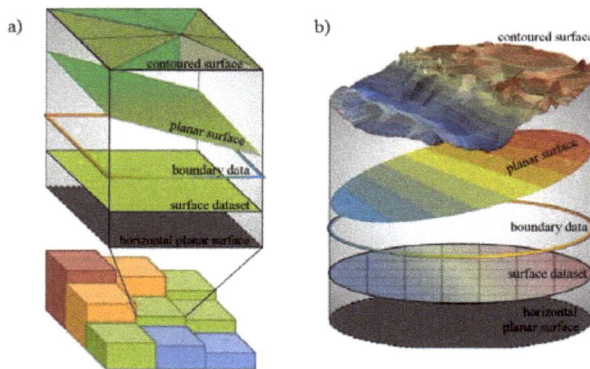

Figure 4. Components required to calculate Arc-Chord Ratio (ACR) (a) for a bathymetric surface using a moving 3 × 3 window, and (b) for an area of interest using a triangulated irregular network. Reprinted by permission from Cherisse Du Preez: Springer Nature Landscape Ecology, 2015 [24].

Surface Area to Planar Area (Slope-corrected): The second method of calculating the *Arc-Chord Ratio* is exposed through the **Surface Area to Planar Area** tool described above. By selecting the "Correct Planar Area for Slope" option, the tool will divide the contoured area by the cell area as projected onto the plane of the local slope. The result is a surface where each cell represents the local ACR rugosity.

2.1.4. Distribution Moments

In addition to specialized morphometrics, BTM computes summary statistics on neighborhoods of data. It supports both circular or square neighborhoods, internally using the neighborhood tools provided in ArcGIS. These measurements provide a spatially aggregated summary of the values within the neighborhood, and are a measure of biotic assemblage structure [45]. These are provided in BTM with the **Calculate Metrics (Depth Statistics)** tool. **Calculate Metrics (Depth Statistics)** exposes summary statistics currently available in ArcGIS with the **Focal Statistics** tool, as well as several statistics that have not yet been implemented in ArcGIS.

Mean is the average water depth in the neighborhood. This simple filter is frequently used to filter noise from the data collection process, but also can provide a smoothed surface which accounts for large scale variations in the DTM, and when subtracted from the DTM, provides a residual of fine scale morphologies [46]. *Standard deviation* captures the local dispersion and is a measure of rugosity, or bathymetric roughness [40]. BTM leverages the SciPy library [47] to provide two additional measures: *Interquartile range* is the difference between the 75th and 25th percentiles of the data, and is a measure of dispersion more robust against outliers than standard deviation. *Kurtosis* measures the weight of the tails of the distribution relative to the overall distribution [21]. BTM computes unbiased kurtosis using Fisher's definition [48], or *excess kurtosis*, which subtracts 3 from the raw value to give 0.0 for a normal distribution. In a geographic context, it can describe the relative dominance of hills or valleys in the neighborhood. Values much less than zero show a dominance of hills, values much greater than zero show a dominance of valleys, and values approach zero for normally distributed surfaces [49].

2.1.5. Multiscale Analysis

Most terrain attributes depend on one of two cases: either the window of analysis is fixed to a 3×3 neighborhood, or the user must decide on an appropriate scale of analysis to match the features of interest. However, many metrics may contain valuable information at multiple scales. Here, BTM implements the same moving window approach, but allows the metrics to be computed over a variety of scales.

The **Compare Scales of Analysis** tool samples a user-selected 200×200 neighborhood of cells in a bathymetry surface and calculates a user-specified statistic. The statistic is then computed at 20 neighborhood sizes within a specified range, and the 20 result grids are presented in a single image for visual comparison, as shown in Figure 5. This image can help users qualitatively understand which scales of analysis identify benthic features and processes of interest. Focal statistics available to use for comparison include *median*, *minimum*, *maximum*, and *percentile*, which require a percentile value used for filtering to be specified.

Some tools included in BTM rely on a fixed focal neighborhood size (such as **Slope**, **Aspect**, **SAPA** and **SAPA (Slope-corrected)**), but the balance of the tools provide an opportunity to explore the impact of scale on the results. While the **Compare Scales of Analysis** tool offers a generalized comparison of scales, the **Calculate Metrics at Multiple Scales** tool generates grids of one or more of the following at any number of scales: *mean, standard deviation, variance, kurtosis, interquartile range*, and *VRM*. This provides an automated solution to generating full grids of distribution moments and rugosity across multiple scales for use in a multiscale model, or simply for the purpose of investigating the impact of scale on the results.

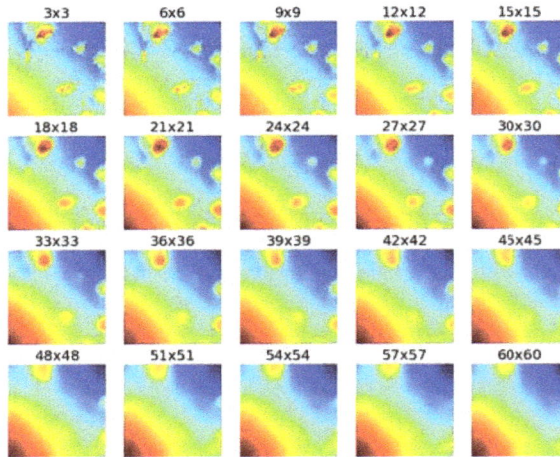

Figure 5. Output from **Compare Scales of Analysis** tool, with *Median* calculated for focal neighborhoods ranging from 3×3 to 60×60. Fine scale detail is lost by the 21×21 neighborhood, and only broad trends remain at the 48×48 scale.

2.2. Software Architecture

While software has played an important role in geomorphometry for decades [50], more recently data-intensive science using software has become the dominant approach to science [51]. This change has made many more scientists in charge of writing code in order to solve their work. Domain scientists have deep knowledge within their field, but their work requires using so many different software components that a deep knowledge of software architecture is often impractical. Scientists and students alike want tools which are simple to use, create reproducible results, and can be easily integrated into their existing workflows, which typically include GIS in an important capacity. Benefits of ArcGIS include a familiar interface, extensive data format support, and capabilities that can be easily extended with Python, the most widely taught programming language at the undergraduate level [52]. To facilitate these needs, in 2012 BTM was rewritten from a closed source Visual Basic Application extension into an open source Python toolbox and Add-in for ArcGIS.

BTM attempts to serve both newcomers and experienced scientists in marine geomorphometry. For newcomers, BTM is easy to use through a graphical interface, includes extensive documentation on each tool, and provides a detailed tutorial on getting started with the software. BTM has been widely used in educational settings, often as an introduction to performing marine geomorphometry analysis. For researchers, BTM is often used as a stepping stone in addressing a broader analysis question, and its extensibility is important, so that it can be composed with other tools in a reproducible workflow. Where the earlier release of BTM provided a single mode of interaction via a wizard-based tool, BTM can now be used in four different contexts (see Table 2).

Table 2. Benthic Terrain Modeler Interfaces.

Interface	Role
Graphical menu	Direct interaction (see Figure 6)
Python toolbox	Direct and scripted interaction
Command-line	Reproducible, scalable, programmatic
Jupyter Notebooks	Teaching and exploration

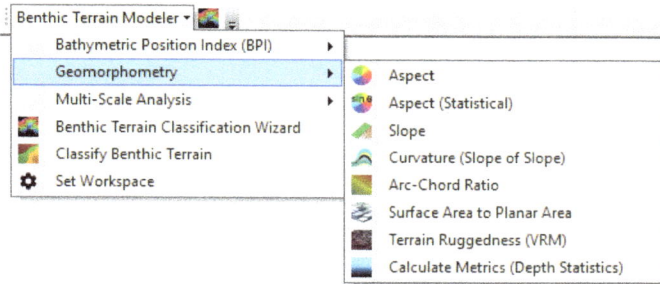

Figure 6. The BTM Python Add-in, which provides a simple graphical interface for accessing the analytical tools provided.

This flexibility means that the software remains easy to use, but can grow with the faculties of researchers as their use deepens. Beyond BTM, recent improvements to the software architecture allow users to utilize the extensive scientific computing stack in Python, which ships with ArcGIS and can be further extended with open source modules for specialized operations, as BTM does with the NetCDF, SciPy and NumPy libraries. Similarly, ArcGIS can now be integrated directly with the R statistical language, which is a logical next step for many researchers when moving to robust statistical models, and we discuss some of the implications of this change in the results section.

BTM also utilizes many best practices in its development. Its source code is publicly available on GitHub [53], and the software is extensively tested across many platforms to ensure its correct function. It also includes numerical validation tests to ensure that changes to the software continue to produce consistent results. While these are best practices in software development, they are less frequently observed in specialized scientific code.

Many excellent analytical tools now exist for geomorphological applications, including those of SAGA [54] and GRASS [55]. ArcGIS is fortunate to have numerous extensions for marine geomorphology, which in addition to BTM include ArcGeomorphometry for computing terrain attributes against a quadratic surface [28], the TASSE toolbox for carefully selected relevant terrain attributes [56], the Surface Gradient and Geomorphometric Modeling toolbox [57], multidirectional texture analysis via MADTools [58], and MGET [59] which ties together many aspects of marine analysis, including data collection and statistical modeling with R.

2.3. Example Study Site

Buck Island Reef National Monument (BIRNM) is a protected area northeast of Saint Croix, U.S. Virgin Islands. Established as a protected area in 1961, both it and its surrounding seascape are well documented; the site has had extensive study of its ecology by NOAA scientists [36,60,61]. Here, our focus is not on producing a specific new map for a location, but to highlight the analytical workflows possible using BTM and its allied tools, and Buck Island provides an excellent testing ground. The reef area has extensive coverage of both MBES and LiDAR (including reflectance values), which we use for our analysis. We restricted our efforts to the LiDAR covered shallow depth locations (depth $\leq$ 50 m), which include an additional LiDAR reflectance product that can be used in other packages to derive additional terrain attributes [62], here used in our covariate comparison.

Few studies of marine geomorphometry provide the data they analyzed, hampering reproducibility and standardized testing datasets that can be used across multiple studies. Buck Island was selected in part because of its full availability of datasets used, which are hosted by NOAA [63]. Additionally, the BTM generated covariates used in the analysis are also available [64]. Broader sharing of marine geomorphometry data will help this nascent field continue to grow, and benefit analytical approaches which require looking beyond individual datasets.

A breadth of literature covers in-depth analysis techniques for habitat mapping, often building up sophisticated statistical models using machine learning techniques like random forest classification [45,65]. This literature is valuable, and illuminates in-depth a single study system with the best available data and models to maximize understanding of the location and its broader modeling implications. For this work, we have targeted the common prerequisites to such a task: the production of a bethic zone mapping and interpretation of a compelling set of covariates derived from bathymetry of the generated covariates. BTM can solve both of these problems directly, and with minor additional work in R, the analytical workflow can be adopted with minimal effort. Demonstrating this workflow, and how it fits into a broader analytical task is the primary goal.

3. Results

3.1. Classified Benthnic Zone Mapping

Since its initial release, BTM has supported the generation of seafloor classification maps using only a bathymetric surface and a classification dictionary. Traditionally, the bathymetric surface was used to generate two grids of BPI (at two different scales) and a grid of slope prior to feeding these intermediate artifacts into the **Classify Benthic Terrain** tool, in which a classification dictionary was manually created to generate a map of benthic zones for the study area, or stored as XML. The newest release of BTM includes the **Run All Model Steps** tool, which condenses this workflow into a single step, and accepts both CSV and Excel files as classification dictionaries.

To demonstrate BTM's classification mapping capabilities, a subset of the LiDAR bathymetry dataset shown in Figure 7 was classified using the **Run All Model Steps** tool. The original dataset was clipped to obtain a smaller study area that highlights the benthic structures immediately surrounding Buck Island.

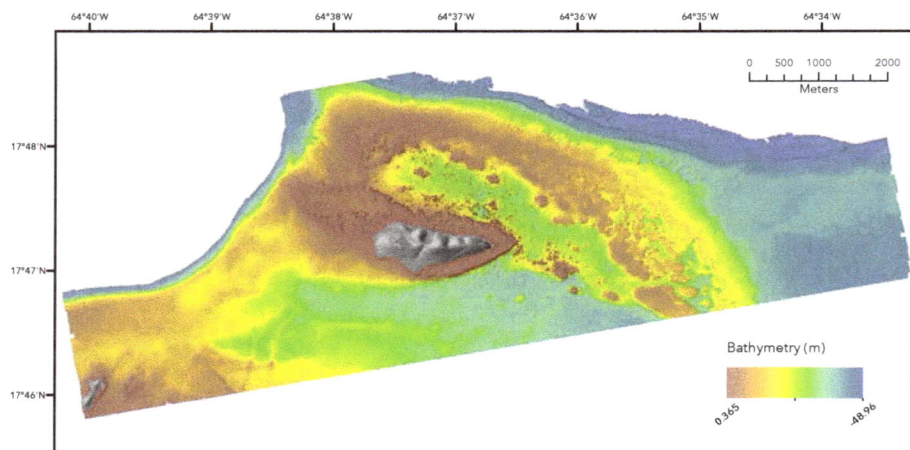

Figure 7. Buck Island National Marine Reserve LiDAR bathymetry (3 m resolution) coverage.

Typical users of BTM should carefully build a unique classification dictionary for each study area by considering the context of the input data and the goals of the analysis. For each benthic zone or habitat in the table, upper and lower bounds of broad scale BPI, fine scale BPI, slope, and depth should be chosen by considering some or all of the following:

- Scale and resolution of the input data
- Scale of focal operations used to calculate BPI and slope

- Previous studies of the area of interest
- Typical values observed for the benthic zone of interest
- The range of values in each classification artifact

There is no universally applicable approach to creating a classification dictionary for use in BTM. In each case, users will need to utilize a different combination of the above resources in conjunction with professional judgment and perhaps an iterative refinement of the values used. A more rigorous approach can be obtained by using a generalized linear model (GLM) to extract the key variables as in Dunn et al. [15].

However, the purpose of this example is not to provide an accurate mapping of benthic zones in BINMR, but rather to demonstrate the typical steps that would be taken within BTM to accomplish that goal. In that context, the classification dictionary shown in Table 3 was created with an alternative approach to that recommended above.

Table 3. Classification table used for creation of Figure 8. Missing values indicate that the bound is not applicable to the benthic zone.

Class	Zone	Broad BPI (Lower)	Broad BPI (Upper)	Fine BPI (Lower)	Fine BPI (Upper)	Slope (Lower)	Slope (Upper)	Depth (Lower)	Depth (Upper)
1	Reef Crest	−87	541	−907	2229		11	−2	
2	Back Reef	−87	541	−459	885		52	−6	
3	Bank or Shelf	−402	699	−1355	4469		74	−18	
4	Reef Flat	−9	541	−459	1333		67	−8	
5	Channel	−323	384	−907	1333		53	−12	−1
6	Fore Reef	−323	541	−1355	2229		61	−14	
7	Lagoon	−166	463	−459	1333		49	−7	
8	Salt Pond	−9	−9	−11	−11		53	−7	
9	Shoreline Intertidal	−9	384	−11	436		19	−2	

Two BPI grids (broad and fine scale) and a slope grid were created using the clipped bathymetry dataset and the parameters listed in Table 4. Each grid was then overlayed with polygon features, each representing a benthic zone described in a 2012 benthic habitat map of the study area produced by the NOAA/NOS/NCCOS/CCMA Biogeography Branch [61]. Using the Zonal Statistics as Table tool provided with ArcGIS, the maximum and minimum values of each grid within each polygon were summarized into the classification dictionary and then revised. This method is only relevant for creating a classification dictionary to demonstrate software functionality and should not be used otherwise.

Table 4. Parameters used for benthic classification in **Run All Model Steps** tool.

Parameter	Value
Bathymetry Raster:	LiDAR bathymetry (clipped)
Broad-scale BPI Inner Radius:	25
Broad-scale BPI Outer Radius:	250
Fine-scale BPI Inner Radius:	5
Fine-scale BPI Outer Radius:	25
Classification Dictionary:	See Table 3

The **Run All Model Steps** tool was then executed in ArcGIS using the parameters summarized in Table 4. The result is shown in Figure 8.

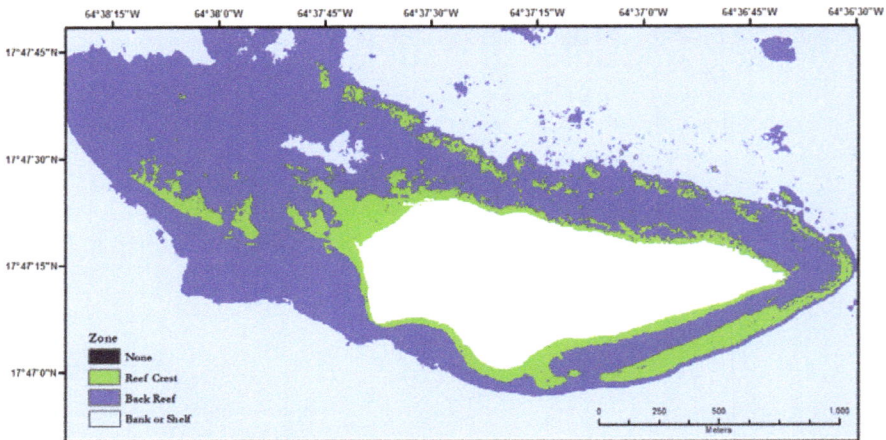

Figure 8. A habitat classification map created for a subregion of Buck Island National Marine Reserve using the Run All Model Steps tool. The white zone represents terrain above sea level.

3.2. Integrating R Statistical Analysis

The ease of calculating BTM covariates is further enhanced by the ability to utilize them in predictive models or other analyses. The direct link between ArcGIS and the statistical programming language R, allows for the direct transfer of data to either software and the result of which is ready for immediate use. This functionality enables the creation of maps, the aggregation and wrangling of data, the calculation of covariates, the examination of the relationships between covariates, the building of predictive models, the analysis of diagnostic measures and charts, and more. Ultimately, the bridge enables the ability to utilize the needed tools or functions to answer the questions at hand and to create efficient and reproducible workflows. To demonstrate this, the bridge will be used to transfer BTM data to R to consider a dimensionality reduction method and to analyze the results.

After all desired BTM covariates have been derived, simultaneously working in ArcGIS and R with the same data is possible using the R-ArcGIS bridge [66]. The R-ArcGIS bridge consists of the R package, **arcgisbinding**, which provides functions for reading, writing, converting, and manipulating spatial data between ArcGIS and R. The advantages of the **arcgisbinding** package compared to packages like **rgdal** are most noticeable when considering the breath of data that can be transferred and when coordinated data manipulation is needed. For example, the package can read and write to any data source that exists within ArcGIS. This includes vector data stored in formats such as shapefiles, file geodatabase, or a URL for a feature service, and any supported raster data types, including complex types like mosaic datasets. Additionally, in the **arcgisbinding** package, the same functions used to read in data, can also be used to perform actions like creating custom subsets and selections, reprojecting both vector and raster data on the fly, and resampling and adjusting the extent of raster data. All of the above mentioned functionality is contained in the functions *arc.open* or *arc.select* if working with vector data, as shown, or *arc.open* and *arc.raster* if working with raster data.

```
# Load library containing R-ArcGIS bridge functionality
library(arcgisbinding)
# Check connection between R and ArcGIS has been established
arc.check_product()

input_gdb <- "C:/ArcGIS/Projects/BTM/BTM.gdb"
feature_class <- "Field_AAData_ClassifiedLocations"

# Establish pointer to desired data's stored location
arc_locations <- arc.open(file.path(input_gdb, feature_class))

# Convert from an ArcGIS data type to an R data frame object
df_locations <- arc.select(arc_locations)

# Convert from an R data frame object to a spatial data frame object
# from the R package sp
spdf_locations <- arc.data2sp(df_locations)
```

Once this data is in its desired format, any of the functions or packages in R can be used. For example, since covariates typically used in Benthic Terrain Modeling are highly correlated, extracting the linear components of each predictor to alleviate multicollinearity prior to modeling, might be of interest. Principal component analysis is one such method for accomplishing this, by quantifying how much variance is explained by each covariate, but also by providing a useful metric for dimension reduction and building predictive models.

```
# Remove NAs
df_locations <- na.omit(df_locations)

# Creation of training/testing datasets
smp_size <- floor(0.80 * nrow(df_locations))

# Randomly select observation numbers to ensure sample is randomly selected
train_ind <- sample(seq_len(nrow(df_locations)), size = smp_size)

# Subset the original data based on the randomly selected observation
# numbers to make the training data set
df_locations_train <- df_locations[train_ind, ]
# Subset the original data on the remaining observations to make the
# testing data set
df_locations_test <- df_locations[-train_ind, ]

# Make predictor variable into a factor
df_locations_train$D_STRUCT <- as.factor(df_locations_train$D_STRUCT)

# Separate response from the covariates
btm.train_covariates <- df_locations_train[, 2:17]
btm.train_response <- df_locations_train[, 1]

# Apply Principal Component Analysis
btm.train_pca <- prcomp(btm.train_covariates,
                        center = TRUE,
                        scale. = TRUE)
```

Results of the PCA analysis can be explored using functions like *summary()* (Figure 9) and *plot()* (Figure 10) to determine the proportion of variance explained by each component and which components together are able to explain up to 95 percent of the variance. From this point, results could be used to determine the most influential covariates which could then be used to construct a predictive model and assess model fit.

```
> summary(btm.train_pca)
Importance of components:
                          PC1    PC2     PC3     PC4     PC5    PC6     PC7     PC8     PC9     PC10
Standard deviation      2.4356 1.6903 1.26275 1.09419 1.0159 0.93868 0.87839 0.83530 0.66985 0.57986
Proportion of Variance  0.3708 0.1786 0.09966 0.07483 0.0645 0.05507 0.04822 0.04361 0.02804 0.02101
Cumulative Proportion   0.3708 0.5493 0.64897 0.72380 0.7883 0.84337 0.89159 0.93520 0.96324 0.98426
                          PC11   PC12    PC13    PC14    PC15    PC16
Standard deviation      0.40476 0.24426 0.13309 0.10206 0.01644 0.002306
Proportion of Variance  0.01024 0.00373 0.00111 0.00065 0.00002 0.000000
Cumulative Proportion   0.99450 0.99822 0.99933 0.99998 1.00000 1.000000
```

Figure 9. Output from the *summary()* function on the results from the principal component analysis to explain the proportion of variance explained by each selected covariate.

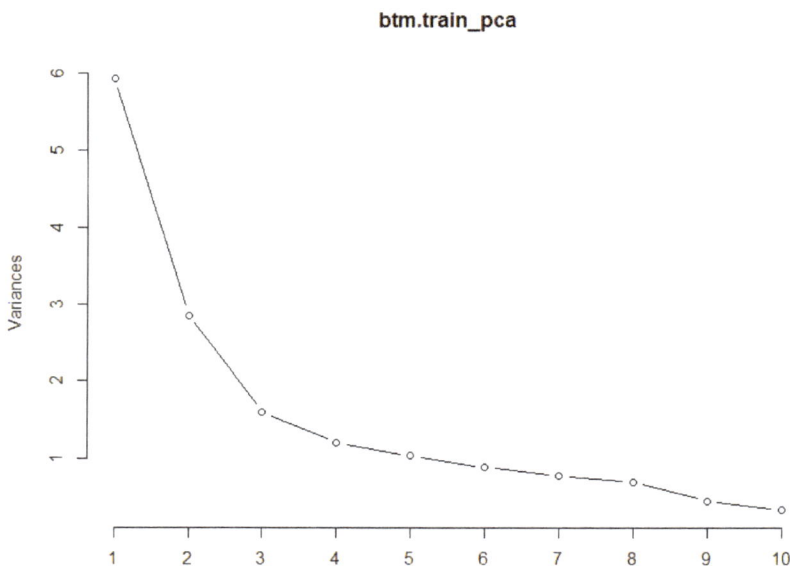

Figure 10. Output from the *plot()* function on the results from the principal component analysis which show the components that explain 95 percent of the variance.

This example is just one possibility of the functionality between R and ArcGIS. The ease of transferring data back and forth, with the ability to perform a variety of statistical and spatial analyses, coupled with the creation of new covariates and maps allows for in depth exploration of the statistics and the science. Final results can be converted back to ArcGIS data types for the production of final maps or tables and R functionality plots can be integrated into the software through the creation of script tools and models further integrating the two.

3.3. Terrain Attribute Relationships

3.3.1. Slope

We compared Slope differences between the traditional planar form and its new geodesic variant for our example site. In most locations, little variation is seen. High rugosity areas show more marked variation, with valleys between peaks showing the biggest declines, and the edges of steep peaks showing the greatest increases (see Figure 11, Difference pane). Over the full raster, the two layers are highly correlated (Pearson correlation > 0.99). Because of this high correlation, we only relied on planar slope in our broader comparison between covariates.

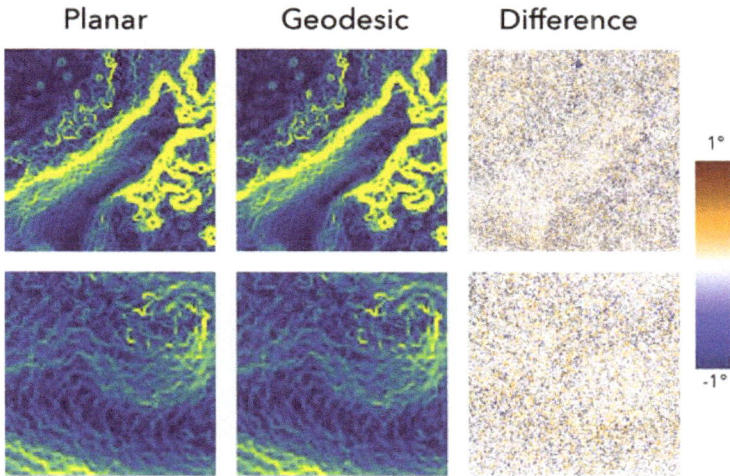

Figure 11. Left: planar slope, S_p, in degrees. Center: geodesic slope, S_g, in degrees. Right: Slope differences, $(S_g - S_p)$, with the displayed range of (1 to -1). Most areas show very small changes in the values, but locations with high rugosity, and steep slopes show localized larger differences. Same inset locations described in Figure 2.

3.3.2. Covariate Selection

For this study, a set of 16 covariates was selected (Table A1, [64]). Following the ordering of Section 2.1, we used BTM to generate three surface gradients (*slope, aspect eastnerness,* and *aspect northness*), four measures of relative position (*broad* and *fine scale BPI,* and *slope of the slope, relative difference to the mean*), three measures of surface roughness (*2D ACR, VRM* and *SAPA*), and four distribution moments (*mean, standard deviation, inner-quartile range* and *kurtosis*), with only sampling the 3 × 3 derivatives of the multiscale algorithms. The final two covariates are the bathymetry itself, and the reflectance data from the LiDAR returns. In typical applications, the multiscaling algorithms would also be used, to measure the effect of scale on the properties, or to generate a broad suite of covariates to be used in a data reduction algorithm to select the key variables and scales.

Removing colinearity is an important step for many applications of geomorphometry [56], but is missing in most analyses, where frequently derivatives of a single DTM are used without examining the covariation that should be expected when deriving many outputs from a single input. This could be addressed within geomophometry packages, but it is a complex topic that currently resists generalization. Here, we examined the relationships between the layers (Table A2), to illustrate the importance of this step when using tools like BTM, and to examine some relationships between multiple algorithms claiming to capture similar features of DTMs. While this is limited to a single dataset, it provides some insight into some relationships between covariates that marine geomorphological tools produce.

Rugosity, and its relationship to slope, have received attention with multiple algorithms working to decouple the two [22,24]. BTM features two such rugosity algorithms, ACR (or slope-corrected SAPA) and VRM. In our study site, ACR and VRM are very strongly correlated (Pearson: 0.89), and both only moderately correlated with SAPA. In turn, SAPA shows a high correlation to slope, the standard deviation over a 3 × 3 window, and the interquartile range. In practice, standard deviation and the interquartile range are less correlated over larger neighborhoods, but the strong correlation between SAPA and slope shows the limited utility of SAPA as an independent measurement in models which already contain slope as a covariate.

A measure suggested by Lecours et al. [56] based on his evaluation of over 200 covariates is relative difference from mean depth (the value minus the mean divided by the range). With our data this measure showed no meaningful correlations, providing a useful addition. Similarly, Pittman et al. [65] and others have recommended "slope of the slope", executing the slope function a second time on the slope raster, and again we show little correlation with this result to others in the covariates we analyzed, though the term "slope of the slope" can be methodologically confusing because of its conceptual overlap with curvature. Finally, kurtosis showed no meaningful correlations and should be examined as another potentially useful derivative.

3.3.3. Remarks

Here we have shown the basic workflow of using BTM to create benthic zone classifications, to use BTM to create a collection of covariates for later use and interpretation, and how to extend the workflow using R. In practice, BTM is used in all these roles. It may be used to create simple slope and rugosity layers, as was done in McNeil et al. [67]. The software can also be used to generate covariates at multiple scales to aid in understanding the effects of scale on prediction models like Maximum Entropy (MaxEnt) model outputs as in Miyamoto et al. [68]. In use cases like Li et al. [69], the outputs from BTM are used in an unsupervised classification using the ISODATA algorithm to build a validated benthic classification map. While BTM has been providing new algorithms for rugosity for many years, it is still common for users to rely on earlier methods like SAPA [70], likely because of its well-known association with capturing rugosity. It will take time, and a greater exposure of the benefits of slope-decoupled roughness for this to change. More studies which explicitly examine a large gamut of covariates like [56] are needed to improve our understanding of the relationships between the derived covariates, which usually ultimately derive from a single input dataset.

4. Discussion

The analysis shows how BTM can be used to produce semi-automated classifications of locations, using easy to use tools, and how to incorporate BTM and the terrain attributes it produces into a larger analysis context. This case study is intended to be simple with reproducible data, and the authors hope that reuse of publicly available well documented datasets will become more common in the community to facilitate collaboration. An accessible but powerful analysis framework is available by combining BTM and ArcGIS with the broader scientific software ecosystem in Python and R. As marine geomorphology matures as a discipline, adapting to rich and accessible tools provides a key way for both students and researchers to perform rigorous analysis. BTM has grown in part to adapt to these needs, and we illustrate some of the new approaches it has incorporated to help with this process.

While improvements accumulate, there remain gaps in the analytical tools available that deserve deeper investment. Implementations measuring fractal dimension [71] remain rare, and could be aided by being more broadly accessible. Multiscale analysis, while partially captured in BTM and other tools, could be greatly improved by the incorporation of characteristic scales [72] to formalize understanding on the role of scale. Instead of relying on cell-based analysis and resampling, object-based models provide a better potential basic unit for analysis [73]. Similarly, the data capture process primarily observes point based observations, and analyzing these directly provides another avenue forward [74]. Terrestrial geomorphology still enjoys advantages in consistent global datasets which form a common benchmark for analysis, but as the process for capturing and processing marine data continue to improve, this goal should also be attainable in the marine domain. Continued improvements in the marine geomorphology research community will be well served with broadly available and well understood tools available for researchers, and we hope to allow BTM to continue serving in this capacity.

Acknowledgments: Institutional support provided by: Esri, NOAA Costal Service Center, Massachusetts Office of Coastal Zone Management, and Oregon State University. Thanks to Cherisse Du Preez, Matt Pendleton, Jen Boulware, Lisa Wedding, Pat Iampietro and Ellen Hines for their contributions. Thanks to our anonymous reviewers, they have greatly improved the text.

Author Contributions: N.S. and S.W. conceived and designed the analysis; N.S., M.P. and S.W. performed the analysis; N.S., M.P. and S.W. analyzed the data; D.W., N.S., M.P. and S.W. wrote the paper.

Conflicts of Interest: The authors declare no conflict of interest.

Abbreviations

The following abbreviations are used in this manuscript:

AOI	Area of interest
BIRNM	Buck Island Reef National Monument
BPI	Bathymetric Position Index
BTM	Benthic Terrain Modeler
DTM	Digital terrain model
GIS	Geographic information system
IQR	Interquartile range
LiDAR	Light Detection and Ranging
MBES	Multibeam Ecosounder
NOAA	National Oceanic and Atmospheric Administration
POBF	Plane of best fit
SAPA	Surface area to planar area
TIN	Triangulated irregular network
VRM	Vector ruggedness measure

Appendix A.

Appendix A.1. Derived Parameters

Covariance is defined as:

$$cov_{ij} = \frac{\sum_{k=1}^{N}(Z_{ik}-\mu_i)(Z_{jk}-\mu_j)}{N-1} \tag{A1}$$

where
k denotes the position of the current raster cell;
Z is the value of the cell;
i and j are the two rasters being compared;
μ is the mean of the raster;
N is the total number of cells.

The Pearson correlation coefficient between rasters i and j is:

$$corr_{ij} = \frac{cov_{ij}}{\sigma_i\sigma_j} \tag{A2}$$

where (σ_i, σ_j) is the standard deviation of rasters i and j.

Table A1. Mapping of the environmental covariates to their layer names.

Layer	Environmental Covariate
1	Aspect easterness
2	Aspect northerness
3	Broad scale BPI standardized (60–80)
4	Fine scale BPI standardized (10–15)
5	Interquartile range (IQR)
6	Bathymetry (LiDAR)
7	Kurtosis (Pearson)
8	Mean
9	Relative difference to mean
10	Standard deviation
11	Reflectance (LiDAR)
12	Slope (planar)
13	Slope of the slope (max Δ in slope)
14	Rugosity (ACR)
15	Rugosity (SAPA)
16	Rugosity (VRM)

Table A2. Upper triangle is pairwise covariance, lower triangle is pairwise correlation. Pearson correlation coefficients **bold** for *very strong* correlations, *italic* for *strong* correlations [75], layer labels in Table A1.

Layer	1	2	3	4	5	6	7	8	9	10	11	12	13	14	15	16
1		−0.001505	−0.000064	−0.000070	−0.000074	−0.001187	−0.000116	−0.001187	−0.000151	−0.000113	0.000753	−0.000293	−0.001150	−0.000004	−0.000014	−0.000005
2	−0.02035		−0.000174	−0.000018	0.000274	−0.005216	−0.000269	−0.005231	0.000012	0.000360	0.002170	0.000938	0.003334	0.000013	0.000045	0.000013
3	−0.00781	−0.02005		0.000123	0.000072	0.000495	−0.000025	0.000460	0.000110	0.000102	−0.000443	0.000236	0.000704	0.000010	0.000036	0.000022
4	−0.01952	−0.00471	0.29382		0.000070	−0.000161	−0.000008	−0.000159	0.000063	0.000099	−0.000181	0.000223	0.000489	0.000016	0.000040	0.000031
5	−0.02094	0.0727	0.17456	0.38188		−0.000080	−0.000094	−0.000078	0.000012	0.000235	−0.000157	0.000551	0.001311	0.000020	0.000083	0.000033
6	−0.0371	−0.15321	0.13208	−0.09758	−0.04917		−0.000071	0.014664	−0.000356	−0.000091	−0.005869	−0.000263	−0.001478	−0.000001	0.000015	0.000014
7	−0.01123	−0.02448	−0.02069	−0.0156	−0.17901	−0.01488		−0.000068	−0.000030	−0.000077	−0.000073	−0.000219	0.000277	−0.000001	−0.000013	0.000002
8	−0.03705	−0.15354	0.12279	−0.09619	−0.0476	**0.99459**	−0.0142		−0.000357	−0.000091	−0.005891	−0.000261	−0.001482	−0.000001	0.000015	0.000014
9	−0.01078	0.00082	0.06726	0.0872	0.01686	−0.05514	−0.01445	−0.05526		0.000016	−0.000172	0.000040	−0.000146	0.000002	0.000005	0.000002
10	−0.02384	0.07124	0.18349	0.40352	**0.96982**	−0.0415	−0.10875	−0.04173	0.01651		−0.000540	−0.002241	−0.000031	−0.000086	−0.000055	−0.000055
11	0.03378	0.0915	−0.16976	−0.15727	−0.13822	−0.57182	−0.02204	−0.57346	−0.0384	−0.15457		−0.000540	−0.002241	−0.000031	−0.000086	−0.000048
12	−0.02606	0.0785	0.17926	0.38571	**0.96182**	−0.05086	−0.1308	−0.0505	0.01777	**0.99042**	−0.14996		−0.000540	0.000066	0.000245	0.000106
13	−0.0055	0.06997	0.13414	0.21164	0.57397	−0.07169	0.04158	−0.0718	−0.0162	*0.61259*	−0.15597	*0.63653*		0.000182	0.000415	0.000301
14	−0.00734	0.01546	0.10146	0.38474	0.48963	−0.00152	0.02906	−0.00171	0.00986	0.54408	−0.11896	0.50955	0.35024		0.000012	0.000013
15	−0.00395	0.02177	0.16034	0.40183	**0.83457**	0.01642	−0.04372	0.01634	0.01338	**0.85408**	−0.13833	*0.78137*	0.33215	0.54067		0.000020
16		0.0102	0.15266	0.48178	0.51791	0.02527	0.0124	0.02513	0.01487	0.5628	−0.13889	0.52841	0.37625	**0.89434**	0.5639	

References

1. Halpern, B.S.; Walbridge, S.; Selkoe, K.A.; Kappel, C.V.; Micheli, F.; D'Agrosa, C.; Bruno, J.F.; Casey, K.S.; Ebert, C.; Fox, H.E.; et al. A global map of human impact on marine ecosystems. *Science* **2008**, *319*, 948–952.
2. Wright, D.J.; Heyman, W.D. Introduction to the Special Issue: Marine and Coastal GIS for Geomorphology, Habitat Mapping, and Marine Reserves. *Mar. Geodesy* **2008**, *31*, 1–8.
3. Becker, J.J.; Sandwell, D.T.; Smith, W.H.; Braud, J.; Binder, B.; Depner, J.; Fabre, D.; Factor, J.; Ingalls, S.; Kim, S.H.; et al. Global Bathymetry and Elevation Data at 30 Arc Seconds Resolution: SRTM30_PLUS. *Mar. Geodesy* **2009**, *32*, 355–371.
4. Sayre, R.; Wright, D.; Aniello, P.; Breyer, S.; Cribbs, D.; Frye, C.; Vaughan, R.; Van Esch, B.; Stephens, D.; Harris, P.; et al. Mapping EMUs (Ecological Marine Units)–the creation of a global GIS of distinct marine environments to support marine spatial planning, management and conservation. In Proceedings of the Fifteenth International Symposium of Marine Geological and Biological Habitat Mapping (GEOHAB 2015), Bahia, Brazil, 1–8 May 2015; p. 21.
5. Pittman, S.J. *Seascape Ecology*; John Wiley & Sons: Hoboken, NJ, USA, 2017.
6. Brock, J.C.; Wright, C.W.; Clayton, T.D.; Nayegandhi, A. LIDAR optical rugosity of coral reefs in Biscayne National Park, Florida. *Coral Reefs* **2004**, *23*, 48–59.
7. García-Alegre, A.; Sánchez, F.; Gómez-Ballesteros, M.; Hinz, H.; Serrano, A.; Parra, S. Modelling and mapping the local distribution of representative species on the Le Danois Bank, El Cachucho Marine Protected Area (Cantabrian Sea). In *Deep-Sea Research Part II: Topical Studies in Oceanography*; Elsevier: Amsterdam, The Netherlands, 2014; Volume 106, pp. 151–164.
8. Brown, C.J.; Smith, S.J.; Lawton, P.; Anderson, J.T. Benthic habitat mapping: A review of progress towards improved understanding of the spatial ecology of the seafloor using acoustic techniques. *Estuar. Coast. Shelf Sci.* **2011**, *92*, 502–520.
9. Bishop, M.P.; James, L.A.; Shroder, J.F.; Walsh, S.J. Geospatial technologies and digital geomorphological mapping: Concepts, issues and research. *Geomorphology* **2012**, *137*, 5–26.
10. Passalacqua, P.; Belmont, P.; Staley, D.M.; Simley, J.D.; Arrowsmith, J.R.; Bode, C.A.; Crosby, C.; DeLong, S.B.; Glenn, N.F.; Kelly, S.A.; et al. Analyzing high resolution topography for advancing the understanding of mass and energy transfer through landscapes: A review. *Earth-Sci. Rev.* **2015**, *148*, 174–193.
11. Lecours, V.; Dolan, M.F.; Micallef, A.; Lucieer, V.L. A review of marine geomorphometry, the quantitative study of the seafloor. *Hydrol. Earth Syst. Sci.* **2016**, *20*, 3207–3244.
12. Lundblad, E.R.; Wright, D.J.; Miller, J.; Larkin, E.M.; Rinehart, R.; Naar, D.F.; Donahue, B.T.; Anderson, S.M.; Battista, T. A Benthic Terrain Classification Scheme for American Samoa. *Mar. Geodesy* **2006**, *29*, 89–111.
13. Wright, D.; Pendleton, M.; Boulware, J.; Walbridge, S.; Gerlt, B.; Eslinger, D.; Sampson, D.; Huntley, E. ArcGIS Benthic Terrain Modeler (BTM), v. 3.0, Environmental Systems Research Institute, NOAA Coastal Services Center, Massachusetts Office of Coastal Zone Management. Available online: https://esriurl.com/5754 (accessed on 27 February 2018).
14. Young, M.; Ierodiaconou, D.; Womersley, T. Forests of the sea: Predictive habitat modelling to assess the abundance of canopy forming kelp forests on temperate reefs. *Remote Sens. Environ.* **2015**, *170*, 178–187.
15. Dunn, D.; Halpin, P. Rugosity-based regional modeling of hard-bottom habitat. *Mar. Ecol. Prog. Ser.* **2009**, *377*, 1–11.
16. Verfaillie, E.; Degraer, S.; Schelfaut, K.; Willems, W.; Van Lancker, V. A protocol for classifying ecologically relevant marine zones, a statistical approach. *Estuar. Coast. Shelf Sci.* **2009**, *83*, 175–185.
17. Diesing, M.; Coggan, R.; Vanstaen, K. Widespread rocky reef occurrence in the central English Channel and the implications for predictive habitat mapping. *Estuar. Coast. Shelf Sci.* **2009**, *83*, 647–658.
18. Horn, B.K. Hill shading and the reflectance map. *Proc. IEEE* **1981**, *69*, 14–47.
19. Olaya, V. Basic land-surface parameters. *Geomorphometry Concepts Softw. Appl.* **2009**, *33*, 141–169.
20. Florinsky, I.V. An illustrated introduction to general geomorphometry. *Prog. Phys. Geogr.* **2017**, *41*, 723–752.
21. Zwillinger, D.; Kokoska, S. *CRC Standard Probability and Statistics Tables and Formulae*; CRC Press: Boca Raton, FL, USA, 1999.
22. Sappington, J.M.; Longshore, K.M.; Thompson, D.B. Quantifying Landscape Ruggedness for Animal Habitat Analysis: A Case Study Using Bighorn Sheep in the Mojave Desert. *J. Wildl. Manag.* **2007**, *71*, 1419–1426.

23. Jenness, J.S. Calculating landscape surface area from digital elevation models. *Wildl. Soc. Bull.* **2004**, *32*, 829–839.
24. Du Preez, C. A new arc–chord ratio (ACR) rugosity index for quantifying three-dimensional landscape structural complexity. *Landsc. Ecol.* **2015**, *30*, 181–192.
25. Weisstein, E.W. MathWorld: Moore Neighborhood. Available online: http://mathworld.wolfram.com/MooreNeighborhood.html (accessed on 27 February 2018).
26. Evans, I.S. An integrated system of terrain analysis and slope mapping. *Z. fur Geomorphol.* **1980**, *36*, 274–295.
27. Wood, J. The Geomorphological Characterisation of Digital Elevation Models. PhD Thesis, University of Leicester, Leicester, UK, 1996.
28. Rigol-Sanchez, J.P.; Stuart, N.; Pulido-Bosch, A. ArcGeomorphometry: A toolbox for geomorphometric characterisation of DEMs in the ArcGIS environment. *Comput. Geosci.* **2015**, *85*, 155–163.
29. Dolan, M.F.; Lucieer, V.L. Variation and Uncertainty in Bathymetric Slope Calculations Using Geographic Information Systems. *Mar. Geodesy* **2014**, *37*, 187–219.
30. Ligas, M.; Banasik, P. Conversion between Cartesian and geodetic coordinates on a rotational ellipsoid by solving a system of nonlinear equations. *Geodesy Cartogr.* **2011**, *60*, 145–159.
31. Lecours, V.; Devillers, R.; Simms, A.E.; Lucieer, V.L.; Brown, C.J. Towards a framework for terrain attribute selection in environmental studies. *Environ. Model. Softw.* **2017**, *89*, 19–30.
32. Davies, A.J.; Wisshak, M.; Orr, J.C.; Murray Roberts, J. Predicting suitable habitat for the cold-water coral *Lophelia pertusa* (Scleractinia). In *Deep-Sea Research Part I: Oceanographic Research Papers*; Elsevier: Amsterdam, The Netherlands, 2008; Volume 55, pp. 1048–1062.
33. Smith, N.; Van der Walt, S. A Better Default Colormap for Matplotlib. Available online: https://bids.github.io/colormap/ (accessed on 27 February 2018).
34. Wilson, M.F.J.; O'Connell, B.; Brown, C.; Guinan, J.C.; Grehan, A.J. Multiscale Terrain Analysis of Multibeam Bathymetry Data for Habitat Mapping on the Continental Slope. *Mar. Geodesy* **2007**, *30*, 3–35.
35. Lucieer, V.; Pederson, H. Linking morphometric characterisation of rocky reef with fine scale lobster movement. *ISPRS J. Photogramm. Remote Sens.* **2008**, *63*, 496–509.
36. Pittman, S.J.; Hile, S.D.; Jeffrey, C.F.G.; Caldow, C.; Kendall, M.S.; Monaco, M.E.; Hillis-Starr, Z. Fish assemblages and benthic habitats of Buck Island Reef National Monument (St. Croix, U.S. Virgin Islands) and the surrounding seascape: A characterization of spatial and temporal patterns. *NOAA Tech. Memo. NOS NCCOS* **2008**, *71*, 1–80.
37. Zevenbergen, L.W.; Thorne, C.R. Quantitative analysis of land surface topography. *Earth Surf. Process. Landf.* **1987**, *12*, 47–56.
38. Weiss, A.D. Topographic Position and Landforms Analysis. In Proceedings of the Esri User Conference, San Diego, CA, USA, 26–30 June 2000; p. 1.
39. Iampietro, P.; Kvitek, R. Quantitative seafloor habitat classification using GIS terrain analysis: Effects of data density, resolution, and scale. In Proceedings of the 22nd Annual ESRI User Conference, San Diego, CA, USA, 9–13 July 2002; pp. 8–12.
40. Lecours, V. On the Use of Maps and Models in Conservation and Resource Management (Warning: Results May Vary). *Front. Mar. Sci.* **2017**, *4*, 1–18.
41. Riley, S.; Degloria, S.; Elliot, S. A Terrain Ruggedness Index that Quantifies Topographic Heterogeneity. *Int. J. Sci.* **1999**, *5*, 23–27.
42. Friedman, A.; Pizarro, O.; Williams, S.B.; Johnson-Roberson, M. Multi-Scale Measures of Rugosity, Slope and Aspect from Benthic Stereo Image Reconstructions. *PLoS ONE* **2012**, *7*, doi:10.1371/journal.pone.0050440.
43. Hobson, R. Surface roughness in topography: Quantitative approach. In *Spatial Analysis in Geomorphology*; Methuen: London, UK, 1972; pp. 221–245.
44. Du Preez, C.; Curtis, J.M.; Clarke, M.E. The structure and distribution of benthic communities on a shallow seamount (Cobb Seamount, Northeast Pacific Ocean). *PLoS ONE* **2016**, *11*, e0165513.
45. Wedding, L.M.; Friedlander, A.M.; McGranaghan, M.; Yost, R.S.; Monaco, M.E. Using bathymetric lidar to define nearshore benthic habitat complexity: Implications for management of reef fish assemblages in Hawaii. *Remote Sens. Environ.* **2008**, *112*, 4159–4165.
46. Grohmann, C.H.; Smith, M.J.; Riccomini, C. Multiscale analysis of topographic surface roughness in the Midland Valley, Scotland. *IEEE Trans. Geosci. Remote Sens.* **2011**, *49*, 1200–1213.

47. Jones, E.; Oliphant, T.; Peterson, P. SciPy: Open Source Scientific Tools For Python. Available online: http://www.scipy.org/ (accessed on 10 January 2018).

48. DeCarlo, L.T. On the meaning and use of kurtosis. *Psychol. Methods* **1997**, *2*, 292, doi:10.1037/1082-989X.2.3.292.

49. Bartels, M.; Wei, H.; Mason, D.C. DTM generation from LIDAR data using skewness balancing. In Proceedings of the 18th IEEE International Conference on Pattern Recognition, Hong Kong, China, 20–24 August 2006; Volume 1, pp. 566–569.

50. Mark, D.M. Geomorphometric parameters: A review and evaluation. *Geogr. Ann. Ser. A Phys. Geogr.* **1975**, 165–177, doi:10.1080/04353676.1975.11879913.

51. Hey, T.; Tansley, S.; Tolle, K. *The Fourth Paradigm: Data-Intensive Scientific Discovery*; Microsoft Research: Washington, DC, USA, 2009.

52. Guo, P. Python Is Now the Most Popular Introductory Teaching Language at Top U.S. Universities. Available online: https://cacm.acm.org/blogs/blog-cacm/176450 (accessed on 27 February 2018).

53. Walbridge, S.; Slocum, N. GitHub: Benthic Terrain Modeler. Available online: https://github.com/EsriOceans/btm (accessed on 27 February 2018).

54. Conrad, O.; Bechtel, B.; Bock, M.; Dietrich, H.; Fischer, E.; Gerlitz, L.; Wehberg, J.; Wichmann, V.; Böhner, J. System for automated geoscientific analyses (SAGA) v. 2.1.4. *Geosci. Model Dev.* **2015**, *8*, 1991–2007.

55. Neteler, M.; Bowman, M.; Landa, M.; Metz, M. GRASS GIS: A multi-purpose Open Source GIS. *Environ. Model. Softw.* **2012**, *31*, 124–130.

56. Lecours, V.; Brown, C.J.; Devillers, R.; Lucieer, V.L.; Edinger, E.N. Comparing selections of environmental variables for ecological studies: A focus on terrain attributes. *PLoS ONE* **2016**, *11*, 1–18.

57. Evans, J.; Oakleaf, J.; Cushman, S. An ArcGIS Toolbox for Surface Gradient and Geomorphometric Modeling, Version 2.0-0. Available online: http://evansmurphy.wix.com/evansspatial (accessed on 3 January 2018).

58. Trevisani, S.; Rocca, M. MAD: Robust image texture analysis for applications in high resolution geomorphometry. *Comput. Geosci.* **2015**, *81*, 78–92.

59. Roberts, J.J.; Best, B.D.; Dunn, D.C.; Treml, E.A.; Halpin, P.N. Marine Geospatial Ecology Tools: An integrated framework for ecological geoprocessing with ArcGIS, Python, R, MATLAB, and C++. *Environ. Model. Softw.* **2010**, *25*, 1197–1207.

60. Kendall, M.; Miller, T.; Pittman, S. Patterns of scale-dependency and the influence of map resolution on the seascape ecology of reef fish. *Mar. Ecol. Prog. Ser.* **2011**, *427*, 259–274.

61. Costa, B.M.; Tormey, S.; Battista, T.A. *Benthic Habitats of Buck Island Reef National Monument*; NOAA National Centers for Coastal Ocean Science: Silver Spring, MD, USA, 2012.

62. Zavalas, R.; Ierodiaconou Daniel, D.; Ryan, D.; Rattray, A.; Monk, J. Habitat classification of temperate marine macroalgal communities using bathymetric LiDAR. *Remote Sens.* **2014**, *6*, 2154–2175.

63. Battista, T.; Costa, B. Buck Island Reef National Monument. Available online: https://products.coastalscience.noaa.gov/collections/benthic/e93stcroix/ (accessed on 10 January 2018).

64. Walbridge, S. Buck Island, Covariate Geodatabase. Available online: https://figshare.com/articles/Buck_Island_-_Covariate_Stack/5946463 (accessed on 27 February 2018).

65. Pittman, S.; Costa, B.; Battista, T. Using lidar bathymetry and boosted regression trees to predict the diversity and abundance of fish and corals. *J. Coast. Res.* **2009**, 27–38, doi:10.2112/SI53-004.1.

66. Pavlushko, D.; Pobuda, M.; Walbridge, S.; Kopp, S.; Aydin, O.; Krivoruchko, K.; Janikas, M. R-ArcGIS Bridge, Environmental Systems Research Institute. Available online: https://r-arcgis.github.io (accessed on 27 February 2018).

67. McNeil, M.A.; Webster, J.M.; Beaman, R.J.; Graham, T.L. New constraints on the spatial distribution and morphology of the Halimeda bioherms of the Great Barrier Reef, Australia. *Coral Reefs* **2016**, *35*, 1343–1355.

68. Miyamoto, M.; Kiyota, M.; Murase, H.; Nakamura, T.; Hayashibara, T. Effects of Bathymetric Grid-Cell Sizes on Habitat Suitability Analysis of Cold-water Gorgonian Corals on Seamounts. *Mar. Geodesy* **2017**, *40*, 205–223.

69. Li, D.; Tang, C.; Xia, C.; Zhang, H. Acoustic mapping and classification of benthic habitat using unsupervised learning in artificial reef water. *Estuar. Coast. Shelf Sci.* **2017**, *185*, 11–21.

70. D'Antonio, N.L.; Gilliam, D.S.; Walker, B.K. Investigating the spatial distribution and effects of nearshore topography on Acropora cervicornis abundance in Southeast Florida. *PeerJ* **2016**, *4*, e2473.

Geosciences **2018**, *8*, 94

71. Mark, D.M.; Aronson, P.B. Scale-dependent fractal dimensions of topographic surfaces: An empirical investigation, with applications in geomorphology and computer mapping. *J. Int. Assoc. Math. Geol.* **1984**, *16*, 671–683.

72. Perron, J.T.; Kirchner, J.W.; Dietrich, W.E. Formation of evenly spaced ridges and valleys. *Nature* **2009**, *460*, 502–505.

73. Deng, Y.X.; Wilson, J.P. The Role of Attribute Selection in GIS Representations of the Biophysical Environment. *Ann. Assoc. Am. Geogr.* **2006**, *96*, 47–63.

74. Brasington, J.; Vericat, D.; Rychkov, I. Modeling river bed morphology, roughness, and surface sedimentology using high resolution terrestrial laser scanning. *Water Resour. Res.* **2012**, *48*, doi:10.1029/2012WR012223.

75. Evans, J. *Straightforward Statistics for the Behavioral Sciences*; Brooks/Cole Publishing Company: Pacific Grove, CA, USA, 1996.

geosciences

MDPI

Article

Geomorphometric Characterization of Pockmarks by Using a GIS-Based Semi-Automated Toolbox

Joana Gafeira [1,*] , Margaret F. J. Dolan [2] and Xavier Monteys [3]

1   British Geological Survey (BGS), The Lyell Centre, Research Avenue South, Edinburgh EH14 4AP, UK
2   Geological Survey of Norway (NGU), Postal Box 6315 Torgarden, NO-7491 Trondheim, Norway;
     Margaret.Dolan@ngu.no
3   Geological Survey Ireland (GSI), Beggars Bush, Haddington Road, Dublin 4, Ireland; Xavier.Monteys@gsi.ie
*   Correspondence: jdlg@bgs.ac.uk; Tel.: +44-131-650-0422

Received: 8 March 2018; Accepted: 24 April 2018; Published: 27 April 2018

Abstract: Pockmarks are seabed depressions developed by fluid flow processes that can be found in vast numbers in many marine and lacustrine environments. Manual mapping of these features based on geophysical data is, however, extremely time-consuming and subjective. Here, we present results from a semi-automated mapping toolbox developed to allow more efficient and objective mapping of pockmarks. This ArcGIS-based toolbox recognizes, spatially delineates, and morphometrically describes pockmarks. Since it was first developed, the toolbox has helped to map and characterize several thousands of pockmarks on the UK continental shelf, especially within the central North Sea. This paper presents the latest developments in the functionality of the toolbox and its adaptability for application to other geographic areas (Barents Sea, Norway, and Malin Deep, Ireland) with varied pockmark and seabed morphologies, and in different geological settings. The morphometric characterization of vast numbers of pockmarks allows an unprecedented statistical analysis of their morphology. The outputs from the toolbox provide an objective, quantitative baseline for combining this information with the geological and oceanographical knowledge of individual areas, which can provide further insights into the processes responsible for their development and their influence on local seabed conditions and habitats.

Keywords: pockmarks; automated-mapping; ArcGIS; Glaciated Margin; North Sea; Malin Basin; Barents Sea

1. Introduction

First reported by offshore Nova Scotia [1], pockmarks are depressions formed in soft sediments at the seabed by fluid flow processes. Since then, the known occurrences of pockmarks have increased dramatically. The rise of the documented occurrences of pockmarks is largely thanks to technical improvements in marine geophysical survey equipment, particularly of multibeam echo-sounders (MBES), along with a more widespread use of such data for seabed mapping. Pockmarks have been found worldwide, from water depths of less than 10 m in estuaries (e.g., [2]) to over 3000 m in offshore canyons (e.g., [3]). However, the distribution of known pockmarks is currently still geographically biased, with more occurrences reported in economically developed areas of the world where high-resolution geophysical surveys have been undertaken.

Although the first accounts described these features as concave crater-like depressions [1,4], the complexity and diversity of morphologies possible have gradually become evident. Pockmarks have sizes recorded across four magnitudes, although most are between 10 and 200 m diameter and 1–25 m deep [5]. They can occur in both random and non-random distributions controlled by the underlying geology due to, for example, the presence of faults (e.g., [6]) or buried channels (e.g., [7]). The variability of size, spatial distribution, and geometry results from their development depending on a variety of parameters such as fluid type, flow fluxes, thickness and nature of near bottom sediments, underlying structure, and lithology.

Most of the initial descriptions of these features were based on low penetration seismic profiles (mainly Boomer system) and/or from sidescan sonar data (e.g., [1,4,8]). Presently, pockmarks are predominantly mapped manually using seabed digital terrain models (DTM) created from MBES data (e.g., [6]), generally a time-consuming task. In addition, delineating their individual boundaries is subjective and consistency of criterion is hard to achieve. To address this, the British Geological Survey (BGS) developed a semi-automated mapping toolbox. This ArcGIS-based BGS Seabed Mapping Toolbox recognizes, spatially delineates, and morphometrically describes seabed features including pockmarks [9], coral mounds [10], and other confined features. The toolbox is embedded in ESRI® ArcGIS, a geographic information system (GIS) widely used in the field of marine geology, allowing users to work in a familiar and integrated mapping environment. Furthermore, the scripts developed use standard ESRI® algorithms that increase the clarity of the steps taken during delineation and characterization processes. With this approach, human interaction and expert knowledge is still part of the mapping process but is limited to restricting criteria for feature mapping. This allows multiple mapping exercises to be performed with the same criterion, improving comparisons across different areas or quantification of seabed changes over time. The tools also allow an extensive morphological characterization of the mapped features with a fraction of the effort that would be required using manual techniques.

The consistent characterization of vast numbers of pockmarks with multiple morphological characteristics allows an unprecedented statistical analysis of their morphology. Combining this statistical analysis with the geological and oceanographic knowledge of individual areas provides insights into the processes responsible for their development and the influence of local seabed conditions. The application of this method to different datasets, over a wider range of water depths, seabed sediments types, and geological settings, as reported in this study, significantly increases the understanding of the formation, evolution, and preservation of these common, but still poorly understood, seabed features.

This work is the result of a semi-formal collaboration, established in 2015, between national seabed mapping programmes in Norway (MAREANO, www.mareano.no), Ireland (INFOMAR, www.infomar.ie), and the UK (MAREMAP, www.maremap.ac.uk). This collaboration has facilitated the exchange of data and methods for the development of various aspects of geological mapping of the seafloor. Mapping of pockmarks was quickly identified as an objective of common interest to the geological surveys within the three mapping programmes, and one for which quantitative, objective methods were not yet readily available. Testing this mapping approach originally developed for UK seabed data in other geological settings data from the Barents Sea (Norway) and Malin Basin (Ireland) will lead to further developments of the toolbox. In this paper, we explore the applicability of this practical and effective mapping approach, fully integrated within the most used GIS software, and illustrate the potential of the use of morphometric parameters to characterize pockmarks.

2. Materials and Methods

2.1. Study Areas and Datasets

A total of 20 MBES datasets were used in this study. These comprise: (1) 18 site survey datasets from the North Sea (UK) held by the British Geological Survey (BGS) and forming part of the MAREMAP data repository; (2) a dataset from the Malin Basin (Ireland) acquired by INFOMAR; and (3) a dataset from the Barents Sea (Norway) acquired by the MAREANO programme (Figure 1).

The selected datasets represent not just a diversity of geological settings, but also a range of data resolutions and quality, and present different mapping challenges. A resolution range from 1 to 10 m and pockmarks from 20 m to almost 800 m wide provided an opportunity to assess the impact of cell size and feature size in the mapping results. The dataset from Malin Deep provided a chance to test the impact of both regional topography and the presence of MBES acquisition artefacts on the delineation and characterization of pockmarks. The Barents Sea example, where pockmarks are particularly prolific, allowed us to explore the efficiency of the tools where vast numbers of seabed features are present.

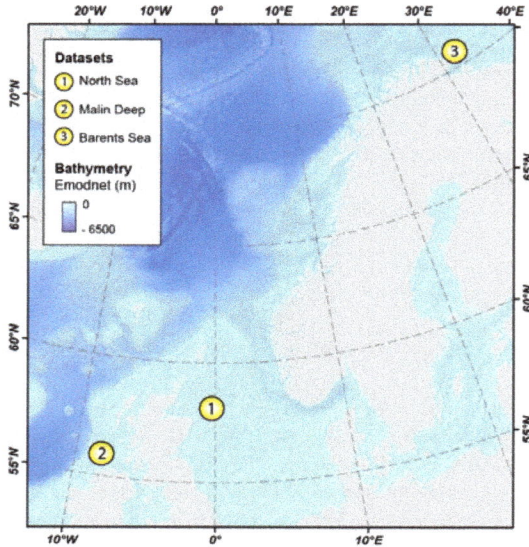

Figure 1. Location map of the three study areas used in this work. Base layer: EMODnet Bathymetry map.

2.1.1. Witch Ground Basin–North Sea (UK)

In the northern part of the North Sea, especially across the Witch Ground Basin, pockmarks can be found in high numbers (Figure 2). The Witch Ground Basin is an extensive area of muds reaching water depths greater than 150 m. It was a depocentre for fine-grained sedimentation at the end of the Weichselian glaciation, when sediments were deposited very rapidly, creating a thick sequence of very soft muds [11]. Seismic profiles often show acoustic blanking within the shallow section of the Quaternary sequence underlying the Witch Ground Formation, which suggests that shallow gas is trapped at selected horizons [12]; such accumulations support the hypothesis that the pockmarks found in this basin were formed by gas escape at irregular intervals since deglaciation [13]. In addition, since the seabed in the Witch Ground Basin has remained essentially unchanged by erosion or sedimentation once the sea level stabilized after the last glaciation, the pockmarks represent the cumulative effects of gas-escape activity over a period of at least 8000 years [14].

The vast majority of the pockmarks mapped within the Witch Ground Basin are less than 3 m deep [9], with an area of approximately 2000 to 4000 m^2. In cross-section, these pockmarks are mainly V-shaped (Figure 2d), with a few of them being U-shaped and very rarely W-shaped with a degree of asymmetry, whereas in plan-view, they are generally circular to elongate. Besides the vast number of unit pockmarks, several unusually large pockmarks were found in the Witch Ground Basin area. The largest of them, and among the largest pockmarks known globally, is the western pockmark of the Scanner Pockmark Complex (Figure 2a,b) [15].

The BGS has gathered numerous MBES datasets over the last decades in UK territorial waters, especially from the North Sea. These include the 18 multibeam datasets used by [9] in their first study using the semi-automatic mapping approach (Figure 2). These datasets were collected for the Strategic

Environment Assessment programme (SEA programme) and data were collected as part of site surveys for exploration wells commissioned by different operators. Due to the purpose for which the data were acquired, these datasets tend to be of high resolution (of 2 m for most) but cover small areas, ranging from less than 5 km² up to 36 km². In total, the 18 datasets cover an area of 306 km².

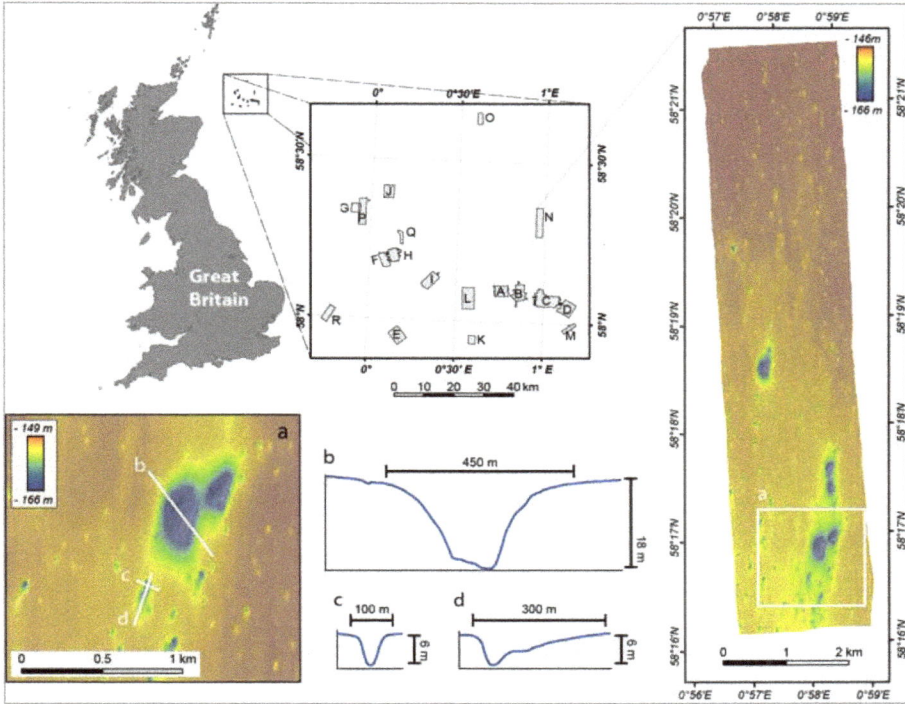

Figure 2. Location map of the 18 multibeam datasets (MAREMAP) in the North Sea and the bathymetric map of the SEA2 Box 4 dataset. See Table S1 for more information on individual survey areas. (**a**) Detailed view of the SEA2 Box 4 showing the Scanner Pockmark Complex (SPC) and surrounding pockmarks; (**b**) NW-SE profile across Western Scanner Pockmark; (**d**) WNW-ESE profile across a pockmark southwest of SPC, showing a symmetric transverse profile; (**c**) NNE-SSW profile along the same pockmark, showing a markedly asymmetric longitudinal profile.

2.1.2. Malin Basin (Ireland)

The Malin Shelf lies immediately north of Ireland and west of Scotland, with typical depths between 100 and 150 m. Seabed morphology is rocky and irregular in the east, while to the west, it is relatively sandy and smooth. The Malin shelf shallow geology is characterized by glacial diamictons, muds, and sands [16]. The Malin Deep pockmark field lies in the outer Malin shelf, approximately 70 km offshore northwest of the Malin Head. The area covers approximately 1000 km², extending from 7°45′ W to 8°20′ W and from 55°45′ N to 56° N (Figure 3). Water depths range from 140 m to 182 m, in the Malin Deep (c. 55°55′ N; 8°14′ W). The pockmark field lies in a basin characterized by a smooth and soft seabed composed of fine-grained marine sediments, ranging from fine sands to silts. The thickness of the Quaternary deposits varies N-S from 175 to 125 m [17]. Both acoustic and electromagnetic evidence indicates the presence of fluids within these deposits [18].

Geochemical analysis of sulphate profiles indicates that gas from the shallow reservoir has been migrating upwards [18].

The Malin Deep dataset covers an area of c. 865 km^2 (Figure 3). The Geological Survey of Ireland and the Marine Institute acquired this dataset in 2003, as part of the Irish National Seabed Survey (INSS), the precursor to the INFOMAR mapping programme. Data were acquired on-board the R.V. Celtic Explorer using a Kongsberg-Simrad EM1002 multibeam echosounder, with an operational frequency of 93–98 kHz. Bathymetric data cleaning was performed on-board and statistical analysis of the data indicates a vertical accuracy of <40 cm across the region. Resulting bathymetric terrain models were gridded at 5 m× 5 m.

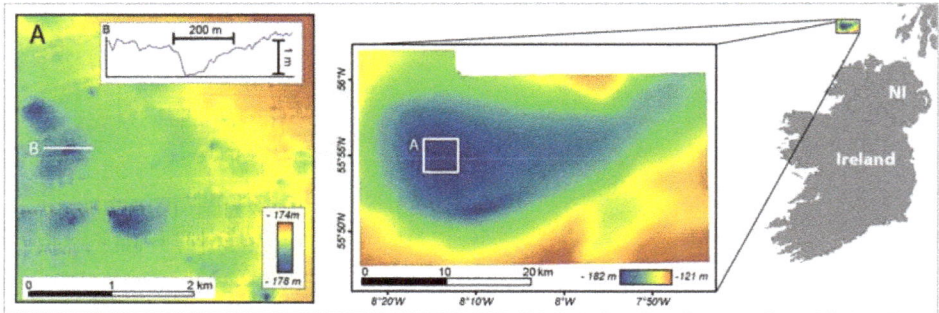

Figure 3. Location map of the Malin Deep dataset (Ireland) and bathymetric map obtained by INFOMAR. (**A**) Detailed view of the central area; (**B**) Bathymetric profile across of the pockmark indicated in (**A**).

2.1.3. Barents Sea (Norway)

Pockmarks are a frequent occurrence in the Barents Sea, occurring far beyond the area investigated in this study. The distribution and likely origins of the pockmarks in the SW Barents Sea and Finnmark fjords, northern Norway, have been recently documented as part of a broad-scale study by [19], where the study area investigated here falls under their description for the "Easternmost Norwegian sector". The study area considered here was classified by [19] as having a high density (300–800 per km^2) of pockmarks; however, the individual pockmarks were not delineated in that study. The authors showed that in the Barents Sea, pockmarks occur in areas where soft glaciomarine and marine sediments were deposited after the ice margin retreated [19]. The distribution of pockmarks thereby reflects the distribution of soft, fine-grained postglacial deposits in the SW Barents Sea. They also conclude that the pockmarks formed due to the melting of gas hydrates. It is suggested that this process probably started c. 14,500 cal. years ago, after the ice cap had melted and the bottom water temperature and thus the seabed temperature had increased due to the inflow of warm Atlantic water.

The Barents Sea dataset provided by MAREANO covers an area of 100 km^2 near the Norway-Russian border (Figure 4). The data were collected by MMT under contract to the Norwegian Hydrographic Service as part of the MAREANO seabed mapping programme in 2011 using a Kongsberg EM710 MBES on the survey vessel M/V Franklin. Data are gridded at a 5 m resolution, a standard output for MAREANO bathymetric mapping. The data are of good quality and do not exhibit any significant artefacts from data acquisition or processing.

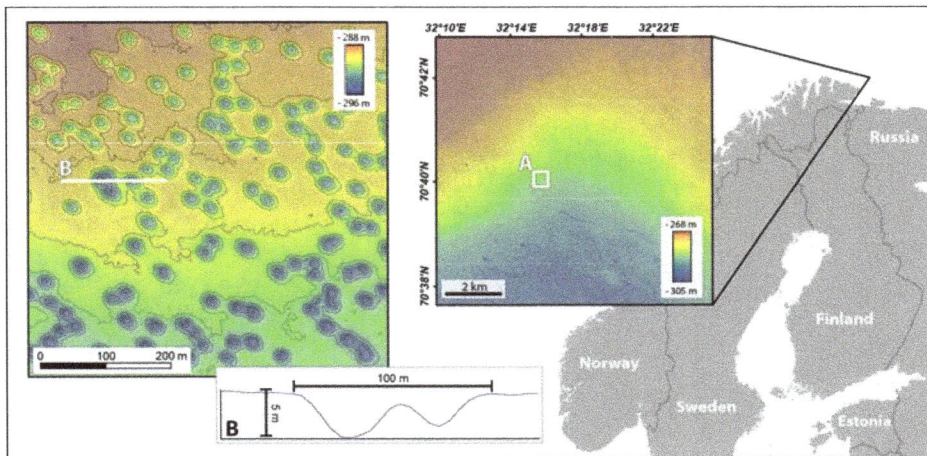

Figure 4. Barents Sea dataset (Norway). Location map of the Barents Sea dataset (Norway) and bathymetric map obtained by MAREANO. (**A**) Detailed view of the bathymetric map (contour lines every 1 m) and localization of profile (**B**), across two pockmarks.

2.2. Manual Mapping

Marine geological mapping methods have evolved substantially over recent decades, alongside diversification in the uses for this mapping. However, manual mapping of seabed features still represents a huge component of the effort to map and understand the seabed. Almost invariably, manual mapping of seabed features will involve the use of a GIS. It allows the creation, visualization, and analysis of DTMs, hillshade (from one or more directions), and several other rasters derived from the bathymetry data (e.g., slope, aspect, and curvature). GIS also provides a convenient platform for manual, expert-driven digitization of seabed features, based on the analysis of multiple layers of information. The definition of the limits of the geomorphic features is based on expert judgement complemented with classification schemes, but is often subjective.

The example of manual mapping of pockmarks included in this study is based on the 5 m DTM and slope maps derived from the Malin Basin dataset. Seabed depressions shallower than 0.5 m were dismissed, as this approaches the vertical accuracy of the data (c. 40 cm). This area is particularly challenging due to the artefacts present in the dataset.

2.3. Pixel-Based Calculation of Terrain Attributes as a Basis for Semi-Automated Mapping

Pixel-based analyses of bathymetric DTMs are often used to produce derived terrain attributes that serve as an aid to the manual digitization of features (as in the Malin Deep region). However, these derived terrain attributes can also be used for automatic pixel-based mapping, based on expert-defined threshold values before manual fine-tuning.

Here, we explore some terrain attributes, derived from pixel-based analyses, related to slope, curvature, and relative position in order to examine their potential for delineating pockmarks. Other terrain attributes relating to orientation and terrain variability may be useful for describing the nature of the pockmarks but are not so directly relevant to their delineation.

The pixel-based terrain analysis methods presented as part of this study were conducted on SEA2 Box4 of the Witch Ground Basin datasets. This area is particularly challenging due to the different shapes and sizes of the pockmarks. Therefore, it is a good example for examining how well various methods are able to delineate the pockmarks. The methods and analysis scales used (Table 1) were selected to provide examples of the various approaches rather than conducting an exhaustive analysis of the area. A selection of the results is presented in Section 3.1, where results from the BGS Seabed Mapping Toolbox are shown for comparison.

Table 1. Summary of pixel-based approaches to pockmark mapping in SEA2 Box 4, Witch Ground Basin.

Terrain Attribute	Analysis Scale Used (Pixel $n \times n$ Neighbourhood, Except BPI (BTM Toolbox)	Method
Single scale slope	$n = 3$	Standard 3×3 curvature from ArcGIS 10.4.1 Spatial Analyst [20]
Multiple scale slope	$n = 5, 9, 15$	GRASS module r.param.scale via QGIS 2.18.11 based on the methods of [21].
Multiple scale Bathymetric Position Index (BPI)	Inner radius 1, 3 Outer radius 5, 15	Fine_scale BPI using the BTM toolbox [22]. Annulus neighbourhood with selected inner and outer radius. Produces integer output.
Multiple scale BPI (modified)	$n = 5, 9, 15$	This modified version of BPI implemented in ArcGIS Spatial Analyst raster calculator produces a floating-point grid. The computation is similar to that presented by [23] but uses a rectangular neighbourhood to facilitate using the same neighbourhood as other analyses.
Curvature	$n = 3$	Standard, 3×3 curvature from ArcGIS 10.4.1 Spatial Analyst [24].
Multiple scale minimum curvature Multiple scale profile curvature	$n = 5, 9, 15$ $n = 5, 9, 15$	Multiple scale curvature calculated using GRASS module r.param.scale via QGIS 2.18.11 which is based on the methods of [21].
Multiple scale feature classification	$n = 9, 15, 27, 51$	Multiple scale feature classification calculated using GRASS module r.param.scale via QGIS 2.18.11 from r.param.scale [21].

2.4. BGS Seabed Mapping Toolbox

The BGS-developed tools in the BGS Seabed Mapping Toolbox run individual Python scripts that use a sequence of pre-existing ArcGIS geoprocessing tools. The toolbox includes (1) data preparation tools; (2) feature delineation tools; and (3) characterization tools.

2.4.1. Data Preparation

The basic input to the BGS Seabed Mapping Toolbox is a DTM, either obtained from a multibeam echosounder dataset or from another dataset that can be used to generate a DTM of the seabed or buried surface (e.g., 3D seismic—[25]). In datasets strongly affected by artefacts, when using the semi-automated mapping toolbox, pockmarks may be incorrectly delineated and spurious values may be captured during their characterization. This was the case for the dataset available for the Malin Basin, where acquisition artefacts markedly affect this dataset (Figure 5). Artificial vertical reliefs (corrugations) of up to 50 cm are detected systematically across the dataset due to tidal shifts in the lines overlap and vessel motion-related artefacts across the swath. These are comparable in magnitude to the vertical relief of some of the pockmarks. For that reason, it was necessary to smooth the initial bathymetric surface. The ArcGIS *Focal Statistics* tool was used to smooth the original surface. The smoothed surface was then used as the input surface for both the pockmark delineation and characterization tools. The *Focal Statistics* tool performs a neighbourhood operation that computes an output raster where the value for each output cell is a function of the values of all the input cells that are in a specified neighbourhood around that location.

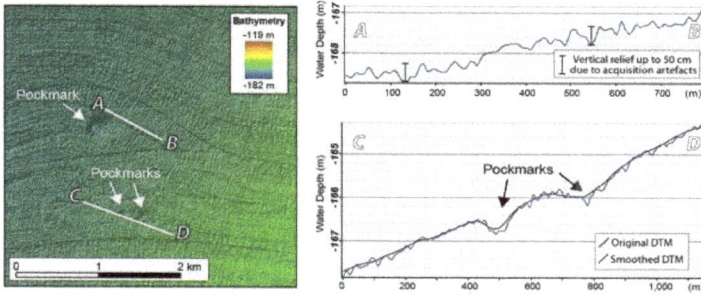

Figure 5. Left: Detail of the Malin Basin dataset, illustrating the impact of artefacts on the dataset. Right: (Top) Bathymetric profile from (**A**) to (**B**). (Down) Profile from (**C**) to (**D**) across the original dataset and the smoothed bathymetry.

The overall regional morphology and the presence of overlapping morphological features can also affect the ability of the delineation tools to correctly map pockmarks. This can be addressed by using the BGS-developed *Filter-based Clip Tool*. This tool automatically identifies and clips areas of special interest where seabed features are likely to be present. The use of this tool is particularly useful in a setting like the Malin Basin (Figure 6), where the pockmarks occur within a regional basin. During the first delineation, the entire small basin was erroneously mapped as a pockmark since it is also a confined depression (Figure 6A). The *Filter-based Clip Tool* identifies and outlines areas of vertical relief changes, which are then used to clip the original DTM excluding the data from zones of smooth bathymetry (Figure 6B–D). Using the clipped DTM as the input to the delineation tool, this tool is now capable of delineating individual pockmarks within the small basin (Figure 6E).

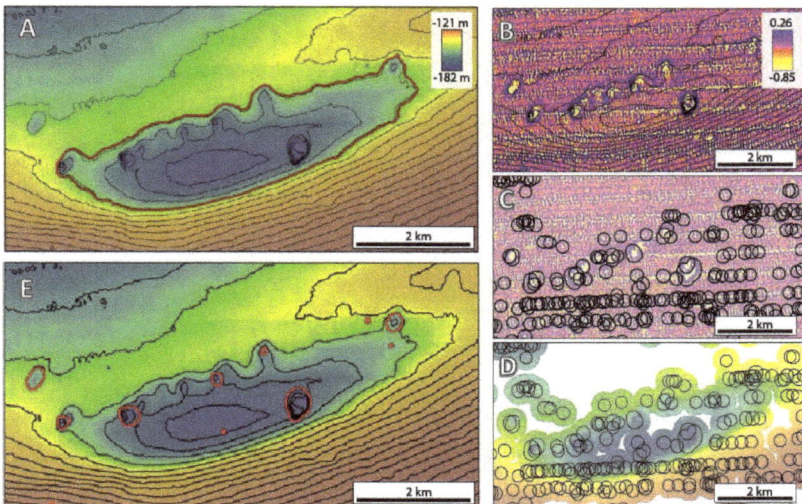

Figure 6. (**A**) Example of a small basin initially mapped by the Delineation Tool as a confined depression (contour lines every 1 m); (**B**) Detail of the raster obtained by applying a High Pass filter followed a Low Pass filter; (**C**) Black polygons delineate areas of negative values within the filtered dataset (in yellow), which can correspond to either pockmarks or artefacts; (**D**) Clipped bathymetric dataset based on the delineated polygons; (**E**) The result of running the Delineation Tool with a clipped bathymetry.

2.4.2. Feature Delineation

As explained in [9], pockmarks are generally represented by confined depressions in a DTM, therefore it is possible to employ hydrological algorithms such as the Fill function in ArcGIS Spatial Analyst to define what would be the lowest elevation on the rim of a sink depression (i.e., the overflow point if the depression was being filled). In fact, "Fill" is one of the key algorithms of a sequence of steps used by the *Feature Delineation [Bathy]* tool as described by [9] and is used here to delineate pockmarks. Experience has shown that in certain regional settings, using the rim of the confined depression related to the presence of the pockmark may result in an underestimation of its size, especially in areas with steep slopes. To address this issue, the alternative *Feature Delineation [Derived]* tool was created, which can use the derived terrain attribute Bathymetric Position Index (BPI) calculated from the bathymetry data using *Benthic Terrain Modeler* toolbox [22] as the input, instead of the bathymetry data. BPI measures whether a certain location is higher or lower than the surrounding seabed by comparing the depth of each pixel with the mean depth of neighbouring pixels within a user-defined neighbourhood (inner and outer radii). The BPI value obtained for any pixel, however, depends on both the regional setting and the neighbourhood used in the BPI calculation. The values of BPI are also sensitive to data resolution. This alternative approach using BPI rather than bathymetry therefore introduces unavoidable sources of inconsistencies on the criterion used to map pockmarks in different study areas and is only recommended for use in more local studies.

Due to the limitations of the BPI approach mentioned above, in this study involving data from different regions, the pockmark delineation was done directly from the bathymetric data. Five values must be defined to run the *Feature Delineation [Bathy]* tool; these are the *Cut-off Vertical Relief*, *Minimum Vertical Relief*, *Minimum Width*, *Minimum Width/Length Ratio*, and *Buffer Distance*. The *Cut-off Vertical Relief* defines the contour line that will be used to delineate the features. The *Minimum Vertical Relief*, *Minimum Width*, and *Minimum Size Ratio* define which features will be mapped; only the features that present dimensions above the specified thresholds will be delineated. The *Cut-off Vertical Relief* and the *Minimum Vertical Relief* can be the same value. The defined *Buffer Distance* compensates for the fact that the delineation process is based on the feature's internal contour line corresponding to the *Cut-off Vertical Relief* threshold. This parameter should approximate the distance, in plan-view, from the internal contour line delineated based on the *Cut-off Vertical Relief* to the actual rim of the feature. The greater the value of *Cut-off Vertical Relief*, the greater the *Buffer Distance*. Figure 7 illustrates the different mapping results obtained by choosing different *Cut-off Vertical Relief* and *Buffer Distance* values.

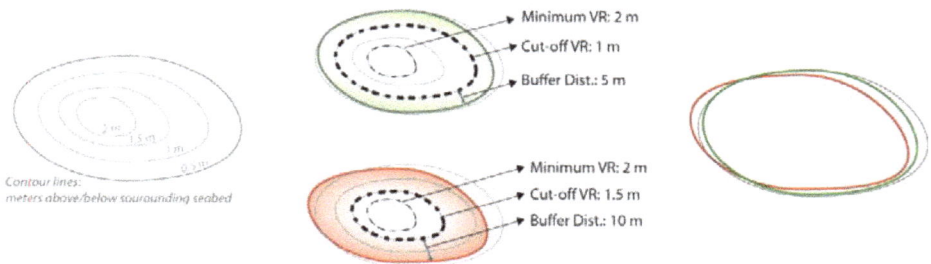

Figure 7. Example of the delineated outline depending on the choice of the different values of Cutoff Vertical Relief (cut-off VR) and Buffer Distance. Note that the green outline better represents the pockmarks' limits than the red outline.

The North Sea pockmarks were mapped as part of [9] using the first version of the *Feature Delineation [Bathy]* tool, which did not require the *Minimum Vertical Relief* and used *Minimum Area* instead of the *Minimum Width* as one of the threshold values. The *Minimum Width* later replaced the

Minimum Area threshold because it was found to be more user-friendly and easier to relate to the resolution of the dataset used as input. The thresholds used to map the North Sea pockmarks were *Cut-off Vertical Relief* (then referred to as *Minimum Depth*) of 0.5 m and *Minimum Area* of 100 square metres. Table 2 presents the respective values used for Malin Deep and Barents Sea study areas.

Table 2. Table of thresholds used for three study areas: Cut-off Vertical Relief (Cut-off VR), Minimum Vertical Relief (Min VR), Minimum Width (Min Width), Minimum Size Ratio (Min W/L), and Buffer Distance. Note that North Sea's pockmarks were mapped with the first version of the delineation tool that used a different set of thresholds.

Study Area	Cut-off VR	Min VR	Min Width	Min W/L	Buffer Distance
North Sea	0.5 m	—	—	0.2	7.5 m
Malin Basin	0.2 m	0.2 m	40 m	0.2	80 m
Barents Sea	0.75 m	0.75 m	4 m	0.2	10 m

The output of the tool is a polygon shapefile that delineates the mapped features (pockmarks). The shapefile attribute table contains the following fields (1) Area; (2) Perimeter; (3) VRelief; (4) MBG_Width; (5) MBG_Length; (6) MBG_Orient; and (7) MBG_W_L. Area and Perimeter describe the geometry of each delineated feature. The VRelief provides the vertical relief measure for each delineated feature. The MBG_Width, MBG_Length, and MBG_Orient describe the Minimum Bounding Geometry (MBG) envelope that contains each delineated feature. MBG_W_L describes the aspect ratio of these envelope polygons. Jorge et al. [26] assessed the use of different automated methods to measure longitudinal bedform's morphometry. Although these authors focus on subglacial positive-relief bedform (e.g., drumlins), their conclusions are still relevant to the morphometry measurements of pockmarks. They established that the MBG approach provides the most suitable measurements, from the tested methods, for both Orientation and Length, only showing a wider range of errors for the measurement of the Longitudinal Asymmetry, which is not used in this study.

The output shapefiles from the delineation tool require visual assessment as part of a semi-automatic workflow. Visual assessment of the polygons can be performed by overlaying the generated shapefile onto both the original bathymetric data and derived surfaces, such as the slope map. This allows a check on the mapping results and assessment of the need to manually edit sporadic polygons and/or to add features that were missed by the automated method. Additionally, the visual assessment can be complemented by an analysis of the values reported in the table of attributes.

2.5. Morphometric Analysis

In addition to the morphological attributes extracted by the characterization tool, extra morphological size and shape ratios can also be calculated. This includes the Vertical Relief to Area (VR/A) ratio. To characterize the profile of the pockmarks, we also define a Profile Indicator (PI) by the following equation:

$$PI = \frac{MinWD - MeanWD}{MinWD - MaxWD} \tag{1}$$

This morphologic PI ratio can help to distinguish between depressions with a V-shape profile, more typical of single pockmarks, and the depressions with a U-shape profile, more common on complex pockmarks. V-shape pockmarks will generate lower values of PI compared to the pockmarks with a U-shape (Figure 8).

Figure 8. Schematic representation of two different type of pockmarks profiles and how the position of the mean will lead to lower or higher PI values.

3. Results

3.1. Witch Ground Basin—North Sea

A total of 4146 pockmarks, deeper than 50 cm, were mapped in the Witch Ground Basin. Here, we present the general trends and measures for the whole basin, but [9] summarizes the descriptive measurements for each of the individual site survey areas.

The highest pockmark density occurs within the survey site Roisin in the centre of the basin, with a pockmark density of almost 30 pockmarks per square kilometre (Figure 9). The density of pockmarks decreases from the centre of the basin where water depths exceed 150 m, to less than 5 per km^2 on the edge of the basin where water depths are around 120 m. However, more than the changes in water depth, the number of pockmarks seems to be controlled by the thickness of the very soft late glacial sediments within which the pockmarks are developed. This is greatest in the centre of the basin but thins towards the edge.

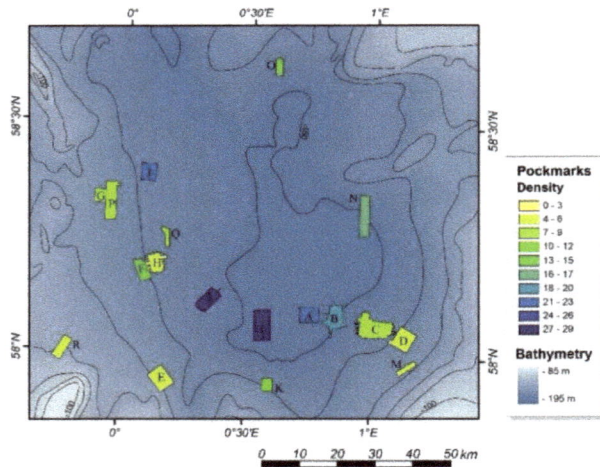

Figure 9. Pockmark density across the Witch Ground Basin. Survey sites colour coded according to the pockmark density observed. The regional bathymetry shows deeper areas (−150 m) in a darker blue and shallower areas (−90 m) in light blue [9].

The mean vertical relief of the pockmarks is 1.82 m, with most pockmarks between 1 and 2.4 m deep (Q1 and Q3, respectively). However, in SEA2 Box 4, there are five pockmarks deeper than 12 m and one of them reaches almost 18 m deep (Figure 10A). These unusually large pockmarks, in UK license block 15/25, have long been known as sites of active seepage [27–29]. None of the datasets exhibit a significant variation of vertical relief with water depth (Figure 10B).

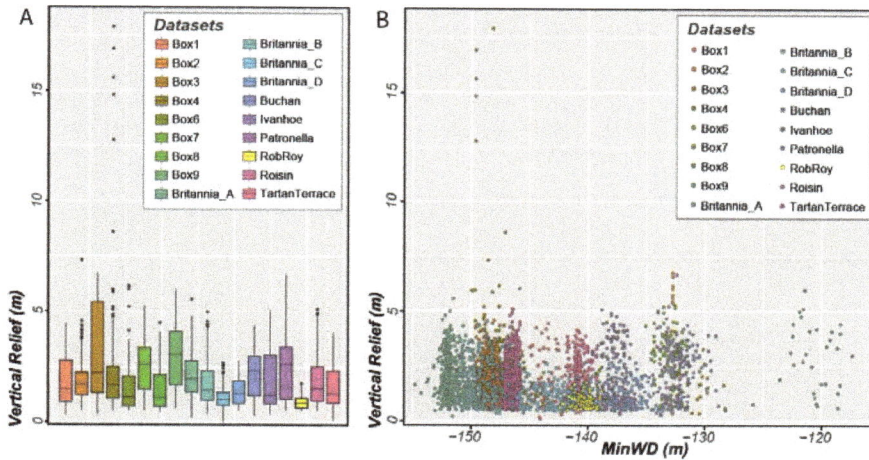

Figure 10. (**A**) Box plot of the pockmarks' vertical relief over individual study areas mapped in the Witch Ground Basin; (**B**) Vertical relief as a function of Minimum Water Depth.

The mean area of the pockmarks is 3222 m^2, with half of the pockmarks between 1960 (Q1) and 5385 m^2 (Q3). The pockmark area does not follow a trend from the centre to the edge of the basin. However, it does show a strong correlation between vertical relief and pockmark area (Figure 11). With the exception of the Rob Roy dataset, all the other datasets studied in the Witch Ground Basin show a marked trend where pockmarks with higher vertical reliefs have greater areas. The reason why the Rob Roy dataset does not present the same trend as the rest of the dataset will be discussed later.

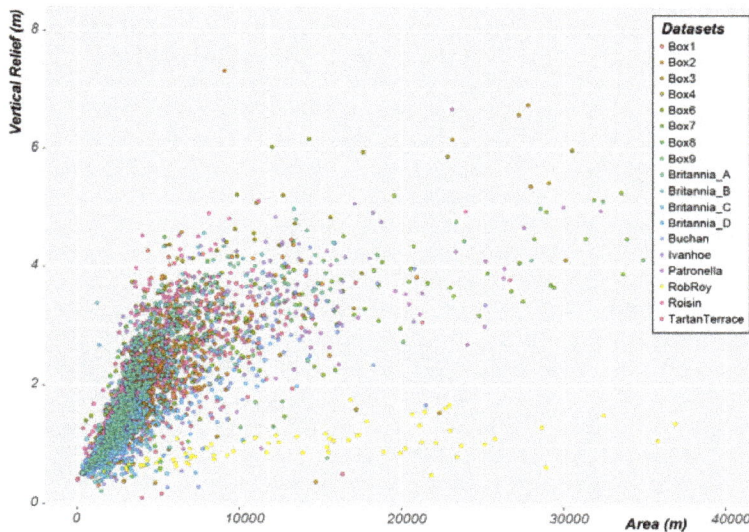

Figure 11. Vertical Relief as a function of Area in pockmarks with vertical relief lower than eight metres within the Witch Ground Basin. Note that the unusually large pockmarks existent in the Witch Ground Basin are not displayed in this figure.

Comparison between the Results from the BGS Semi-Automatic Approach and from the Pixel-Based Analysis

As shown in Figure 12, the *Feature Delineation [Bathy]* tool successfully delineates pockmarks across a range of sizes and with different sizes and shapes. Results from pixel-based analyses are more varied in their ability to delineate the various morphologies of the pockmarks in this area. Some examples are shown in Figure 13.

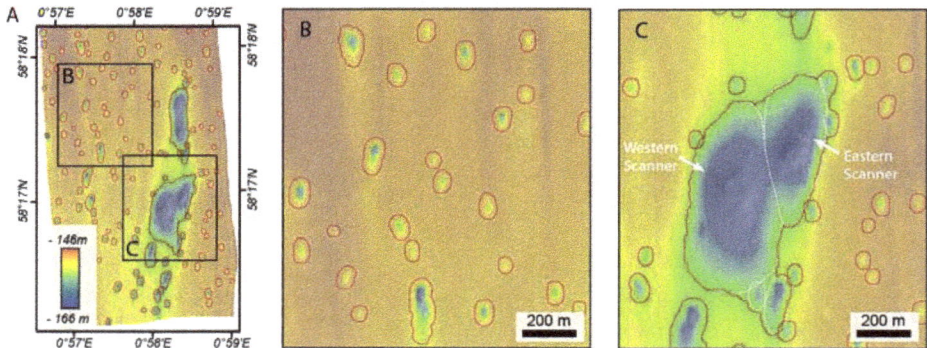

Figure 12. Examples of the initial results from BGS toolbox for SEA 2 Box 4—the pockmarks delineated by the tool are shown in red over a colour-shaded relief map from the southern part of Box 4. Resolution 6 m. Note that manual editing was later required to split the Scanner Pockmark Complex (SPC) into Western Scanner, Eastern Scanner, and adjacent pockmarks.

The first terrain attribute tested was slope, initially tested using standard 3×3 pixel analysis and Horn's algorithm [20] in ArcGIS' Spatial Analyst. Using a colour ramp classified by natural breaks (Figure 13A), we see that all the pockmarks delineated by the BGS tool are highlighted by the slope map. The difficulty lies in finding a slope cut-off value that would delimit pockmarks of different morphologies, and in representing the entire pockmark. For instance, some pockmarks in SEA2 Box 4 present an asymmetric profile (Figure 2c), with steep northern slopes facing very elongated southern slopes, and a delineation based on slope would tend to delineate a crescent shape whilst missing the areas of very gentle slope. Testing larger analysis windows and using methods for generating multiple scale slopes (Table 1) has shown that increasing the analysis window is of limited value as the extent of the pockmarks can become overestimated. As with all raster outputs of pixel-based analyses, it is important to be conscious of the influence of colour ramp choice on visual interpretation.

Both marine geoscientists and marine biologists have used BPI widely [10,30]. BPI can be computed using the BTM toolbox [22]. For this study, only fine-scale BPI was calculated, testing the inner and outer radii indicated in Table 2. Figure 13B shows fine-scale BPI with inner radius 1 (i.e., 2 m) and outer radius 15 (i.e., 30 m). The areas with negative BPI give a good delineation of the larger pockmarks, including the base. However, using this neighbourhood, we fail to capture the smaller, shallower, pockmarks, due to the integer rounding inherent in the tool. Decreasing the outer radius to try to capture the smaller pockmarks is not successful and results in fewer pockmarks being detected overall. Increasing the inner radius has a negligible effect on the delineation.

The small, shallow pockmarks can be detected with a modified version of BPI without integer rounding (Figure 13C, Table 2). This result is more successful at highlighting the small pockmarks as well as most of the larger ones, but it also detects artefacts in the MBES data, which are of a similar magnitude to the BPI values within parts of the pockmarks.

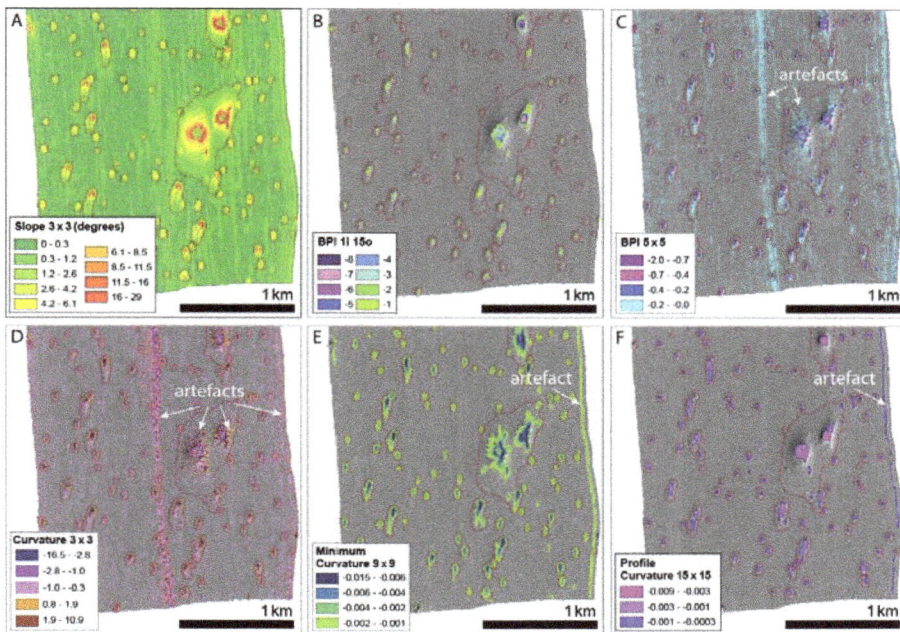

Figure 13. Examples of pixel-based terrain attributes from the southern part of SEA2 Box 4. Background data—greyscale shaded relief image of bathymetry data. Pockmark features delineated by the BGS Mapping toolbox are shown in red for reference. All terrain attributes except (**B**) are coloured according to natural breaks classification and no colour is used for irrelevant classes, where appropriate, to aid visualisation. (**A**) Slope calculated in ArcGIS Spatial Analyst using standard 3 × 3 rectangular neighbourhood; (**B**) BPI (integer) using the BTM toolbox using annulus neighbourhood with an inner radius of 1 and an outer radius 5; (**C**) BPI (floating point) calculated using a 5 × 5 rectangular window—negative BPI values correspond to negative features (depressions) in the seabed, including pockmarks, positive features (BPI value > 0) are not shown. Note that values in the range −0.2–0 (cyan) highlight numerous small artefacts in the data in addition to the shallower portions of pockmarks; (**D**) Curvature calculated using ArcGIS Spatial Analyst using standard 3 × 3 rectangular neighbourhood—values from −0.3 to 0.6 (close to flat) are not coloured. (**E**) Minimum Curvature calculated using a 9 × 9 rectangular neighbourhood—values greater than −0.001 are not coloured; (**F**) Profile curvature calculated using a 15 × 15 rectangular neighbourhood—values greater than −0.0003 are not coloured (flat and positive features). Note that visual interpretation of the results, and the extent to which features and artefacts show up is very dependent on the colour ramp used. These figures are intended as examples only.

Like BPI, measures of curvature also highlight positive and negative features of the terrain. We first tested standard, 3 × 3 curvature in ArcGIS' Spatial Analyst [24] with the results shown in Figure 13D. This, like Figure 13C, shows up artefacts in data in addition to pockmarks, but in this case, it is even more difficult to separate the artefacts and the pockmarks from the curvature values. It is clear that larger, alternate analysis scales are required that overlook the artefacts and find the pockmarks.

We have also tested two measures of curvature that can be generated at multiple scales in the GRASS module r.param.scale (Table 1). Minimum curvature (Figure 13E) should find the inflexion point in the bathymetry surface corresponding to the pockmark and is reasonably successful in capturing entire pockmarks spanning a range of sizes, although some artefacts are also highlighted.

Profile curvature [21] describes the rate of change of slope along a profile of the surface and can be useful in highlighting convex or concave slopes in the bathymetry surface. This appears to be one of the most successful pixel-based approaches to delineating the various sizes of pockmarks in the study area; however, we note that artefacts at the eastern edge of the dataset are also detected.

Various properties of curvature are combined in feature classification. We tested the feature classification from r.param.scale [21]. Here, we show only the 'pit' class output which identifies depressions in the surface and seems well matched to the detection of pockmarks (Figure 14). Experimenting with analysis windows at multiple scales, it is clear that the 9×9 analysis window fails to separate out pockmarks from artefacts in the data. A larger analysis window of 15×15 successfully detects the small-medium size pockmarks but fails to capture the larger ones. Consequently, we see that larger analysis windows of 27×27 and 51×51 cells find the larger and largest pockmarks, respectively, but overlook the small ones.

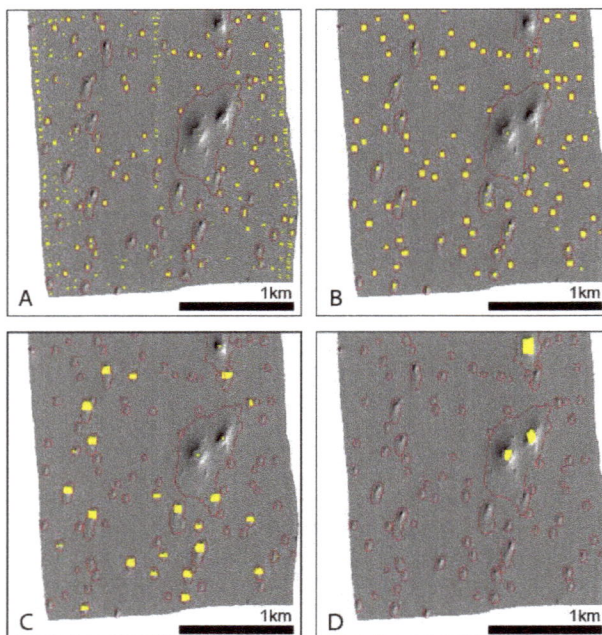

Figure 14. Examples of feature classification using different analysis window size. Background data - greyscale shaded relief image of bathymetry data. Pockmark features delineated by the BGS Mapping toolbox are shown in red for reference. (**A**) Pit features (yellow) from feature analysis using 9×9 rectangular neighbourhood; (**B**) Pit features (yellow) from feature analysis using 15×15 rectangular neighbourhood; (**C**) Pit features (yellow) from feature analysis using 27×27 rectangular neighbourhood; (**D**) Pit features (yellow) from feature analysis using 51×51 rectangular neighbourhood.

It appears that some of the pixel-based methods tested here can give good results where pockmarks are of a relatively uniform size. Whereas, multiple scale analysis can be employed to detect features of different sizes if required. It is challenging, however, to find a single scale of analysis and analysis type that detects all pockmarks and delineates the entire feature. Further, with the exception of the feature classification, the user must determine a suitable cut-off value in the terrain attribute (BPI, curvature etc.) to use as the limit of the pockmark. If such a value can be used, then these raster outputs of pixel-based terrain analysis can be used as a basis for conversion to vector features (shapefiles), should these be required for the application in question. Following conversion to polygon features, area attributes can be applied to the pockmarks. However, the areas may not be a consistent representation of all complete pockmarks due to the dependence on the raster cut-off value, which may well be a compromise between pockmarks of different morphometry. Manual editing of the polygons may therefore be a necessary expert-driven step in the mapping process. Pixel-based terrain analysis, based on the type of methods illustrated here, does not provide any estimate of pockmark depth or volume. Methods such as Geomorphons [31], which are designed to simultaneously identify landform elements across a wider range of scales, may be more suitable, although they face similar limitations in terms of the metrics they provide, unless combined with other methods.

3.2. Malin Basin

Due to the artefacts present in this dataset, it was necessary to smooth the bathymetry before applying the delineation tool (Section 2.5). As a consequence of this smoothing, the vertical relief of the pockmarks was most likely underestimated and shallow pockmarks were not detected, or where detected, their area may well be underestimated (Figure 15).

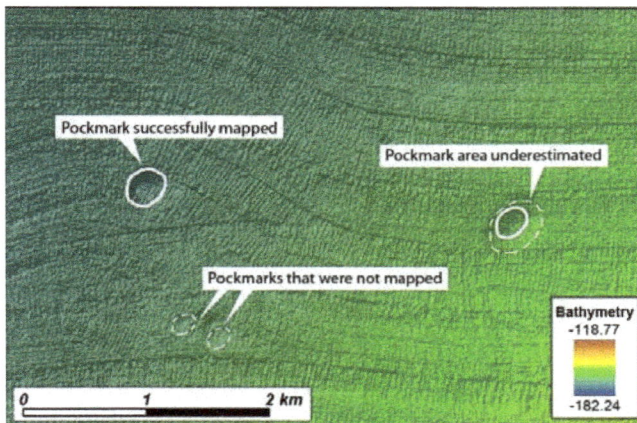

Figure 15. Detail of the original Malin Deep dataset and the mapping results based on the smoothed bathymetry.

Nevertheless, after smoothing and clipping the bathymetry using the *Filter-based Clip Tool*, the *Delineation Tool* identified 150 pockmarks with a vertical relief greater than 20 cm (Figure 16) within the Malin Basin. Pockmarks were mapped in water depths ranging from 126 to 177 m, but most of the pockmarks are found in water depths below 167 m (Figure 16B). The largest mapped pockmark is at least 5.5 m deep, but 75% of the pockmarks are shallower than 60 cm, and the mean vertical relief is 36 cm (Figure 17). We also remark that all the pockmarks with vertical relief higher than one metre occur at water depths deeper than 168 m (Figure 16A). The pockmark area varies from 2000 m^2 to almost 303,000 m^2, with a mean value of 32,073 m^2. Whereas, the pockmark length varies from the smallest pockmark with less than 50 m to the largest with 785 m long. They tend to be quite concentric, with a mean size ratio of 0.83.

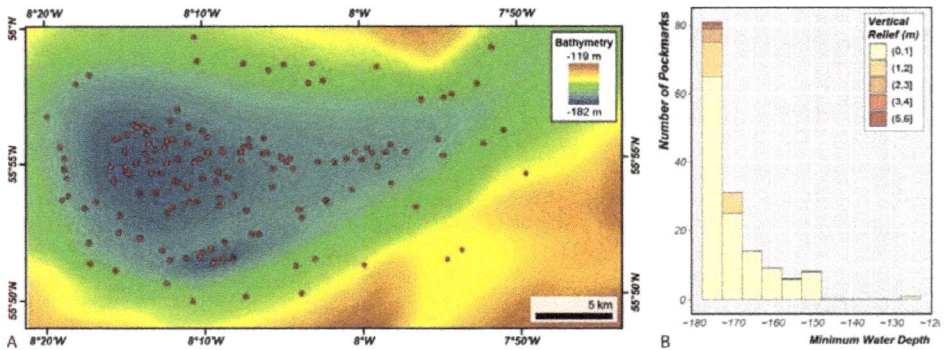

Figure 16. (**A**) Location map of the 150 pockmarks mapped with the BGS Seabed Mapping Toolbox (red dots); (**B**) Vertical Relief as a function of Area for the pockmarks mapped in the Malin Basin.

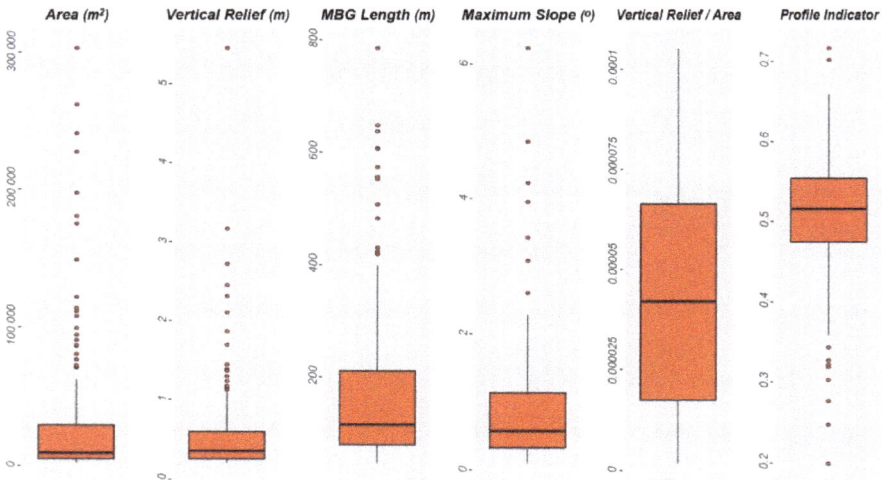

Figure 17. Boxplots for Area, MBG Length, and Vertical Relief for pockmarks mapped in the Malin Basin.

Comparison between the Results from the BGS Semi-Automatic Approach and from Manual Mapping

Previously, 214 depressions deeper than 0.5 m had been mapped manually [32]. When comparing the manual mapping with the automated mapping, it is evident that fewer pockmarks were identified by the *Delineation Tool*, especially in the steeper areas (Figure 18). Almost 80% of the pockmarks that were not mapped automatically are located flanks of the Malin Deep, where the smoothing of the data has a stronger impact on the automatic recognition of shallow pockmarks. However, within the centre of the basin, the delineation tool detected several pockmarks that were not manually mapped. Some of the missed pockmarks are deeper than 0.5 m and therefore would fulfil the requirements used for the manual mapping. This illustrates the ability of the automatic tools to recognise subtle features that can escape the attention of expert examination.

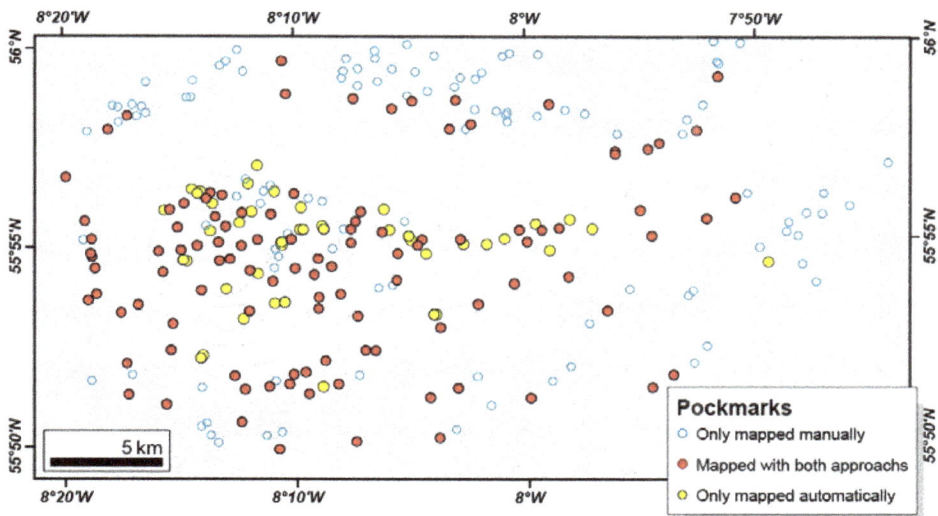

Figure 18. Malin Deep's pockmarks distribution based on both manual and semi-automatic mapping.

Based on the manual mapping, the pockmarks were described as being between 40 and 850 m wide, and up to 8.5 m deep, with a mean length of 124 m and mean vertical relief of 1.02 m [32]. These measurements present higher vertical relief than the values extracted by the semi-automatic method. We believe that this difference also results from the need to smooth, in this case, the bathymetry before applying the *Feature Delineation [Bathy]* tool.

3.3. Barents Sea

The Barents Sea dataset is characterized by the presence of vast numbers of pockmarks. These are the main topographic features present, but there are also a few iceberg ploughmarks, which can be a kilometre long. After applying the *Filter-based Clip Tool*, to avoid the erroneous delineation of ploughmarks as pockmarks, the *Feature Delineation [Bathy]* tool identified and delineated more than 35,000 pockmarks in two hours. These were then characterized by using the *Feature Description* tool.

The pockmarks are found in water depths ranging from 255 to 305 m, with half of all pockmarks located between 275 and 291 m. The mean pockmark vertical relief is 2.2 m, but some can be up to 7 m deep. The pockmark area varies from 538 m^2 to more than 11,000 m^2, but most are less than 1643 m^2 (Q3). Their length measure, using the MBG envelope, varies from just 25 m to more than 220 m. They tend to be concentric with a mean MBG_Width/MBG_Length ratio of 0.88. Figure 19 shows the relationship between vertical relief and area for the pockmarks mapped in the Barents Sea dataset. Almost 80% of the pockmark polygons followed a distinct distribution compared to the remaining pockmarks, with a higher VR:A ratio (Figure 19). By selecting these, it became evident in GIS that these polygons corresponded to the single pockmarks (Figure 19), whereas the polygons with a lower VR:A ratio corresponded to pockmarks with a more complex geometry and with multiple possible venting points.

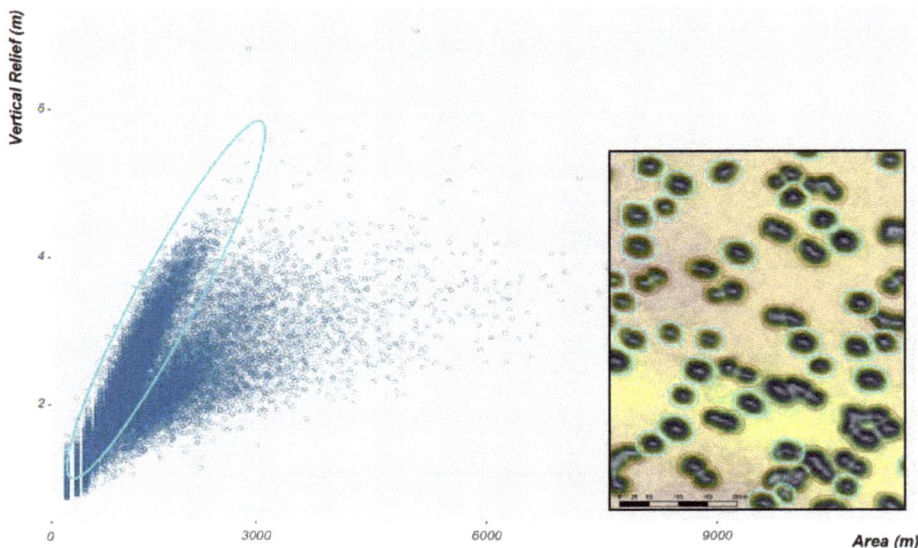

Figure 19. (**Left**) Vertical Relief as a function of Area in pockmarks mapped in the Barent Sea's dataset. Selection of the pockmarks with the higher ratio of VR/A (in cyan); (**Right**) Detail of the DTM showing the pockmark delineation obtained, with the polygons with the higher ratio of VR/A highlighted on the map view.

By separating these two populations, using the following function:

$$\text{Vertical Relief} = (0.0022 \times \text{Area}) - 1.25 \tag{2}$$

it was possible to study their morphologic characteristics separately. Table 3 shows some of the metrics extracted from these two types of pockmarks.

Table 3. Number of depressions mapped (N), mean area (μ Area), mean width (μ Width), mean length (μ Length), mean of the ratio between width and length (μ W/L), and vertical relief (μ VR) from the depressions classified as single and complex pockmarks, respectively.

	N	μ Area	μ Width	μ Length	μ W/L	μ VR
Single Pockmarks	28,014	1223	15.5	19.4	0.82	2.12
Complex Pockmarks	7223	2386	24.6	45.4	0.5	2.5

The mean area of the complex pockmarks is almost twice the mean area of the single pockmarks and the larger of the complex pockmarks can be almost one order of magnitude larger than the single pockmarks (Figure 20). However, the mean vertical relief of these features shows only a slight increase from 2.12 to 2.5, from the single to the complex pockmarks group, respectively (Figure 20). The mean value for the profile indicator for the complex pockmarks is 0.38, whereas the mean PI value for the single pockmarks is 0.33 (Figure 20). These values indicate that the single pockmarks will have profiles closer to the V-shape compared to the complex pockmarks.

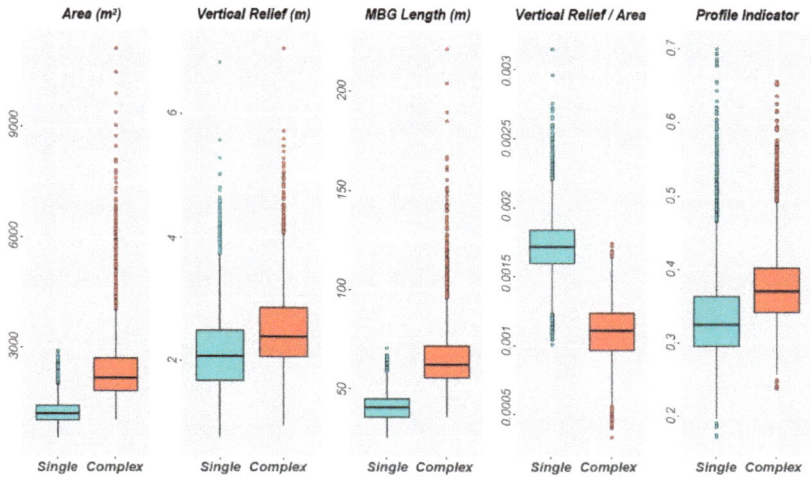

Figure 20. Box plot for the Area, Vertical Relief, MBG Length Vertical Relief/Area ratio, and PI measured for both the single and complex pockmarks mapped.

The stacked histogram in Figure 21A shows that, contrary to the other two regions, the pockmarks with the highest vertical relief tend to occur in deeper waters. This is not the result of a preferential occurrence of complex pockmarks in deeper waters as both types of pockmarks exhibit a trend of increase in vertical relief with water depth (Figure 21B). However, it should be noticed that changes in water depth in itself should not be the reason for this trend. Differences in the depositional environment linked to the water depth (e.g., thickness of soft, fine-grained postglacial deposits) are expected to be the primary parameter driving the observed increase of vertical relief with water depth. Other factors linked to the gas release could also be responsible for such a trend.

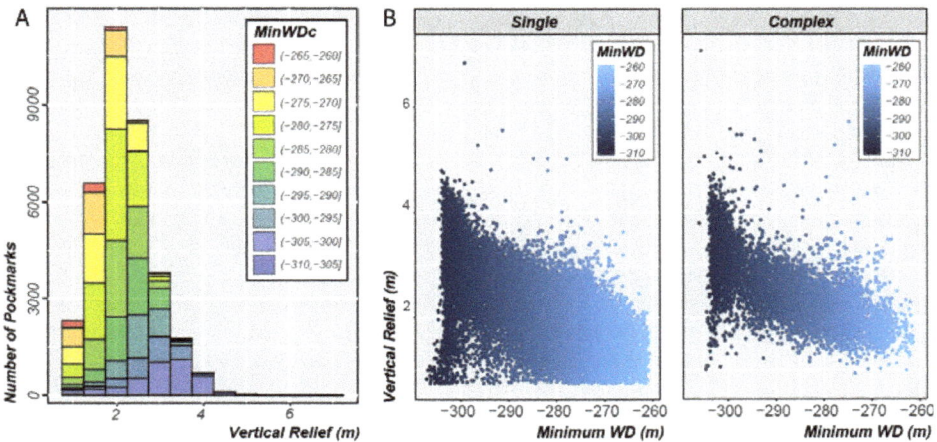

Figure 21. (**A**) Stacked histogram of the Vertical Relief values for all mapped pockmarks in the Barents Sea by minimum water depths classes; (**B**) Internal Vertical Relief versus Minimum Water Depth plot of both single and complex pockmarks.

3.4. Geomorphometric Comparison between the Different Areas

Due to the systematic and consistent mapping approach adopted, the morphological attributes used in this study to describe the pockmarks can be employed to compare the morphology and trends between different areas. For example, we see a clear distinction in the vertical relief of pockmarks from the three study areas (Figure 22).

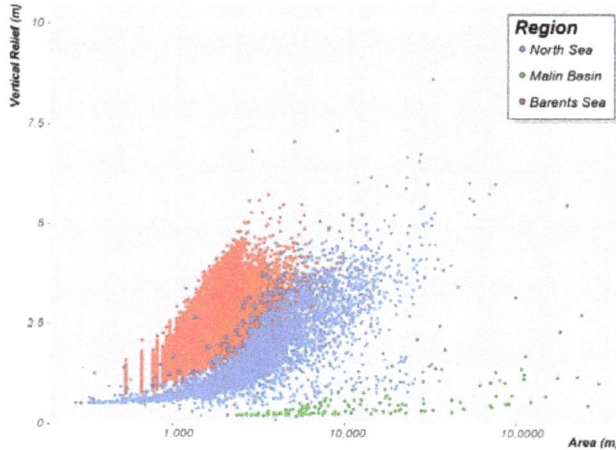

Figure 22. Vertical Relief as a function of Area comparing pockmarks mapped in the three studied regions (Barents Sea: red dots, North Sea: blue dots; and Malin Basin; green dots). Note that the five outlier values from the Witch Ground Basin are not presented here to facilitate better display of the general trends in the remaining data.

The North Sea study area is the one where the pockmarks exhibit the highest vertical relief. However, these usually large pockmarks are just five outliers to the general trend that exceed 12 m, whilst the mean value of Vertical Relief is only 1.8 m. The Barents Sea dataset shows the highest Vertical Relief mean value (μ VR = 2.2 m), which is almost four times higher than the mean value calculated for the Malin Basin dataset (μ VR = 0.6 m). Nonetheless, the Malin Basin shows the highest Area mean value (μ A = 32,073 m^2); it is one order of magnitude bigger than the mean values for the other two datasets (North Sea: 4798 m^2 and Barents Sea: 1462 m^2).

All the study areas show a positive correlation between vertical relief and area, controlled by the angle of repose. However, the strength of this correlation and the ratio between these two morphometric variables varies from region to region (Figure 23). The pockmarks from the Barents Sea present the highest VR/A mean value—μ VR/A = 1.592 $\times$ 10^{-3}; where the single pockmarks have μ VR/A = 1.720 $\times$ 10^{-3} and complex pockmarks μ VR/A = 1.095 $\times$ 10^{-3}. Nevertheless, even the complex pockmarks that tend be less deep than their single pockmark counterparts (i.e., with the same vertical relief), tend to have higher VR/A values than the pockmarks from the North Sea (μ VR/A = 0.623 $\times$ 10^{-3}). The Malin Basin pockmarks present much lower VR/A values (μ VR/A = 0.044 $\times$ 10^{-3}) but, as mentioned earlier, their vertical relief is significantly affected by the smoothing applied to the dataset, leading to artificially lower VR/A values. The regional correlation coefficients calculated for the three study areas are, respectively, 0.72, 0.57, and 0.67 for the North Sea, the Barents Sea, and the Malin Basin (Figure 23). The correlation is even stronger if looking either to individual study areas within the Witch Ground Basin—R $\approx$ 0.85 for most survey areas, or individual type of pockmarks—e.g., R = 0.94 for the single pockmarks within the Barents Sea—the highest observed correlation coefficient in this study (Figure 24).

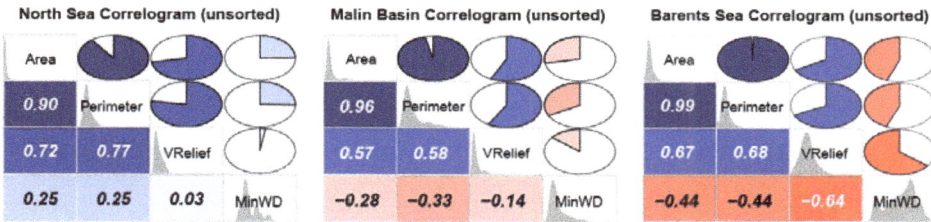

Figure 23. Unsorted correlograms showing the relationships between key pockmark metrics for the three studied regions.

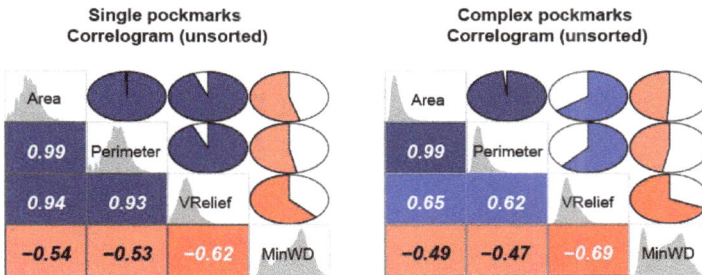

Figure 24. Unsorted correlograms showing the different relationships between key pockmark metrics for the two types of pockmarks found in the Barents Sea. Note the higher value of correlation between Area and Vertical Relief for the single pockmarks.

4. Discussion

It has been widely shown that human-cognitive approaches, i.e., manual mapping, have a number of limitations such as scale bias, azimuth bias, detection bias, and operator bias e.g., [33,34]. These limitations have encouraged new research into automated and semi-automated GIS-based mapping routines across many sub-disciplines of quantitative geomorphology [35]. In various fields of terrestrial geomorphology, pixel-based analysis techniques have been used with much success to discretize landform elements or the separable constituents of landforms, such as ridges or peaks [36,37]. However, integrating landform elements to characterize individual landforms has proven more problematic and requires subjective operator decision inputs (e.g., [38]). More recently, object-oriented approaches for automated mapping have become favoured in the terrestrial geomorphological community, whereby remote sensing imagery is segmented into meaningful objects, whose characteristics are assessed through spatial, spectral, and temporal scales [36–38].

As highlighted by [39], the availability of high-quality seabed DTMs and quantitative analyses of these data for delineation of seabed features is much more recent. We are aware of several attempts among the seabed mapping community to delineate individual pockmarks using pixel-based analysis of bathymetric data to produce terrain attributes that highlight the negative features (e.g., slope, curvature). However, there are few published studies, which may be due to the limited success of these approaches. A promising feature extraction method, based on kernel matching and machine learning, was developed by [40], but their approach, developed on synthetic data with limited testing on real-world data, does not seem to have made the crossover from computer science method development to further use in applied seabed mapping. Although other feature extraction methods such as [21] or [31] have been applied to other seabed geomorphic features, to our knowledge, they have not been applied to the delineation of pockmarks. The methods of [21] are among those tested here, but the application of the Geomorphon approach of [31] which is starting to gain attention for marine applications (e.g., [41,42]) has not yet been tested for pockmark delineation. Future applications of this method may benefit from the developments to the method reported by [42], especially where the pockmarks also have a characteristic MBES backscatter signature. Object-Based Image Analysis (OBIA) is currently gaining momentum in the seabed mapping community for seabed classification and delineation of features (e.g., [43–45]). The approach seems suitable for application to the delineation of features such as pockmarks, but we are not aware of any published studies.

These methods vary in complexity depending on the data available for the seabed classification and the approach used but are all much more complex than the approach presented here. The BGS Seabed Mapping Toolbox provides a simple extension to ArcGIS functionality that gives the user the possibility to extract morphologic information of a vast number of pockmarks from multiple surveys in a systematic and consistent way. All that is required from the user is the informed definition a limited number of thresholds. This is a significant advantage within the context of a national scale mapping programme, since it would be challenging for multiple interpreters to maintain consistent criteria throughout the laborious process of manual mapping. Such standardisation may also be difficult to achieve for pixel- or object-based classification, and none of the alternative methods identified so far address the need for morphological characterisation in addition to delineation. The BGS toolbox, although simpler than some potential alternative methods, provides both delineation and characterisation of the pockmarks. This appears to be well matched to the needs of national mapping programmes (see also Section 4.4).

Regardless of the mapping approach used for the delineation of the bedforms, it is crucial to understand how the resolution and data quality will influence morphometric studies of bedforms. Although this has been stressed in certain fields of terrestrial geomorphology that rely on the use of DTMs, relatively little attention has been paid to this to date in marine geomorphological studies. As a result, there is currently a lack of studies assessing how seabed mapping and morphometric characterisation are influenced by variations in the grid cell size or data quality. These types of studies could advise on the optimum resolution depending on the dimensions of the features studied, as well as establishing protocols for dealing with datasets where the fidelity of the DTM is sub-optimal.

4.1. Importance of Pockmarks Geomorphometric Characterization

Although it is widely accepted that pockmarks are a superficial expression of fluid flow, their formation mechanism is still elusive. Quantitative morphological data can be an important source of information in addressing this question. As we have demonstrated, quantitative analysis of pockmark morphology can provide valuable insights into the factors that control their formation and development. Supported by the wider availability of high resolution bathymetry data and appropriate DTM-based methodologies for morphometric analysis, we hope that more studies linking morphological trends to geological processes will be forthcoming in the literature, such as the one in this Special Issue by [46]. We hope the results of such studies will also be carried forward in validating numerical models of the development of seabed features.

Further, we note that with better understanding the factors controlling the morphology and spatial distribution of pockmarks, their morphometric characteristics could be used as proxies to subsurface conditions, local hydrodynamics, or fluid flow regime. As mentioned in Section 2.1, [9] shows that in the Witch Ground Basin, the thickness of very soft late glacial sediments seems to control the number of pockmarks. In this area, pockmark density maps can be used to predict variation in the thickness of the Witch Ground Formation. However, in areas of the South Western Barents Sea, according to [17], pockmark distribution seems to be almost independent of the Quaternary sediment thickness, although the authors do report an inferred correlation between pockmark size and the thickness of fine-grained deposits. This reflects the complex development of these seabed features and the fact that morphometric characteristics of pockmarks extracted from bathymetry data as proxies to subsurface conditions or flow regimes are somewhat restricted to variations within a certain regional setting.

Even if the actual development of pockmarks is not yet totally understood, there is evidence that lateral collapse of the pockmarks sidewall is a key process for the widening of the pockmarks (e.g., [47]). Therefore, it is valid to conclude that the geotechnical characteristics of the sediments affected will have a crucial role in the ratio between vertical relief developed and area affected by the seepage. Areas with sediment packages comprised of stiffer material will sustain steeper slopes and show higher rations between vertical relief and area. Therefore, a gradual and systematic mapping of the pockmarks, complemented with data on the local geotechnical characteristics would be an invaluable resource that could facilitate the use of this morphometric ratio as an indicator of the sediment proprieties.

Multibeam datasets can provide bathymetric data over vast areas of the seabed more economically than any sub-surface acoustic system. It would seem advantageous for both the scientific community and any commercial sector that requires seabed infrastructures to utilize the bathymetric data for many more quantitative analyses of seabed morphometry than are currently in common use.

4.2. Impact of Data Resolution and Quality

Although the BGS semi-automated mapping approach provided robust results for most areas, it was affected by the resolution and quality of the bathymetric dataset used. This is particularly evident when comparing the results between the different datasets from the Witch Ground Basin. The results obtained for the Rob Roy survey are a good example of the impact of using a dataset with an insufficient resolution for morphological studies. The lower resolution of the data from the Rob Roy dataset, a 10 m grid, prevented both the correct identification and delineation of the pockmarks at the seabed. This is evident by comparing the pockmark density of the adjacent survey site, Ivanhoe, which has a pockmark density of 13.8 pockmarks per km^2, with the pockmark density within the Rob Roy dataset, which is only 4.81 (Figure 9). As noted by [9], this apparent abrupt reduction of pockmark occurrence between these adjacent study areas does not appear to be controlled by: (1) marked changes in the underlying geology; (2) distinct fluid flow regime; or (3) variation in the nature of the seabed sediments. It can best be explained by the lower resolution of the Rob Roy bathymetric dataset. The lower resolution also affected the morphometric characterization of the pockmarks, in particular the measurements of vertical relief. The vertical relief mean value defined was merely 0.87 m, whereas the mean value for the adjacent area, Ivanhoe, is 1.22 m. The Rob Roy pockmarks are described as having significantly lower vertical relief values than any other study area within the North Sea. No significant changes in the morphometry of the pockmarks between the datasets with 1, 2, and 5 m resolution were evident.

This comparison suggests an optimum resolution (i.e., the coarsest resolution, providing efficiency for measurement, computing time and data storage, at which detail is not sacrificed) of at least 5 m for the dimensions of the pockmarks found in the North Sea. However, to accurately define the optimum resolution, multiple resolution datasets from the same area should be used to assess the effect of resolution on the morphometry measurement of pockmarks. That analysis goes beyond the scope of this work, but as seabed mapping continues to integrate morphometric analyses of bedforms based on digital data, it is crucial to understand how bathymetric data resolution influences morphometric variables related to bedforms.

The dataset from the Malin Deep provides a valuable insight into the potential impact of bathymetric data quality on the detection of pockmarks. Whilst reasonable results may be obtained by employing methods such as smoothing to reduce the effect of the artefacts, it is hard to overcome the inherent limitations of the dataset entirely and it should be acknowledged that there are quality issues affecting the results besides those related to data resolution. Lecours et al. [48] recently showed how heave, pitch, roll, and timing artefacts in MBES bathymetry impact habitat maps and species distribution models which utilise bathymetry data and derived terrain attributes as predictor variables. Their results show that artefacts can lead to misleading and counterintuitive results. For similar reasons to those given for data resolution, the assessment of data quality is equally important for the delineation and quantitative characterisation of bedforms.

4.3. Impact of Regional Morphology

The Malin Basin dataset has demonstrated that the mapping output of the BGS *Feature Delineation [Bathy]* tool under represents small features located on slopes. These features are not distinguished using bathymetric data alone, and when identified, their vertical relief is frequently underestimated. For that type of setting, a semi-automated approach based on the BPI, as presented by [49], could be more effective. These features are not distinguished using bathymetric data alone, and when identified, their vertical relief is frequently underestimated. However, it should be noted that for statistical comparison, the mapping based on the BPI would not provide the same consistency of delineation criterion as it is based on a relative defined value depending on the surrounding neighbourhood.

4.4. Mapping Strategies at Multiple Scales

The output of a quantitative tool like the BGS mapping toolbox is well suited to providing information on several levels of detail to suit the particular mapping purpose and scale of map product. The present study has only illustrated the implementation of the toolbox on relatively small datasets, but the method is suitable for application to larger datasets subject to computing resources and/or with tiling of the bathymetric data.

Whilst some investigations may require the delineation of individual pockmarks on detailed maps at a 1:20,000 scale or finer, others rather use this information as a basis for providing a summarized quantification of properties of the pockmarks. For example, in producing maps at a 1:100,000 scale or greater, the MAREANO programme is interested in mapping polygons indicating "pockmark areas" to provide information for management authorities on the location and extent of these terrain features, which may be linked to particular habitats. This is currently done by expert judgement, but the output of the BGS toolbox provides a quantitative indication of pockmark density that will be useful for defining objective criteria for defining the limits of a pockmark field. Further, summary statistics on the size and shape of the pockmarks can be attributed to such polygons.

Depending on the scale used, it can be impractical, or even pointless, to represent individual pockmarks in a map. In the context of regional mapping, the use of density contours is often the best approach to capture the number and spatial distribution of these seabed features. Figure 25 shows the type of pockmark density map that can be generated for the Barents Sea study area based on the outputs from the BGS Seabed Mapping Toolbox. This figure was generated using the point shapefile that represents the deepest point in each polygon. The density values range from 60 to almost 600 pockmarks per square kilometre, and the density shading highlights geographical trends in the distribution.

Even in a situation where the end-user does not require the pockmark density distribution displayed on the map and only the outline of pockmark field is needed, there are advantages to using a density contour line to outline the limits of a pockmark field. This approach increases the efficiency of pockmarks fields' delineation, improves the consistency of the mapping, and prevents any discrepancies between the mapping of adjacent areas. In addition, subjectivity issues related to manually defining the limits of a pockmark field are greatly reduced by using a more automatic approach.

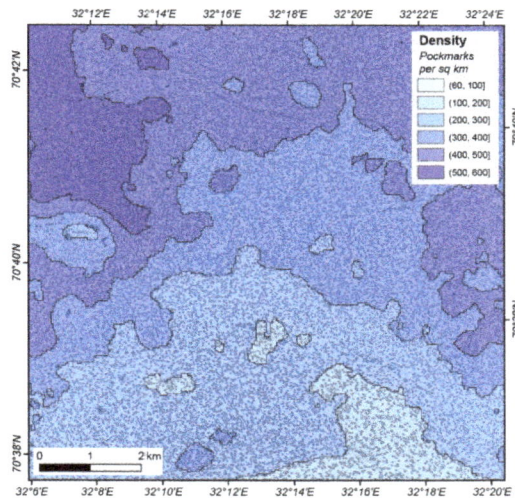

Figure 25. Pockmark density and distribution map of the Barents Sea study area, obtained using the mapping results from the BGS Seabed Mapping Toolbox.

5. Conclusions

This study explored the potential of a practical and effective semi-automatic approach for mapping pockmarks that is fully integrated within commonly used GIS software. A total of 39,533 pockmarks were mapped, across three different geological settings within the European Glaciated Margin, each of which can be considered a test case study for that setting. The results from the semi-automatic mapping exercise using the BGS Seabed Mapping Toolbox were compared to other mapping approaches, demonstrating that the semi-automatic mapping approach presented here is a valid and useful alternative. Our approach overcomes the subjectivity intrinsic to manual delineation of pockmarks, and is also considerably quicker. The results of this study indicate that this approach is more flexible than pixel-based delineation methods when dealing with pockmarks of varying sizes. Moreover, this approach incorporates the automatic extraction of the morphometric attributes of the pockmarks. This information facilitates an unprecedented statistical analysis of their morphology and the analysis of spatial trends within the pockmark fields, providing insights into the processes responsible for their development and the influence of local seabed conditions. Whilst the study has focussed on the delineation of individual pockmarks, we have also shown how our quantitative methodology, once applied to larger datasets, can provide a convenient and objective basis for summarizing pockmark distribution at broader map scales in a consistent way, increasing the relevance of the mapping results to multiple end users.

Supplementary Materials: The following are available online at http://www.mdpi.com/2076-3263/8/5/154/s1. Table S1: describes the datasets used in the Witch Ground Basin study area.

Author Contributions: J.G. and M.D. conceived the idea of the study and wrote the paper. J.G. developed the BGS Seabed Mapping Toolbox and carried out the semi-automatic mapping of the three study areas. M.D. performed the pixel-based analyses of the SEA2: box 4. X.M. was responsible for the manual mapping of the pockmarks within the Malin Basin study area. J.G. performed the statistical analyses and morphometric comparison.

Acknowledgments: Bathymetry data from the Barents Sea were acquired and provided by the Norwegian Mapping Authority Hydrographic Service as part of the MAREANO programme and are used here under a Creative Commons Attribution 4.0 International Public Licence. Bathymetry data for the Malin Shelf was provided by the Geological Survey of Ireland and the Marine Institute through the INFOMAR programme, funded by the Irish Government, Department of Communications, Climate Action and the Environment (DCCAE). Amerada

Hess, ConocoPhillips and Talisman Energy are thanked for providing site survey reports to the national offshore database and allowing data from these reports to be used in this regional study. The DTI Strategic Environment Assessment programme is thanked for providing data from 8 surveys. Joana Gafeira publishes with the permission of the Executive Director of the British Geological Survey (NERC). We thank Leif Rise and the two anonymous reviewers for their thoughtful comments that considerably improved the manuscript.

Conflicts of Interest: The authors declare no conflict of interest.

References

1. King, L.H.; MacLean, B. Pockmarks on the Scotian Shelf. *Bull. Geol. Soc. Am.* **1970**, *81*, 3141–3148. [CrossRef]
2. Martínez-Carreño, N.; García-Gil, S. The Holocene gas system of the Ría de Vigo (NW Spain): Factors controlling the location of gas accumulations, seeps and pockmarks. *Mar. Geol.* **2013**, *344*, 1–19. [CrossRef]
3. Olu-Le Roy, K.; Caprais, J.C.; Fifis, A.; Fabri, M.C.; Galéron, J.; Budzinsky, H.; Le Ménach, K.; Khripounoff, A.; Ondréas, H.; Sibuet, M. Cold-seep assemblages on a giant pockmark off West Africa: Spatial patterns and environmental control. *Mar. Ecol.* **2007**, *28*, 115–130. [CrossRef]
4. Hovland, M.; Judd, A.G.; King, L.H. Characteristic features of pockmarks on the North Sea Floor and Scotian Shelf. *Sedimentology* **1984**, *31*, 471–480. [CrossRef]
5. Pilcher, R.; Argent, J. Mega-pockmarks and linear pockmark trains on the West African continental margin. *Mar. Geol.* **2007**, *244*, 15–32. [CrossRef]
6. Dandapath, S.; Chakraborty, B.; Karisiddaiah, S.M.; Menezes, A.; Ranade, G.; Fernandes, W.; Naik, D.K.; Prudhvi Raju, K.N. Morphology of pockmarks along the western continental margin of India: Employing multibeam bathymetry and backscatter data. *Mar. Pet. Geol.* **2010**, *27*, 2107–2117. [CrossRef]
7. Gay, A.; Lopez, M.; Cochonat, P.; Sultan, N.; Cauquil, E.; Brigaud, F. Sinuous pockmark belt as indicator of a shallow buried turbiditic channel on the lower slope of the Congo basin, West African margin. *Subsurf. Sediment Mobilization* **2003**, *216*, 173–189. [CrossRef]
8. Harrington, P.K. Formation of pockmarks by pore-water escape. *Geo-Mar. Lett.* **1985**, *5*, 193–197. [CrossRef]
9. Gafeira, J.; Long, D.; Diaz-Doce, D. Semi-automated characterisation of seabed pockmarks in the central North Sea. *Near Surf. Geophys.* **2012**, *10*, 301–312. [CrossRef]
10. De Clippele, L.H.; Gafeira, J.; Robert, K.; Hennige, S.; Lavaleye, M.S.; Duineveld, G.C.A.; Huvenne, V.A.I.; Roberts, J.M. Using novel acoustic and visual mapping tools to predict the small-scale spatial distribution of live biogenic reef framework in cold-water coral habitats. *Coral Reefs* **2017**, *36*, 255–268. [CrossRef]
11. Sejrup, H.P.; Haflidason, H.; Aarseth, I.; King, E.; Forsberg, C.F.; Long, D.; Rokoengen, K. Late Weichselian glaciation history of the northern North Sea. *Boreas* **1994**, *23*, 1–13. [CrossRef]
12. Barkan, R.; Chaytor, J.D.; Andrews, B.D. Size distributions and failure initiation of submarine and subaerial landslides. *Earth Planet. Sci. Lett.* **2009**, *287*, 31–42. [CrossRef]
13. Long, D. Devensian late-glacial gas escape in the central North Sea. *Cont. Shelf Res.* **1992**, *12*, 1097–1110. [CrossRef]
14. Judd, A. Pockmarks in the UK Sector of the North Sea. 2001. Available online: https://assets.publishing.service.gov.uk/government/uploads/system/uploads/attachment_data/file/197337/TR_SEA2_Pockmarks_Dist.pdf (accessed on 26 April 2018).
15. Gafeira, J.; Long, D. *Geological Investigation of Pockmarks in the Scanner Pockmark SCI Area*; JNCC Report No 570; Joint Nature Conservation Committee (JNCC): Peterborough, UK, 2015.
16. Stoker, M.S.; Hitchen, K.; Graham, C.C. *The Geology of the Hebrides and West Shetland Shelves, and Adjacent Deep-Water Areas*; United Kingdom Offshore Reginal Report; HMSO: Richmond, UK, 1993; p. 149.
17. Evans, D.; Whittington, R.J.; Dobson, M.R. *Tiree Sheet 56 N—08 W Quaternary Geology*; British Geological Survey: Edinburgh, UK, 1987.
18. Szpak, M. Chemical and Physical Dynamics of Marine Pockmarks with Insights into the Organic Carbon Cycling on the Malin Shelf and in the Dunmanus Bay. Ireland. Ph.D. Thesis, Dublin City University, Dublin, UK, 2012.
19. Rise, L.; Bellec, V.K.; Ch, S.; Bøe, R. Pockmarks in the southwestern Barents Sea and Finnmark fjords. *Nor. J. Geol.* **2014**, *94*, 263–282. [CrossRef]
20. Horn, B.K.P. Hill Shading and the Reflectance Map. *Proc. IEEE* **1981**, *69*, 14–47. [CrossRef]
21. Wood, J. The Geomorphological Characterisation of Digital Elevation Models. Ph.D. Thesis, University of Leicester, Leicester, UK, 1996.

22. Wright, D.J.; Pendleton, M.; Boulware, J.; Walbridge, S.; Gerlt, B.; Eslinger, D.; Sampson, D.; Huntley, E. *ArcGIS Benthic Terrain Modeler (BTM)*, v. 3.0; Environmental Systems Research Institute, NOAA Coastal Services Center, Massachusetts Office of Coastal Zone Management. Available online: http://esriurl.com/5754 (accessed on 26 April 2018).

23. Wilson, M.F.J.; O'Connell, B.; Brown, C.; Guinan, J.C.; Grehan, A.J. Multiscale Terrain Analysis of Multibeam Bathymetry Data for Habitat Mapping on the Continental Slope. *Mar. Geodesy* **2007**, *30*, 3–35. [CrossRef]

24. Zevenbergen, L.W.; Thorne, C.R. Quantitative analysis of land surface topography. *Surf. Process. Landf.* **1987**, *12*, 47–56. [CrossRef]

25. Geldof, J.; Gafeira, J.; Contet, J.; Marquet, S. GIS Analysis of Pockmarks From 3D Seismic Exploration Surveys. In Proceedings of the Offshore Technology Conference, Houston, TX, USA, 5–8 May 2014; pp. 1–10.

26. Jorge, M.G.; Brennand, T.A. Semi-automated extraction of longitudinal subglacial bedforms from digital terrain models—Two new methods. *Geomorphology* **2017**, *288*, 148–163. [CrossRef]

27. Hovland, M.; Sommerville, J.H. Characteristics of two natural gas seepages in the North Sea. *Mar. Pet. Geol.* **1985**, *2*, 319–326. [CrossRef]

28. Dando, P.R.; Austen, M.C.; Burke, R.A.; Kendall, M.A.; Kennicutt II, M.C.; Judd, A.G.; Moore, D.C.; O'Hara, S.C.M.; Schmaljohann, R.; Southward, A.J. Ecology of a North Sea pockmark with an active methane seep. *Mar. Ecol. Prog. Ser.* **1991**, *70*, 49–63. [CrossRef]

29. Judd, A.; Long, D.; Sankey, M. Pockmark formation and activity, UK block 15/25, North Sea. *Bull. Geol. Soc. Den.* **1994**, *41*, 34–49. [CrossRef]

30. Walbridge, S.; Slocum, N.; Pobuda, M.; Wright, D.J. Unified Geomorphological Analysis Workflows with Benthic Terrain Modeler. *Geosciences* **2018**, *8*, 94. [CrossRef]

31. Jasiewicz, J.; Stepinski, T.F. Geomorphons-a pattern recognition approach to classification and mapping of landforms. *Geomorphology* **2013**, *182*, 147–156. [CrossRef]

32. Monteys, X.; Hardy, D.; Doyle, E.; Garcia-Gil, S. Distribution, morphology and acoustic characterisation of a gas pockmark field on the Malin Shelf, NW Ireland. In *Symposium OSP-01, Proceedings of the 33rd International Geological Congress, Oslo, Norway, 6–14 August 2008*; Curran Associates, Inc.: New York, NY, USA, 2008.

33. Smith, M.J.; Wise, S.M. Problems of bias in mapping linear landforms from satellite imagery. *Int. J. Appl. Earth Obs. Geoinf.* **2007**, *9*, 65–78. [CrossRef]

34. Smith, M.J.; Clark, C.D. Methods for the visualisation of digital elevation models for landform mapping. *Earth Surf. Process. Landf.* **2005**, *30*, 885–900. [CrossRef]

35. Bishop, M.P.; James, L.A.; Shroder, J.F.; Walsh, S.J. Geospatial technologies and digital geomorphological mapping: Concepts, issues and research. *Geomorphology* **2012**, *137*, 5–26. [CrossRef]

36. Reuter, H.I.; Wendroth, O.; Kersebaum, K.C. Optimisation of relief classification for different levels of generalisation. *Geomorphology* **2006**, *77*, 79–89. [CrossRef]

37. MacMillan, R.A.; Shary, P.A. Landforms and Landform Elements in Geo-morphometry. *Dev. Soil Sci.* **2009**, *33*, 227–254. [CrossRef]

38. Wieczorek, M.; Migoń, P. Automatic relief classification versus expert and field based landform classification for the medium-altitude mountain range, the Sudetes, SW Poland. *Geomorphology* **2014**, *206*, 133–146. [CrossRef]

39. Lecours, V.; Dolan, M.F.J.; Micallef, A.; Lucieer, V.L. A review of marine geomorphometry, the quantitative study of the seafloor. *Hydrol. Earth Syst. Sci.* **2016**, *20*, 3207–3244. [CrossRef]

40. Harrison, R.J.P.; Bellec, V.K.; Mann, D.; Wang, W. A new approach to the automated mapping of pockmarks in multi-beam bathymetry. In Proceedings of the 2011 18th IEEE International Conference on Image Process (ICIP), Brussels, Belgium, 11–14 September 2011; pp. 1–4.

41. Di Stefano, M.; Mayer, L. An Automatic Procedure for the Quantitative Characterization of Submarine Bedforms. *Geosciences* **2018**, *8*, 28. [CrossRef]

42. Masetti, G.; Mayer, L.; Ward, L. A Bathymetry- and Reflectivity-Based Approach for Seafloor Segmentation. *Geosciences* **2018**, *8*, 14. [CrossRef]

43. Diesing, M.; Thorsnes, T. Mapping of Cold-Water Coral Carbonate Mounds Based on Geomorphometric Features: An Object-Based Approach. *Geosciences* **2018**, *8*, 34. [CrossRef]

44. Diesing, M.; Green, S.L.; Stephens, D.; Lark, R.M.; Stewart, H.; Dove, D. Mapping seabed sediments: Comparison of manual, geostatistical, object-based image analysis and machine learning approaches. *Cont. Shelf Res.* **2014**, *84*, 107–119. [CrossRef]

45. Lucieer, V.; Lamarche, G. Unsupervised fuzzy classification and object-based image analysis of multibeam data to map deep water substrates, Cook Strait, New Zealand. *Cont. Shelf Res.* **2011**, *31*, 1236–1247. [CrossRef]

46. Sánchez-Guillamón, O.; Fernández-Salas, L.; Vázquez, J.-T.; Palomino, D.; Medialdea, T.; López-González, N.; Somoza, L.; León, R. Shape and Size Complexity of Deep Seafloor Mounds on the Canary Basin (West to Canary Islands, Eastern Atlantic): A DEM-Based Geomorphometric Analysis of Domes and Volcanoes. *Geosciences* **2018**, *8*, 37. [CrossRef]

47. Gafeira, J.; Long, D. *Geological Investigation of Pockmarks in the Braemar Pockmarks SCI and Surrounding Area*; JNCC Report No. 571; Joint Nature Conservation Committee (JNCC): Peterborough, UK, 2015.

48. Lecours, V.; Devillers, R.; Edinger, E.N.; Brown, C.J.; Lucieer, V.L. Influence of artefacts in marine digital terrain models on habitat maps and species distribution models: A multiscale assessment. *Remote Sens. Ecol. Conserv.* **2017**, *3*, 232–246. [CrossRef]

49. Picard, K.; Radke, L.C.; Williams, D.K.; Nicholas, W.A.; Siwabessy, J.P.; Floyd, H.; Gafeira, J.; Przeslawski, R.; Huang, Z.; Nichol, S. Origin of high density seabed pockmark fields and their use in inferring bottom currents. *Geosciences* **2018**, in press.

geosciences

MDPI

Article

Mapping of Cold-Water Coral Carbonate Mounds Based on Geomorphometric Features: An Object-Based Approach

Markus Diesing * and **Terje Thorsnes**

Geological Survey of Norway, Postal Box 6315 Torgarden, NO-7491 Trondheim, Norway; terje.thorsnes@ngu.no
* Correspondence: markus.diesing@ngu.no; Tel.: +47-7390-4309

Received: 14 December 2017; Accepted: 20 January 2018; Published: 23 January 2018

Abstract: Cold-water coral reefs are rich, yet fragile ecosystems found in colder oceanic waters. Knowledge of their spatial distribution on continental shelves, slopes, seamounts and ridge systems is vital for marine spatial planning and conservation. Cold-water corals frequently form conspicuous carbonate mounds of varying sizes, which are identifiable from multibeam echosounder bathymetry and derived geomorphometric attributes. However, the often-large number of mounds makes manual interpretation and mapping a tedious process. We present a methodology that combines image segmentation and random forest spatial prediction with the aim to derive maps of carbonate mounds and an associated measure of confidence. We demonstrate our method based on multibeam echosounder data from Iverryggen on the mid-Norwegian shelf. We identified the image-object mean planar curvature as the most important predictor. The presence and absence of carbonate mounds is mapped with high accuracy. Spatially-explicit confidence in the predictions is derived from the predicted probability and whether the predictions are within or outside the modelled range of values and is generally high. We plan to apply the showcased method to other areas of the Norwegian continental shelf and slope where multibeam echosounder data have been collected with the aim to provide crucial information for marine spatial planning.

Keywords: cold-water coral; carbonate mound; habitat mapping; spatial prediction; image segmentation; geographic object-based image analysis; random forest; accuracy; confidence

1. Introduction

Cold-water coral (CWC) reefs are complex three-dimensional habitats that provide niches for many species and hence are notable for their high biodiversity [1]. Such reefs might also function as nurseries for fish larvae [2] and fish habitat [3], although such claims are contested [4]. Moreover, CWC are of importance as archives of palaeo-environmental changes due to their longevity, cosmopolitan distribution and banded skeletal structure [1].

At the same time, human activities threaten these ecosystems in at least three ways [1]: (i) physical damage through demersal fishing [5–8], (ii) potential impacts from hydrocarbon drilling [9,10] and (iii) likely effects of ocean acidification on calcifying reef fauna [11–14]. Therefore, knowledge on the spatial distribution of CWC reefs is vital for effective marine spatial planning and conservation. Deriving the necessary spatial information is typically attempted by spatial distribution modelling for selected framework-forming CWC species [15,16]. However, data on CWC species' presence and absence are often sparse and important predictor variables frequently only exist at a coarse spatial resolution. Consequently, the resulting maps of CWC species distributions are frequently too coarse for many management applications.

The geological products of CWC reefs are 'cold-water coral carbonate mounds', positive topographic features that owe their origin, partially or entirely, to the framework-building capacity of

CWCs [17]. In the following, we will refer to these structures as carbonate mounds. Such carbonate mounds are identifiable from multibeam echosounder (MBES) data; however, the large number of carbonate mounds that can be seen in MBES data makes manual interpretation and mapping a tedious process.

In Norway, 'bioclastic sediments' associated with carbonate mounds have been mapped as part of the MAREANO (Marine AREA database for NOrwegian coast and sea areas) seabed mapping programme. Such maps give a good approximation of the occurrence of carbonate mounds [18]. However, for many applications including marine spatial planning and conservation, these products are too generalised as they do not pinpoint individual mounds, nor do they convey the confidence in the map products in a spatially explicit manner.

The objectives of this research are therefore to develop a quick, repeatable and validated methodology that allows mapping of carbonate mounds with high spatial detail and that expresses the confidence in the mapped outputs in a spatially explicit way.

2. Materials and Methods

2.1. Site

Iverryggen is situated on the mid-Norwegian shelf approximately 75 km west of the Norwegian coast. It is a nearly 20 km long N–S trending ridge (Figure 1), 7 km wide and up to 120 m high [19]. The Iverryggen ridge is interpreted as the product of glacitectonic processes, where compressive subglacial stresses from a westward ice-sheet during the Late Weichselian glaciation caused thrusting and build-up of the ridge. The stresses were induced by basal freezing close to a thin ice-sheet margin. The seabed sediments on the ridge are dominated by gravelly deposits on the highest, central parts of the ridge, with sand and gravelly sand in the fringe areas. Sandy mud and gravelly sandy mud dominate around the ridge [20]. CWC reefs on Iverryggen have been known to fishermen since bottom-trawling started in the area in the 1990s. Following scientific investigations in the late 1990s that indicated extensive trawl damage to the reefs, the area was closed to bottom trawling in 2000 [21].

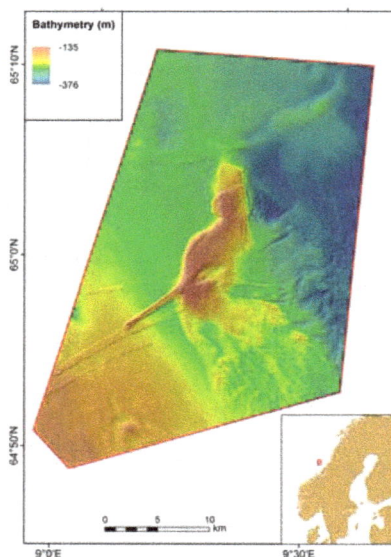

Figure 1. Map showing bathymetry in the study site Iverryggen. Inset map shows the location of the study site off the coast of Norway.

2.2. Data

MBES data were collected and processed by the Norwegian Mapping Authority, Division Hydrographic Service, in April 2012 with a Kongsberg EM710 system onboard the Norwegian survey vessel Hydrograf. The surveyed area has a size of 918 km^2. The raw MBES data were processed with software Neptune and a digital elevation model (DEM) with a grid cell size of 5 m produced. Bathymetric derivatives were calculated from the 3 × 3 Lee-filtered DEM as 32-bit floating point raster files (Table 1). Additionally, the bathymetric position index with a radius of 3 pixels (bpi3) was transformed to an 8-bit raster image used for image segmentation (see below). All layers were projected to UTM Zone 33 North.

Table 1. Bathymetric derivatives utilised in this study.

Derivative	Description	Abb.	Unit	Reference
Slope	The maximum slope gradient	slope	°	[22]
Roughness	The difference between minimum and maximum of a cell and its eight neighbours.	rgh	m	[22]
Vector ruggedness measure (VRM)	The variation in three-dimensional orientation of grid cells within a neighbourhood. A radius of 3 pixels was used.	vrm3	-	[23]
Curvature	Rate of change of slope. Profile (PR) curvature is measured parallel to maximum slope; plan (PL) curvature is measured perpendicular to slope.	Curv curvPL curvPR	-	[22]
Bathymetric position index (BPI)	Vertical position of a cell relative to its neighbourhood. Radii of 3, 5, 10 and 25 pixels were used.	bpi3 bpi5 bpi10 bpi25	m	[24]
Zero-mean bathymetry	A moving mean filter with a rectangular neighbourhood cf 25 m by 25 m was applied to the bathymetry layer. The resulting smoothed bathymetry was subtracted from the bathymetry layer.	0mean	m	[25]

A pre-existing shapefile showing bioclastic sediments [18] was used as a spatial constraint. This was extracted from [20]. The layer was derived by visual analysis of MBES data and digitisation and is depicting areas where bioclastic sediments, including those associated with carbonate mounds, are likely to exist. The following analysis was restricted to areas highlighted as bioclastic sediments and a 25-m buffer around these.

2.3. Research Strategy

The general research strategy consisted of three major elements and is outlined in Figure 2: geographic object-based image analysis (GEOBIA) [26,27], visual classification of sample objects and modelling the spatial distribution of carbonate mounds. Initially, image segmentation was carried out to derive image objects (polygons) from suitable raster derivative layers. The aim was to form objects that were either complete mounds (especially smaller ones) or parts of these (e.g., top, slope, foot). In a subsequent step, candidate mounds were defined by applying a simple rule-based classification. These candidate mounds would ideally contain all true mounds, but unavoidably also a certain number of features that have similar characteristics, but are not carbonate mounds. A wide range of object features was extracted for these candidate mounds and used for further analysis.

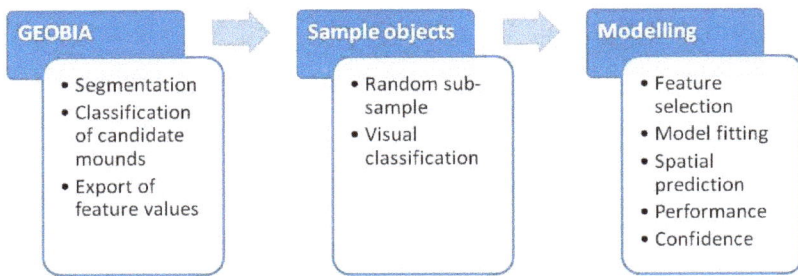

Figure 2. General work flow.

A randomly selected subset of the candidate mounds was visually inspected and classified into objects that either belonged to a carbonate mound or not. This dataset was subsequently split into objects used for feature selection and model fitting and those used to assess the performance and accuracy of the model. Finally, the model was applied to all image objects and mapping confidence was assessed.

2.4. Image Segmentation

Image segmentation is a vital step in GEOBIA, which is a two-step approach consisting of segmentation and classification. Segmentation is the process of complete partitioning of an image into non-overlapping polygons in image space [28] on the basis of homogeneity [27]. The aim of the segmentation is to divide the image into meaningful polygons of variable sizes, based on their spectral and spatial characteristics. The resulting image objects can be characterised by various features such as layer values (mean, standard deviation etc.), geometry (extent, shape etc.), image texture and many others. Classification is then based on user-specified combinations of these image object features or by applying statistical methods of spatial prediction.

Segmentation was carried out using the multi-resolution segmentation algorithm in eCognition 9.2.0. This algorithm is an optimisation procedure, which locally minimises the average heterogeneity of image objects for a given resolution of image objects. Starting from an individual pixel, it consecutively merges pixels until a certain threshold, defined by the scale parameter, is reached. The scale parameter is an abstract term that determines the maximum allowable heterogeneity for the resulting image objects.

The object heterogeneity, to which the scale parameter refers, is defined by the 'composition of homogeneity' criterion. This criterion defines the relative importance of 'colour' (pixel value) versus shape of objects. If high weight is given to colour then the object boundaries will be predominantly determined by variations in colour of the image. Further on, the shape criterion has contributions from smoothness and compactness, both of which can be weighted. A high value for smoothness will lead to smoother boundaries of the objects. High values of compactness will increase the overall compactness of image objects.

Image segmentation was carried out on the BPI3 8-bit layer and the buffered bioclastic sediments shapefile, using the multi-resolution segmentation algorithm with a scale factor of 10, shape of 0.1 and compactness of 0.5. The selection of the segmentation parameters was made after trialling different combinations and assessing the results visually.

2.5. Classification of Candidate Mounds

Candidate mounds were classified with the zero-mean bathymetry (Table 1) $\geq$0.05 m within buffered areas of bioclastic sediments.

2.6. Export of Feature Values

Image objects classified as candidate mounds were exported as a shapefile together with 24 image-object features (Table 2). These included object mean values of derivatives, maximum pixel values of selected derivatives and geometry features (extent and shape features). Documentation of the eCognition rule-set (RuleSetDocu.txt) is provided as supplementary material.

Table 2. Image-object features extracted for further analysis. Abbreviations of extent and shape features given in brackets.

Feature Type	Features
Object mean value (Mean_...)	slope, rgh, vrm3, curv, curvPL, curvPR, bpi3, bpi5, bpi10, bpi25, 0mean
Maximum pixel value (Max_...)	slope, rgh, vrm3, curvPL, bpi3, 0mean
Extent	Area (Area_Pxl), border length (Border_len), length-width ratio (LengthWidt)
Shape	Asymmetry, compactness (Compactnes), elliptic fit (Elliptic_f), main direction (Main_direc)

2.7. Defining Sample Objects

One thousand image objects were randomly selected using the NOAA Sampling Design Tool (https://www.arcgis.com/home/item.html?id=ecbe1fc44f35465f9dea42ef9b63e785) in ArcGIS 10.4.1. Based on shaded relief and bpi3 maps, image objects were manually classified as belonging to a carbonate mound or otherwise. In total 573 image objects indicated the presence of carbonate mounds and 427 image objects the absence of mounds. The selected sample objects were split into training and test data with a ratio of 2:1 using a stratified random approach to ensure the same class frequencies in both training and test data (Table 3). Subsequent feature selection and model fitting was conducted with the training data, while the test set was used to estimate model performance. The training data set (Iverryggen_training_data.txt) is provided as supplementary material.

Table 3. Number of sample objects in training and test set.

	Training	Test	Sum
Presence (1)	382	191	573
Absence (0)	285	142	427
Sum	667	333	1000

2.8. Feature Selection

The feature selection approach consisted of two steps: Initially, all relevant features were identified with the Boruta algorithm. Subsequently, the set of predictors was reduced to uncorrelated features.

The Boruta variable selection wrapper algorithm [29] was employed for the identification of all relevant predictor features. Wrapper algorithms identify relevant features by performing multiple runs of predictive models, testing the performance of different subsets [30]. The Boruta algorithm creates copies of all features and randomises them. These so-called shadow features are added to the predictor feature data set and the random forest algorithm (see below) is run to compute feature importance scores for predictor and shadow features. The maximum importance score among the shadow features (MZSA) is determined. For every predictor feature, a two-sided test of equality is performed with the MZSA. Predictor features that have a feature importance score significantly higher than the MZSA are deemed important. Likewise, predictor features that have a variable importance score significantly lower than the MZSA are deemed unimportant. Tentative features have a variable importance score that is not significantly different from the MZSA. The Boruta algorithm was run with a maximum of 500 iterations and a p-value of 0.05. Only important features were retained for further analysis. A further feature selection step removed correlated features from the set of predictors used in the final model. Out of any two features with a correlation coefficient above 0.7, only the feature with the higher importance measured as the mean decrease in accuracy (see below) was retained in the final set of predictors. The variance inflation factor (VIF) was used as an indicator to assess the remaining degree of multicollinearity in the data. The R script (FeatureSelection.r) is provided as supplementary material.

2.9. Model Fitting

The random forest (RF) prediction algorithm [31] was chosen as the modelling tool for the analysis because it has shown high predictive performance in a number of domains [32–35]. RF can be used without extensive parameter tuning, it can handle a large number of predictor variables, is insensitive to the inclusion of some noisy/irrelevant features, makes no assumptions regarding the shape of distributions of the response or predictor variables [36] and is therefore suitable for this analysis. The RF is an ensemble technique that 'grows' many classification trees. Two elements of randomness are introduced: Firstly, each tree is constructed from a bootstrapped sample of the training data. Secondly, only a random subset of the predictor features is used at each split in the tree building process. This has the effect of making every tree in the forest unique. The underlying principle of the technique is that although each tree in the forest may individually be a poor predictor and that any two trees could give different answers, by aggregating the predictions over a large number of uncorrelated trees, prediction variance is reduced and accuracy improved ([37], p. 316). For binary classifications, probability predictions are initially derived based on the fraction of votes given for a specified class by the ensemble of trees. These are subsequently dichotomised by applying suitable thresholds (see below).

Two parameters can be specified when fitting a RF model. These are the number of trees to grow (n_{tree}) and the number of features available for splitting (m_{try}). We varied n_{tree} between 500 and 5000 (steps of 500) to investigate the effect on the accuracy and stability of the model. Varying n_{tree} had little influence on model accuracy but model stability increased with the number of trees. We selected n_{tree} = 2000 as a compromise between stability and processing time. For m_{try} the default value, equal to the square root of the number of predictor variables (rounded down), was chosen.

RF also provides a relative estimate of predictor variable importance. This is measured as the mean decrease in accuracy associated with each variable when it is assigned random but realistic values and the rest of the variables are left unchanged. The worse a model performs when a predictor is randomised, the more important that predictor is in predicting the response variable. The mean decrease in accuracy was left unscaled, i.e., it was not divided by the standard deviation of the differences in prediction errors, and is reported as a fraction ranging from 0 to 1.

The randomForest package [38], executed via the Marine Geospatial Ecology Tools v08a.67 [39], was used for the implementation of the model. The resultant Rdata file (rf2000_2.Rdata) and model statistics (rf2000_2_model_stats.txt) are provided as supplementary materials.

2.10. Spatial Prediction

The selected RF model was applied to the whole dataset to predict the probability of carbonate mound presence for each image object. For the dichotomisation of the probabilistic outputs, a threshold was calculated from the training data by maximising the Youden index [40].

2.11. Model Performance

Model performance of presence-absence predictions was assessed with the receiver-operating characteristic (ROC) curve by plotting sensitivity against 1—specificity for various thresholds. Sensitivity is the amount of true presence predictions as a proportion of the total number of presence observations (Table 4 and Equation (2)). Specificity is the amount of true absence predictions as a proportion of the total number of absence observations (Table 4 and Equation (3)). The area under the

ROC curve, or AUC, was used to assess the model performance. An AUC value of 0.5 is indicating discrimination ability no better than random and a value of 1 is indicative of perfect discrimination.

Table 4. Contingency table for presence-absence predictions and selected associated accuracy metrics.

	Observed Absence	Observed Presence
Predicted absence	True negative (TN)	False negative (FN)
Predicted presence	False positive (FP)	True positive (TP)

We assessed the accuracy of the dichotomised predictions with three widely applied metrics, which can be derived from a contingency table (Table 4), namely the overall accuracy (percent classified correctly, PCC), sensitivity and specificity. PCC is the proportion of correctly classified presences and absences. The accuracy metrics were calculated with the caret package [41] in R. The R script (MapAccuracy.r) is provided as supplementary material.

$$PCC = \frac{(TN + TP)}{(TN + FN + FP + TP)} \tag{1}$$

$$Sens = \frac{TP}{(FN + TP)} \tag{2}$$

$$Spec = \frac{TN}{(TN + FP)} \tag{3}$$

2.12. Confidence

The predicted probabilities were also employed to assign a confidence score to the outputs similar to reference [42]: To indicate high confidence in the absence of carbonate mounds, a sensitivity value of 0.95 was selected. In other words, we requested that 95% of the predicted presence of mounds is correct based on the test set. Likewise, to indicate high confidence in the presence of mounds, a specificity value of 0.95 was required, equal to 95% of absence of mounds correctly classified based on the test set. The choices for the selected values of sensitivity and specificity are somewhat arbitrary here. However, they were selected for illustrative purposes and will be discussed in greater detail later on. Additionally, we considered whether the predictions were within or outside the modelled range of values of the RF model. Scores of 0 and 1 were given for low and high confidence for both aspects of the predictive model and summed to give total confidence scores of 0 (low confidence), 1 (medium confidence) or 2 (high confidence). The Presence Absence [43] and ROCR [44] packages were used in R to calculate thresholds for required values of sensitivity and specificity and create diagnostic plots. The R script (BinaryClassification.r) and the input data file (rf2000_2_export_for_PresenceAbsence.txt) are provided as supplementary materials.

2.13. Representation on Maps

Since carbonate mounds are small features compared to the size of the study area, we represent them in an aggregated form by calculating the fraction of mound area per 1 km^2 block. We also display the mean total confidence score in a similar way.

3. Results

3.1. Feature Selection

The Boruta feature selection reduced the number of predictor features from 24 to 22 (Figure 3). Two features were deemed unimportant (Asymmetry, Main_direc). The subsequent correlation analysis reduced the number of predictor features to six. All VIFs were below 3 indicating a low degree of multicollinearity in the remaining predictors (Table 5). Mean_curPL, Mean_rgh, Mean_curPR, Area_Pxl, LengthWidt and Mean_bpi25 were used to build the final model.

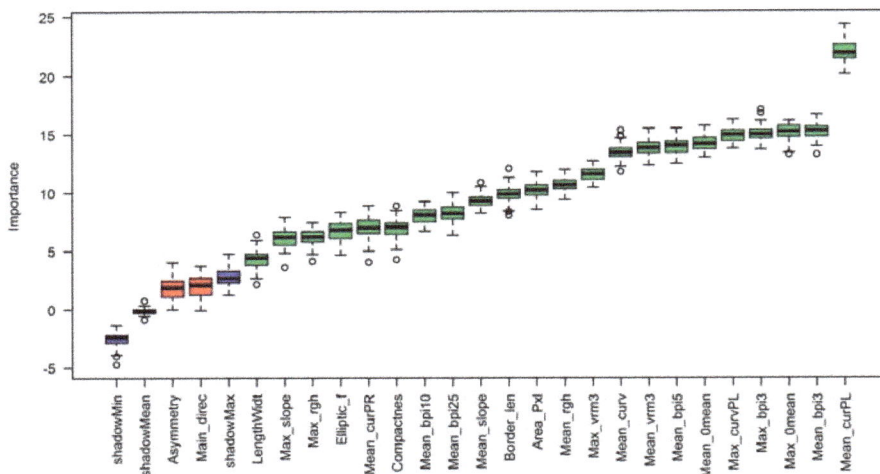

Figure 3. Results of the Boruta feature selection analysis. Green colours indicate important features, red colours indicate unimportant features. Shadow features are shown in blue.

Table 5. Variance inflation factors for the selected features.

Feature	VIF
Mean_curPL	2.654873
Mean_rgh	1.9249
Mean_curPR	2.250794
LengthWidt	1.133754
Area_Pxl	1.340249
Mean_bpi25	2.28189

3.2. Model Performance and Confidence

The model had an AUC = 0.91, which indicates a well-performing model with excellent discriminatory ability (Figure 4, top left). The probability threshold for maximising the Youden index was 0.619. The selected threshold also coincided with a maximum in PCC (Figure 4, top right). Probabilities above this threshold were classified as presence of carbonate mounds. Accuracy statistics associated with this threshold are as follows: 84.4% of objects were correctly classified (PCC = 0.844; 95% confidence intervals: (0.8003, 0.8811)). This is significantly higher ($p < 2 \times 10^{-16}$) than the no information rate (NIR = 0.574), which was taken to be the largest class percentage in the data. The model had a high sensitivity (Sens = 0.827) and specificity (Spec = 0.866).

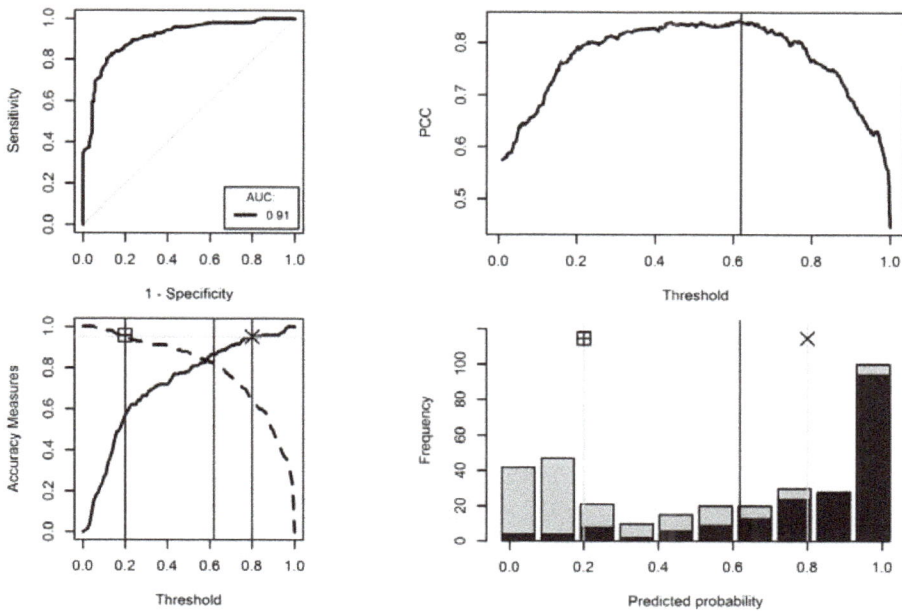

Figure 4. Summary of model performance: **Top left**—ROC curve (bold black line) indicating AUC = 0.91. The grey diagonal indicates AUC = 0.5 (no discrimination ability). **Top right**—PCC versus selected threshold: The threshold selected by maximising the Youden index (0.619) gave the highest PCC. **Bottom left**—Sensitivity (dashed line) and specificity (solid line) versus selected threshold: Requesting Sens = 0.95 and Spec = 0.95 gave thresholds of 0.2 and 0.8, respectively. **Bottom right**—Histogram showing the frequency of presence (black) and absence (grey) observations against the predicted probability.

Probability thresholds for Sens = 0.95 and Spec = 0.95 were 0.20 and 0.80, respectively (Figure 4, bottom left). These were employed to typify confidence in the predictions: Confidence is high for probabilities below 0.20 (high confidence in absence) and equal to or greater than 0.80 (high confidence in presence) as indicated by the low proportion of incorrectly classified objects seen in Figure 4 (bottom right). Confidence is low for intermediate probability values. In 1.5% of the cases were the predictions outside the range of modelled values, leading to low confidence in this aspect of the model. Predictions were inside the modelled range for 98.5% of the cases and high confidence was associated with these predictions. Total confidence was high in 67.42%, medium in 32.13% and low in 0.45% of the cases.

3.3. Spatial Representation

Figure 5 shows an example detail of the results with predicted carbonate mounds mapped with three levels of total confidence. These are overlaid on a hillshade image of the bathymetry.

The relative mound area (Figure 6) varies between 0% and close to 18%; however, the distribution is uneven with mound areas below 1% in more than half of the blocks (and not including blocks with a mound area equal to 0%). Mounds occur predominantly in a NNW-SSE trending zone. Confidence in the predictions is high overall.

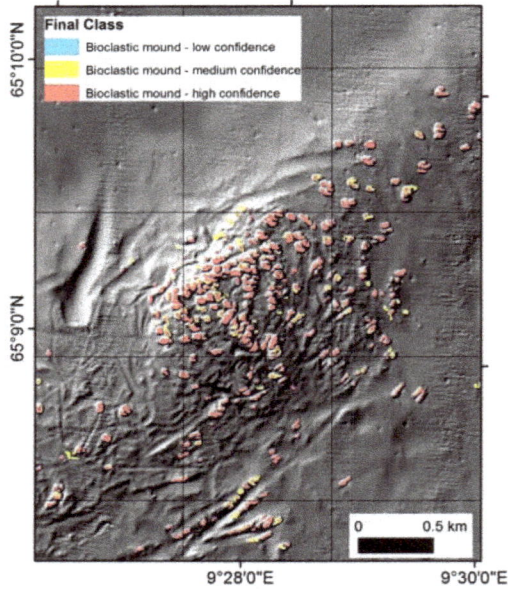

Figure 5. Left—Example detail of mapping results showing predicted occurrence of carbonate mounds and associated total confidence overlaid on a hillshade image of the bathymetry. Also visible is the 1 km by 1 km raster used to aggregate data on carbonate mounds.

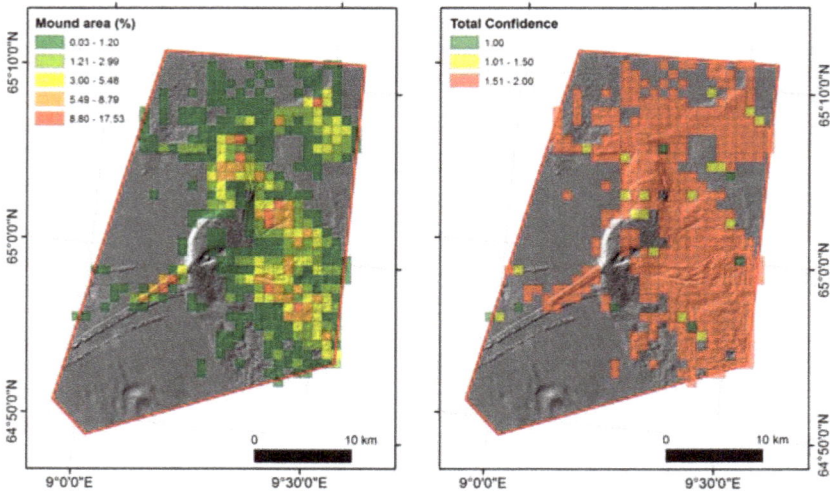

Figure 6. Left—Carbonate mound area per 1 km² block expressed as percentage of mound area per block. **Right**—Average total confidence per 1 km² block. Only blocks that were occupied by at least one carbonate mound are shown.

3.4. Predictor Feature Importance

Mean_curvPL was by far the most important predictor feature with a mean decrease in accuracy of approximately 15% (Figure 7). Other features were significantly less important with mean decreases in accuracy below 5%. However, it should be noted that other features that have been removed from the model due to high correlation with those features included in the model may also be useful predictors for carbonate mounds. Based on the feature importance scores of the Boruta analysis (Figure 3), the most likely candidates are Mean_bpi3, Max_0mean, Max_bpi3, Max_curvPL, Mean_0mean, Mean_bpi5, Mean_vrm3 and Mean_curv.

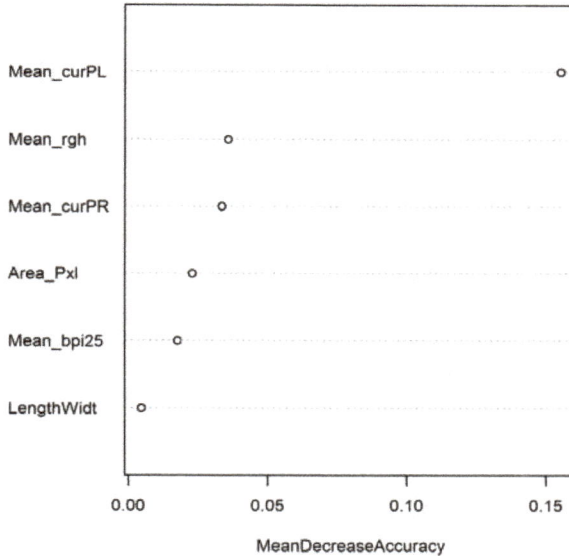

Figure 7. Predictor feature importance expressed as the mean decrease in accuracy when a feature is randomised. Larger decreases mean that the feature is more important. The mean decrease in accuracy ranges from 0 to 1.

4. Discussion

We have demonstrated how carbonate mounds can be mapped from derivatives of MBES bathymetry in an efficient and validated way with high accuracy by combining image segmentation with presence-absence spatial predictions. This approach is reasonably quick; the most time-consuming steps being data preparation (calculation of derivatives etc.) and classification of sample objects, both of which took approximately one working day. All in all, it took about one working week to derive the showcased maps. This is considerably faster than manual interpretation of the MBES data.

Image segmentation was a vital step in the mapping approach for two reasons: First, it allowed selecting sample objects for visual classification. Such a step would have been difficult to exercise based on image pixels. Second, additional features other than image layer values were extracted and utilised in the spatial prediction. The final model did not only include object mean values, but also extent features, which contributed to the final predictions.

We used a pre-existing bioclastic sediments [18] map layer as a spatial constraint for the analysis. Although this might carry the risk of missing some of the mounds present in the area, we judge the chances of this being the case as low. The main benefit of spatially constraining the analysis were that the total number of image objects used in the analysis was markedly smaller (23,602 instead of 83,085 image objects) and, more importantly, the presence-absence ratio was higher. Initial trials without a spatial constraint at a different but similar site showed that it was still possible to derive meaningful results with high accuracy, but more sample objects are required to give a sufficiently high number of mound presences.

The predictive model was trained and tested with samples that were derived by visually assessing a sub-set of image objects. This approach is arguably prone to introducing error due to the incorrect classification of an unknown number of sample objects. However, we caution that ground-truth information derived from underwater stills or video is equally not free of error due to mis-classification [45]. Moreover, ground-truth information is often difficult and costly to obtain, and these factors typically limit the number of samples available for an analysis. Our approach allowed us to select a large amount of sample objects without the additional costs. We believe that the approach was justified as carbonate mounds are typically conspicuous objects on the seabed readily identifiable from MBES data but do not recommend utilising this approach in situations where the target classes are less obvious.

It is important to assess the performance of a model to understand how reliable the predictions are. In this case, discriminatory ability (AUC) and accuracy (PCC, Sens and Spec) of the predictions were generally high. However, model performance statistics are either global or class-specific, giving no information on the accuracy of the prediction in a spatially explicit way [46]. Maps that portray the spatial variation of mapping error or uncertainty can be highly informative: They might be used to guide future ground-truth sampling campaigns or provide managers with information on the reliability of predictions when making decisions. Here, we have adopted and modified a scheme to assess the confidence in spatial predictions developed for mapping bedrock outcrops on the United Kingdom's continental shelf [42]. The confidence in the presence of carbonate mounds is high where probabilities are 0.8 or higher and the predictions are within the modelled range of values of the RF model. Medium confidence in the presence of carbonate mounds means that one of the two criteria is not fulfilled, and low confidence is assigned when both criteria are not fulfilled, but probabilities are higher than 0.619. In this way, we express confidence in the predictions in a way that is simple to understand; yet, confidence is based on objective and measurable criteria rather than scoring systems and it is spatially explicit.

As mentioned earlier, the choice of sensitivity and specificity values of 0.95 was somewhat arbitrary and used for illustrative purposes. Preferably, the selection of those values should be informed by the objectives of the mapping task. Increasing the value for the required specificity will reduce the number of false positives, i.e., the number of image objects incorrectly classified as carbonate mounds will be reduced. For example, increasing the required specificity from 0.95 to 0.99 will increase the probability threshold from 0.8 to 0.96. Consequently, the number of image objects classified accordingly will be reduced from 10,107 to 5868. This might be useful when a high level of certainty in the presence of carbonate mounds is required, for example when a field survey is planned, and targeted stations should correspond to carbonate mounds as frequently as possible.

This approach allows mapping carbonate mounds from MBES data with high accuracy and confidence. Carbonate mounds are the geological product of CWCs [17]; however, it is important to note that the cover of such mounds with live corals can vary dramatically. In fact, mounds might exhibit live coral cover approaching 100% or be completely devoid of live corals, even in close proximity, as shown by detailed results from the Mingulay Reef Complex west of Scotland [47]. Although it has been demonstrated that the likelihood of live coral presence can be spatially predicted [47], it will still require detailed ground-truthing, for example with video or stills images, to establish whether or not carbonate mounds are occupied by live corals and to what degree.

In this case study, Mean_curPL was by far the most important predictor for the presence of carbonate mounds. However, other features relating to terrain variability, curvature and relative position are expected to be suitable predictors for carbonate mounds as well, but were strongly correlated with Mean_curPL and hence removed from the analysis. The comparatively small size of the mounds in relation to the resolution of the terrain data means that small neighbourhood sizes of 3 and 5 were the most relevant. We recommend that other studies attempting to predict carbonate mounds should consider features of terrain variability, curvature and relative position [22] at a relevant scale that reflects both the size of the mounds to be mapped and the resolution of the data utilised.

Future work will include a comparison of the results presented here with ground-truth information from underwater video tows. This information might also be used to add another layer of confidence: Image objects that coincide with interpreted video frames and agree with the interpretation might be given a higher confidence score [42,48]. In a next step, the developed methodology will be applied to areas of the Norwegian continental shelf and slope where MBES data have been collected. Previous research has shown that carbonate mounds are identifiable from MBES data in areas of flat seabed, iceberg ploughmarks and glacial lineations, while slide scar environments and bedrock outcrops limit the ability to detect carbonate mounds [18]. To what extent the presented method is capable of mapping carbonate mounds in such challenging environments remains to be tested. Further work might include the extraction of additional morphometric variables such as the maximum height of a mound above the surrounding seabed and the volume of a mound. Such metrics could be used to derive the amount of inorganic carbon stored in carbonate mounds on the Norwegian seabed, information that is critical for a better understanding of the importance of CWC reefs as carbonate sinks [49] and possible impacts of climate change (warming and acidification) on those carbon stores [12].

Supplementary Materials: The following are available online at http://www.mdpi.com/2076-3263/8/2/34/s1, RuleSetDocu.txt: Documentation of the eCognition rule-set (Chapters 2.4–2.6); Iverryggen_training_data.txt: Model training data set containing observed classes (carbonate mound presence and absence) and predictor feature values; FeatureSelection.r: R script for feature selection (Chapter 2.8); rf2000_2.Rdata: Output from MGET random forest model fitting (Chapter 2.9); rf2000_2_model_stats.txt: Model performance summary and confusion matrix as provided by MGET; MapAccuracy.r: R script for calculation of accuracy statistics (Chapter 2.11); rf2000_2_export_for_PresenceAbsence.txt: Text file containing observed classes (test data) and predicted probabilities; BinaryClassification.r: R script to derive thresholds for selected sensitivity and specificity and model performance plots (Chapter 2.12).

Acknowledgments: This study was funded by the Geological Survey of Norway. The bathymetry data were made available by the Norwegian Mapping Authority, Division Hydrographic Service, funded by the MAREANO programme.

Author Contributions: Markus Diesing and Terje Thorsnes conceived and designed the study; Markus Diesing performed the analysis; Markus Diesing analysed the data; Markus Diesing and Terje Thorsnes wrote the paper.

Conflicts of Interest: The authors declare no conflict of interest. The founding sponsors had no role in the design of the study; in the collection, analyses, or interpretation of data; in the writing of the manuscript, and in the decision to publish the results.

References

1. Roberts, J.M.; Wheeler, A.J.; Freiwald, A. Reefs of the Deep: The Biology and Geology of Cold-Water Coral Ecosystems. *Science* **2006**, *312*, 543–547. [CrossRef] [PubMed]

2. Baillon, S.; Hamel, J.-F.; Wareham, V.E.; Mercier, A. Deep cold-water corals as nurseries for fish larvae. *Front. Ecol. Environ.* **2012**, *10*, 351–356. [CrossRef]

3. Costello, M.J.; McCrea, M.; Freiwald, A.; Lundälv, T.; Jonsson, L.; Bett, B.J.; van Weering, T.C.E.; de Haas, H.; Roberts, J.M.; Allen, D. Role of cold-water Lophelia pertusa coral reefs as fish habitat in the NE Atlantic. In *Cold-Water Corals and Ecosystems*; Freiwald, A., Roberts, J.M., Eds.; Springer: Berlin/Heidelberg, Germany, 2005; pp. 771–805. ISBN 978-3-540-27673-9.

4. Auster, P.J. *Are deep-water corals important habitats for fishes? In Cold-Water Corals and Ecosystems*; Freiwald, A., Roberts, J.M., Eds.; Springer: Berlin/Heidelberg, Germany, 2005; pp. 747–760. ISBN 978-3-540-27673-9.

5. Althaus, F.; Williams, A.; Schlacher, T.; Kloser, R.J.; Green, M.; Barker, B.; Bax, N.; Brodie, P. Impacts of bottom trawling on deep-coral ecosystems of seamounts are long-lasting. *Mar. Ecol. Prog. Ser.* **2009**, *397*, 279–294. [CrossRef]

6. Hall-Spencer, J.; Allain, V.; Fosså, J.H. Trawling damage to Northeast Atlantic ancient coral reefs. *Proc. R. Soc. Lond. Ser. B Biol. Sci.* **2002**, *269*, 507–511. [CrossRef] [PubMed]

7. Fosså, J.H.; Mortensen, P.B.; Furevik, D.M. The deep-water coral Lophelia pertusa in Norwegian waters: Distribution and fishery impacts. *Hydrobiologia* **2002**, *471*, 1–12. [CrossRef]

8. Huvenne, V.A.I.; Bett, B.J.; Masson, D.G.; Le Bas, T.P.; Wheeler, A.J. Effectiveness of a deep-sea cold-water coral Marine Protected Area, following eight years of fisheries closure. *Biol. Conserv.* **2016**, *200*, 60–69. [CrossRef]

9. Fisher, C.R.; Hsing, P.-Y.; Kaiser, C.L.; Yoerger, D.R.; Roberts, H.H.; Shedd, W.W.; Cordes, E.E.; Shank, T.M.; Berlet, S.P.; Saunders, M.G.; et al. Footprint of Deepwater Horizon blowout impact to deep-water coral communities. *Proc. Natl. Acad. Sci. USA* **2014**, *111*, 11744–11749. [CrossRef] [PubMed]

10. Purser, A.; Thomsen, L. Monitoring strategies for drill cutting discharge in the vicinity of cold-water coral ecosystems. *Mar. Pollut. Bull.* **2012**, *64*, 2309–2316. [CrossRef] [PubMed]

11. Guinotte, J.M.; Orr, J.; Cairns, S.; Freiwald, A.; Morgan, L.; George, R. Will human-induced changes in seawater chemistry alter the distribution of deep-sea scleractinian corals? *Front. Ecol. Environ.* **2006**, *4*, 141–146. [CrossRef]

12. Hoegh-Guldberg, O.; Poloczanska, E.S.; Skirving, W.; Dove, S. Coral Reef Ecosystems under Climate Change and Ocean Acidification. *Front. Mar. Sci.* **2017**, *4*, 158. [CrossRef]

13. Maier, C.; Watremez, P.; Taviani, M.; Weinbauer, M.G.; Gattuso, J.P. Calcification rates and the effect of ocean acidification on Mediterranean cold-water corals. *Proc. R. Soc. Lond. Ser. B Biol. Sci.* **2012**, *279*, 1716–1723. [CrossRef] [PubMed]

14. Büscher, J.V.; Form, A.U.; Riebesell, U. Interactive Effects of Ocean Acidification and Warming on Growth, Fitness and Survival of the Cold-Water Coral Lophelia pertusa under Different Food Availabilities. *Front. Mar. Sci.* **2017**, *4*, 101. [CrossRef]

15. Davies, A.J.; Wisshak, M.; Orr, J.C.; Murray Roberts, J. Predicting suitable habitat for the cold-water coral Lophelia pertusa (Scleractinia). *Deep Sea Res. Part I Oceanogr. Res. Pap.* **2008**, *55*, 1048–1062. [CrossRef]

16. Davies, A.J.; Guinotte, J.M. Global Habitat Suitability for Framework-Forming Cold-Water Corals. *PLoS ONE* **2011**, *6*, e18483. [CrossRef] [PubMed]

17. Wheeler, A.J.; Beyer, A.; Freiwald, A.; de Haas, H.; Huvenne, V.A.I.; Kozachenko, M.; Olu-Le Roy, K.; Opderbecke, J. Morphology and environment of cold-water coral carbonate mounds on the NW European margin. *Int. J. Earth Sci.* **2007**, *96*, 37–56. [CrossRef]

18. Bellec, V.K.; Thorsnes, T.; Bøe, R. Mapping of Bioclastic Sediments—Data, Methods and Confidence. Available online: http://www.ngu.no/upload/Publikasjoner/Rapporter/2014/2014_006.pdf (accessed on 10 November 2017).

19. Thorsnes, T.; Bellec, V.; Baeten, N.; Plassen, L.; Bjarnadóttir, L.; Ottesen, D.; Dolan, M.; Elvenes, S.; Rise, L.; Longva, O.; et al. The Seabed - Marine Landscapes, Geology and Processes. Available online: http://mareano.no/resources/images/2015/chapter-7.pdf (accessed on 10 November 2017).

20. Bjarnadóttir, L.R.; Ottesen, D.; Bellec, V.; Lepland, A.; Elvenes, S.; Dolan, M.; Bøe, R.; Rise, L.; Thorsnes, T.; Selboskar, O.H. Geologisk Havbunnskart, KART 65000900, Mai 2017. M 1: 100 000. 2017. Available online: http://www.ngu.no/upload/Publikasjoner/Kart/Maringeologi/Geologisk_havbunnskart_65000900.pdf (accessed on 10 December 2017).

21. Buhl-Mortensen, P.; Buhl-Mortensen, L.; Dolan, M. Bottom Habitats and Fauna. Available online: http://mareano.no/resources/images/2015/chapter-7.pdf (accessed on 10 November 2017).

22. Wilson, M.F.J.; O'Connell, B.; Brown, C.; Guinan, J.C.; Grehan, A.J. Multiscale Terrain Analysis of Multibeam Bathymetry Data for Habitat Mapping on the Continental Slope. *Mar. Geodesy* **2007**, *30*, 3–35. [CrossRef]

23. Sappington, J.M.; Longshore, K.M.; Thompson, D.B. Quantifying Landscape Ruggedness for Animal Habitat Analysis: A Case Study Using Bighorn Sheep in the Mojave Desert. *J. Wildl. Manag.* **2007**, *71*, 1419–1426. [CrossRef]

24. Lundblad, E.R.; Wright, D.J.; Miller, J.; Larkin, E.M.; Rinehart, R.; Naar, D.F.; Donahue, B.T.; Anderson, S.M.; Battista, T. A Benthic Terrain Classification Scheme for American Samoa. *Mar. Geodesy* **2006**, *29*, 89–111. [CrossRef]

25. Krämer, K.; Holler, P.; Herbst, G.; Bratek, A.; Ahmerkamp, S.; Neumann, A.; Bartholomä, A.; van Beusekom, J.E.E.; Holtappels, M.; Winter, C. Abrupt emergence of a large pockmark field in the German Bight, southeastern North Sea. *Sci. Rep.* **2017**, *7*, 5150. [CrossRef] [PubMed]

26. Hay, G.J.; Castilla, G. Geographic Object-Based Image Analysis (GEOBIA): A new name for a new discipline. In *Object-Based Image Analysis: Spatial Concepts for Knowledge-Driven Remote Sensing Applications*; Blaschke, T., Lang, S., Hay, G.J., Eds.; Springer: Berlin/Heidelberg, Germany, 2008; pp. 75–89.

27. Blaschke, T.; Hay, G.J.; Kelly, M.; Lang, S.; Hofmann, P.; Addink, E.; Queiroz Feitosa, R.; van der Meer, F.; van der Werff, H.; van Coillie, F.; et al. Geographic Object-Based Image Analysis—Towards a new paradigm. *J. Photogramm. Remote Sens.* **2014**, *87*, 180–191. [CrossRef] [PubMed]

28. Schiewe, J. Segmentation of high-resolution remotely sensed data—Concepts, applications and problems. In Proceedings of the Symposium on Geospatial Theory, Processing and Applications, Ottawa, ON, Canada, 9–12 July 2002.

29. Kursa, M.; Rudnicki, W. Feature selection with the Boruta Package. *J. Stat. Softw.* **2010**, *36*, 1–11. [CrossRef]

30. Guyon, I.; Elisseeff, A. An Introduction to Variable and Feature Selection. *J. Mach. Learn. Res.* **2003**, *3*, 1157–1182.

31. Breiman, L. Random Forests. *Mach. Learn.* **2001**, *45*, 5–32. [CrossRef]

32. Prasad, A.M.; Iverson, L.R.; Liaw, A. Newer classification and regression tree techniques: Bagging and random forests for ecological prediction. *Ecosystems* **2006**, *9*, 181–199. [CrossRef]

33. Huang, Z.; Siwabessy, J.; Nichol, S.L.; Brooke, B.P. Predictive mapping of seabed substrata using high-resolution multibeam sonar data: A case study from a shelf with complex geomorphology. *Mar. Geol.* **2014**, *357*, 37–52.

34. Diesing, M.; Kröger, S.; Parker, R.; Jenkins, C.; Mason, C.; Weston, K. Predicting the standing stock of organic carbon in surface sediments of the North–West European continental shelf. *Biogeochemistry* **2017**, *135*, 183–200. [CrossRef]

35. Hasan, R.C.; Ierodiaconou, D.; Monk, J. Evaluation of Four Supervised Learning Methods for Benthic Habitat Mapping Using Backscatter from Multi-Beam Sonar. *Remote Sens.* **2012**, *4*, 3427–3443. [CrossRef]

36. Cutler, D.; Edwards, T.; Beards, K.; Cutler, A.; Hess, K.; Gibson, J.; Lawler, J. Random Forests for classification in Ecology. *Ecology* **2007**, *88*, 2783–2792. [CrossRef] [PubMed]

37. James, G.; Witten, D.; Hastie, T.; Tibshirani, R. *An Introduction to Statistical Learning*; Springer: Berlin/Heidelberg, Germany, 2013; ISBN 9781461471370.

38. Liaw, A.; Wiener, M. Classification and regression by randomForest. *R News* **2002**, *2*, 18–22. [CrossRef]

39. Roberts, J.J.; Best, B.D.; Dunn, D.C.; Treml, E.A.; Halpin, P.N. Marine Geospatial Ecology Tools: An integrated framework for ecological geoprocessing with ArcGIS, Python, R, MATLAB, and C++. *Environ. Model. Softw.* **2010**, *25*, 1197–1207. [CrossRef]

40. Perkins, N.J.; Schisterman, E.F. The inconsistency of "optimal" cutpoints obtained using two criteria based on the receiver operating characteristic curve. *Am. J. Epidemiol.* **2006**, *163*, 670–675. [CrossRef] [PubMed]

41. Kuhn, M. Building Predictive Models in R Using the caret Package. *J. Stat. Softw.* **2008**, *28*, 1–26. [CrossRef]

42. Downie, A.L.; Dove, D.; Westhead, R.K.; Diesing, M.; Green, S.; Cooper, R. *Semi-Automated Mapping of Rock in the North Sea*; JNCC Report 592; JNCC: Peterborough, UK, 2016. [CrossRef]

43. Freeman, E.A.; Moisen, G. PresenceAbsence: An R Package for Presence Absence Analysis. *J. Stat. Softw.* **2008**, *23*, 1–31. [CrossRef]

44. Sing, T.; Sander, O.; Beerenwinkel, N.; Lengauer, T. ROCR: Visualizing classifier performance in R. *Bioinformatics* **2005**, *21*, 3940–3941. [CrossRef] [PubMed]

45. Rattray, A.J.; Ierodiaconou, D.; Monk, J.; Laurenson, L.; Kennedy, P. Quantification of Spatial and Thematic Uncertainty in the Application of Underwater Video for Benthic Habitat Mapping. *Mar. Geodesy* **2014**, *37*, 315–336. [CrossRef]

46. Diesing, M.; Mitchell, P.; Stephens, D. Image-based seabed classification: What can we learn from terrestrial remote sensing? *ICES J. Mar. Sci.* **2016**, *73*, 2425–2441. [CrossRef]

47. De Clippele, L.H.; Gafeira, J.; Robert, K.; Hennige, S.; Lavaleye, M.S.; Duineveld, G.C.A.; Huvenne, V.A.I.; Roberts, J.M. Using novel acoustic and visual mapping tools to predict the small-scale spatial distribution of live biogenic reef framework in cold-water coral habitats. *Coral Reefs* **2017**, *36*, 255–268. [CrossRef]

48. Lillis, H. *A Three-Step Confidence Assessment Framework for Classified Seabed Maps*; JNCC Report 591; JNCC: Peterborough, UK, 2016.
49. Titschack, J.; Baum, D.; De Pol-Holz, R.; López Correa, M.; Forster, N.; Flögel, S.; Hebbeln, D.; Freiwald, A. Aggradation and carbonate accumulation of Holocene Norwegian cold-water coral reefs. *Sedimentology* **2015**, *62*, 1873–1898. [CrossRef]

geosciences

MDPI

Article

A Bathymetry- and Reflectivity-Based Approach for Seafloor Segmentation

Giuseppe Masetti * , Larry Alan Mayer and Larry Guy Ward

Center for Coastal and Ocean Mapping/National Oceanic and Atmospheric Administration (NOAA) Joint Hydrographic Center, University of New Hampshire, Durham, NH 03824, USA; lmayer@ccom.unh.edu (L.A.M.); lgward@ccom.unh.edu (L.G.W.)
* Correspondence: gmasetti@ccom.unh.edu; Tel.: +1-603-862-3452

Received: 13 November 2017; Accepted: 5 January 2018; Published: 8 January 2018

Abstract: A robust and flexible technique to segment seafloor acoustic mapping data by analyzing co-located bathymetric digital elevation models and acoustic backscatter mosaics is presented. The algorithm first uses principles of topographic openness, pattern recognition, and texture classification to identify geomorphic elements of the seafloor or "area kernels", and then derives the final seafloor segmentation by merging or splitting the kernels based on principles of similarity and multi-modality. The output is a collection of homogeneous, non-overlapping seafloor segments of consistent morphology and acoustic backscatter texture. Each labeled segment is enriched by a list of derived, physically-meaningful attributes that can be used for subsequent task-specific analysis.

Keywords: Acoustic applications; object segmentation; seafloor; underwater acoustics

1. Introduction

Numerous studies have described approaches to seafloor segmentation and classification using acoustic backscatter data from multibeam sonar [1–9] or, alternatively, seafloor bathymetry [10–15]. However, there are few studies that have offered general methods for using a machine-focused approach to combine and use the information found in co-located bathymetric digital elevation models (DEMs) and acoustic mosaics [16–22]. Modern multibeam sonars and processing software now typically produce geo-located bathymetry and backscatter mosaic products, thus offering the opportunity to treat both data sets together [21,23]. This paper explores a methodology to combine both bathymetry and backscatter data to automatically segment the seafloor.

The proposed method attempts to mimic the approach taken by a skilled analyst assuming that, when called upon to manually segment a seafloor area, the analyst initially evaluates the context surrounding the area and attempts to take full advantage of both bathymetric and reflectivity products rather than focusing on small-scale geomorphometric variability (e.g., local rugosity). The result is a bathymetry- and reflectivity-based estimator for seafloor segmentation (BRESS) that mimics the positive aspects of the segmentation process as performed by a skilled analyst (e.g., the use of context and multiple inputs) but avoids the inherent deficiencies (subjectivity, processing time, lack of reproducibility). The initial phase of the algorithm performs a segmentation of the DEM surface through the identification of its seafloor "geoform" elements (i.e., contiguous regions of similar morphology, e.g., valleys or edges). These elements represent "area kernels" (regions of consistent morphological type) whose backscatter is then analyzed to derive final seafloor segments by merging or splitting the kernels based on the principles of similarity and multi-modality. The output of BRESS is a collection of homogeneous, non-overlapping seafloor segments. Each labeled segment is enriched by a list of derived, physically-meaningful attributes that can be used for task-specific analysis (e.g., habitat mapping, backscatter model inversion, or change detection).

2. Methods

2.1. Area Kernels Based on Landform Classification

The BRESS algorithm starts by performing a preliminary bathymetry-derived segmentation, that is, a segmentation of the DEM surface through the identification of its seafloor geomorphological (geoform) elements. From the perspective of the final task of the algorithm (a collection of homogeneous, non-overlapping seafloor segments based on both morphology and backscatter), this preliminary step usually provides a general over-segmentation of the area (i.e., a homogenous area is split into more than one segment), although localized cases of under-segmentation (a single segment covers a non-homogenous area) are also possible since only the bathymetry is analyzed in the first step.

Common methods to segment the seafloor into geoform elements use differential geometry and geo-morphometric proxies that are derived locally from the DEM by calculating a variable combination of first and second derivatives [24]. These methods offer several approaches for how the geo-morphometric variables are used (mainly, cell-based or object-based), with respect to the selected units of classification (landforms, landform elements, and physiographic units), and the adopted classifier (based on expert knowledge or driven by a machine learning algorithm) [25,26]. However, well-known limitations (particularly the sensitivity to the selected scale) of these methods have impact on the stability of the overall segmentation algorithm [27–29]. Due to this scale-dependent limitation, we have taken a different approach that uses the principles of topographic openness, pattern recognition and texture classification to create 'area kernels' (regions of consistent morphological type), evolving this approach from the innovative concepts introduced by [30,31].

To derive geoform elements, we have adapted the concepts of Local Ternary Pattern (LTP) [32] and texton [33] to trigger the preliminary bathymetry-derived segmentation. In image processing, the LTP represents an evolution of the Local Binary Pattern (LBP) texture descriptor introduced by [34]. LBP labels the pixels of an image by thresholding the neighborhood of each pixel and characterizing it with a binary indicator (i.e., a "+" or a "-" that describes the relative value of the pixel with respect to its neighborhood). Despite its simplicity, LBP has been applied successfully to tasks such as texture classification, face recognition, and background modeling [32]. LTP adds a third (neutral) level to the original two possible LBP levels of contrast variation against the central cell. Thus, each direction of evaluation of LTP has three possible states: "+", "0", and "-".

Together with LTP, the BRESS algorithm borrows concepts from texton theory. This theory proposes a model of how humans perceive texture [35]. Based on this theory, the human brain, when in the pre-attentive mode, does not process complex forms and yet in parallel, without effort or scrutiny, can easily recognize differences in a few local conspicuous features (i.e., the textons) over the entire human visual field [33]. Following a similar approach, we apply a preliminary segmentation that is obtained by extracting bathymetric features (directly derived from LTP evaluation for each DEM node) and then apply a seafloor geoform classification scheme to identify the constituent regions of connected nodes with homogeneous bathymorphologic characteristics. These regions represent preliminary, bathymetry-only derived segments and, based on the fact that they will be used to derive the final segments, we call them "area kernels". In this context, we have created a bathymorphologic texton. For similarity with the blend word of "geomorphon" adopted in [31], we use the term of "bathymorphon" as the bathymorphologic archetype to label the nodes of a bathymetric DEM.

The bathymorphon is derived from the LTP to capture the local morphologic context of each DEM node. In the bathymorphological realm, the neutral level of the LTP (a "flat") identifies the application-specific absence of meaningful slope variation in comparison to "shoal" and "deep" levels (for positive and negative slopes, respectively). In principle it is possible to identify more than just these three levels (i.e., "flat", "shoal", and "deep"), however, our observations support those of [31,34] in concluding that a ternary solution is able to capture enough of the data structure to appropriately describe the morphologic variation while keeping the overall approach relatively simple.

In order to capture bathymorphologic elements at the desired scales, the algorithm supports the definition of a search annulus contained between an internal radius (r_i) acting like a low-pass filter, and an external radius (r_e) that limits the extent of the spatial analysis. Although the node neighborhood may be potentially evaluated in any range of directions, exploratory tests have shown that limiting the analysis to eight directions (d_n) (the four cardinal directions and the four main inter-cardinal directions) provides a good working tradeoff between computation efficiency and stability of the retrieved information (Figure 1).

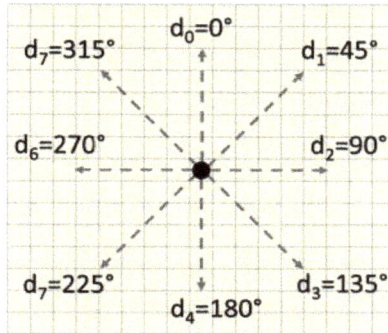

Figure 1. Selected eight directions (d_n) surrounding a DEM node in solid black. Each direction is evaluated independently by only taking into account the neighborhood nodes between the search annulus identified by the internal (r_i) and external radii (r_e).

Adopting the described analytic schema, bathymorphons can only fall into a finite number of configurations based on the ternary nature of the LTP and the eight selected neighbor directions (Figure 2). Having three possible states for each of the eight directions, the number of possible LTP values is 6561 (3^8). However, given the symmetry of many of these configurations, the actual number of unique bathymorphon classes, after having evaluated all the possible rotations and mirroring operations, is 498 [31]. Based on evidence from preliminary tests, some of these morphological seafloor types are very common, while others describe quite rare forms (an example of bathymorphon distribution is presented in Figure 3).

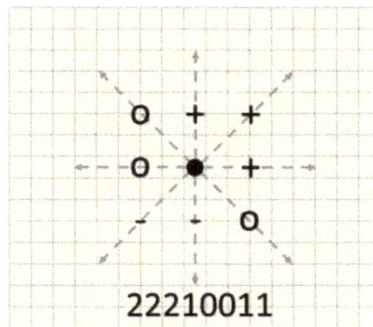

Figure 2. Example of LTP evaluation surrounding a DEM node whereas a "+" is used to identify a "shoal", a "0" for "flat", and a "-" for "deep" configurations, respectively. Following the convention described in the text, those levels are translated in a ternary value: a ternary digit ("-" = 0, "0" = 1, "+" = 2) for each of the eight directions (the "22210011" code in the example provided). The ternary value will then be reduced to one of the possible 498 bathymorphon classes.

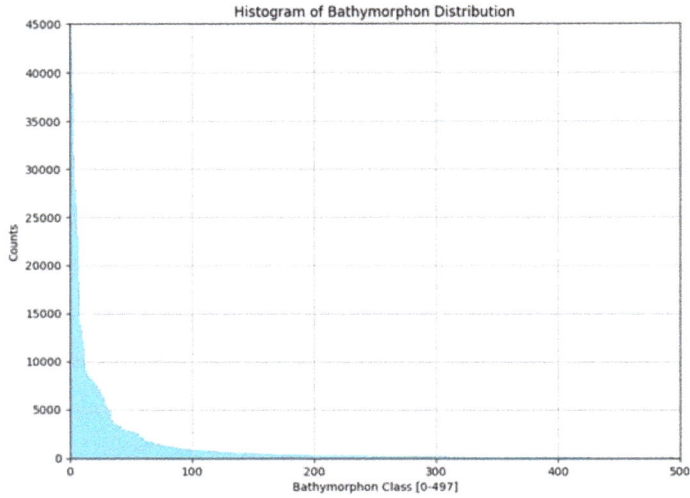

Figure 3. Histogram providing the distribution of the bathymorphon classes for the testing dataset described in the Results section.

The evaluation on the assigned ternary level ($L_{r_i,r_e}^{d_n}$) per direction is based on the line-of-sight principle: two straight lines are passed connecting each DEM node to the "visible" highest and lowest node in each of the eight directions identified in Figure 1. The line-of-sight principle is implemented by using a user-defined parametric angular flatness threshold (α) and the difference between the zenith ($\varphi_{r_i,r_e}^{d_n}$) and the nadir ($\psi_{r_i,r_e}^{d_n}$) angles (Figure 4) as defined in Equation (1):

$$L_{r_i,r_e}^{d_n} \begin{cases} 2 & \text{if } \psi_{r_i,r_e}^{d_n} - \varphi_{r_i,r_e}^{d_n} > \alpha \\ 1 & \text{if } \left| \psi_{r_i,r_e}^{d_n} - \varphi_{r_i,r_e}^{d_n} \right| < \alpha \\ 0 & \text{if } \psi_{r_i,r_e}^{d_n} - \varphi_{r_i,r_e}^{d_n} < -\alpha \end{cases} \tag{1}$$

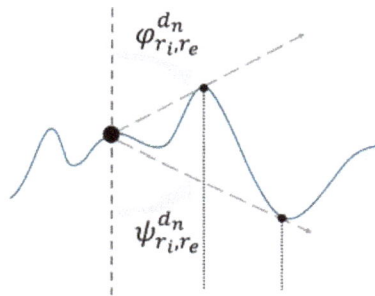

Figure 4. A simplified, two-dimensional representation of DEM nodes (represented by black dots) and the underline the real surface (blue line). In particular, the large black dot represents the currently evaluated node, while the smaller dots are nodes in its surrounding, along the same direction. The line-of-sight principle is adopted to define the node openness in terms of zenith ($\varphi_{r_i,r_e}^{d_n}$) and nadir ($\psi_{r_i,r_e}^{d_n}$) angles. Both angles are defined to be always positive, within a range from $0°$ to $180°$. The difference among the two angles is used to define the ternary level in each node direction.

Using α and the described angular difference, each node is assigned to a bathymorphon class that expresses the degree of dominance or enclosure of a node location on an irregular surface (i.e., the openness) [30], at the user-identified scale (i.e., the search annulus). This line-of-sight approach has the advantage of reducing complications derived from having to manage the range of spatial scales that is often a limitation in approaches based on differential geometry.

As demonstrated in [31], the bathymorphons can be grouped into a relatively small (ten) number of landform classes that capture most the relevant morphologic relationships related to landform description. In our case we believe that the majority of the information can be captured by mapping the generated bathymorphons into six possible seafloor geoform classes, each representing a common bathymorphologic element, using the lookup table provided in Table 1. These six classes thus represent a simplification of the lookup table proposed in [31]. Pits and peaks present in [31] have been merged with valleys and ridges, respectively; and, the concave- and convex-slope cases describe in [34] redistributed among the slope, shoulder, and footslope classes provided in Table 1.

Table 1. Lookup table adopted to map the bathymorphons (listed on the left inset together with their abbreviations) to the six seafloor form classes of interest for this step of the segmentation. Given the possibility of having a neutral level (a "flat"), the number of "shoals" and "deeps" surrounding the node point may vary between zero and eight. The header row and column provide the total number of positive and negative levels (respectively) for the eight directions surrounding each node. FL = Flat, RI = Ridge, SH = Shoulder, SL = Slope, FS = Footslope, VL = Valley.

+ / −	0	1	2	3	4	5	6	7	8
0	FL	FL	FL	FS	FS	VL	VL	VL	VL
1	FL	FL	FS	FS	FS	VL	VL	VL	-
2	FL	SH	SL	SL	SL	VL	VL	-	-
3	SH	SH	SL	SL	SL	SL	-	-	-
4	SH	SH	SH	SL	SL	-	-	-	-
5	RI	RI	RI	SL	-	-	-	-	-
6	RI	RI	RI	-	-	-	-	-	-
7	RI	RI	-	-	-	-	-	-	-
8	RI	-	-	-	-	-	-	-	-

The lookup table step usually greatly reduces the complexity of the segmentation by at least one order of magnitude. The next step is the creation of area kernels, i.e., connected regions of common bathymorphon class. The area kernels are created by applying the connected components labeling algorithm (specifically, the Block Based with Decision Trees algorithm described in [36]) to the seafloor form-classified grid creating a bathy-morphometric map. This step of the algorithm scans the created map with bathymorphon classes, and groups its nodes into components based on node connectivity: all nodes in a connected component share same bathymorphon class and are connected with each other. Once all groups have been determined, each node is labeled according to the component it was assigned to.

The algorithm adopted for the definition of the area kernels can be summarized in the following four main steps applied to each node of the grid:

- Calculation of the ternary value based on neighborhood and search annulus.
- Reduction (by mirroring and rotating) of the ternary value to one of the 498 bathymorphon classes.
- Assignment of each bathymorphon to one of the six seafloor geoform classes through a user-modifiable lookup table (Table 1).
- Creation of the area kernels by clustering all the connected nodes within the same geoform class.

After preliminary tests, the described approach highlighted two possible distortions in the classification: nodes close to the edge of the surface, and in case of the selection of a large external radius for the search annulus. Two optional corrections have thus been identified and introduced: a "Node-on-the-Edge" correction that classifies only nodes that have a minimum number of valid ternary levels (default value adopted for this parameter is 6); and, an "extended-form" correction that stops, after a given distance, the effects of the angular flatness threshold on the resulting calculated elevation used to identify a "flat".

2.2. Derivation of Seafloor Segments

In the second and final phase, the algorithm then evaluates each area kernel within the context of the backscatter mosaic much as an experienced analyst would use the backscatter to understand the context of given morphological regions. The region of each area kernel in the bathy-morphometric map is evaluated using only the co-located pixels in the acoustic backscatter mosaic. This evaluation is performed both in isolation (to assess the requirement subdividing into smaller area kernels), and by pairwise comparison to other area kernels of the same seafloor form type (to cluster area kernels with similar characteristics). Thus, the area kernels identified in the bathymetric realm are split or merged based on the intensity-level distribution of the pixels in the corresponding region in the mosaic. To maximize the robustness of this operation and avoid the introduction of biases, the acoustic backscatter mosaic has to be created using the best practices for radiometric and geometric corrections (e.g., the effect of local slope) and for normalization to minimize the angular dependency and possible artifacts in the collected reflectivity data [5,23].

A normalized (45°), slope-corrected (by deriving the true incidence angle from DEM) intensity-level histogram for each individual area kernel is created and then analyzed to identify the possible presence of multiple modes (Figure 5). The process first identifies the index of the bin of the peak by taking the first order differences, then enhances the resolution of the peak detection by using Gaussian fitting, centroid computation [37,38]. The multi-modal detection is tailored by adopting two customizable parameters: A is the amplitude threshold given as percentage of the total number of elements (that is, the peaks with amplitude lower than the given threshold are ignored) and D represents the minimum distance between peaks. These parameters can be used to improve the algorithm robustness in case of the presence of small artifacts in the acoustic mosaic. The identification of more than one peak triggers the execution of a simple k-means algorithm [39], an unsupervised learning algorithm commonly used to solve the well-known clustering problem. This algorithm is used here to split the nodes belonging to a specific area kernel into a number of clusters defined by the identified peaks.

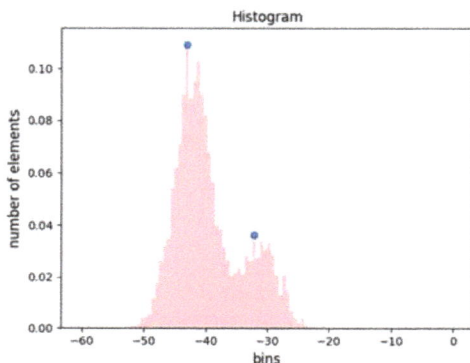

Figure 5. Each individual area kernel is analyzed to detect the presence of multi-modal distribution. The figure shows an example with a bi-modal distribution (the two peaks are represented as blue dots). The bin values represent decibel values, but the same approach can be adopted by adapting the bins to dimensionless digital numbers. The number of elements is normalized.

The evaluation in the pairwise comparison is performed through a simplified histogram comparison approach [40,41]. The percentage of intersection (*I*) between a pair of histograms is adopted as a criterion to evaluate whether the area kernel belonging to the same seafloor form type (e.g., a valley) has the same textural characteristics in the mosaic to be judged as representative of the same seafloor segment (Figure 6). The value used as the merging threshold varies based on the specific task that the algorithm is adopted for; however, initial tests have identified a validity range between 50% and 80%.

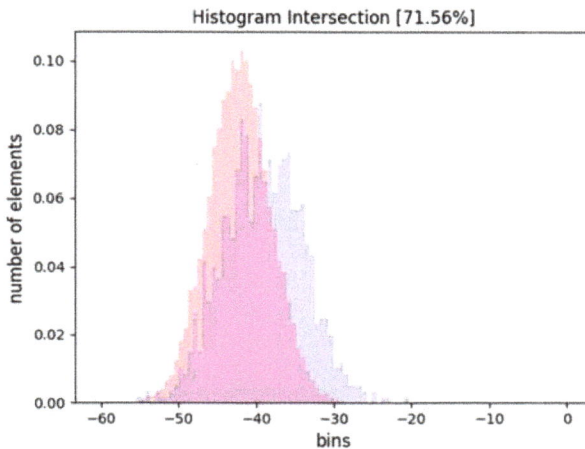

Figure 6. The histogram intersection is used as a criterion to compare pair of area kernels. The figure shows an example whereas the two normalized histograms overlap for more than 71% and are, thus, considered to come from seafloor with similar acoustic response.

As a rule of thumb, backscatter data collected with modern multibeam echo sounders, typically produces a mosaic with resolution of 2–3 times the resolution of the corresponding bathymetric DEM. In order to apply the two described steps to area kernels with statistics that are stable, only area kernels of at least 10 nodes (corresponding at about 40 pixels on the mosaic) are evaluated. Area kernels smaller than this size are currently staged as unclassified.

3. Results

In order to test the described bathymetric and reflectivity-based estimation method for seafloor segmentation, we applied it to a well-studied bedform field in the mouth of the Piscataqua River, a well-mixed estuary between New Hampshire and Maine, USA. The area was selected because it has been the subject of numerous previous mapping efforts including studies of bedform migration [42], automated segmentation [10], and seafloor scattering models [43,44]. The study area is centered on a shallow, sandy sediment region, determined by multiple means (sonars, divers, and video observations) to be a rippled sand-wave field composed of largely medium to coarse sand and fine shell hash, surrounded by bedrock and gravelly channel sediments [10,42,44,45]. The bedform field is a persistent, elongated feature with its major axis oriented north-south along the main channel axis of the lowermost part of the Piscataqua River estuary [42].

The bathymetric DEM and the acoustic backscatter mosaic was constructed using data collected with a dual-head Kongsberg (Kongsberg, Norway) EM3002D multibeam echosounder (operating at 300-kHz central frequency and installed on the University of New Hampshire's *R/V Coastal Surveyor*). Positioning and attitude data were collected from an Applanix (Richmond Hill, Ontario, Canada) POS/MV system with integrated real-time kinematic GPS. The bathymetric data were processed and gridded at a 0.5-m resolution using QPS Qimera (version 1.5.5) processing software (Quality Positioning Services BV, Zeist, The Netherland), while the backscatter mosaic was generated using Center for Coastal and Ocean Mapping (Durham, NH, USA)'s in-house research code [5,46] at 0.13-m resolution (Figure 7); backscatter is presented normalized to a 45 degree angle of incidence. Both products were created using a UTM 18N/WGS84 cartographic projection and stored in a portable ASCII grid format.

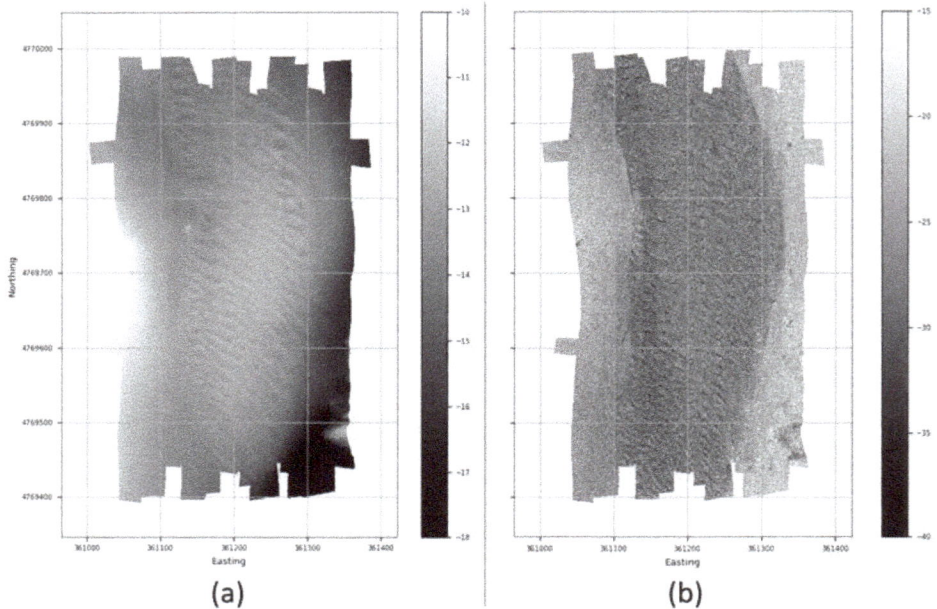

(a)   (b)

Figure 7. Test data inputs collected over a bedform field located in the mouth of the Piscataqua River, USA. (**a**) Bathymetric DEM with a 50-cm resolution. (**b**) Acoustic backscatter mosaic assembled at 13-cm resolution.

The BRESS algorithm is implemented mainly in C++11, with a graphical user interface created in Python. The implementation creates all the outputs (the final products and several optional intermediate layers) with the same shape and projection of the original bathymetric DEM. Although not required by the adopted set of input data, the code implementation is able to identify inputs with different projections and re-project the mosaic input to match the projection of the bathymetric input.

An example of the processing steps is shown in Figure 8. Figure 8a shows the initial LTP values calculated for each node in the DEM (with: $r_i = 3$ nodes, $r_e = 10$ nodes and $= 1°$). Using this parameter set, the length scale of detectable features is between 1.5 and 5 m. Figure 8b shows the reduction of the original 6561 LPT values to the six seafloor geoform classes described in Table 1. All six of the geoform types are present in the dataset, and this representation shows a clear differentiation between the central rippled sand-wave field and the surrounding regions that matches nicely the manually-derived delineation of physiographic differences in the area [47,48]. For some applications, this simple segmentation based on bathymetry alone may be adequate. However, if an application requires information on sediment type (e.g., grain size), the bathymetrically-derived segmentation may not suffice. For instance, a flat seafloor that has various sediment types will likely end in a single seafloor geoform.

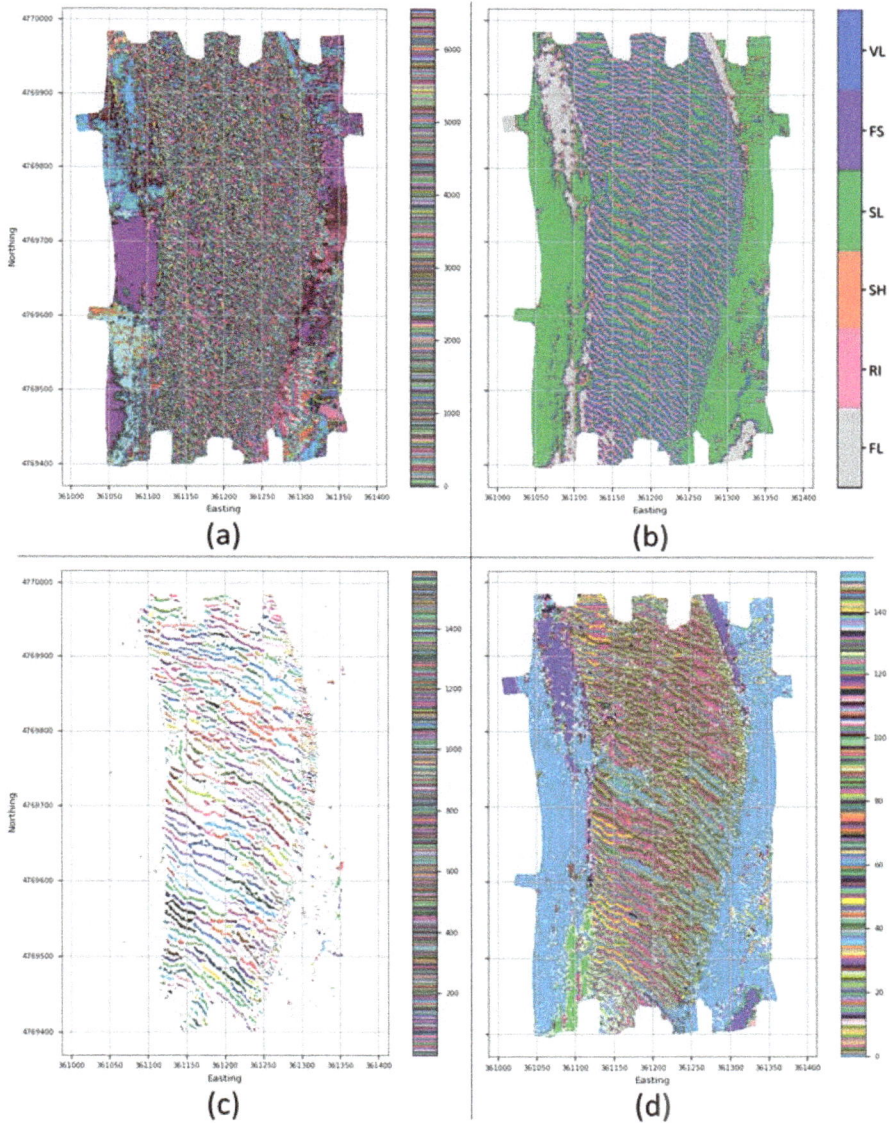

Figure 8. The four major steps of the BRESS algorithm. (**a**) A ternary value is associated to each node of the DEM. (**b**) The 6561 (3^8) possible ternary values are first reduced to the 498 bathymorphons, then mapped to six geoform classes of interest (abbreviations defined in Table 1). (**c**) The area kernels are created by clustering all the connected nodes within the same geoform class (valley in the shown examples). Each area kernel has assigned a random color. (**d**) The output seafloor segments are generated by analyzing each area kernel in isolation, to assess the requirement of its sub-division in smaller area kernels, and by pairwise comparison to other area kernels of the same geoform type.

To bring in the context of the backscatter mosaic, the area kernels are then generated by clustering all the connected nodes within the same geoform class (valleys are shown in Figure 8c). Finally, each area kernel is analyzed in isolation to assess the need for sub-division into smaller area kernels ($A = 0.02\%$ *of total elements*, $D = 10\ dB$), and by pairwise comparison to other area kernels of the same seafloor form type ($I = 60\%$) to generate the final segmentation in Figure 8d. The final segmentation provides the advantage of also capturing the effects of areas of different seafloor reflectivity (i.e., it shows whether the connected geoform segments (valleys in our example) are homogenous or segmented with respect to backscatter). By analyzing the different labels assigned within the same type of geoforms, it is now possible to identify detailed patterns of variability. By stopping the analysis at the area kernels in Figure 8b, the presence of these patterns would have been otherwise ignored.

Figure 9 shows the impact of using different values for the search annulus for a portion of the study area presented in Figure 8. The portion area in Figure 9a presents a larger number of segments when compared to Figure 9b as the result of increasing the range of the search annulus (from $r_i = 3$ *nodes*, $r_e = 10$ *nodes* to $r_i = 6$ *nodes*, $r_e = 20$ *nodes*), thus, by capturing only geoforms at a scale larger than 3 m. Although decreased, the effect of reduction in output complexity by increasing the search annulus is also present in Figure 9c. By changing the search parameters, the BRESS algorithm can be tuned for different usages and scenarios.

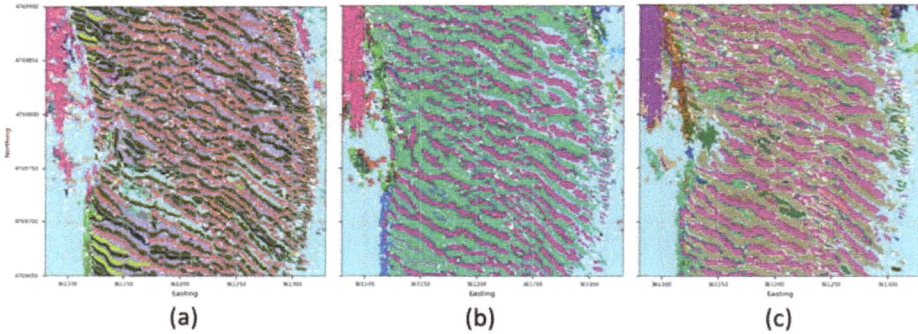

(a)   **(b)**   **(c)**

Figure 9. The three insets show, on a central portion of the study area, the effects of adopting search annuli of increasing size: (**a**) $r_i = 3$ *nodes*, $r_e = 10$ *nodes*; (**b**) $r_i = 6$ *nodes*, $r_e = 20$ *nodes*; and (**c**) $r_i = 9$ *nodes*, $r_e = 30$ *nodes*. A direct effect of such variation is a reduction on the number of output segments for the whole study area: (a) 154, (b) 118, and (c) 96.

Finally, an exploratory evaluation of the advanced discrimination capabilities of the algorithm is presented in Figure 10. The region shown in Figure 10 is the rippled sand-wave field who's central region is generally characterized as medium sand and that has been shown to be stable over the years [42,47]. Figure 10 shows just the "valley" and "ridge" bathymorphons in the area, in this case, the troughs and crests of the sand waves. The analysis of the backscatter of the valleys and the ridges shows that they vary in their reflectivity behavior in a spatially-consistent manner. For instance, the cluster of yellow (for valleys) and orange (for ridges) segments present in Figure 10 in the south-west region of the sand wave area (pointed by the red arrow) versus the cluster of blues (for valleys) and dark green (for ridges) segments in the central region (light green arrow). These clusters appear to correlate with the variations in the percentage of gravel and shells based on the limited ground-truth datasets available (Figure 10).

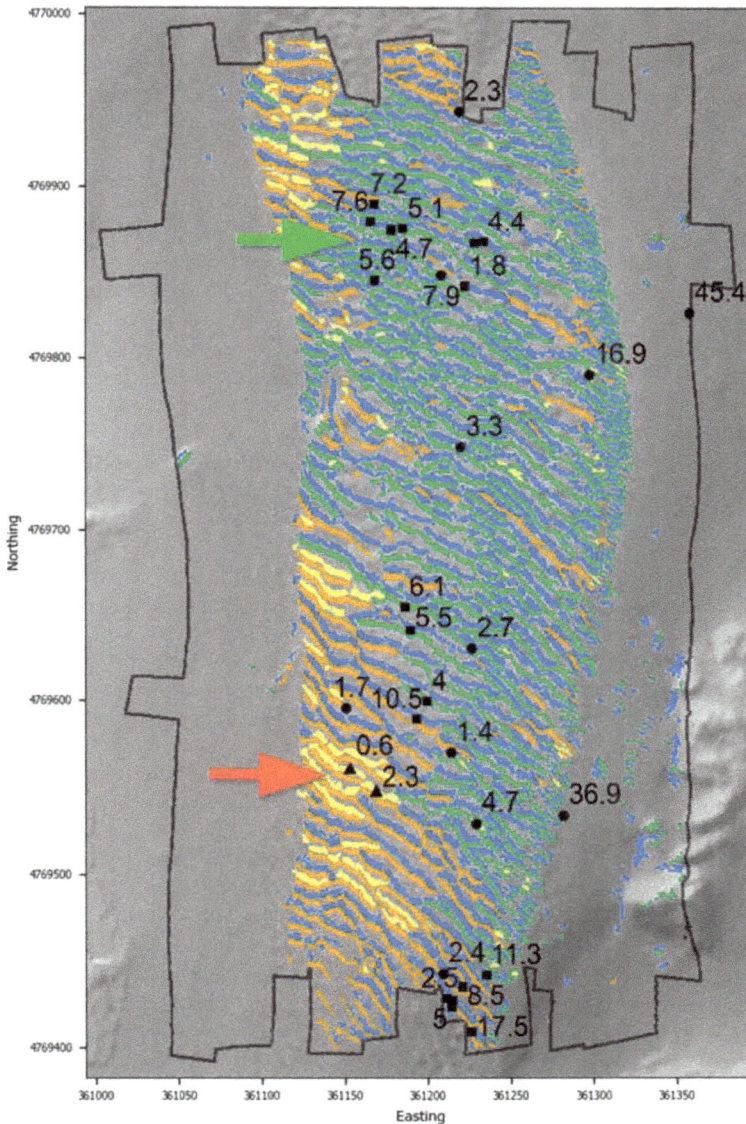

Figure 10. Algorithm segmentation output for valley (in yellow and blue) and ridge (in orange and green) geoforms compared with collected samples. The numerical values shown represent the percentage of gravel in the retrieved sediments. The samples were collected by three different studies that are represented using different symbol shapes: the circles for [42]; the squares for [44]; and the triangles for [45]. Shaded bathymetric relief is shown as the background and the survey polygon of the whole test area is represented with a gray line. The light green arrow and the red arrow point to areas with relatively high and low percentages of gravel and shells, respectively.

Although this correlation is promising, it is based on limited data collected for other purposes. With this new analysis, we can now (and will) test the discrimination capabilities of the algorithm by carefully designing a sampling plan to ground-truth the different segment types or test the fit-for-purpose for field-specific applications (e.g., the evaluation of seabed habitat maps presented in [49]). In fact, the collection of additional ground-truth values will provide means to evaluate the efficacy of the BRESS method in comparison with other available methods: for example, the Object Based Image Analysis applied to both MBES bathymetry and acoustic backscatter [19,20] or the combined use of the terrain attributes obtained with Benthic Terrain Modeler [14,50], and a parallel classification of the acoustic backscatter.

From a theoretical point of view, the BRESS method presents the key advantage of having both physically meaningful and statistically intuitive (intermediate and final) steps. In other words, the processing steps can be translated into relatively simple questions (e.g., what geoform class does this DEM node belong to? Which range of scales is the target of the analysis? By looking at the reflectivity content, is a specific pair of area kernels similar?). This intuitiveness should facilitate its adoption and proper use, while other existing methods often require quite abstract evaluations (e.g., the tuning of "magic numbers" used as parameters) or deep statistical insights on the study area. The amount of local, specific knowledge often required by methods like [19] and [50] makes them difficult to be ported, with a consistent success classification rate, to different study areas. The evaluation of the BRESS method in comparison with those methods should also consider such factors.

4. Conclusions

BRESS offers a novel approach for the quantitative analysis of ocean mapping data. The incentive for developing this method was a desire to create an automated, scale-independent, robust and computationally efficient technique to segment the seafloor for a range of research applications. The method differs from standard approaches that tend to use either bathymetry or backscatter data independently in that it attempts to emulate the approach of a skilled analyst by using the full context of both the bathymetry and the backscatter (Figure 11 summarizes the algorithm workflow).

From a bathymetric perspective, the adoption of the concept of grid openness (using thresholds of flatness) makes the algorithm self-adaptive within the identified search annulus, with a clear gain in computational efficiency [31]. On the acoustic reflectivity side, the derivation of the seafloor segments based on local normalized histograms makes the method robust in the presence of the many possible artifacts that could be present in an acoustic backscatter mosaic [46,51]. The optional corrections, "Node-on-the-Edge" and "Extended-Form", improve the overall algorithm robustness for cases in which the general approach was shown to have possible weaknesses.

The output of BRESS is a collection of homogeneous, non-overlapping seafloor segments. Each labeled segment is enriched by a list of derived, physically-meaningful attributes that can be used for subsequent task-specific analysis. The ability to natively perform a multi-scale analysis through a prescribed search annulus mitigates the risk of mismatch of spatial scales between measurements and their interpretation. Although the method cannot overcome the limitations that result from the inherent resolution of the system used for data collection [52], recent developments in acoustic mapping systems are currently achieving an unprecedented high-resolution view of the seafloor at a broad range of spatial scales. In particular, modern multibeam echosounders may produce continuous coverage depth measurements and co-located, high-quality reflectivity measurements that reveal well-defined texture patterns.

The described method is able to identify patterns of seafloor topography representing areas of homogenous geomorphological feature types and seafloor textures remotely sensed by acoustic devices. Given the relevance of this information to the spatial distributions of habitats, we believe that the method has a potential application for habitat mapping. Although habitat delineations can be done manually, robust automated procedures, like the BRESS method, offer increased speed, efficiency, more objectivity, and reproducible map products. Another possible application is to improve the understanding of seafloor stability, particularly important in the coastal environment [53]. In fact, the quality of the identification of valleys and ridges within a defined range of scales makes BRESS outputs a good candidate for use in the calculation of bedform migration rates. Finally, another potential application is its adoption as a source of acoustic themes for seafloor characterization by backscatter model inversion [54].

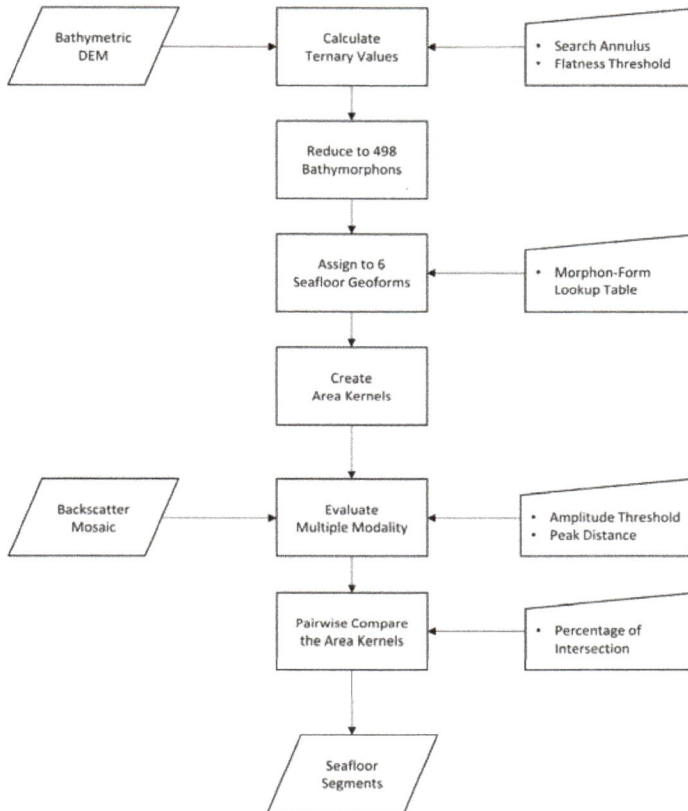

Figure 11. Flowchart identifying the main inputs, outputs, and processing steps of the BRESS algorithm.

Acknowledgments: This work was supported by the NOAA grant NA15NOS4000200.

Author Contributions: Giuseppe Masetti and Larry Alan Mayer conceived and designed the algorithm; and Larry Guy Ward contributed to its improvement and to the analysis of the results.

Conflicts of Interest: The authors declare no conflict of interest. The founding sponsors had no role in the design of the study; in the collection, analyses, or interpretation of data; in the writing of the manuscript; or in the decision to publish the results.

References

1. De Moustier, C.; Alexandrou, D. Angular dependence of 12-khz seafloor acoustic backscatter. *J. Acoust. Soc. Am.* **1991**, *90*, 522–531. [CrossRef]
2. Hughes Clarke, J.E.; Mayer, L.A.; Wells, D.E. Shallow-water imaging multibeam sonars: A new tool for investigating seafloor processes in the coastal zone and on the continental shelf. *Mar. Geophys. Res.* **1996**, *18*, 607–629. [CrossRef]

3.   Augustin, J.; Dugelay, S.; Lurton, X.; Voisset, M. Applications of an image segmentation technique to multibeam echo-sounder data. In Proceedings of the OCEANS'97, MTS/IEEE Conference Proceedings, Halifax, NS, Canada, 6–9 October 1997; pp. 1365–1369.

4.   Rzhanov, Y.; Fonseca, L.; Mayer, L. Construction of seafloor thematic maps from multibeam acoustic backscatter angular response data. *Comput. Geosci.* **2012**, *41*, 181–187. [CrossRef]

5.   Fonseca, L.; Mayer, L. Remote estimation of surficial seafloor properties through the application angular range analysis to multibeam sonar data. *Mar. Geophys. Res.* **2007**, *28*, 119–126. [CrossRef]

6.   Lamarche, G.; Lurton, X.; Verdier, A.L.; Augustin, J.M. Quantitative characterisation of seafloor substrate and bedforms using advanced processing of multibeam backscatter—Application to cook strait, New Zealand. *Cont. Shelf Res.* **2011**, *31*, S93–S109. [CrossRef]

7.   De, C.; Chakraborty, B. Model-based acoustic remote sensing of seafloor characteristics. *IEEE Trans. Geosci. Remote Sens.* **2011**, *49*, 3868–3877. [CrossRef]

8.   Kloser, R.J.; Penrose, J.D.; Butler, A.J. Multi-beam backscatter measurements used to infer seabed habitats. *Cont. Shelf Res.* **2010**, *30*, 1772–1782. [CrossRef]

9.   Masetti, G.; Calder, B. Remote identification of a shipwreck site from mbes backscatter. *J. Environ. Manag.* **2012**, *111*, 44–52. [CrossRef] [PubMed]

10.   Cutter, G.R.; Rzhanov, Y.; Mayer, L.A. Automated segmentation of seafloor bathymetry from multibeam echosounder data using local fourier histogram texture features. *J. Exp. Mar. Biol. Ecol.* **2003**, *285*, 355–370. [CrossRef]

11.   Malinverno, A. Segmentation of topographic profiles of the seafloor based on a self-affine model. *IEEE J. Ocean. Eng.* **1989**, *14*, 348–359. [CrossRef]

12.   Wilson, M.F.J.; O'Connell, B.; Brown, C.; Guinan, J.C.; Grehan, A.J. Multiscale terrain analysis of multibeam bathymetry data for habitat mapping on the continental slope. *Mar. Geod.* **2007**, *30*, 3–35. [CrossRef]

13.   Dolan, M.; Thorsnes, T.; Leth, J.; Al-Hamdani, Z.; Guinan, J.; Van Lancker, V. Terrain Characterization from Bathymetry Data at Various Resolutions in European Waters—Experiences and Recommendations. Available online: https://www.ngu.no/en/publikasjon/terrain-characterization-bathymetry-data-various-resolutions-european-waters-experiences (accessed on 4 January 2018).

14.   Lundblad, E.R.; Wright, D.J.; Miller, J.; Larkin, E.M.; Rinehart, R.; Naar, D.F.; Donahue, B.T.; Anderson, S.M.; Battista, T. A benthic terrain classification scheme for american samoa. *Mar. Geod.* **2006**, *29*, 89–111. [CrossRef]

15.   Zarai, M.; Boudraa, A.O.; Garlan, T.; Thibaud, R.; Ray, C. Seafloor characterization by bathymetric image segmentation. In Proceedings of the OCEANS, Shanghai, China, 10–13 April 2016; pp. 1–4.

16.   Pirtle, J.L.; Weber, T.C.; Wilson, C.D.; Rooper, C.N. Assessment of trawlable and untrawlable seafloor using multibeam-derived metrics. *Methods Oceanogr.* **2015**, *12*, 18–35. [CrossRef]

17.   Lawrence, E.; Hayes, K.R.; Lucieer, V.L.; Nichol, S.L.; Dambacher, J.M.; Hill, N.A.; Barrett, N.; Kool, J.; Siwabessy, J. Mapping habitats and developing baselines in offshore marine reserves with little prior knowledge: A critical evaluation of a new approach. *PLoS ONE* **2015**, *10*, e0141051. [CrossRef] [PubMed]

18.   Chakraborty, B.; Menezes, A.; Dandapath, S.; Fernandes, W.A.; Karisiddaiah, S.; Haris, K.; Gokul, G. Application of hybrid techniques (self-organizing map and fuzzy algorithm) using backscatter data for segmentation and fine-scale roughness characterization of seepage-related seafloor along the western continental margin of india. *IEEE J. Ocean. Eng.* **2015**, *40*, 3–14. [CrossRef]

19.   Lucieer, V.; Lamarche, G. Unsupervised fuzzy classification and object-based image analysis of multibeam data to map deep water substrates, cook strait, new zealand. *Cont. Shelf Res.* **2011**, *31*, 1236–1247. [CrossRef]

20.   Lacharité, M.; Brown, C.J.; Gazzola, V. Multisource multibeam backscatter data: Developing a strategy for the production of benthic habitat maps using semi-automated seafloor classification methods. *Mar. Geophys. Res.* **2017**, 1–16. [CrossRef]

21.   Brown, C.J.; Blondel, P. Developments in the application of multibeam sonar backscatter for seafloor habitat mapping. *Appl. Acoust.* **2009**, *70*, 1242–1247. [CrossRef]

22.   Tang, Q.H.; Zhou, X.H.; Liu, Z.C.; Du, D.-W. Processing multibeam backscatter data. *Mar. Geod.* **2005**, *28*, 251–258. [CrossRef]

23.   Lurton, X.; Lamarche, G.; Brown, C.; Lucieer, V.; Rice, G.; Schimel, A.; Weber, T. *Backscatter Measurements by Seafloor-Mapping Sonars: Guidelines and Recommendations*; A collective report by members of the GeoHab Backscatter Working Group; GeoHab Backscatter Working Group: Salvador da Bahia, Brazil, May 2015.

24. Lecours, V.; Dolan, M.F.; Micallef, A.; Lucieer, V.L. A review of marine geomorphometry, the quantitative study of the seafloor. *Hydrol. Earth Syst. Sci* **2016**, *20*, 3207–3244. [CrossRef]

25. Saadat, H.; Bonnell, R.; Sharifi, F.; Mehuys, G.; Namdar, M.; Ale-Ebrahim, S. Landform classification from a digital elevation model and satellite imagery. *Geomorphology* **2008**, *100*, 453–464. [CrossRef]

26. Bishop, M.P.; James, L.A.; Shroder, J.F.; Walsh, S.J. Geospatial technologies and digital geomorphological mapping: Concepts, issues and research. *Geomorphology* **2012**, *137*, 5–26. [CrossRef]

27. Lecours, V.; Devillers, R.; Schneider, D.C.; Lucieer, V.L.; Brown, C.J.; Edinger, E.N. Spatial scale and geographic context in benthic habitat mapping: Review and future directions. *Mar. Ecol. Prog. Ser.* **2015**, *535*, 259–284. [CrossRef]

28. Grohmann, C.H.; Smith, M.J.; Riccomini, C. Multiscale analysis of topographic surface roughness in the midland valley, Scotland. *IEEE Trans. Geosci. Remote Sens.* **2011**, *49*, 1200–1213. [CrossRef]

29. Wu, J. Effects of changing scale on landscape pattern analysis: Scaling relations. *Landsc. Ecol.* **2004**, *19*, 125–138. [CrossRef]

30. Yokoyama, R.; Shirasawa, M.; Pike, R.J. Visualizing topography by openness: A new application of image processing to digital elevation models. *Photogramm. Eng. Remote Sens.* **2002**, *68*, 257–266.

31. Jasiewicz, J.; Stepinski, T.F. Geomorphons—a pattern recognition approach to classification and mapping of landforms. *Geomorphology* **2013**, *182*, 147–156. [CrossRef]

32. Liao, W.H. Region description using extended local ternary patterns. In Proceedings of the 2010 20th International Conference on Pattern Recognition (ICPR), Istanbul, Turkey, 23–26 August 2010; pp. 1003–1006.

33. Julesz, B. A brief outline of the texton theory of human-vision. *Trends Neurosci.* **1984**, *7*, 41–45. [CrossRef]

34. Ojala, T.; Pietikäinen, M.; Mäenpää, T. Gray scale and rotation invariant texture classification with local binary patterns. In Proceedings of the 6th European Conference on Computer Vision Dublin, Dublin, Ireland, 26 June–1 July 2000; pp. 404–420.

35. Alvarez, S.; Vanrell, M. Texton theory revisited: A bag-of-words approach to combine textons. *Pattern Recognit.* **2012**, *45*, 4312–4325. [CrossRef]

36. Grana, C.; Borghesani, D.; Cucchiara, R. Optimized block-based connected components labeling with decision trees. *IEEE Trans. Image Process.* **2010**, *19*, 1596–1609. [CrossRef] [PubMed]

37. Sezan, M.I. A peak detection algorithm and its application to histogram-based image data reduction. *Comput. Vis. Graph. Image Process.* **1990**, *49*, 36–51. [CrossRef]

38. Abdel-Aal, R.E. Comparison of algorithmic and machine learning approaches for the automatic fitting of gaussian peaks. *Neural Comput. Appl.* **2002**, *11*, 17–29. [CrossRef]

39. MacQueen, J. Some methods for classification and analysis of multivariate observations. In Proceedings of the fifth Berkeley Symposium on Mathematical Statistics and Probability, Oakland, CA, USA, 21 June–18 July 1965; pp. 281–297.

40. Barla, A.; Odone, F.; Verri, A. Histogram intersection kernel for image classification. In Proceedings of the 2003 International Conference on Image Processing—ICIP 2003, Barcelona, Spain, 14–17 September 2003; p. III-513.

41. Boughorbel, S.; Tarel, J.-P.; Boujemaa, N. Generalized histogram intersection kernel for image recognition. Proceeding of the 2005 12th IEEE International Conference on Image Processing (ICIP 2005), Genova, Italy, 11–14 September 2005; p. III-161.

42. Felzenberg, J.A. Detecting Bedform Migration from Gigh-Resolution Multibeam Bathymetry in Portsmouth Harbor, New Hampshire, USA. Master's Thesis, University of New Hampshire, Durham, NH, USA, September 2009.

43. Weber, T.C.; Ward, L.G. Observations of backscatter from sand and gravel seafloors between 170 and 250 khz. *J. Acoust. Soc. Am.* **2015**, *138*, 2169–2180. [CrossRef] [PubMed]

44. Weber, T.C.; Ward, L. High-frequency seafloor scattering in a dynamic harbor environment: Observations of change over time scales of seconds to seasons. *J. Acoust. Soc. Am.* **2016**, *140*, 3348–3349. [CrossRef]

45. Nifong, K.L. Sedimentary Environments and Depositional History of a Paraglacial, Estuarine Embayment and Adjacent Inner Continental Shelf: Portshmouth Harbor, New Hampshire, USA. Master's Thesis, University of New Hampshire, Durham, NH, USA, December 2016.

46. Masetti, G.; Calder, B.R.; Hughes Clarke, J.E. Methods for artifact identification and reduction in acoustic backscatter mosaicking. In Proceedings of the U.S. Hydro 2017 Conference, Galveston, TX, USA, 20–23 March 2017.

47. Ward, L.G. *Sedimentology of the Lower Great Bay/Piscataqua River Estuary*; Final Report, Revision 1; Jackson Estuarine Laboratory, University of New Hampshire: Durham, NH, USA, 1995.

48. Ward, L.G.; Zaprowski, B.J.; Trainer, K.D.; Davis, P.T. Stratigraphy, pollen history and geochronology of tidal marshes in a gulf of maine estuarine system: Climatic and relative sea level impacts. *Mar. Geol.* **2008**, *256*, 1–17. [CrossRef]

49. Diesing, M.; Green, S.L.; Stephens, D.; Lark, R.M.; Stewart, H.A.; Dove, D. Mapping seabed sediments: Comparison of manual, geostatistical, object-based image analysis and machine learning approaches. *Cont. Shelf Res.* **2014**, *84*, 107–119. [CrossRef]

50. *Arcgis Benthic Terrain Modeler (BTM)*; version 5.1; NOAA Office for Coastal Management: Charleston, SC, USA, 2012.

51. Hughes Clarke, J.E.; Li, M.Z.; Sherwood, C.R.; Hill, P.R. Optimal use of multibeam technology in the study of shelf morphodynamics. In *Sediments, Morphology and Sedimentary Processes on Continental Shelves: Advances in Technologies, Research and Applications*, 1st ed.; International Association of Sedimentologists: Gent, Belgium, 2013; pp. 3–28.

52. Lurton, X.; Eleftherakis, D.; Augustin, J.M. Analysis of seafloor backscatter strength dependence on the survey azimuth using multibeam echosounder data. *Mar. Geophys. Res.* **2017**. [CrossRef]

53. Nittrouer, J.A.; Allison, M.A.; Campanella, R. Bedform transport rates for the lowermost Mississippi River. *J. Geophys. Res. Earth Surf.* **2008**, *113*, 1–16. [CrossRef]

54. Fonseca, L.; Brown, C.; Calder, B.; Mayer, L.; Rzhanov, Y. Angular range analysis of acoustic themes from stanton banks ireland: A link between visual interpretation and multibeam echosounder angular signatures. *Appl. Acoust.* **2009**, *70*, 1298–1304. [CrossRef]

geosciences

MDPI

Article

An Automatic Procedure for the Quantitative Characterization of Submarine Bedforms

Massimo Di Stefano [1,2,*] and Larry Alan Mayer [1,2]

[1]   Department of Earth Science, University of New Hampshire, Durham, NH 03824, USA;
     lmayer@ccom.unh.edu
[2]   Center for Coastal and Ocean Mapping/National Oceanic and Atmospheric Administration (NOAA) Joint
     Hydrographic Center, University of New Hampshire, Durham, NH 03824, USA
*   Correspondence: distefano@ccom.unh.edu; Tel.: +1-508-292-4078

Received: 22 November 2017; Accepted: 10 January 2018; Published: 21 January 2018

Abstract: A model for the extraction and quantitative characterization of submarine landforms from high-resolution digital bathymetry is presented. The procedure is fully automated and comprises two parts. The first part consists of an analytical model which extracts quantitative information from a Digital Elevation Model in the form of objects with similar parametric characteristics (terrain objects). The second part is a rule-based model where the terrain objects are reclassified into distinct landforms with well-defined three dimensional characteristics. For the focus of this work, the quantitative characterization of isolated dunes (height greater than 2 m) is used to exemplify the process. The primary metrics used to extract terrain objects are the flatness threshold and the search radius, which are then used by the analytical model to identify the feature type. Once identified as dunes, a sequence of spatial analysis routines is applied to identify and compute metrics for each dune including length, height, width, ray of curvature, slope analysis for each stoss and lee side, and dune symmetry. Dividing the model into two parts, one scale-dependent and another centered around the shape of the landform, makes the model applicable to other submarine landforms like ripples, mega-ripples, and coral reefs, which also have well-defined three-dimensional characteristics.

Keywords: geomorphometry; GIS; spatial scale; spatial analysis; terrain analysis

---

1. Introduction

There is evidence that the shape of the earth, at many scales, directly affects the spatial distribution of life. This connection between landscapes and organisms is bidirectional, where "the land is shaped by life, [and] life is shaped by the landscapes it inhabits" [1,2]. This two-way interaction between species and environment, including geomorphology, is an important object of study in ecology and is at the basis of niche theory as well as the foundation of the concept of habitat [3].

Mapping of seafloor habitat is an essential element to understanding the ecology of the benthic zone. By acquiring knowledge of seafloor characteristics and benthic community structure, habitat mapping is a useful tool for many purposes ranging from the study of marine biodiversity, design of marine protected areas, and planning marine fishing reserves [4]. In particular, habitat mapping is crucial for the development of marine resource management plans maintaining a sustainable fishing industry and gauging the performance of existing management plans.

The linkage between abiotic and biotic components of a specific set of species in a well-defined environmental setting is what is used in ecology to identify a particular habitat (or biotope). Species distribution modeling is a complex and multidisciplinary scientific art that studies still poorly understood interactions. It involves multidimensional environmental analyses, where models built on top of "explanatory variables" (predictors) are tested against a few detailed and accurate observations (ground truth). Those observations can be species presence/absence, species abundance, species assemblage, etc.

Much work has been done to relate geomorphology and biological composition. A relationship between physical properties of the seafloor and the occurrence of certain species assemblages has been demonstrated in some studies (e.g., [5,6]), but such linkage is still poorly understood, and it is the object of ongoing research in marine ecology. Many studies (e.g., [4], Table 5.2) report seafloor morphology and sediment type as the most important variables explaining the spatial distribution of benthic species assemblages. Such habitat surrogates can be extracted from bathymetry and acoustic backscatter mosaics derived by Multibeam Echosounder (MBES) data [7,8]. Thus, depending on the scale of the analysis, MBES is one of the most valuable tools to study seafloor habitat.

This study focuses on the application of quantitative models of Digital Bathymetric Models (DBM) to characterize morphology by landform extraction. The quantitative approaches described are the theoretical foundation of Digital Terrain Analysis (DTA).

The quantitative description of a form, *morphometry*, applied to the Earth's surface is what is known, in physical geography, as *geomorphometry* [9]. Geomorphometry can be divided into *general geomorphometry*, which analyzes the land surface as a ontinuous surface described by local attributes, and *specific geomorphometry* which relates to a specific surface's features and deals with their quantitative description. *Specific geomorphometry* can then be defined as the process of subdivision of the terrain into landforms based on a segmentation process and then the characterization of each segment based on their relationship to one another [10]. DTA allows the partitioning of the landscape into objects with a distinct parametric representation resulting from a standard combination of processes, materials, and conditions. Such objects (or relief features) are often called landforms. Each landform exhibits a predictable range of visual and physical characteristics [11] and the analysis of such characteristics is the object of the study of *specific geomorphometry*.

Unfortunately, approaches for classifying and segmenting land surfaces into useful units are often tuned to model local processes with explicit assumptions and limitations that only apply to the local conditions [12]. Variables such as grid resolution and size of the neighborhood operators have a large influence on the terrain derivatives used by landform recognition models [13] affecting their ability to detect all the terrain features that characterize a landscape. Moreover, to correlate landforms with geological processes, a specific set of rules is usually implemented, which will work only for that particular process. Hence, a unique, generic terrain-classification framework that will work for any geological process is difficult, if not impossible, to achieve.

The objective of this paper is to develop an approach to analyzing digital bathymetric models in order to identify and quantitatively describe specific bedforms with the final goal of achieving a better understanding of the geological and physical settings of the area that can be used for habitat characterization and other studies.

Presented is an approach for detection, extraction and quantitative characterization of a particular seascape. The approach first makes use of specific morphometry techniques to extract terrain features including ridges, slopes, summits, spurs, etc. from high-resolution digital bathymetry and identify the feature. Then, it uses these terrain features to extract and quantitatively characterize specific submarine bedforms. We used large, isolated sand dunes as the example bedform in this paper.

This study is part of ongoing research which aims to characterize the benthic habitats of the Great South Channel (GSC) through the use of the Coastal and Marine Ecological Classification Standard (CMECS). The Great South Channel is characterized by the widespread presence of large sand dunes, varying in scale and type. One of the requirements of the CMECS is the characterization of major geomorphic and structural characteristics of the seafloor at various levels of detail, which are described in the CMECS *geoform component* [14], and which motivated the development of the present method. The seafloor characterization tool presented in this paper will be used to identify and quantitatively characterize these particular bedforms and their associated biotopes.

2. Methods

2.1. Area of Study

The analysis in this study focuses on an area located on the western boundary of Georges Bank, 50 nautical miles east of Nantucket Island in the proximity of the west side of the Great South Channel at the southern border of the Gulf of Maine (Figure 1). Georges Bank is a relatively shallow (less than 200 m) bank composed mostly of glacial debris transported from the continent in the late Pleistocene [15] and constantly reworked by strong M2 tidal currents [16]. Due to the interaction of a complex current system and its bathymetry, Georges Bank is considered one of the most productive shelf ecosystems in the world [16–21] with a primary production that reaches 400–500 gCm^2 yr^{-1} [22].

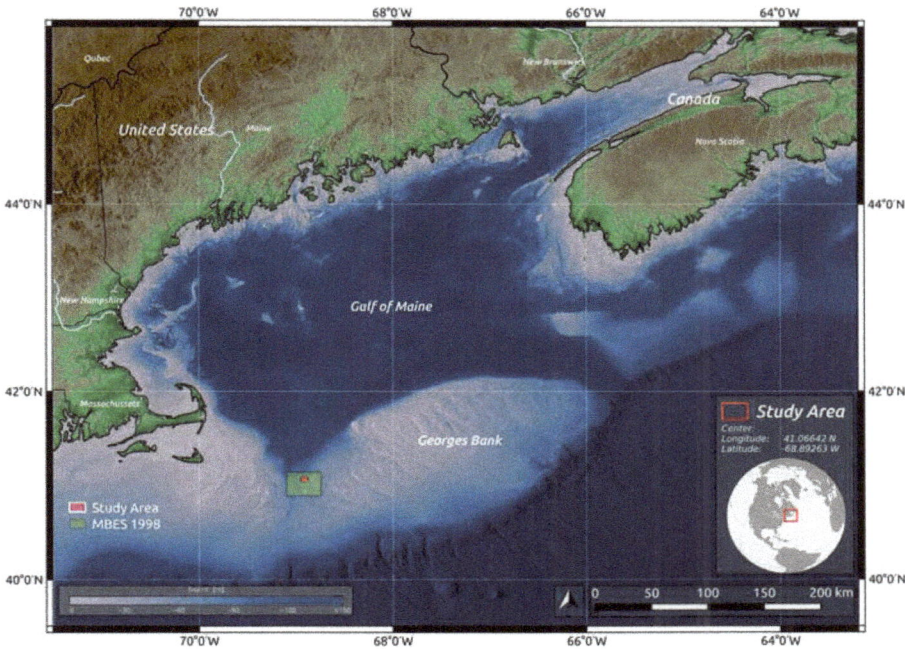

Figure 1. Location of the area of study (red circle), USGS MBES survey (1998) [23] (green area). The shaded relief used as background is derived from the 3 arc second digital elevation model of the Gulf of Maine, source: United States Geological Survey (USGS)—Coastal and Marine program, Woods Hole (MA).

The area is characterized by major topographic features formed by glacial and postglacial processes [23]. The east side of the DBM presents a cluster of giant dunes in a sediment-rich area, whereas the west side is characterized by large and isolated dunes (Figure 2) in a sediment-starved area, with coarser sediments and presence of boulders probably resulting from ice-rafted debris. The whole area is subject to a strong tidal regime with a fast and strong southward outflow and a weaker but longer northward inflow. In addition, the area is occasionally affected by strong storm currents. All these processes maintain the seabed in a status of dynamic equilibrium with active sediment transport.

2.1.1. Survey Description

The study area covered approximately 3 km^2, with an average depth of 90 m, gently deepening towards the west with an absolute difference in depth of 30 m. Due to the strong tidal regime, all fine sediment in the area is washed away by currents, leaving the seabed covered by a surficial layer of coarse-grained sand and gravel. In such an environment, well-distinguished bedforms such as ripples, mega-ripples, and sand dunes are common (Figure 2).

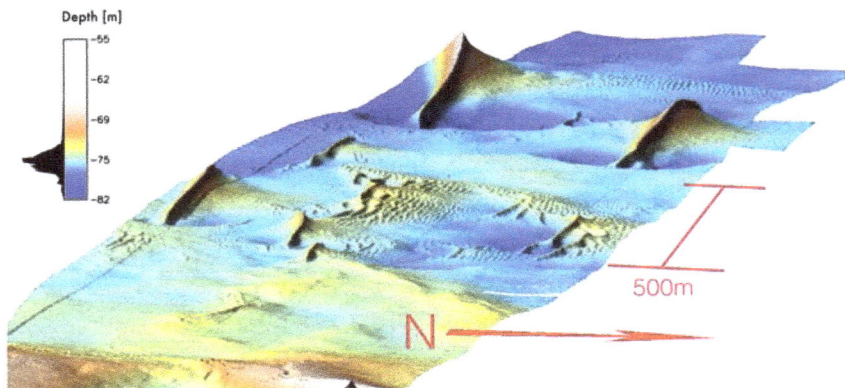

Figure 2. Details of the morphology of the study area (elevation exaggeration 3×).

The data included the acquisition of high-resolution multibeam bathymetry co-registered with a continuous ribbon of still stereographic photos of the seafloor along with the acquisition of a number of water column parameters and physical and biological samples.

2.1.2. Survey Methods

High-resolution multibeam sonar bathymetry and acoustic backscatter data were collected with a Reson Seabat 7125 SVP2 (Teledyne Reson, Slangerup, Denmark). The multibeam echosounder was installed on one of the three drop keel frames of the University of Delaware research vessel, R/V *Hugh R. Sharp*. The sonar was operated at a frequency of 400 kHz and a swath width of 150 degrees. The system was integrated with an Applanix Pos-MV 320 V5 GPS Positioning System which provided accurate altitude, heading, heave, position and velocity of the vessel. This information was used to correct the data for motion artifacts. The survey consisted of eleven parallel track lines running in a east-west direction with a spacing of 20 m. Continuous high-resolution still photographic imagery of the seafloor was collected in the survey area with Prosilica Giga Ethernet stereo cameras mounted on the HabCam system [24]. The co-registration of images and acoustic data was possible by deploying an Ultra Short Baseline (USBL) underwater positioning system, which provided accurate tracking of the HabCam. The USBL Transducer was installed on the second drop down keel frame of the R/V Hugh R. Sharp, with a known offset from the POS/MV. The Beacon/Pinger was mounted on the top frame of the HabCam, in the proximity of the mounting hook of the fiber optic cable. The system was calibrated at the beginning of the cruise by deploying a beacon on the seafloor at a depth of 60 m and performing a static calibration using the four cardinal point scheme, which was possible thanks to the dynamic positioning (DP) capabilities of the R/V Hugh R. Sharp. The four cardinal point scheme calibration method was indicated by USBL manufacturer and is considered the most appropriate calibration method [25]. All the mapping and sampling equipment on the R/V Sharp were synchronized with a time server for accurate time-stamping of the data logs. The survey data also included sediment samples from three box-core stations and biological observations including species, number of individuals, as well as sex and size from a 1.5 km long dredge tow in proximity of the surveyed area. The imaging and physical samples were not used in this study.

2.2. Data Processing

The raw MBES data produced by a Reson Seabat 7125 were acquired with the Hypack software suite (version: 2016) and processed using QPS Qimera software (version: 1.5). The processing steps consisted of:

- Accounting for the refraction effects due to changes in sound speed in the water column. For this purpose, CTD vertical profiles were acquired by the HabCam every two hours, for a total of three sound speed profiles.
- Accounting for vertical offsets due to tide effect by referencing the dataset to the WGS84 ellipsoid and by using GPS and IMU navigation to reprocess the data.
- Removal of outliers by data filtering and manual cleaning.

For gridding, the CUBE (Combined Uncertainty and Bathymetric Estimator) algorithm [26] was used. The CUBE settings were chosen by following the recommendations from the Field Procedures Manual, NOAA, Office of Coast Survey April 2010. The final product consisted of a bathymetric grid with a 1 m×1 m cell size (Figure 3).

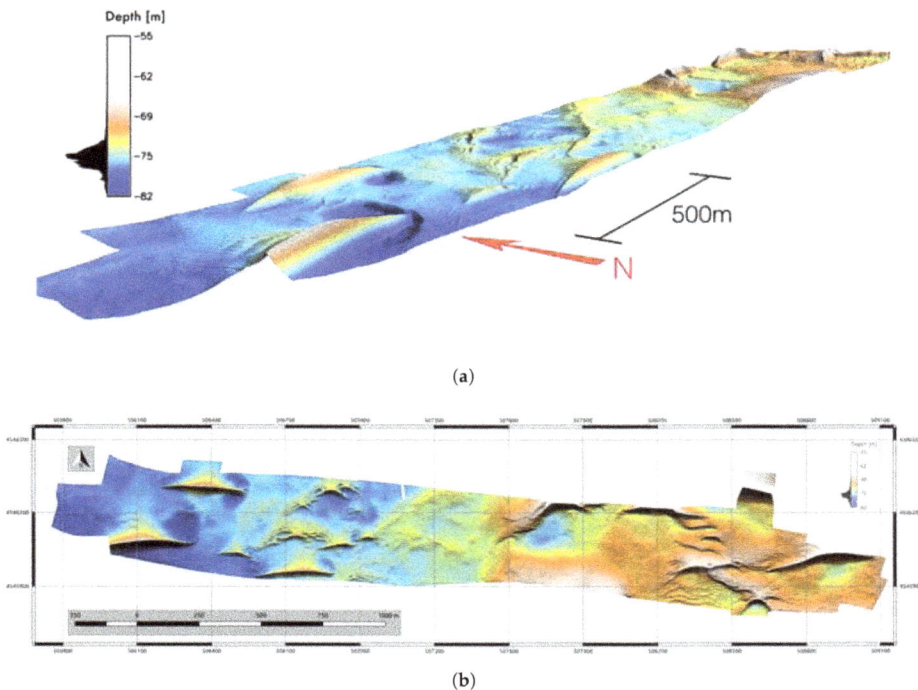

(a)

(b)

Figure 3. Digital Bathymetry produced from the MBES data set., cell size 1 m×1 m. (**a**) Sout West 3D perspective, vertical exaggeration 3×; (**b**) Shaded relief, Azimuth of the sun in degrees to the east of north: 0°, Altitude of the sun in degrees above the horizon: 50°; Factor for exaggerating relief: 3×.

2.3. Model Descriptions

The Spatial Analytic routine here proposed is composed of two main parts:

1. Terrain Feature Extraction (TFE): Where a series of terrain parameters is computed from a digital terrain model
2. Geospatial rule-based model (GRM): The identification and quantitative description of specific bedforms (e.g., sand dunes)

For the first part of the method (TFE), three different approaches for the automatic extraction of the main morphometric characteristics (terrain features) from the digital bathymetry (DBM) were compared. After evaluation of the different TFE models, the most suitable approach was determined and the outputs from this approach were used as input into the second part of the methodology (GRM).

2.3.1. Terrain Feature Extraction (TFE)

The first two approaches to feature extraction (Models Aand B) are both based on differential geometry principles [27–29]. In the first approach (Model A) terrain features are determined analyzing the slope and curvature of each DBM cell by running a fixed size focal window operator on the whole surface. Model A is implemented in the GRASS GIS (version: 7.3) [30] module *r.param.scale* which determines terrain features (namely, planar, pits, channels, pass, ridges and peaks) directly from a DBM by a single run of the model.

In the second approach, Model B, a series of terrain parameters (slope, elevation ratio and curvatures) are first extracted from the DBM and then terrain features are determined by a cell-wise unsupervised classification of these terrain parameters. The unsupervised clustering consists of two steps. First, a modified K-Means algorithm (*i.cluster* available in GRASS GIS) is used to group the terrain parameters into a user-specified number of clusters. Then cluster means and covariances are given as input to a maximum-likelihood discriminant analysis classifier to determine to which class each cell of the terrain has the highest probability of belonging (*i.maxlik* available in GRASS GIS).

A completely different approach is used in Model C which consists of applying the concept of geomorphologic phenotypes (geomorphon) [31] which is integrated into GRASS GIS. This model is based on pattern recognition principles and does not involve the derivation of terrain parameters. To determine and map landforms, the model first extracts and then labels areas characterized by a topographic pattern. Then, the pattern is compared (by matching) with a series of common landform descriptors: flat, peak, ridge, shoulder, spur, slope, hollow, foot-slope, valley, and pit [32]. For each cell, Model C performs an elevation difference between a focus pixel and pixels at a known distance (search radius) in eight principal directions (N, NE, E, SE, S, SW, W, NW). Starting from the east and proceeding counterclockwise, the algorithm produces a ternary operator which identifies a specific topographic pattern (geomorphons) that can be then associated with a particular landform element [32]. The ternary topographic pattern and associated idealized landforms are summarized in Figure 4 (e.g., example local ternary pattern is: [−1, 0, +1] = [lower, same height, higher]).

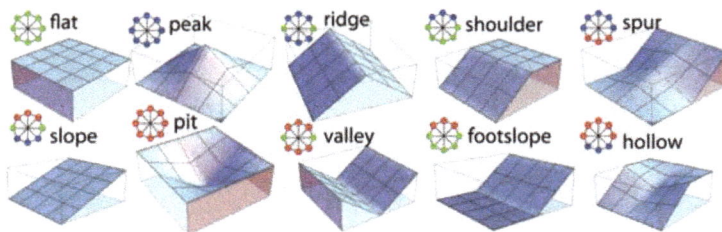

Figure 4. Terrain features represented in *r.geomorphon* (J. Jasiewicz, T. Stępiński, 2012). The rose diagrams for each feature are colored based on the difference in height with the central cell as reference (green: same height, red: higher, blue: lower).

2.3.2. TFE Model Simulation

The three TFE models were evaluated and compared loosely following the evaluation framework described in "Treatise on Geomorphology: Quantitative Modeling of Geomorphology" [33]. All the models include a series of parameters that the operator can tune to control the behavior of their outputs. To investigate the scale-dependency of the TFE models and choose which model was more suitable for the detection of large-scale sand waves, a simulation experiment to analyze their sensitivity to variations in the input parameters was conducted. The list of parameters and the ranges used in the sensitivity analysis is presented in Table 1.

Table 1. Parameters used in the sensitivity analysis of the TFE models.

Parameters	Model A, B
Param Scale Window Operator	$[5, 9, 15, 23]$
Curvature Tolerance	$[0.0001, 0.01, step : 0.001] + [0.01, 0.1, \ step : 0.01]$
Exponent for Distance Weighting	$[0, 4, \ step : 1]$
Vertical Scaling Factor	$[0, 2, \ step : 0.5]$
	Model B
Nearest Neighbor Operator	$[5, 9, 15, 23]$
Number of Classes	$[3, 5, 7, 9]$
Minimum Cluster Size	$[9, 15, 21]$
Number of Iteration	200
	Model C
Outer Search Radius	$[1, 45, \ step : 1]$
Inner Search Radius Threshold	$[1, 45, \ step : 1]$
Flatness Threshold	$[0, 3, \ step : 0.1]$
Flatness Distance	$[0, 10, \ step : 0.5]$

The simulation generated over 10,000 raster layers and was performed using standard consumer hardware by taking advantage of the multiprocessing capabilities of the Python programming language. It was not feasible to manually observe every single output, so a subset was selected for viewing that included the layers corresponding to the parameter combinations at the extreme and middle points of each interval. For the evaluation of the results, the outputs were compared with the desired terrain features (sand dunes) as identified visually from the DBM. When any one parameter combination returned a result that clearly matched the visually identified control features, Figure 5, the corresponding parameters were manually modified to vary around their initial values in order to see which parameter was responsible for detecting the pattern. Model C was selected as the most appropriate of the three; the reasons for this selection will be discussed in the Results section.

2.3.3. Geospatial Rule-Based Model-GRM

The second part of the approach is based on a Geospatial Rule-based Model (GRM) which consists of a series of geospatial processing operations, including map algebra, raster to vector conversion, topological cleaning, and, spatial filtering with the final objective of identifying and quantitatively describing a specific bedform feature, in this case large sand dunes ($height \geq 2$ m). The GRM is centered around the specific bedform and is applied on the output of Model C, the TFE approach that produced the best results.

The GRM rules consisted of a sequence of geospatial analysis operations implemented in two GRASS modules: *r.dunes* and *r.dunes.metrics* which are based on the output of the GRASS GIS module *r.geomorphon*; the entire work-flow is described in Figure 6.

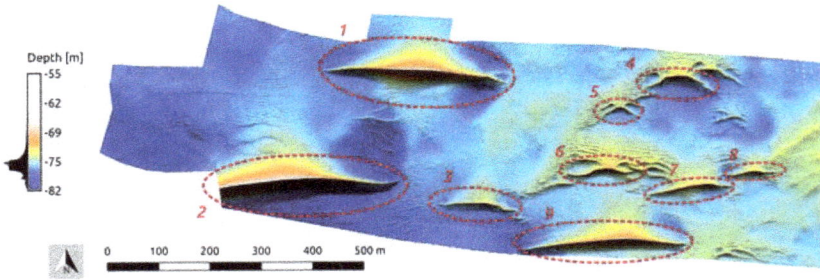

Figure 5. Areas numbered 1–9 indicate visually identified sand waves with *height* ≥ 2 m, used as control features for the selection of model outputs.

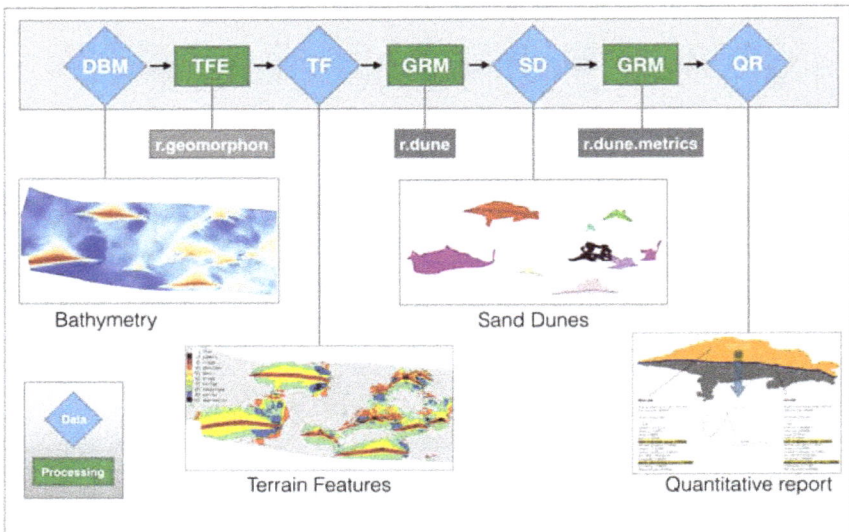

Figure 6. Diagram describing the TFE + GRM approach for the quantitative characterization of sand dunes. The bathymetry used as input is first processed by *r.geomorphon* for the terrain feature extraction (TFE) and then the output is processed by the two GRASS GIS modules developed in this study: *r.dune* to identify sand dunes and and *r.dune.metrics* to produce its quantitative report.

The whole GRM workflow can be divided into three sections:

1. Extract and vectorize sand wave crest (SWC).
2. Extract and vectorize sand wave main body (SW).
3. Identify *lee* and *stoss* side and compute sand wave's metrics.

The sequence of each step are summarized and visually described below:

1. Extract and vectorize sand wave crest for each bedform

   1.1. Extraction of sand wave crest (SWC): From the TFE results (*r.geomorphon*), SWC are identified by extracted by reclassifying the cells with feature type equal to *ridge* and *summit* (feature type class: 2,3) into a new raster feature with category value 1 and setting the remaining cells to null (Figure 7).

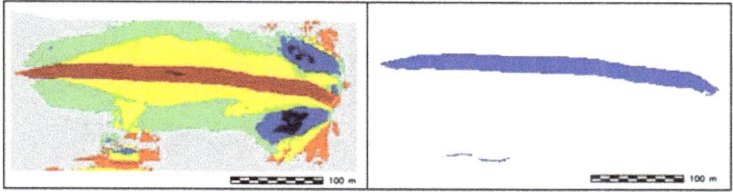

Figure 7. Extraction of sand wave crest (SWC).

1.2. SWC thinning: SWC areas are thinned and reduced to a single pixel width (Figure 8).

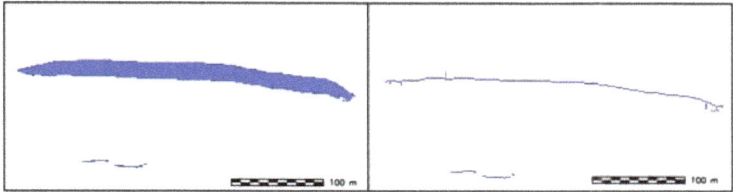

Figure 8. SWC thinning.

1.3. SWC clumping: Each SWC is recategorized by grouping cells that form physically discrete areas into unique categories and assign a distinct color to each raster feature, different colors are assigned to each linear feature (Figure 9).

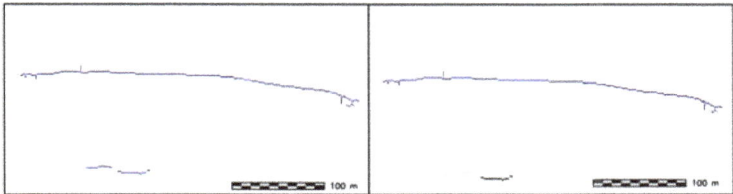

Figure 9. SWC clumping.

1.4. SWC filtering by length: Each feature with same category shorter than a given threshold is removed (Figure 10).

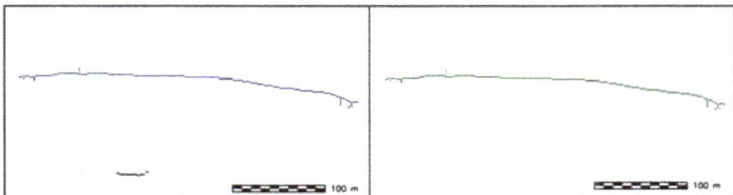

Figure 10. SWC filtering by length.

1.5. SWC vectorization: Conversion from raster to vector to obtain a vector feature of type line representing an approximate sand wave crest (Figure 11).

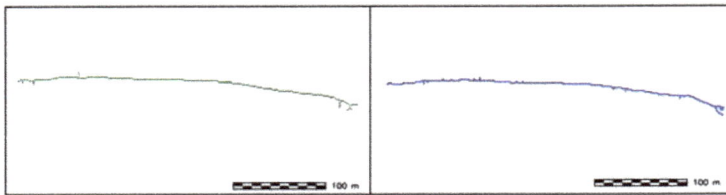

Figure 11. SWC vectorization.

1.6. SWC topological cleaning: Line features are cleaned by removing any dangle (Figure 12).

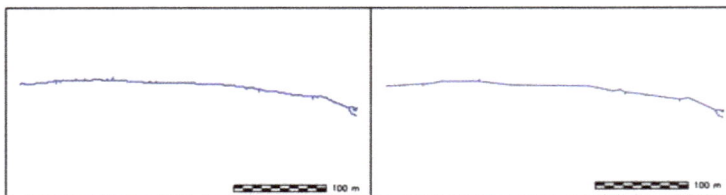

Figure 12. SWC topological cleaning.

1.7. SWC smoothing: A low pass filter is used to obtain a vectorized Sand Wave Crest (Figure 13).

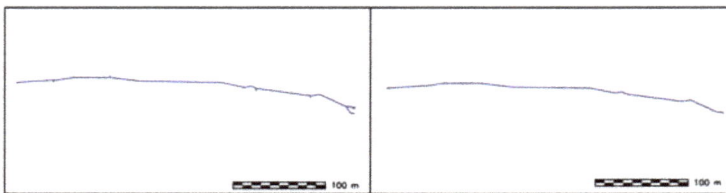

Figure 13. SWC smoothing.

2. Identify areas covered by large scale bedforms

2.1. Sand wave (SW) extraction: From the TFE results (*r.geomorphon*), the entire landform is extracted by reclassifying the cells with feature type equal to summit, ridge, spur, and slope (feature type: class 2, 3, 5, 6) into a new raster feature with category value 1, and setting the remaining cells to null (Figure 14).

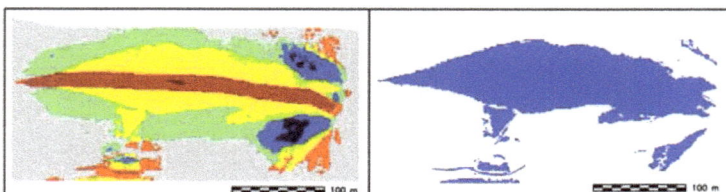

Figure 14. Sand wave (SW) extraction.

2.2. SW clumping: The SW raster map is recategorized by grouping cells that form physically discrete areas into unique categories and assign a distinct color to each raster feature (Figure 15).

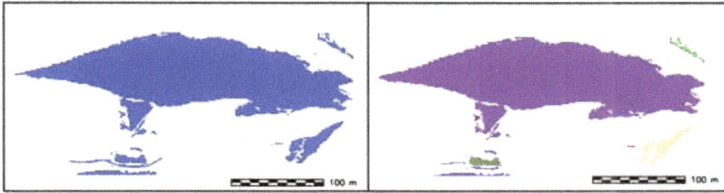

Figure 15. SW clumping.

2.3. SW filtering by area: Each raster feature is reclassified based on its area, all the features having an area smaller than a given threshold are removed (Figure 16).

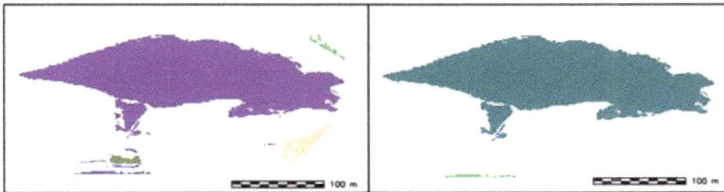

Figure 16. SW filtering by area.

2.4. SW filling: Null cells within the discrete areas are filled with the same category of the surrounding pixels (Figure 17).

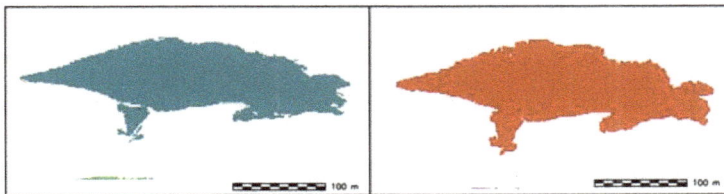

Figure 17. SW filling.

2.5. SW vectorization: Conversion from raster to vector to obtain a vector feature of type polygon representing an approximate sand wave body (Figure 18).

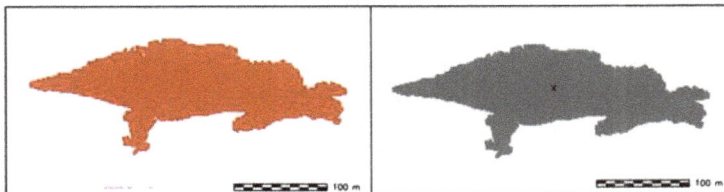

Figure 18. SW vectorization.

3. Identify *lee* and *stoss* side and compute sand wave's metrics

3.1. SW and SWC overlay: The vectorized sand wave crest is overlaid on top of the polygonal area representing the sand wave body (Figure 19).

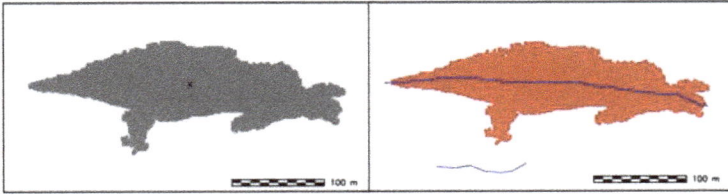

Figure 19. SW and SWC overlay.

3.2. SWC clipping: The portion of the sand wave crest that is not included in the sand wave body is removed (Figure 20).

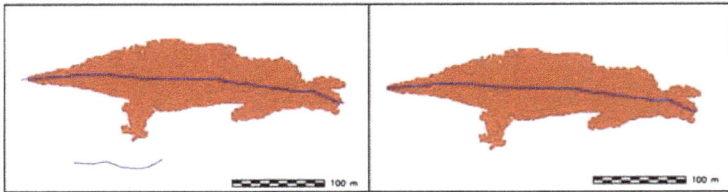

Figure 20. SWC clipping.

3.3. SWC buffering: A polygonal area is created by buffering the sand wave crest (buffer distance equal to the DBM cell size) (Figure 21).

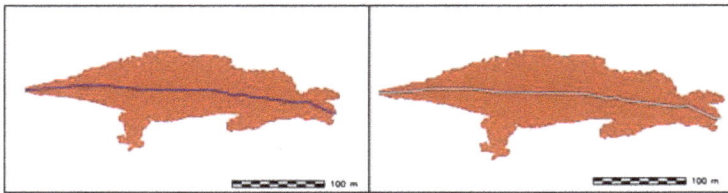

Figure 21. SWC buffering.

3.4. SW splitting: The buffered sand wave crest is used to split the sand wave body into two parts (Figure 22).

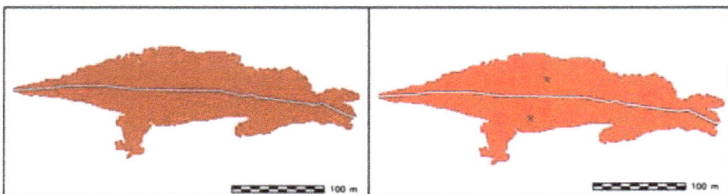

Figure 22. SW splitting.

3.5. Identification of stoss and lee side: the identification is achieved by performing an univariate statistical analysis on the DBM slope, using each sand wave side as a mask. The side with the higher values of the slope is assigned the label of *lee side* and is colored in grey, while the side with the lower values of the slope is assigned the label of *stoss side* (Figure 23).

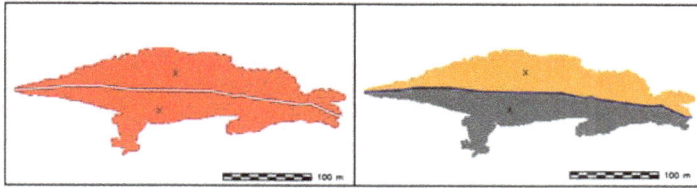

Figure 23. Identification of dune's sides: stoss (colored in gray) and lee (colored in yellow).

3.6. SW height: Derivation of SW height by generating a series of vertical profiles along several transects perpendicular to SWC. The spacing between the equidistant vertical profiles is given by an input parameter and set to 10 m as default value, the length of each profile is also variable by the user (by default is set to be the same length of the sand wave ridge) (Figure 24).

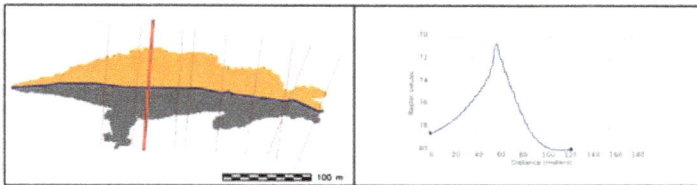

Figure 24. SW height.

3.7. SW width: Sand wave width (horn to horn) is obtained by calculating the distance between the endpoints of the sand wave crest.
3.8. Length: An approximate value for the length of the sand wave is obtained by following the schema in Figure 25.

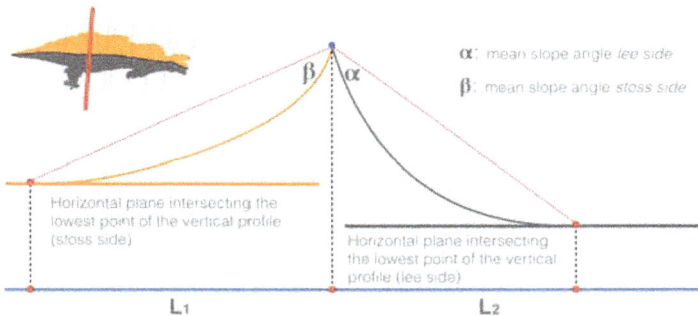

Figure 25. SW length.

Code and documentation to reproduce the whole GRM workflow are available as supplementary material.

3. Results

3.1. TFE Modeling

Using small values of curvature tolerance and with different window size operators, Model A easily isolated large scale features, but was not able to detect important elements of the feature like the sand wave crest (Figure 26).

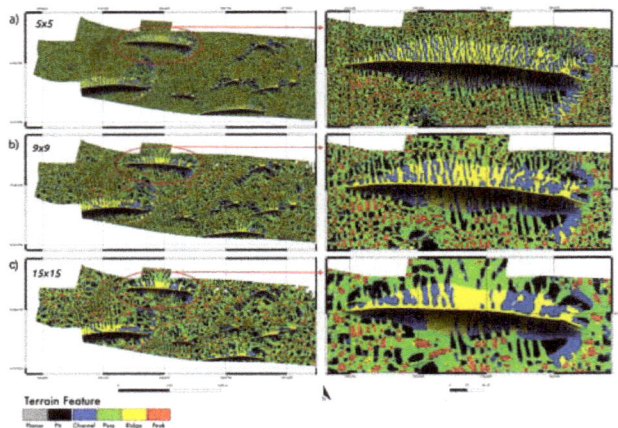

Figure 26. Output details of Model A for small value of curvature tolerance ($ct = 0.0001$) for different sizes (15 (**a**); 9 (**b**); and 5 (**c**)) of the window operator; the other parameters used were: exponent for distance weighting: 0.0; vertical scale factor: 0.5; and slope tolerance: 2.6. The zoomed areas on the right size of the image refer to the bedform labeled "1" in Figure 5.

Sand wave crests and rippled surfaces, on the other hand, could be isolated from the rest of the terrain by using relatively high values of the curvature and slope tolerance tolerance ($ct = 0.0334$, $st = 2.6$), Figure 27a). Model A was also the most sensitive to artifacts in the DBM. The grid presented minor across-track artifacts (likely due to uncompensated vessel motion), which were detected as terrain features when a small (5×5) window operator were used in combination of relatively low values of curvature tolerance, ($ct \leq 0.004$) and slope tolerance, ($st \leq 1.0$), Figure 27b).

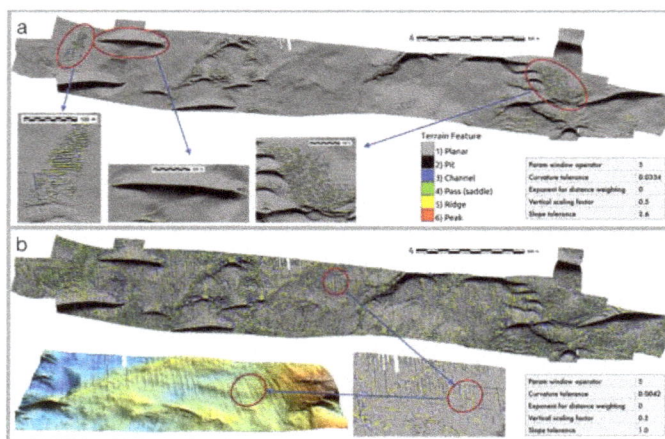

Figure 27. Output details for *Model A*: (**a**) for high value of curvature and slope tolerance ($ct = 0.0334; st = 2.6$), the model depicted rippled areas as well as dune crests; (**b**) for low value of curvature and slope tolerance ($ct = 0.004; st = 1.0$), the model depicted small artifact ($height \leq 20$ cm). In (**b**), the 3D perspective view of the MBES dataset (vertical exaggeration: $3\times$) is illuminated with a light source directed from east to west (along track direction) to highlight the small ($height \leq 20$ cm) across-track artifacts likely due to uncompensated vessel motion

Model B results were more stable (less sensitive) to variation in the parameters used than Model A and more efficient in depicting large scale features (Figure 28). However, the inability of associating the detected feature class to specific terrain features with an established name, and the arbitrary number of user-defined classes made Model B unsuitable for the objective of this paper.

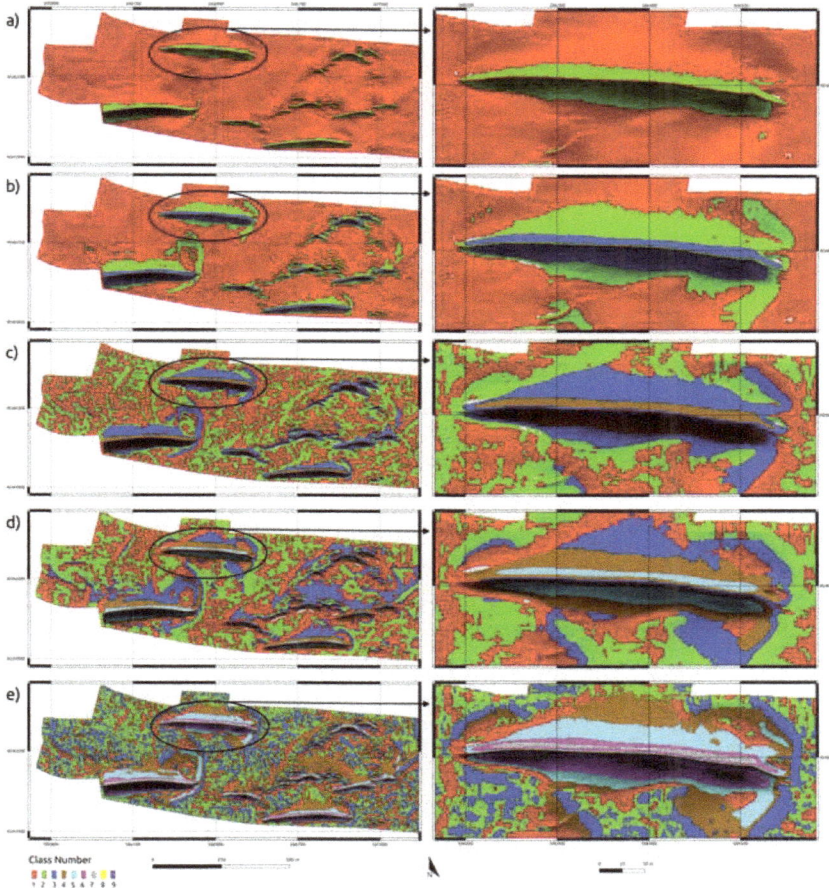

Figure 28. Output details for *Model B*: (**a**) three classes, window operator 5 × 5 cells; (**b**) five classes, window operator 9 × 9 cells; (**c**) five classes, window operator 15 × 15 cells; (**d**) seven classes, window operator 21 × 21 cells; and (**e**) nine classes, window operator 15 × 15 cells. The other parameter are kept constant: slope tolerance ($st = 1$), curvature tolerance ($ct = 0.003$), and vertical scale factor ($zs = 1$). Each color represent a different class. The zoomed area on the right of each panel refer to the bedform labeled "1" in Figure 5.

The *r.geomorphon* approach (Model C) outperformed the methods based on differential geometry. It was able to depict important landform elements at different scales with a single run of the model. Model C presented the most stable results, particularly when the search radius parameter reaches values greater than 30 cells. For almost all combination of parameters, Model C was able to correctly depict all sand wave crests, a key element for sand waves characterization (Figure 29).

The simulation thus showed that, for a particular range of parameters, all three models were able to detect some of the terrain features that could be associated to a specific landform (e.g., isolated crest for sand dunes, concave surface for depressions and patterns of almost parallel slope–ridge–valley in the case of sand ripples) but none could do so in a single run. In particular, Models A and B returned very different results for different input parameters showing a greater sensitivity to parameter changes and thus a greater scale-dependency. Model B results were more stable (less sensitive) to variation in input parameters and most efficient in depicting large scale features, however the lack of an exact labeling for the detected features and the arbitrary number of user-defined classes made the model unsuitable for the objective of this study. Model C produced the most stable outputs particularly when the size of the search radius was 30 cells or more, confirming the scale-flexibility of the model [32]. Thus, Model C was chosen as the most appropriate for extracting the desired bedform feature, in this case large sand dunes; output for the chosen set of parameters is shown in Figure 30.

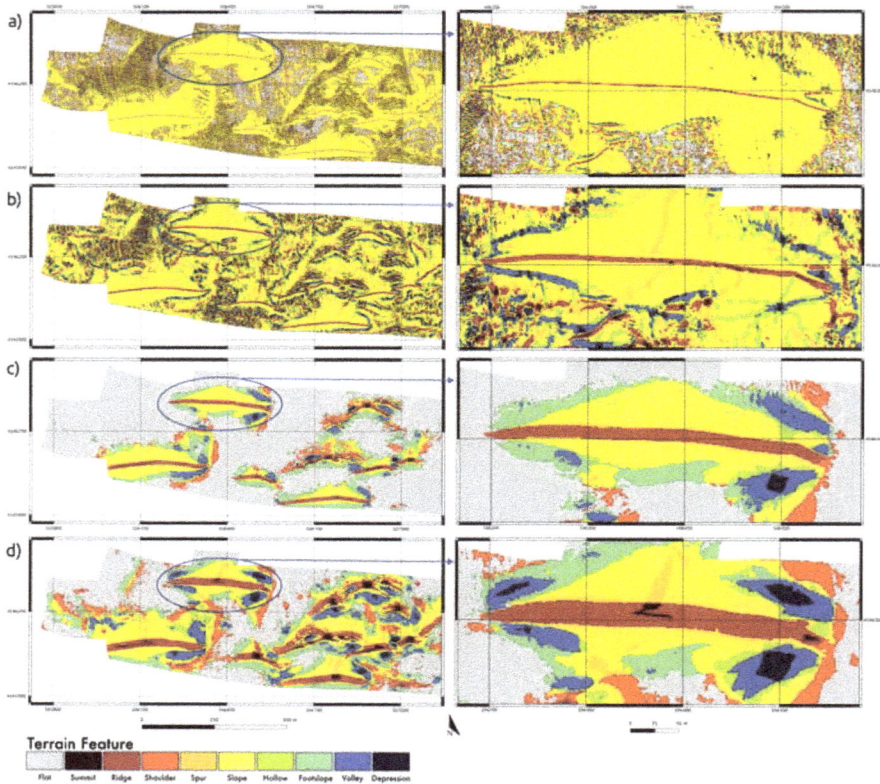

Figure 29. Output details for *Model C, r.geomorphon*: (**a**) *r.geomorphon* default values, outer search radius: 3, inner search radius: 0, flatness threshold: 1, flatness distance: 0; (**b**) outer search radius: 11, inner search radius: 3, flatness threshold: 0.1, flatness distance: 0; (**c**) values chosen to implement the GRM model: outer search radius: 35, inner search radius: 9, flatness threshold: 3.7, flatness distance: 15; and (**d**) outer search radius: 45, inner search radius: 19, flatness threshold:1, flatness distance: 15. The zoomed area on the right of each panel refer to the bedform labeled "1" in Figure 5.

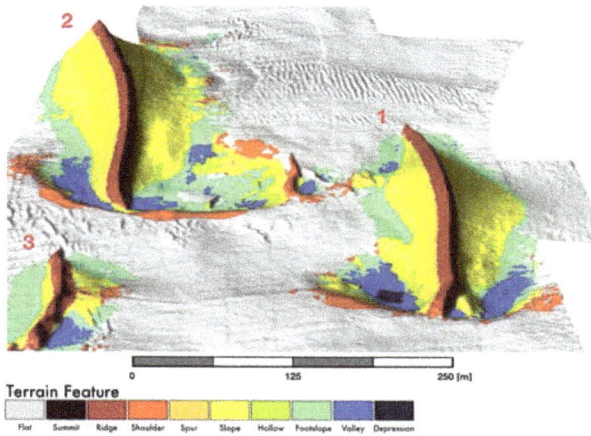

Figure 30. Details of Model C outputs (*r.geomorphon*: outer search radius: 35, inner search radius: 9, flatness threshold: 3.7, flatness distance: 15), draped on top of the DBM, (vertical exaggeration: 3×). The numbers on the top left of each bedform refers to the "bedforms labeled 1, 2, and 3" in Figure 5.

3.2. GRM Modeling

Having selected the most robust approach for feature identification, the rest of the analysis focuses on developing a GRM routine that uses the output of Model C to automatically characterize large (*height* ≥ 2 m) sand waves. A single run of Model C with the appropriate combination of its parameters was able to correctly depict most of the control features chosen to test the models in this paper (Figure 31).

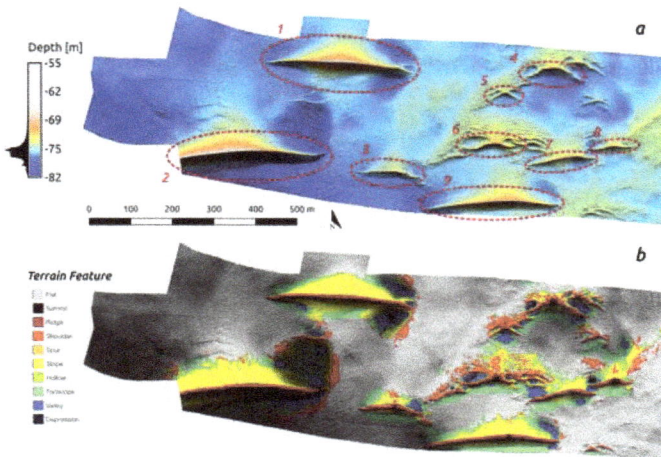

Figure 31. Control units (**a**); and Model C outputs (**b**), *r.geomorphon* parameters used: *Outer search radius*: 35, *Inner search radius*: 9, *Flatness treshold (degrees)*: 3.7, *Flatness distance*: 15. Color images draped over a shaded relief (azimuth of the sun in degrees to the east of north: 0°N, altitude of the sun in degrees above the horizon: 50°).

The model developed in this paper has been implemented in two GRASS GIS modules, (r.dune) and *r.dune.metrics*. In *r.dune*, the output of the TFE model (*r.geomorphon*) is used as input to extract the sand wave and return a vector layer with two features: (1) a polygonal feature representing the area of the bedform; and (2) a line feature that identifies the sand wave crest. In *r.dune.metrics*, the output of *r.dune* is used to query the original DBM and generate a quantitative report for each sand dune. The metrics computed by the module include: length, width, height, vertical profile, identification of stoss and lee side, and slope statistics for each side. The module outputs are summarized in a PDF report for each bedform (Figure 32).

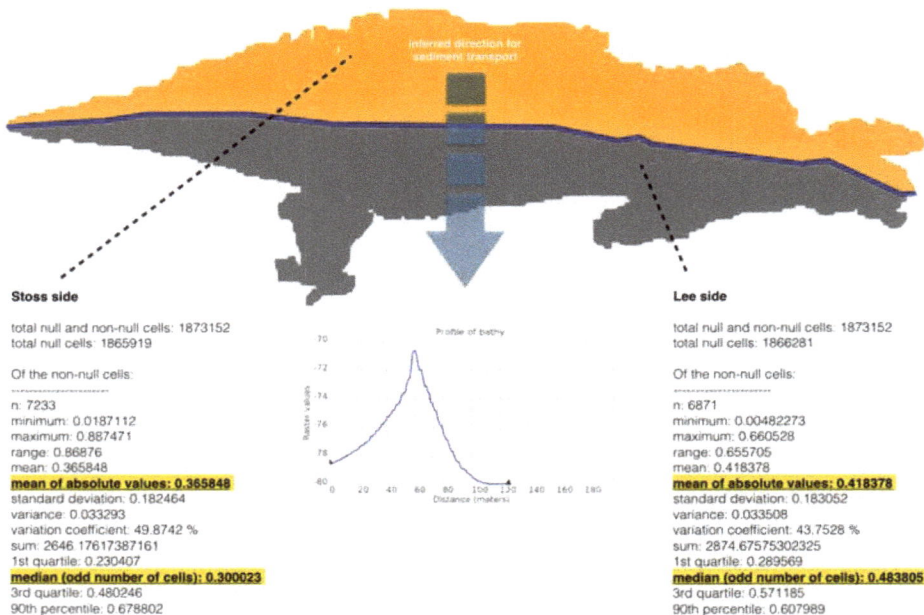

Figure 32. Bedform report, output of *r.dune*. The different colors indicate the stoss (yellow) and lee (gray) sides of the bedform. The tool runs a number of vertical profiles along transects perpendicular to the dune crest (blue). The spacing between transects is controlled by a parameter set (by default) to the DBM cell size in meters, same for the length of the transect which is (by default) set to be equal to the sand crest length (in meters). The vertical profile chosen for the report is the one with the greatest (or maximum) difference in depth.

A report with the main metrics (Length, Width, and Height) for all the bedforms identified is also available as output.

4. Discussion

Quantitative characterization of bedforms can provide a valuable source of information for a number of areas of science and engineering. Engineering applications include sediment transport studies and estimates of sand resource volumes. For ecologists, such a quantitative description can be useful in developing species distribution models by helping to test the hypothesis that morphology has a direct influence on the spatial distribution of benthic species and by relating the distribution of marine life with scale-dependent landforms.

By automatically identifying the areas where specific landforms are located, it is possible to rapidly: (1) compute the extent of those landforms in an objective and reproducible way; and (2) perform a stratified sampling of ecological information (e.g., high-resolution seafloor images, physical samples, etc.) by selecting only samples falling within a specific geomorphic category which will facilitate the definition of biotopes (by annotating species diversity, abundance, etc.) associated with a specific geomorphic category.

The three models presented have both advantages and disadvantages. Model B, by the nature of the unsupervised classification method adopted, needed an *a priori* number of classes, each of which was assigned a dummy label. Although it is possible to tune the number of final clusters so that they are statistically acceptable by varying the *class separation* parameter in *i.cluster*, their labels are still arbitrarily assigned. In contrast, Models A and C assign a specific geomorphological feature label to each object extracted.

A positive aspect of Model B is the ability to accept an undefined number of predictor variables as input (in addition to the terrain derivatives). This last characteristic can be of particular interest when the area of the analysis spans across very different sediment types, which are known to characterize different biotopes [34]. In this case, Model B can take advantage of the valuable information that can be provided by adding acoustic backscatter from MBES as input to the model.

It was observed that, by combining the results of multiple runs of each model using different values of the input parameters, it was possible to identify bedforms at more than one scale in a given area. For example, by identifying and masking out flat areas and large-scale features, and using a high value for the curvature tolerance, Model A performed very well in depicting rippled areas (Figure 10a). This behavior of the models confirms their scale-dependency and validates the idea that a unique and objective terrain classification framework is not easily achieved [29].

As shown in Figure 31, Model C was able to identify large-scale features correctly. In particular, the area covered by each landform was successfully identified and the ridge of each sand wave was correctly depicted. By using these two features, and by knowing the particular label associated with each class, a reclassification routine is enabled for the further quantitative characterization of the bedform.

Once the sand dunes were identified, the GRM approach was used to analyze the asymmetry of the dunes, which in turn can provide evidence of flow direction and speed as well as overall seafloor energy conditions. The dunes mapped in this study were very asymmetric (Figure 33), suggesting that they are mobile [35].

However, in 2017, a new MBES dataset was collected over the same area, which showed the dunes in the same location with an almost identical vertical profile (Figure 34), thus implying that their crests have not migrated over the two-year period. Furthermore, MBES data from 1998 [23] were also available and compared to the dataset used in this paper, confirming that indeed these bedforms have not migrated (Figure 35b).

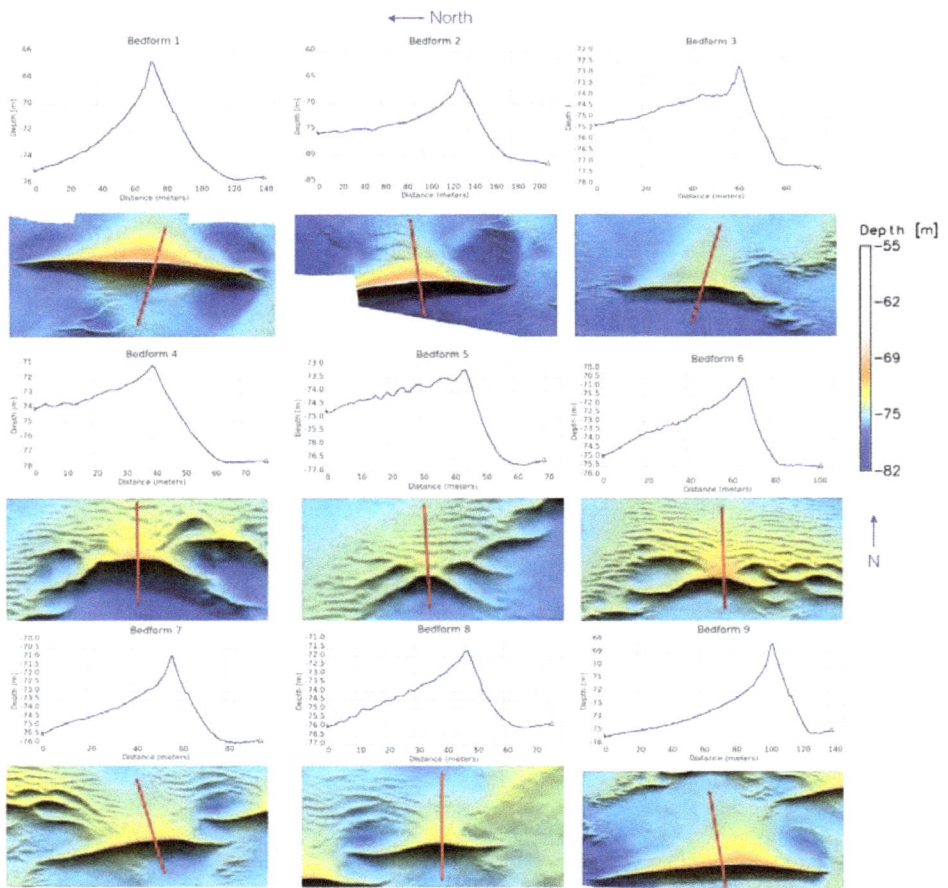

Figure 33. Vertical profiles for the sand waves identified by the *r.dune* module.

From the vertical profiles, it is also possible to observe small differences on the east side of the survey area which is probably due to a larger amount of soft sediment, as can be seen from the acoustic backscatter (BS) collected during the 1998 MBES survey (Figure 36).

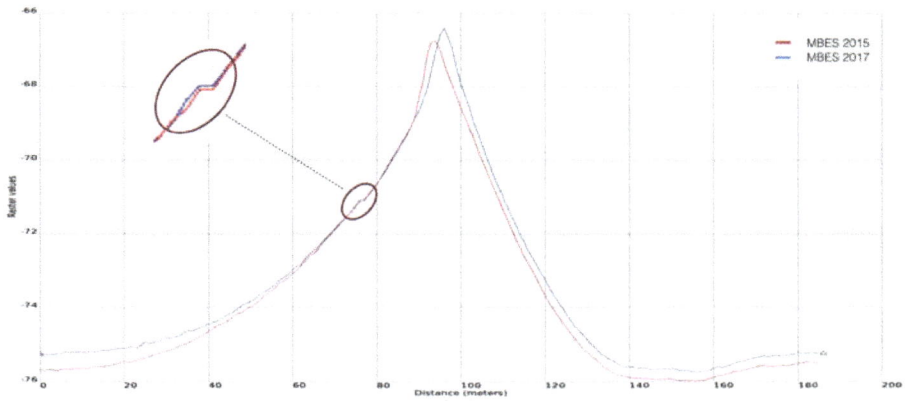

Figure 34. Vertical profile for one of the sand dunes from two different cruises, data collected in 2015 (red) and 2017 (blue). From the zoomed details, on the upper half of the stoss slope, a horizontal step is visible that matches almost perfectly between the two datasets.

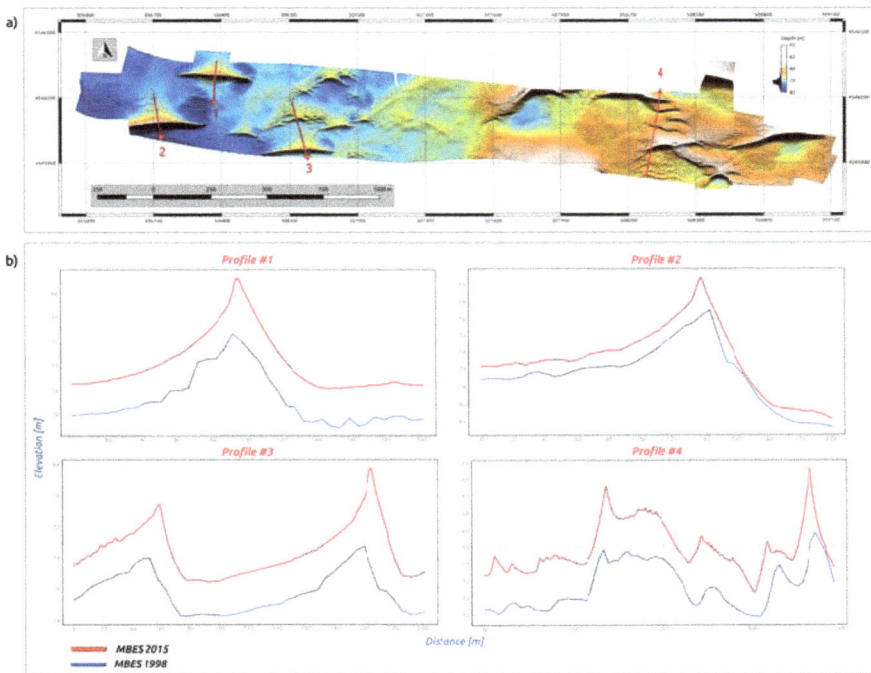

Figure 35. Vertical profiles (**a**, **b**) comparing the two MBES dataset collected in 1998 (blue, cell size: 9m), and 2015 (red, cell size: 1m). Lines numbered 1–4 in (**a**) indicate the transect used to plot the vertical profiles in (**b**). A constant vertical offset of about 1 m can be seen, attributed to differences in accuracy and resolution of the data acquisition systems used in the two cruises.

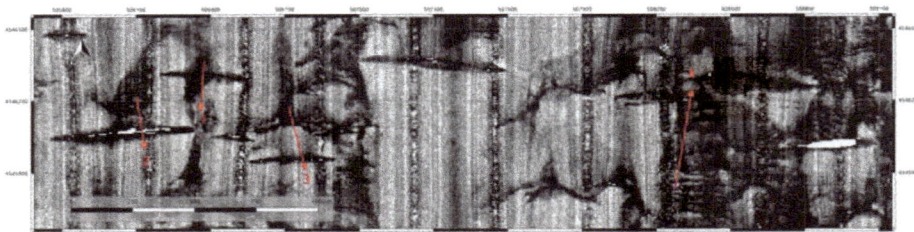

Figure 36. Acoustic Backscatter (BS) collected in 1998 (Simrad Subsea EM 1000 Multibeam Echo Sounder, 95 kHz), dark area low BS. Lines numbered 1–4 indicate the transects used to plot the vertical profiles in Figure 35)

The methods developed here enable the automatic identification and quantification of isolated sand dunes. This allows these bedforms to be rapidly identified, mapped and characterized over large areas, providing valuable information on the distribution and transport of sediment and more generally seabed energy conditions. The method can be easily extended to include other bedforms such as sand ripples and mega-ripples. Moreover, with the development of additional bedform metrics, the GRM method developed here has the potential to further allow the classification of dunes into their types (e.g., barchans versus 'whaleback' dunes). Direct applications of this work extend to the field of ecological science and habitat mapping. For example, in the area described in this study, the HabCam imagery showed a heterogeneous seafloor comprising multiple biotopes (Figure 37).

Figure 37. A selection of HabCam images collected during the 2015 MBES survey that indicate the presence of different biotopes surrounding the bedforms area. Each image covers an area of approximatively 1 m^2.

Given the ability to objectively and consistently define bedform boundaries, it becomes possible to collect habitat information from within (or outside of) these areas and thus relate biotopes and bedforms at various scales. In particular, it will be possible to study habitat patchiness by looking at the correlation, if any, of a particular biotope and its relative position, size and distance from bedforms of a particular type and scale. The isolated bedform objects of this study are a representative feature of the Great South Channel (see Figure 38). The habitat information derived by analyzing the high resolution MBES and imagery dataset can thus be used to extrapolate habitat information over a much wider area where similar "seascape ecological units" can be identified.

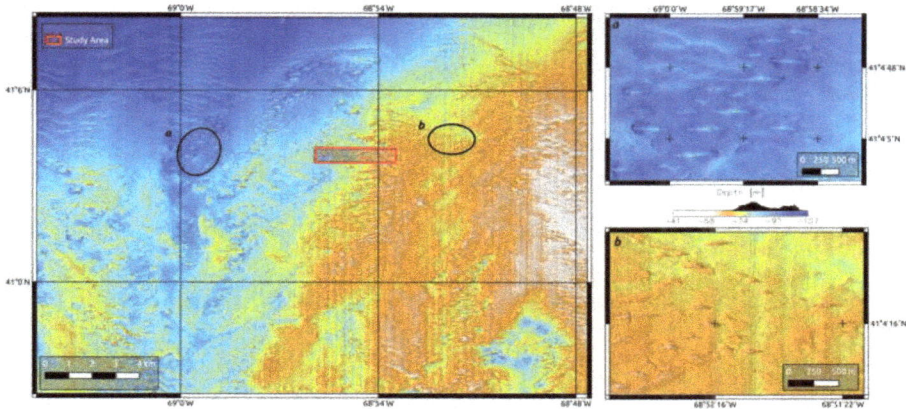

Figure 38. Portion of the USGS bathymetry collected in 1998 [23]. The isolated bedforms object of this study are a representative feature of the Great South Channel seabed. The zoomed areas (**a,b**) show in detail two, of many, areas where it is possible to find isolated dunes.

The adoption of the geomorphon concept in combination with the scripting flexibility of GRASS GIS allows the development of complex and effective GRM routines thus offering a complete platform for automatically detecting a wide range of bedforms. A simple extension to the GRM model presented in this work, is to implement dune classification by type, e.g., barchans vs. whaleback, by computing the planar ray of curvature of the sand wave crest. More complex GRM models can also be implemented for other types of bedforms. By choosing the appropriate TFE model and by tuning its input parameters, it is possible to implement the detection and characterization of other types of bedforms, e.g., rippled areas. For example, using a high value of curvature and slope tolerance (Figure 27a) in the TFE Model A, rippled areas are clearly depicted. One approach to extract the rippled areas for their further quantitative characterization would be to:

1. Set to null all the regions occupied by large scale bedforms and all the *planar* feature class.
2. Vectorize *channel* and *pass* feature classes as vector lines.
3. Split the vector lines into equidistant points.
4. Determine the area occupied by ripples by using a kernel density estimator.

In this way, it is possible to map rippled areas and further develop the model for their allometric characterization (amplitude, wavelength etc.).

5. Conclusions

This study used acoustic data to develop a technique for the quantitative characterization of bedforms. Three models for the extraction of terrain features from a DBM were considered and evaluated in a simulation experiment to study their sensitivity to a range of input parameters including scale dependence. Simulation results were used as a guide to choose a set of parameters more suitable for the extraction of particular scale-dependent terrain features, namely large isolated sand dunes.

From the simulation results, it was shown that, for different parameters, all models were able to accurately depict different terrain features. Such behavior is an indicator that there is no best model, but a combination of them will usually be necessary to achieve a complete multi-scale, quantitative, morphometric characterization of bedforms.

This experiment provided an efficient way to address the scale dependency of digital terrain analysis, enabling the classification of specific landforms by developing ad-hoc (landform/scale specific) reclassification routines. Although only sand waves were considered here, this approach can be readily used in combination with pattern-based geospatial analysis to identify other scale-dependent bedforms, like sand ripples. Through pattern-based geospatial analysis of categorical raster maps, including those generated by TFE models, patterns of terrain features can be defined. These can then be used to identify a given bedform type which in turn can be quantified by the implementation of a GRM model.

The procedure presented here marks a step forward towards the adoption and automation of specific morphometry techniques for the quantitative characterization of bedforms.

Supplementary Materials: Code and documentation to reproduce the results presented in this paper are available at: https://doi.org/10.5281/zenodo.1149563.

Acknowledgments: The survey data were collected during the 2015 NOAA annual federal Sea Scallop Survey on board R/V *Hugh Sharp*, operated by the University of Delaware, and funded by the *Improve a Stock Assessment Program* of the NOAA Northeast National Marine Fisheries Service (NEMFS). The authors would like to thank Burton Shank, for his invaluable help in the survey planning and as facilitator with the other co-PI who generously granted vessel time for data acquisition, James V. Gardner, for his scientific guidance and support, and Richard P. Signell and Page C. Valentine for providing useful insights about the study area. This work was supported by NOAA GRANT NA15NOS400002000.

Author Contributions: Massimo Di Stefano conceived and designed the experiments, conducted the field research and did most of the analysis and manuscript. Larry Mayer discussed concepts and approaches and contributed to the writing of the manuscript.

Conflicts of Interest: The authors declare no conflict of interest.

Abbreviations

The following abbreviations are used in this manuscript:

DEM	Digital Elevation Model
DBM	Digital Batimetry
DTM	Digital Terrain Model
DTA	Digital Terrain Analysis
TPE	Terrain Parameters Extraction
TFE	Terrain Feature Extraction
MBES	Multibeam Echocouder
CTD	Conductivity, Temperature, Depth
USBL	Ultra Short Baseline
R/V	Research Vessel
IMU	Inertial Measurement Unit
GPS	Global Positioning System
GIS	Geographic Information System
GRASS	Geographic Resource Analysis Support System
SW	Sand Wave
SWC	Sand Wave Crest
CCOM	Center For Coastal and Ocean Mapping
WHOI	Woods Hole Oceanographic Institution
USGS	United States Geological Survey
NOAA	National Oceanic and Atmospheric Administration
NEMFS	North East Marine Fisheries Center
CMECS	Coastal and Marine Ecological Classification Standard

References

1. Kruckeberg, A.R. *Geology and Plant Life: The Effects of Landforms and Rock Types on Plants*; University of Washington Press: Seattle, WA, USA, 2002.
2. Porder, S. Coevolution of life and landscapes. *Proc. Natl. Acad. Sci. USA* **2014**, *111*, 3207–3208.
3. Odum, E.P.; Kroodsma, R.L. *Fundamentals of Ecology*; Saunders: Philadelphia, PA, USA, 1976.
4. Harris, P.T.; Baker, E.K. *Seafloor Geomorphology as Benthic Habitat*; Elsevier: Amsterdam, The Netherlands, 2011; pp. 3–22.
5. McArthur, M.A.; Brooke, B.P.; Przeslawski, R.; Ryan, D.A.; Lucieer, V.L.; Nichol, S.; McCallum, A.W.; Mellin, C.; Cresswell, I.D.; Radke, L.C. On the use of abiotic surrogates to describe marine benthic biodiversity. *Estuar. Coast. Shelf Sci.* **2010**, *88*, 21–32.
6. Lecours, V.; Devillers, R.; Simms, A.E.; Lucieer, V.L.; Brown, C.J. Towards a framework for terrain attribute selection in environmental studies. *Environ. Model. Softw.* **2017**, *89*, 19–30.
7. Che Hasan, R.; Ierodiaconou, D.; Laurenson, L.; Schimel, A. Integrating Multibeam Backscatter Angular Response, Mosaic and Bathymetry Data for Benthic Habitat Mapping. *PLoS ONE*, **2014**, *9*, e97339.
8. Brown, C.J.; Smith, S.J.; Lawton, P.; Anderson, J.T. Benthic habitat mapping: A review of progress towards improved understanding of the spatial ecology of the seafloor using acoustic techniques. *Estuar. Coast. Shelf Sci.* **2011**, *92*, 502–520.
9. Thomas, D.S.G. *The Dictionary of Physical Geography*, 4th ed.; John Wiley & Sons: Hoboken, NJ, USA, 2016.
10. Evans, I.S. An integrated system of terrain analysis and slope mapping. *Z. Geomorphol. Suppl.* **1980**, *36*, 274–295.
11. Lillesand, T.M.; Kiefer, R.W. *Remote Sensing and Image Interpretation*; John Wiley & Sons: New York, NY, USA, 1997.
12. Zhou, Q.; Lees, B.; Tang, G. *Advances in Digital Terrain Analysis*; Springer: Berlin/Heidelberg, Germany, 2008.
13. A-Xing, Z.; Burt, J.E.; Smith, M.; Rongxun, W.; Jing, G. The Impact of Neighborhood Size on Terrain Derivatives and Digital Soil Mapping. In *Advances in Digital Terrain Analysis. Lecture Notes in Geoinformation and Cartography*; Springer: Berlin/Heidelberg, Germany, 2008; pp. 333–348.
14. Marine and Coastal Spatial Data Subcommittee Federal Geographic Data Committee. *Coastal and Marine Ecological Classification Standard*; Federal Geographic Data Committee: Reston, VA, USA, 2012.
15. Todd, B.J.; Valentine, P.C. Large submarine sand features and gravel lag sub-strates on Georges Bank, Gulf of Maine. In *Seafloor Geomorphology as Benthic Habitat: GeoHab Atlas of Seafloor Geomorphic Features and Benthic Habitats*, Elsevier: Amsterdam, The Netherlands, 2011; (Chapter 15).
16. Changheng, C.; Beardsley, R.C. Cross-Frontal Water Exchange on Georges Bank: Some Results from an U.S. GLOBEC/Georges Bank Program Model Study. *J. Oceanogr.* **2002**, *58*, 403–420,
17. Riley, G.A. Plankton studies. IV. Georges Bank. Bull. Binghampton Oceanogr. Coll. **1941**, *7*, 1–73. Available online: https://doi.org/10.1093/icesjms/16.3.392 (accessed on 10 November 2017).
18. O'Reilly, J.E.; Evans-Zetlin, C.; Busch, D.A. *Primary Production in Georges Bank*; Backus, R.H., Ed.; MIT Press: Cambridge, MA, USA, 1987; pp. 220–233.
19. Horne, E.P.W.; Loder, J.W.; Harrison, W.G.; Mohn, R.; Lewis, M.R.; Irwin, B.; Platt, T. Nitrate supply and demand at the Georges Bank tidal front. *Scient. Mar.* **1989**, *53*, 145–158.
20. Wiebe, P.H.; Beardsley, R. Physical-biological interactions on Georges Bank and its environs. *Deep Sea Res. Part II Top. Stud. Oceanogr.* **1996**, *43*, 1437–1438,.
21. Franks, P.J.S.; Chen, C. A 3–D prognostic numerical model study of the Georges Bank ecosystem. Part II: biological-physical model. *Deep Sea Res. Part II Top. Stud. Oceanogr.* **2001**, *48*, 457–482, .
22. Cohen, E.B.; Wright, W.R. *Primary Productivity on Georges Bank With an Explanation of Why It Is So High*; National Marine Fisheries Service: Woods Hole, MA, USA, 1979.
23. Valentine, P.C. *U.S. Geological Survey Geologic Investigation Series Map I–2698*; Version 1.0; USGS: Reston, VA, USA, 2002.
24. Howland, J. Development of a Towed Survey System for Deployment by the Fishing Industry. In Proceedings of the OCEANS 2006, Boston, MA, USA, 18–21 September 2006.
25. Philip, D. An Evaluation of USBL and SBL acoustic systems and the optimization of methods of calibration—Part 2. *Hydrographyc J.* **2003**, *109*, 10–20.
26. Calder, B.R.; Mayer, L.A. Automatic processing of high-rate, high-density multibeam echosounder data. *Geochem. Geophys. Geosyst.* **2003**, *4*, doi:10.1029/2002GC000486.

27. *Wood and Snell*; Technical Report EP-214; U.S. Army: Arlington, VA, USA, 1960.
28. Wood, J. The Geomorphological Characterization of Digital Elevation Models. Ph.D. Thesis, University of Leicester, Leicester, UK, 1996.
29. Dragut, L.; Eisank, C.; Strasser, T.; Blaschke, T. A Comparison of Methods to Incorporate Scale in Geomorphometry. In Proceedings of the Geomorphometry 2009, Zurich, Switzerland, 31 August–2 September 2009.
30. GRASS Development Team. Geographic Resources Analysis Support System (GRASS) Software, Version 7.2. 2017. Available online: http://grass.osgeo.org (accessed on 10 November 2017).
31. Stepinski, T.; Jasiewicz, J. Geomorphons—A new approach to classification of landforms. In Proceedings of the Geomorphometry 2011, Redlands, CA, USA, 30 May 2011; pp. 109–112.
32. Jasiewicz, J. Geomorphons—A pattern recognition approach to classification and mapping of landforms. *Geomorphology* **2013**, *182*, 147–156,.
33. Malamud, B.D.; Baas, A.C.W. Nine Considerations for Constructing and Running Geomorphological Models. *Ref. Mod. Earth Syst. Environ. Sci. Treatise Geomorphol.* **2014**, *2*, 6–28.
34. Thorson, G. Some factors influencing the recruitment and establishment of marine benthic communities. *Neth. J. Sea Res.* **1966**, *3*, 267–293.
35. Derek, J.; Henrik, H. *Encyclopedia of Planetary Landforms*; Springer: New York, NY, USA, 2015; pp. 143–148.

geosciences

MDPI

Article

The Morphometry of the Deep-Water Sinuous Mendocino Channel and the Immediate Environs, Northeastern Pacific Ocean

James V. Gardner [iD]

Center for Coastal & Ocean Mapping, University of New Hampshire, Durham, NH 03824, USA;
jim.gardner@unh.edu; Tel.: +1-1-603-862-3473

Received: 7 September 2017; Accepted: 6 November 2017; Published: 29 November 2017

Abstract: Mendocino Channel, a deep-water sinuous channel located along the base of Gorda Escarpment, was for the first time completely mapped with a multibeam echosounder. This study uses newly acquired multibeam bathymetry and backscatter, together with supporting multichannel seismic and sediment core data to quantitatively describe the morphometry of the entire Mendocino Channel and to explore the age and possible causes that may have contributed to the formation and maintenance of the channel. The first 42 km of the channel is a linear reach followed for the next 83.8 km by a sinuous reach. The sinuous reach has a sinuosity index of 1.66 before it changes back to a linear reach for the next 22.2 km. A second sinuous reach is 40.2 km long and the two reaches are separated by a crevasse splay and a large landslide that deflected the channel northwest towards Gorda Basin. Both sinuous reaches have oxbow bends, cut-off meanders, interior and exterior terraces and extensive levee systems. The lower sinuous reach becomes more linear for the next 22.2 km before the channel relief falls below the resolution of the data. Levees suddenly decrease in height above the channel floor mid-way along the lower linear reach close to where the channel makes a 90° turn to the southwest. The entire channel floor is smooth at the resolution of the data and only two large mounds and one large sediment pile were found on the channel floor. The bathymetry and acoustic backscatter, together with previously collected seismic data and box and piston cores provide details to suggest Mendocino Channel may be no older than early Quaternary. A combination of significant and numerous earthquakes and wave-loading resuspension by storms are the most likely processes that generated turbidity currents that have formed and modified Mendocino Channel.

Keywords: seabed mapping; marine geology; submarine topography; marine geomorphology; terrain analysis; multibeam echosounder

1. Introduction

Deep-water sinuous channels are conduits that transport sediments across continental margins and onto abyssal basins, often transporting sands, silts and clays hundreds of kilometers away from the margin. Almost all examples of modern deep-water *sinuous* channels are found on passive margins and are associated with submarine fans (e.g., [1–4], among many others). Mendocino Channel, off the U.S. northern California active continental margin (Figures 1 and 2), is one of the few examples (e.g., Reynisdjup Channel off Iceland and Hikurangi Channels off eastern New Zealand) of a modern deep-water sinuous channel that is not associated with a passive margin or a submarine fan although it also has been a conduit of coarse clastic sediments to a deep-sea basin [5]. During the 1984 mapping of the U.S. Exclusive Economic Zone off the western U.S. continental margin [6,7], a 42 km section of the channel was discovered on the abyssal seafloor at the base of the north side of the Gorda Escarpment segment of Mendocino Ridge, a ridge formed by the transform fault that strikes east from Gorda Ridge and beneath the North American continent. The short section of Mendocino Channel was mapped using GLORIA, a long-range sidescan sonar (Figure 3a) and was investigated on a subsequent cruise that collected widely spaced single-channel seismic-reflection profiles and several box cores within the channel and adjacent levee [5]. The GLORIA sidescan only provides images of 6.5-kHz backscatter but the images show the high-backscatter acoustic response of the channel floor. Unfortunately, the GLORIA system provided no measurements of bathymetry. The channel, called "Mendocino Channel" by Cacchione et al. [5], was traced on the GLORIA images from water depths of ~2450 m to ~2750 m. They suggested, based on NOAA bathymetric charts, that the channel is related to Mendocino and Mattole Canyons. Mendocino Channel is not in a submarine fan setting with associated sinuous distributary channels, such as are found on many submarine fans [1,2,8–20] to cite just a few. Rather, Mendocino Channel is a single channel that trends across a debris apron at the base of the lower continental margin and continues across hemipelagic sediments deposited in a basin setting. The most similar channels to Mendocino Channel and its setting are Reynisdjup Channel off Iceland [21] and Hikurangi Channels off eastern New Zealand [22].

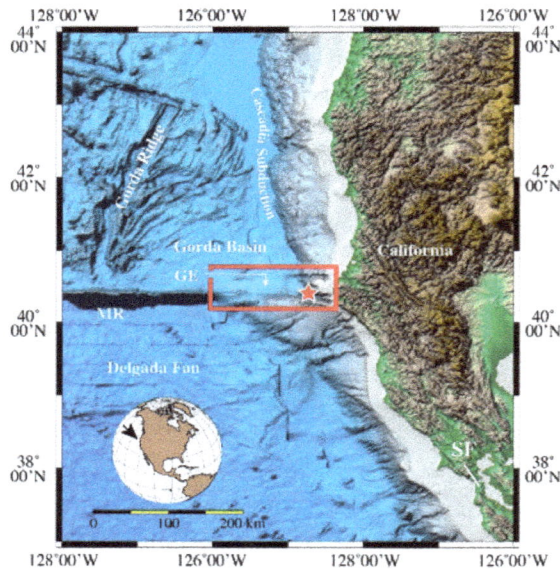

Figure 1. Location of Mendocino Channel (red rectangle) at the base of the northern flank of Gorda Escarpment (GE) in the eastern Pacific Ocean. Red star is Mendocino Triple Junction, MR is Mendocino Ridge and SF is San Francisco, CA. Data from Geomapapp.org v. 3.6.6.

The entire eastern 850 km of Mendocino Ridge was mapped in 2009 with a multibeam echosounder (MBES) [23] and the new bathymetry includes the entire length of Mendocino Channel at a resolution of 40 m/pixel. The channel has many features that resemble those found on fan-related modern deep-water sinuous channels. In addition to resolving these common features, several enigmatic characteristics occur, such as; (1) the channel trends parallel to Gorda Escarpment but is perched several hundred meters above basin depths and descends to basin depths along its length; (2) an initial 48 km-long linear upper reach that makes an abrupt 90° bend towards the Gorda Escarpment but is immediately followed by (3) a 30 km-long section of sinuous channel that parallels the trend of the escarpment and not down slope to the NNW. This section is followed by (4) a 10.5 km reach with a broad flat region that (5) is abruptly diverted 20° and traverses 58.5 km away from the escarpment and out onto the basin. The term "reach" is used here *senso lato* and does not connote the strict subaerial hydraulic or geomorphic definition of the term. The purpose of this study is to use newly acquired multibeam bathymetry and backscatter, together with available supporting data to quantitatively describe the morphometry of the entire Mendocino Channel and to explore the age and possible causes that may have contributed to the formation and maintenance of the channel.

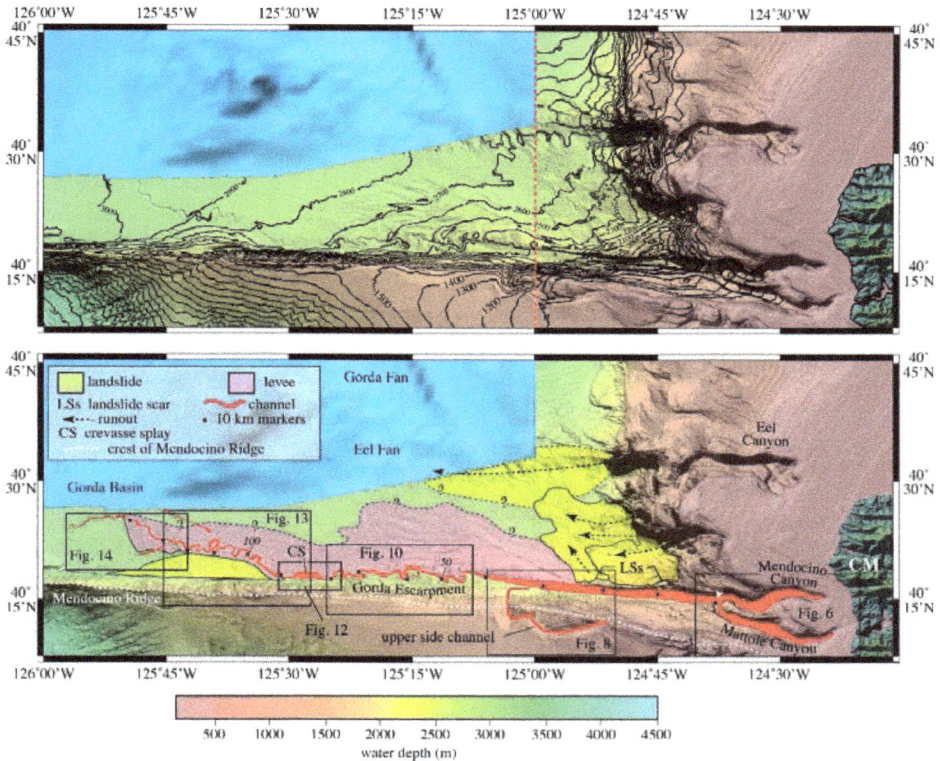

Figure 2. (**Upper**) Overview map view of multibeam bathymetry of the summit and northern flank of Gorda Escarpment, southern-most Gorda Fan and Gorda Basin to the north. Bathymetry east of red dashed line from NOAA Coastal Relief Model (https://ngdc.noaa.gov/mgg/coastal/crm.html) (See text). Contour interval 100 m. (**Lower**) Locations of Mendocino and Mattole Canyon channels and upper side channel (red), landslides (yellow) and levees (purple). Black circles every 10 km mark distances in italics from the junction of Mendocino and Mattole Canyons. White "CM" is Cape Mendocino, Calif., white arrow head is location where Mendocino Channel captured Mattole Channel. LSs is landslide scar. Locations of subsequent figures shown as black rectangles.

At the outset, it should be mentioned that Peakall et al. [24] challenged the analogy often made between subaerial and deep-water sinuous channels based on qualitative planform morphologies. They suggested differences in the fluid mechanics that acted within the two environments would have produced recognizable features. In particular, they stressed the importance of large-scale overbank flows that can produce recognizable morphologic features on submarine channel levees. Consequently, care has been taken to not over-interpret the processes that may have formed and modified Mendocino Channel using subaerial analogies.

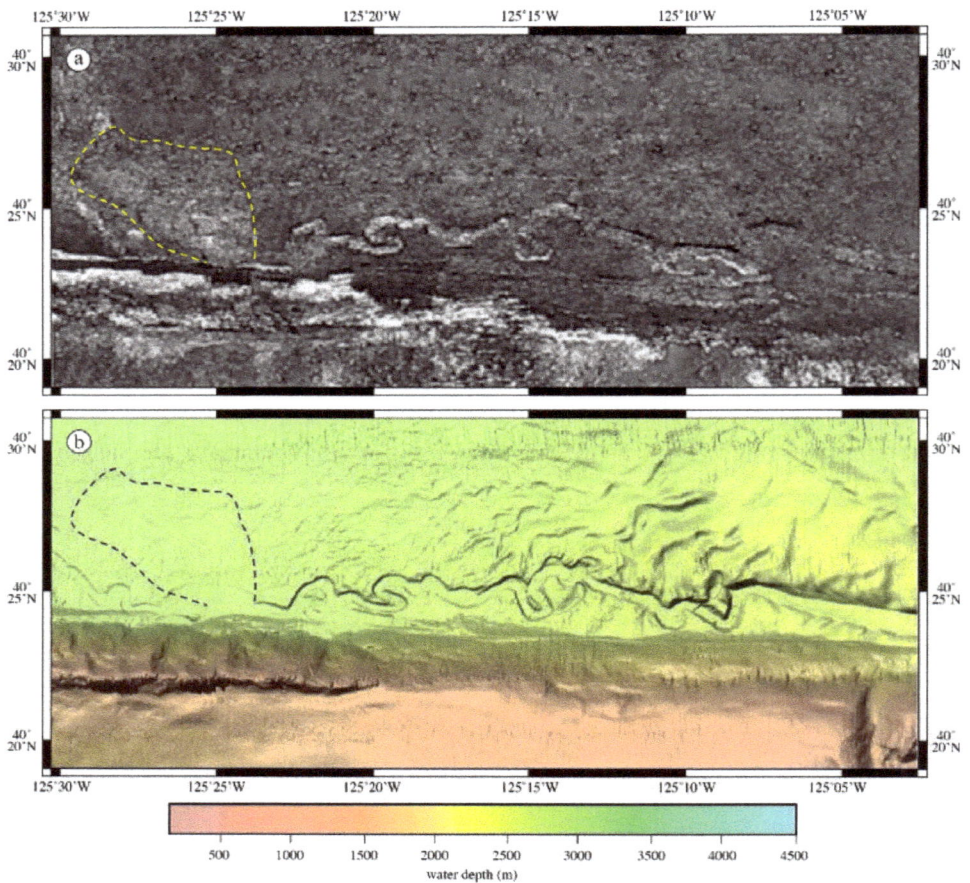

Figure 3. (**a**) GLORIA (6.5-kHz sidescan sonar) backscatter image of the eastern portion of Mendocino Channel, modified from [5]; (**b**) Multibeam bathymetry of same area. Dashed line (yellow and black) outline area of high 6.5-kHz backscatter in area immediately north of crevasse splay (see text for discussion). Location slightly larger than rectangle labeled "Fig. 10" on Figure 2.

2. General Setting

Mendocino Channel occurs at the base of Gorda Escarpment, the north-facing slope of the eastern-most section of Mendocino Ridge. The eastern end of the escarpment is the Mendocino Triple Junction, the point where the Gorda and Pacific Plates meet at the Mendocino Transform Fault. Gorda Escarpment is the eastern section of the transform fault that extends from Gorda Ridge to the west to North America on the east. Mendocino Channel extends westward from the continental margin for more than 148 km and out onto the southernmost Gorda Basin (Figure 2). Mendocino Triple Junction and Transform Fault are the locus of numerous earthquakes with magnitudes greater than M_w5 over the past 100 years whose affects have been felt over the entire region.

The regional bathymetry suggests the southern boundary of Eel Fan is just to the north of the new MBES bathymetry. Tréhu et al. [25] reported on a series of N-S multichannel seismic profiles that cross a section of Gorda Escarpment that includes Mendocino Channel. The profiles (Figure 4) show that Mendocino Channel is perched against the north wall of Gorda Escarpment and has constructed a levee to the north, overlapped in places by landslide deposits and buried in other places by distal sediments of Eel Fan. The channel-levee complex is more than 250 m thick in places and was deposited above a thick deformed hemipelagic sequence that probably represent sediments of Eel Fan as well as landslide deposits.

Mendocino Channel evolves from two canyon heads, Mendocino and Mattole Canyons, that incise the northern California margin [5] (Figure 2). Mattole Canyon channel has been captured by Mendocino Canyon channel 22 km down-canyon and then evolves as a single channel that stretches for at least 148 km to the west-northwest. The channel trends parallel to Gorda Escarpment for 90.5 km before it has been diverted to the NW around a large landslide deposit that originated on the north flank of Gorda Escarpment. At this point, the channel trends NW away from Gorda Escarpment for at least another 48.5 km before the channel turns back on a more westerly trend for an additional 9.4 km. The MBES bathymetry no longer resolves the channel at 148 km down-channel, which is ~114 km east of Gorda Ridge (Figure 1). The average slope of the channel is 0.48° (8.4 m/km) from the capture point to the distal western-most point that is resolved by the multibeam bathymetry.

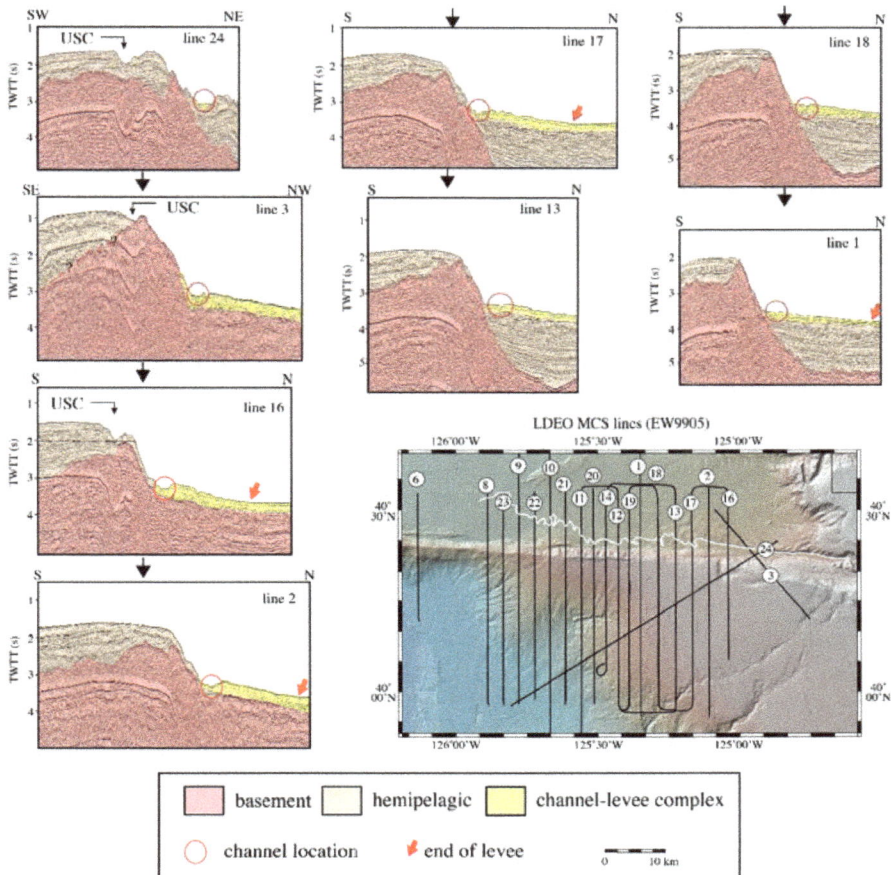

Figure 4. *Cont.*

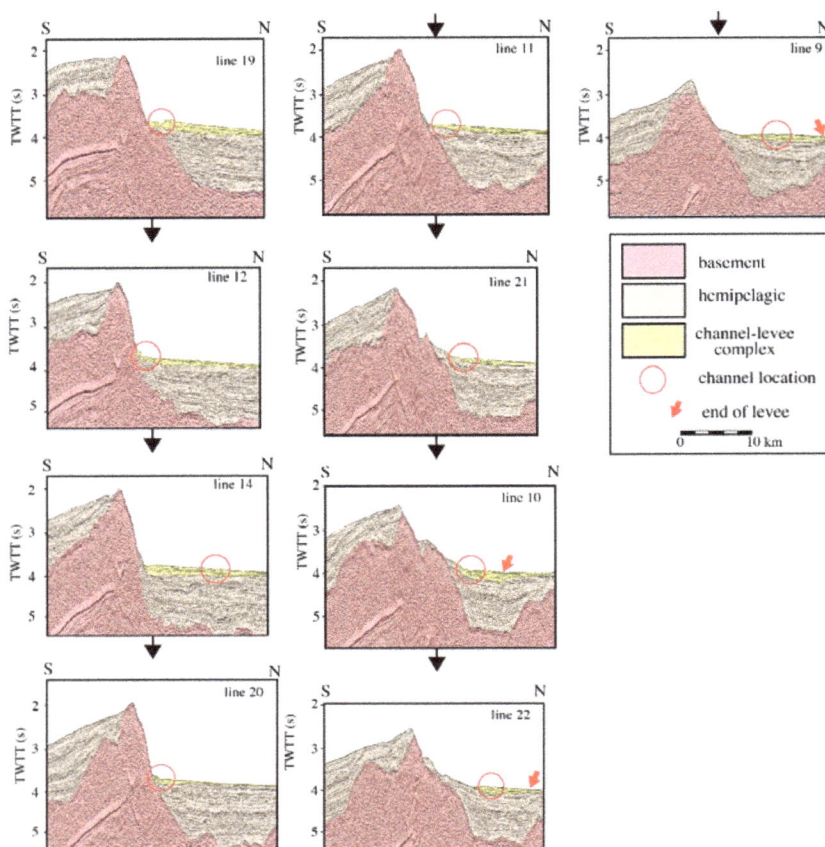

Figure 4. Multichannel seismic-reflection profiles (EW9905) collected by Tréhu et al. [25] and accessed at www.geomapapp.com. TWTT is two-way travel time. Red circle surrounds location of Mendocino Channel; red arrow is location of the distal end of the levee. Large black arrows show the progression of seismic lines from east to west. No channel was resolved on lines 6, 8 and 23. USC is upper side channel (see canyon discussion below).

Water depths of the floor of Mendocino Channel range from 1764 m at the capture point to 3025 m at the western-most extent. Landslides are prominent features on the continental margin adjacent to the eastern 25 km of Mendocino Channel as well as along the north wall of Gorda Escarpment. The California margin immediately north of the eastern reaches of the channel has one area of landslides that originated just north of Eel Canyon, another area of landslides just south of Eel Canyon, a landslide scar on the north levee of Mendocino Channel and a large landslide off the north wall of Gorda Escarpment (Figure 2). The landslide just south of Eel Canyon scattered debris at least 18 km away onto the basin floor. The landslide events are undoubtedly related to the high seismicity of the Mendocino Triple Junction and Transform Fault.

3. Multibeam Echosounder Data

The multibeam bathymetry and acoustic backscatter were collected with the NOAA Ship *Okeanos Explorer* equipped with a hull-mounted Kongsberg Maritime EM302 MBES system. This MBES system transmits a 0.5° wide fore-aft swath and forms up to 864 athwart-ship 1° receive apertures over a maximum swath of 150°. Individual soundings along track are spaced approximately every 20 m, regardless of survey speed.

An Applanix POS/MV 320 version 4 motion reference units (MRU) was used to correct for changes in instantaneous ship heave, pitch, roll and heading. The EM302 system can incorporate transmit beam steering up to ±10° from vertical, and yaw and roll compensation up to ±10°. The MRU was interfaced with a C&C Technologies C-Nav differential-aided GPS (DGPS) receiver that provides real-time correctors to the DGPS position fixes, providing spatial accuracies of <±0.5 m. All horizontal positions were georeferenced to the WGS84 ellipsoid and vertical referencing was to instantaneous sea level.

Water-column sound-speed profiles were calculated from casts of calibrated Sippican model Deep Blue expendable bathythermographs (XBTs) that measured to 760 m maximum water depth. XBT casts were routinely made every 6 h and between scheduled casts whenever the difference between measured sound speed at the transducers differed by more than 0.5 m/s from the sound speed calculated from the XBT value at the depth of the transducer.

The Kongsberg EM302 is capable of simultaneously collecting co-registered full time-series acoustic backscatter along with bathymetry. The backscatter data represent a time series of backscatter measurements across each individual beam footprint on the seafloor. If the received backscatter amplitudes are properly calibrated to the outgoing acoustic signal strength, receiver gains, spherical spreading, and attenuation, then the calibrated backscatter should provide clues about the composition of the surficial seafloor.

A digital terrain model (DTM) was constructed from the MBES bathymetry into a non-projected geographic grid with a resolution of 40 m/pixel using a weighted moving average with a weight diameter of 3 that provides a minimum of 8 of the nearest soundings to influence each grid node of the DTM [26]. The multibeam DTM was combined with a bathymetry grid from the NOAA Coastal Relief Model (CRM) (http://www.ngdc.noaa.gov/mgg/coastal/crm.html) to provide a complete view of the eastern 50 km of the canyon-channel system. The CRM is a compilation of NOAA hydrographic surveys and university single-beam and multibeam bathymetry data gridded at 90 m/pixel. All map and perspective figures in this report were generated from this DTM. Similarly, the MBES backscatter data were gridded at 40 m/pixel and draped over the co-registered MBES bathymetry.

4. Channel Descriptions

Schumm and Brakenridge [27] subdivide fluvial pattern morphologies into (1) straight channels; (2) sinuous-thalweg straight channels; (3) meandering channels; (4) and braided channels. They cite experimental and field studies that have shown that critical thresholds in stream power, gradient and sediment load lead to changes in channel pattern. Although their conclusions may not be relevant to submarine channels because of differences in densities between air and river waters and between oceanic bottom water and turbidity currents, their insights suggest that those three parameters might be important to the formation of submarine channel planforms. Unfortunately, two of those three parameters (stream power and sediment load) are unknown in Mendocino Channel although channel gradient is easily measured from the multibeam bathymetry. Nevertheless, the spirit of their subdivision of channel patterns was used to segment Mendocino Channel.

Mendocino Channel was subdivided into 6 reaches based on the planform geometry. The complete channel includes the following segments: (1) Mendocino and Mattole Canyon channels; (2) USR, an upper straight reach; (3) USinR, an upper sinuous reach; (4) a crevasse splay; (5) LSinR, a lower sinuous reach; and (6) LSR, a lower straight reach. These segments follow in a progression from the continental shelf to the western extent of Mendocino Channel. Each reach has distinct characteristics, as discussed below, that made the subdivision obvious. Although the crevasse splay is not described as a reach, it was identified by planform geometry as such a unique feature that it merited its own subdivision.

Measurements were taken every 0.5 km along the thalweg of the entire length of Mendocino Channel with the junction of Mattole and Mendocino Canyon channels as 0 km. The measured channel parameters include (1) channel floor width; (2) channel top width; (3) channel vertical relief; (4) south-side levee height above the channel floor; (5) north-side levee height above the channel floor; (6) channel cross-sectional area calculated using the formula for a trapezoid; (7) water depth of channel floor and (8) perched height of channel floor above basin depth (Figure 5). Of these measurements, relationships were found with levee heights, channel floor with channel top width, channel cross-section area, water depth of channel floor, perched height of channel floor, all with respect to down-channel distance. Other parameters typically measured in submarine channel meanders were measured (e.g., meander radius of curvature, meander wavelength, meander amplitude) but no statistically significant relationships were found.

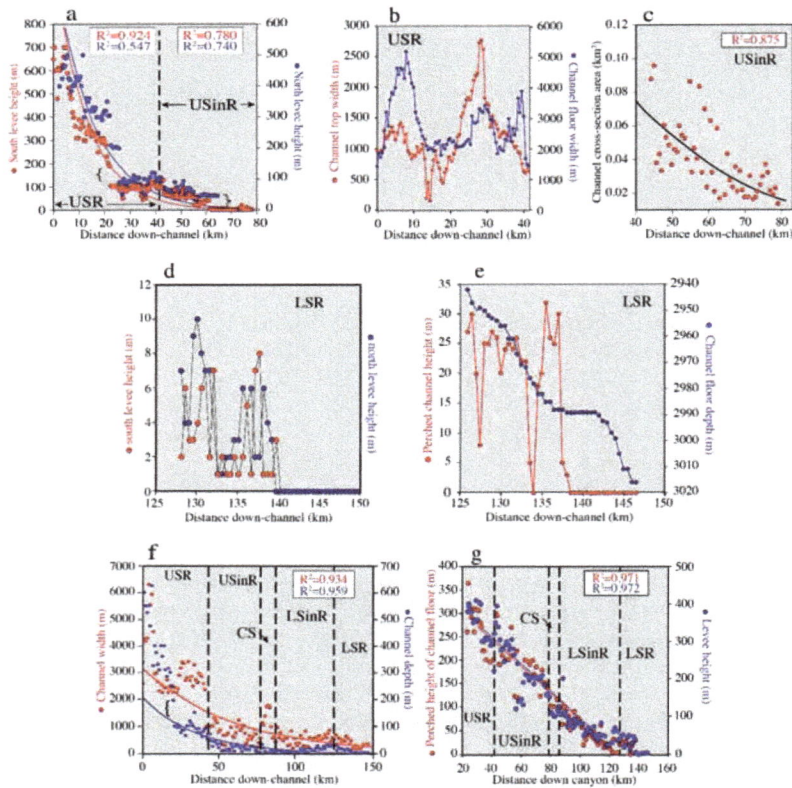

Figure 5. Plots of relationships between (**a**) upper straight reach (USR) and upper sinuous reach (USinR) levee heights vs. distance down-channel. Curly black bracket is gap in USR north levee heights in an otherwise smooth decrease and curly white bracket is gap in USinR north levee heights; (**b**) USR channel floor width and channel top width vs. distance down-channel; (**c**) plot of USinR channel cross-sectional area vs. distance down-channel; (**d**) plot of LSR levee heights vs. distance down-channel; (**e**) plots of LSR perched height above basin depth and depth of channel floor vs. distance down-channel; (**f**) plots of entire Mendocino Channel floor width and channel incision depth vs. distance down-channel. Curly bracket is gap in channel width and depth with down-canyon distance; (**g**) plots of entire Mendocino Channel perched height above basin depths and levee height vs. distance down-channel.

4.1. Mendocino and Mattole Canyon Channels

Mendocino Channel begins where Mendocino Canyon channel captured Mattole Canyon channel at the base of the northern California margin (white arrowhead on Figures 2 and 6). The upper reaches and junction of Mendocino and Mattole Canyons were first illustrated and discussed by H.W. Murray in the late 1930s (www.history.noaa.gov/stories_tales/mendocino.html). Murray's bathymetry was generated from a compilation of more than 80,000 laboriously collected lead-line soundings taken over a 3-year period and provides a remarkably good general overview of the two canyons, although Mendocino Channel was not discovered. Shepard and Dill [28] (their Figure 47) published a version of Murray's map in 1966. The combined MBES and CRM bathymetry provides for the first time an accurate and high-resolution view of the two canyons that shows their complexities. The head of Mendocino Canyon is at a water depth of 109 m (Figure 6) with an overall gradient of 6.33°. Mendocino Canyon channel has a rough and stepped subtle concave-down descent with 7 distinct step-downs that range from 30 to 172 m (Figure 7). The widths of the broad U-shaped canyon valley progressively increase from 1 km at the proximal end to 2.47 km wide at the distal end just east of the capture point. A large 900 m wide, 4 km long terrace stands 200 m above the canyon channel floor on the gentle inside bend of the first bend of the valley (T_a on Figure 6). About 2.5 km down-canyon, a two-stepped tilted terrace occurs on the inside of the next bend (T_b and T_c on Figure 6). The lower terrace (T_c) stands about 350 m above the canyon channel floor and the higher terrace (T_b) stands 730 m above the floor. A flat unpaired terrace (T_d) occurs across the channel from terrace T_c on the south bank (T_d on Figure 6) that stands 355 m above the Mendocino Canyon channel floor. The canyon valley makes a sharp 40° left bend just down-canyon from the T_c terrace and immediately before the capture of Mattole Canyon channel. The upper reaches of Mendocino Canyon channel have no apparent incised thalweg, although any thalweg with less than 0.5 m of relief would be below the vertical resolution of the MBES in these water depths.

The head of Mattole Canyon has a water depth of ~74 m and begins as a gentle swale only 960 m from the present shoreline. The canyon is 26.8 km long with a gentle concave-up surface that has a two-step (40 and 80 m) descent along the reach (Figure 7), with an overall gradient of 4.09°, before it is captured by Mendocino Canyon channel and the two channels evolve into the upper Mendocino Channel at a water depth of 1773 m (Figures 6 and 7). The initial 6.3 km of Mattole Canyon has a southwesterly trend with a broad zig-zag pattern of 3 reaches with lengths of 2.3, 2.1 and 1.9 km from the canyon head. The canyon channel then continues with a 17.3 km linear reach that trends N69W. Mendocino Canyon captured Mattole Canyon at this point and left an abrupt 80 m 16.5° slope down to the floor of the main Mendocino Channel (white arrowhead on Figure 6). The floor of Mattole Canyon descends to the confluence with Mendocino Canyon with numerous small steps mostly with less than 20 m of relief.

The width of the floor of Mattole Canyon channel varies from 515 m at the canyon head to 1155 m at the junction with Mendocino Canyon but the widest section of the canyon floor at 1.65 km occurs in a 3.8 km-long section area centered 12.8 km down canyon. Mattole Canyon has a three-step terrace just above the capture point with heights above the channel floor of 310, 510 and 660 m (T_f, T_g and T_h on Figure 6). The three terraces T_f, T_g and T_h are unpaired with terrace T_d. Two scarps (L_s on Figure 6) occur on the southern wall of Mattole Canyon and the canyon-channel floor adjacent to the scarps is much rougher than the canyon floor farther down.

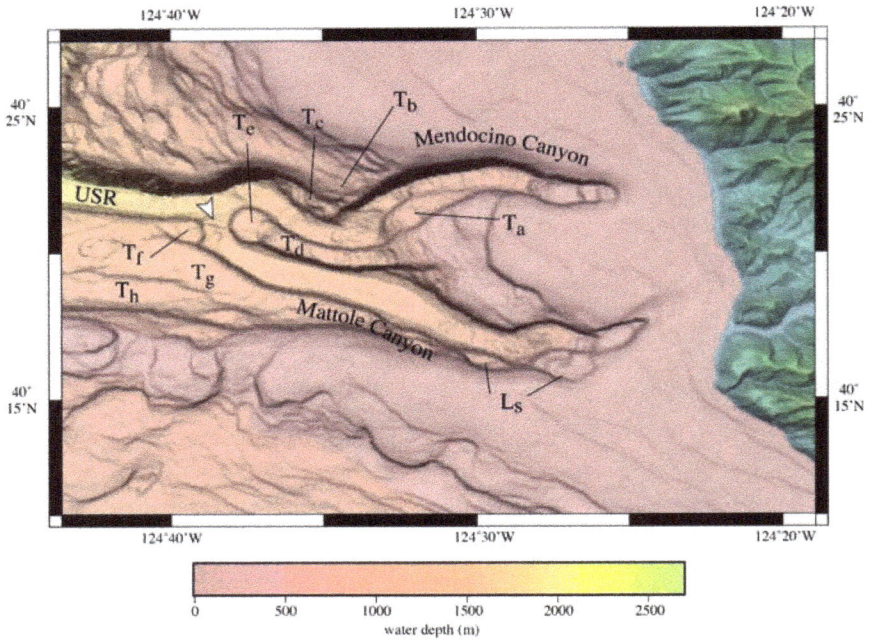

Figure 6. Map view of NOAA Coastal Relief Model bathymetry of Mendocino and Mattole Canyons. Features labeled T_x are 6 individual terraces (see discussion in text) and white arrowhead is an 80 m step-down of the floor of Mattole Channel to the floor of Mendocino Channel at the capture point. L_s points to two scarps on the wall of Mattole Canyon. USR is Upper Straight Reach of Mendocino Channel. See Figure 2 for location.

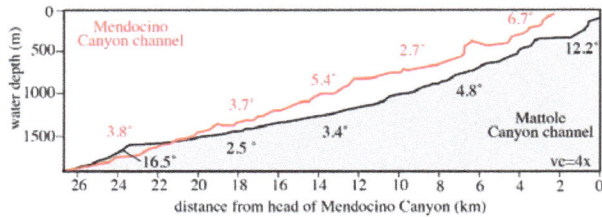

Figure 7. Thalweg profiles of Mendocino Canyon channel (red) and Mattole Canyon channel (black). Note the rough and subtle concave-down profile of Mendocino Canyon channel relative to the smooth but stepped distinct concave-up profile of Mattole Canyon channel.

4.2. Mendocino Channel Upper Straight Reach

The south side of first 12 km of the USR is constrained by the steep (>20°) wall of the Gorda Escarpment whereas the first 12 km of the north side of the USR is flanked by the steep wall of the continental margin and then landslide deposits shed off the California margin (Figures 2 and 8). The channel hugs the north wall of Gorda Escarpment along all but the last 12 km of its length. The upper straight reach extends west in a nearly straight strike for 41.6 km from the capture point of Mendocino and Mattole Canyon channels and ends at a sharp left-hand 90° bend to the south towards the wall of Gorda Escarpment ("c" on Figure 8). The channel floor is 1.1 km wide at the head of the reach and is 4.2 km wide at the top of the channel valley (Figure 5b). The channel floor descends at a smoothly decreasing gradient from 2.83° at its eastern edge to a point 15 km down-channel where the floor is only 145 m wide with a gradient of 1.88°; the top of the channel valley is 2.2 km wide at this point. From this point, the floor of the USR becomes progressively wider down-channel to a maximum floor width of 2.7 km at 31 km down channel and the valley top attains a maximum width of 3.6 km at its western limit. The USR is perched 365 m above general basin depths but slowly descends to 200 m at its down-channel boundary (Figure 5g).

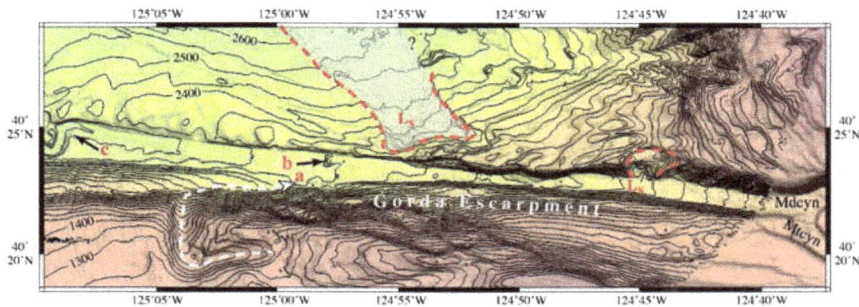

Figure 8. Map view of multibeam bathymetry of the upper straight reach (USR). Ten km black circles mark distances in km in italics from the junction of Mendocino and Mattole Canyons. White dashed arrow shows thalweg of upper side channel (see discussion below), red "a" is the point where the upper side channel enters the USR, red "b" is a large 60 m high mound on the floor of Mendocino Channel (see text) and red "c" is 90° channel bend. "Mdcyn" is lower Mendocino Canyon and "Mtcyn" is lower Mattole Canyon. Red dashed lines outline landslide scarps (Ls) with area of hummocky relief in gray. Isobath interval is 50 m, representative 100 m isobaths are labeled. Isobaths of Gorda Escarpment intentionally not labeled. See Figure 2 for location.

A side channel heads on the eastern summit region of Gorda Escarpment (Figures 8 and 9). The channel was first described by Tréhu et al. [25] and is clearly shown in their seismic data (Figure 4, lines 3, 16 and 24). The new MBES bathymetry shows that the upper side channel does not head in a canyon or even on the margin, but simply appears as a broad shallow swale at the 750 m isobath on a broad high. The first 12.2 km of the upper side channel is relatively straight, but then the channel makes an abrupt series of three near 90° bends before the channel plunges down the north-facing Gorda Escarpment. The upper side channel appears to make yet another 90° bend on a bench about 300 m above the USR floor and traverses east along the bench for another 6.4 km before it enters the USR channel floor at 26.8 km down-channel. At this point, the upper side channel makes an abrupt 120° bend to trend down-channel (west) but no longer can be identified on the MBES bathymetry. However, at the point where the upper side channel enters the USR, the MBES acoustic backscatter abruptly increases from −27 dB to −20 dB and the high backscatter is concentrated along the northern edge of the USR (Figure 9c). From this point on down-channel, the backscatter of the floor of USR spreads out across the channel floor and is consistently between −22 and −20 dB. A 2–5 m high bench occurs on the north side of the channel floor from this point to the end of the reach that reflects sediment from the upper side channel. A large solitary mound sits on the main channel floor at just the point where the side channel reaches the main channel floor ("a" on Figures 8 and 9). The mound has 60 m of relief and basal dimensions of 700 × 940 m.

The head of the USR is perched 300–360 m above the basin water depths and descends 200 m above the basin depths at the end of the reach (Figure 5g). Levees appear at 12 km down-channel on the north bank of the channel and levee (plus undifferentiated landslide and mass transport deposits) heights decrease from 350 m high at 12 km down-channel to 200 m high at 21 km down-channel.

At this point, there is a marked decrease in both north-side and south-side levee heights where the channel has moved beyond the immediate margin and trends out of the zone of landslide and mass transport deposits and onto the proximal basin floor (black curly bracket on Figure 5a). The levees are never higher than ~145 m throughout the remainder of the USR. There is a strong relationship (south levee $R^2 = 0.853$ and north levee $R^2 = 0.830$) of the levee heights with down-channel distance (Figure 5a). Also, there is a rough similarity in trends, but not in absolute values, of the width of the USR channel floor with the width of the USR channel top (Figure 5b). There is no statistical correlation ($R^2 = 0.275$) between the channel floor widths and the down-channel distance. The channel floor width is 825 m at the beginning of the reach and decreases to a minimum of 210 m wide 16 km down-channel but then the channel attains a maximum width of almost 3000 m at 31 km down-channel. The width of the USR channel top decreases from a maximum of 4150 m wide at the beginning of the reach to a minimum of 1800 m wide at a point 20 km down channel. At this point, the channel floor width begins to decrease to a minimum of 100 m wide to the end of the reach. However, the channel top width is 3400 m wide at 32.5 km down-channel but decreases to only 1850 m at 35 km down-channel before increasing to 3690 m wide at 41 km down-channel.

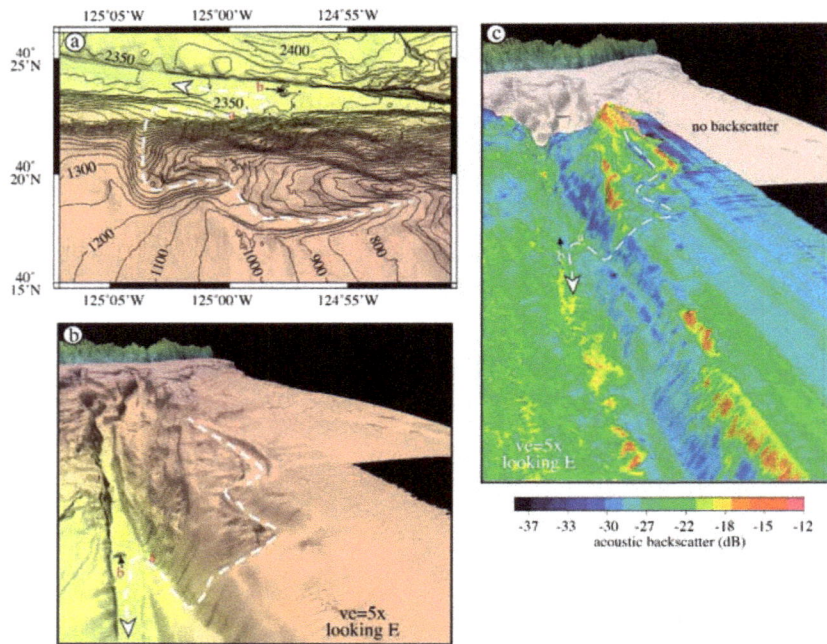

Figure 9. (**a**) Map view of multibeam bathymetry of upper side channel. White dashed line and arrowhead traces the thalweg of the upper side channel. Red "b" points to large mound on upper straight reach (USR) floor. Isobaths of Gorda Escarpment intentionally left unlabeled; (**b**) Perspective view with 5× vertical exaggeration of multibeam bathymetry of the upper side channel, north wall of Gorda Escarpment, and the floor of upper straight reach. Red "b" points to large mound on USR floor; (**c**) Perspective view of multibeam acoustic backscatter draped on the multibeam echosounder (MBES) bathymetry of upper side channel (white dashed line and arrowhead), the north wall of Gorda Escarpment, floor of upper straight and upper sinuous reaches and adjacent levee. White "a" is entry point of upper side channel to Mendocino Channel, white "b" points to large mound on USR floor. Note the high backscatter (red and yellow colors) in Mendocino Channel floor that begins where the upper side channel enters the Mendocino Channel floor and continues down-channel.

4.3. Mendocino Channel Upper Sinuous Reach

The upper sinuous reach (USinR) is a 37.5 km section that begins 42.5 km down-channel and ends at 78.5 km down-channel with an overall gradient of 0.33°. The reach is perched 200 m above basin depths as it trends along the north wall of Gorda Escarpment and continues a westward descent to 140 m above basin depths at the western extent of the reach (Figure 5g). The north bank of USinR is bordered by the continuation of the extensive levee complex found along the upper straight reach but a sudden drop in levee height of 40 m down to basin depths occurs at 65 km down-channel (Figure 5a). The USinR is sinuous with 12 bends, one of which is an abandoned full 190° cut-off meander, whereas other bends turn as little as 20°. The reach has a sinuosity index (ratio of channel length to valley length) of 1.66. The width of the channel floor varies from 22–1107 m with no systematic pattern correlated to distance down-channel. However, there is a strong correlation ($R^2 = 0.875$) of decreased channel cross-section area and distance down the USinR (Figure 5c).

The floor of the USinR is relatively smooth at the vertical resolution of the MBES at these depths (~2 m) although two large mounds occur on the channel floor (b_1 and b_2 on Figure 10a). Mound b_1 stands 18 m high with basal dimensions of 162 × 234 m and is located 4 km down the reach. Mound b_2 is 40 m high with basal dimensions of 266 × 279 m and is located 14 km down channel. Most of the channel floor has high backscatter (−17−−15 dB) (Figure 10b) that is confined within the channel and occurs from a point 58 km down-channel and continues to the western end of the USinR. The middle of the USinR includes a complex series of bends and terraces (Figure 11). Bend A has a flat, level bench (terrace or outer-bank bar [29]) 45 m above the channel floor compared to bend B directly across the channel that has a tilted bench only 3 m above the channel floor. Bend C has a flat, level terrace (or outer-bank bar?) 18 m above the channel floor. Cut-off meander D is a 190° bend but a detailed analysis of the MBES bathymetry shows that a chute channel has broken through so that meander D is in fact a cut-off meander (Figure 11a,b).

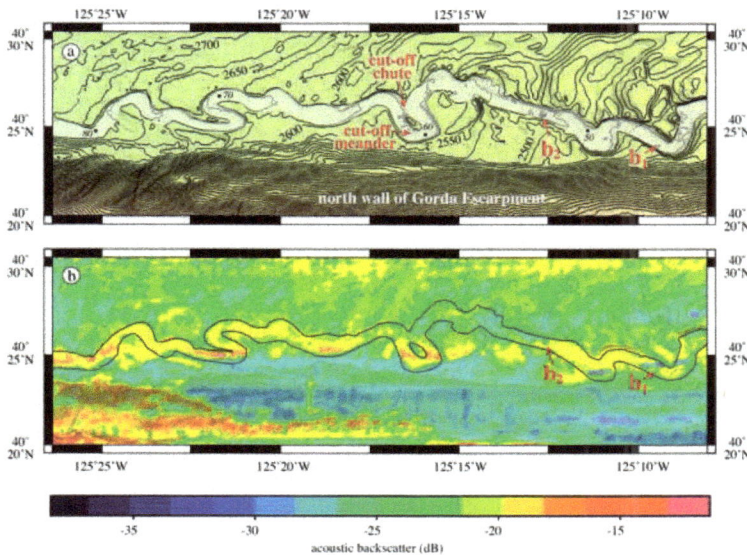

Figure 10. (**a**) Map view of MBES bathymetry and (**b**) backscatter of upper sinuous reach floor (grayed area) and immediate surrounding area. "b_1"and "b_2" are two large mounds on the floor of the reach. A cut-off meander occurs 14 km down-channel. Black circles in (**a**) mark 10 km distances in italics from the junction of Mendocino and Mattole Canyons. See Figure 2 for location. Isobath interval 25 m; isobaths of north wall of Gorda Escarpment intentionally left unlabeled.

A relict cut-off channel and deep depression occurs on the south bank opposite bend A (Figure 11a,b). The depression is 30 m deep at its deepest and its floor is 15 m below the adjacent channel floor at bend A. The depression is separated from the channel floor at bend A by a steep (50°) 200 m wide, 23 m high sill, although the depression is inline with a broad gently curved abandoned channel that begins 18 m above the inner bend of the cut-off meander, and descends to depths 21 m deeper at the center of the depression. The depression and associated abandoned channel are erosional, not depositional, features.

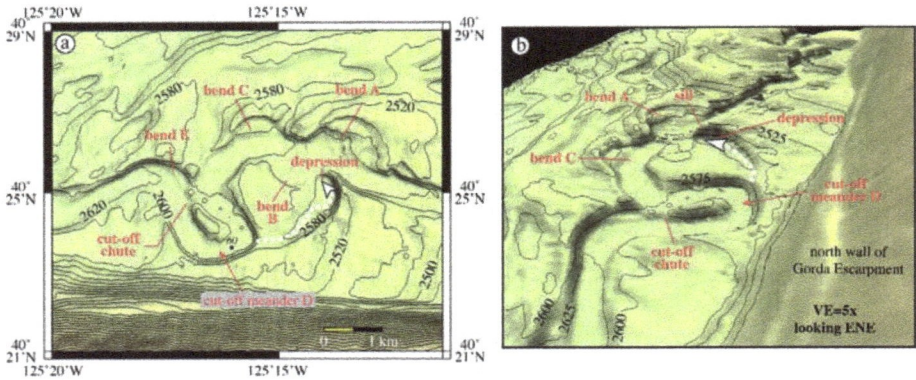

Figure 11. (**a**) Map view of middle of upper sinuous reach (USinR) that shows the cut-off chute that created cut-off meander D and the depression and channel on the southern bank of bend A perched 18 m above the adjacent floor of the USinR. White dashed arrow marks the path of the abandoned channel that leads to the depression. Black circles mark distances in km in italics from the junction of Mendocino and Mattole Canyons. Isobaths of Gorda Escarpment intentionally left unlabeled; (**b**) Perspective view of the same area and features.

4.4. Mendocino Channel Crevasse Splay

A broad conspicuous 5 km-long flat area (4.5 km E-W × 1.7 km N-S) begins at 78.5 km down-channel just beyond the USinR. The area is bounded on the north by a discontinuous low bank. This section has the broadest channel width of the entire length of Mendocino Channel but it continues the perched descent to the west from 140 to 120 m above basin depths while it remains close to the north wall of Gorda Escarpment. The low, broad area resembles a crevasse splay where in the past the channel breached the north bank in several places (Figures 2 and 12). The north-south gradient between the banks is only ~0.02° whereas the east-west gradient is 0.33°, the latter gradient is typical of the USinR immediately up-channel. The north bank varies between 4 and 12 m high except for the suspected breach areas, which are less than 2 m high. High backscatter (−17−−12 dB) decreases in intensity to the north in the broad area of the crevasse splay (Figure 12). Although the high backscatter might reflect debris shed off the escarpment, the location of the broad area of high backscatter is 6.5 km east of the eastern edge of the relief that reflects the large landslide deposit (see below) and nowhere else along the length of Mendocino Channel is high backscatter seen to extend to the north outside the channel.

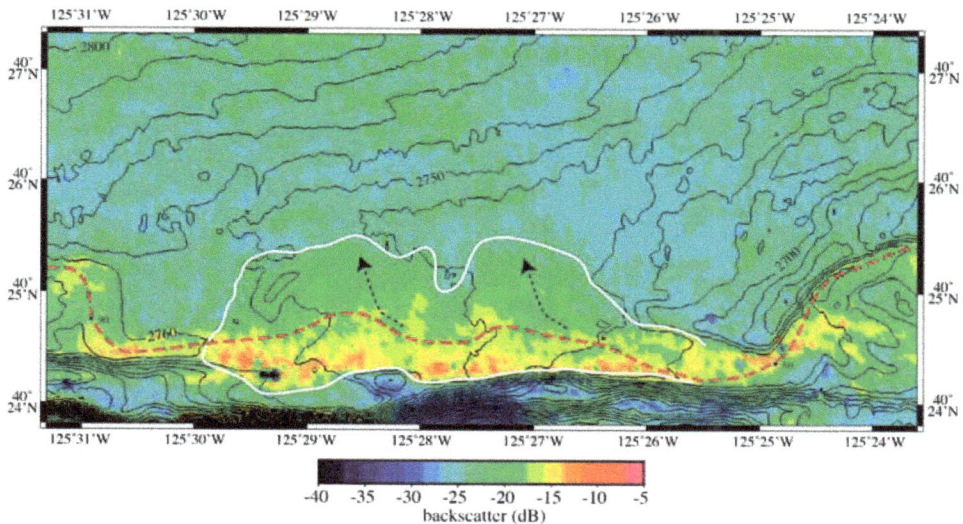

Figure 12. MBES backscatter image of crevasse splay with isobaths. White outline is the extent of the high-backscatter sediment of the crevasse splay that broke out of the channel towards the north (black dashed arrows). Red dashed line is the thalweg of Mendocino Channel. Black circles mark distances in km in italics from the junction of Mendocino and Mattole Canyons. Isobath interval 10 m, isobaths of Gorda Escarpment intentionally left unlabeled. See Figure 2 for location.

4.5. Mendocino Channel Lower Sinuous Reach (Landslide and Beyond Landslide)

The lower sinuous reach (LSinR) includes a complex sinuous channel as well as a large landslide that shed debris off Gorda Escarpment and a channel blockage (Figure 13). The LSinR begins 85.5 km down-channel at a water depth of 2759 m, just beyond the crevasse splay, with a series of tight meanders and ends at 125.7 km down-channel at a water depth of 2944 m. The sinuosity index for LSinR is 1.83. The reach has a gradient of 0.27° with very little resolved relief on the channel floor. However, there is a 5.8 km section of the channel that starts at 94.8 km down-channel (red dashed bracket on Figure 13) that is completely blocked by 3–5 m high rough relief that appears to be debris from the large landslide. The material that blocks the channel has backscatter values of −16–−18 dB, not much different than the backscatter values immediately up-channel in the reach (Figure 13b). The width of the channel floor continues to decrease and water depths continue to increase down-channel (Figure 5f). Levees occur on both sides of the LSinR along almost the entire length. However, the available seismic data do not image levees beyond the LSinR so the full extent of the levees is unknown. The backscatter values are high (−18 to −20 dB) at the beginning of the LSinR but the values decrease to levee values (−26–−30 dB) by 147.6 km down-channel.

As with the other up-channel reaches, the LSinR is perched above basin depths (Figure 5g). The LSinR continues the westward decline, from 120 m above basin depths at the beginning of the reach to less than 20 m at the end of the reach. At the end of the reach, the LSinR has essentially reached basin depths.

The initial west-trending LSinR channel abruptly changes course and trends northwest away from Gorda Escarpment at 89.9 km down-channel because of a large landslide deposit that originated on Gorda Escarpment (Figures 2 and 13). The top of the landslide deposit is 550 m high, stretches 21.7 km along the base against Gorda Escarpment and fans out 4.7 km away from Gorda Escarpment, giving it a minimum volume of ~25 km^3 of talus. There are no resolvable channel obstructions, and other than the area that blocks the channel, no high-backscatter debris within the LSinR that can be confidently attributed to the landslide. Also, there is no hint of a channel on the west side of the landslide deposits; consequently, it appears that the initiation of the landslide predates the formation of Mendocino Channel.

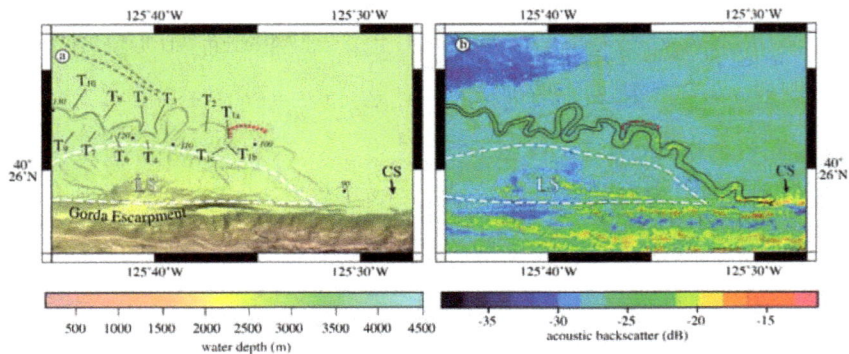

Figure 13. (a) Map view of multibeam bathymetry of the lower sinuous reach and large landslide deposit (white dashed polygon labeled LS) shed off the Gorda Escarpment. Terraces (T) are numbered and discussed in the text. Red dashed bracket is the zone of 3–5 m of relief that presently blocks the channel, discussed in the text. Black hachured lines outline a relict channel. Ten km black circles mark distances in km in italics from the junction of Mendocino and Mattole Canyons; (b) Map view of multibeam backscatter of the lower sinuous reach that shows that the high backscatter from up-channel slowly decreases in intensity with distance down-channel. CS is crevasse splay. See Figure 2 for location.

A series of outer- and inner-bend terraces (T_1 through T_{10} on Figure 13) occurs throughout the LSinR, all at varying heights above the channel floor (Table 1). Some of the terraces are flat in a channel-orthogonal direction whereas other terraces are tilted towards the channel (Table 1). The terraces occur on the outside of channel bends (T_1, T_2, T_3, T_4, T_5, T_7, T_8, T_9, and T_{10}) as well as along linear sections (T_6 and T_8). Paired terraces (similar water depths on adjacent inside and outside bend terraces) include T_7 to T_8 and T_9 to T_{10} whereas all the other adjacent terraces are unpaired.

A subtle 2–3 m deep linear channel strikes NW from the LSinR (black hachured lines in Figure 13). The channel may be related to an older course of Mendocino Channel given that it is in line with the initial eastern straight section of the LSinR. The subtle channel's closest approach to the LSinR at its SE end is only 1.5 km but the floor of the subtle channel is 30 m below the level of the LSinR channel floor and the two are separated by a 15–20 m high ridge. The backscatter values of the linear channel are no different than the levee values in the surrounding area and the multichannel seismic profiles (Figure 4) have no indication of this channel, both of which suggests the subtle channel has been buried by either LSinR levees or distal hemipelagic sediments of Gorda Basin.

Table 1. LSinR terrace water depths and heights above channel floor.

Terrace Number	Terrace Water Depth (m)	Height above Channel (m)	Terrace Orientation
T_1	2846	4	flat
T_2	2857	8	tilt towards channel 1.2°
T_3	2891	15	tilt towards channel 1.6°
T_4	2887	20	tilt towards channel 1.4°
T_5	2897	15	tilt towards channel 1.5°
T_6	2909	21	tilt towards channel 3.2°
T_7	2912	27	tilt towards channel 1.5°
T_8	2914	27	flat
T_9	2924	24	flat
T_{10}	2923	25	flat

4.6. Mendocino Channel Lower Straight Reach (LSR)

The lower straight reach (LSR) begins at 125.7 km down-channel and continues for 23 km before the channel is no longer resolved in the MBES bathymetry or backscatter (Figure 14). The reach has an overall gradient of 0.20° and consists of several relatively gentle bends and five 90° + bends. The LSR strikes west (~276°) for the first 5 km before it makes a broad 90° bend to the north but only stays on that course for 1.1 km before it makes another 90° bend to the west and runs for 2.9 km. At this point, the reach makes the third very broad 90° bend to the north for 3 km before it makes a final 105° turn to the west southwest (~261°) for the remainder of the resolved channel that ends at 148.7 km down channel. The last 105° bend in the reach may reflect the influence of the southern extent of the distal hemipelagic sediments of Gorda Basin or Eel Fan as well as influence from subsurface structure (see basement beneath red arrow on line 9 of Figure 4) that may have steered the LSR away from its NW trend out towards Gorda Basin. The LSR differs from the other reaches in that no terraces are found on either side of any bends. Levee heights have considerable variation down-channel (Figure 5d). The north-side levee is only 4–10 m high and the south-side levee is 2–7 m high until an abrupt decrease in levee height of less than 2 m high at 132.5 km down-channel. The levees then increase in heights to 6–8 m at 136–138 km down-channel and then decrease in height until no bathymetric expression of a levee is found at 140 km down-channel (Figure 5d).

The perched channel height vs. distance down-channel is highly variable for the first 15 km but then at a water depth of ~2990 m the channel has reached basin depths (Figure 5e).

A barely resolved channel trends to the southwest at the eastern beginning of the LSR with a 0.45° gradient. The channel can be followed for 3.9 km and heads just after a sharp bend in the LSR, although it is roughly orthogonal to the LSR trend (Figure 14). The incised depth of the channel ranges from 5 m at its head to less than 2 m at its western limit. The head of the side channel is 725 m WSW of the LSR and descends with a 0.8° gradient from the top of the south wall of LSR. This abandoned channel has no acoustic backscatter signature and appears unrelated to a former course of Mendocino Channel prior to the large landslide because it is not in line with the LSR and is well north of the landslide deposits. It might be a breakout channel from an earlier stage of Mendocino Channel but presently there is a 24 m high south-side levee that separates the eastern-most abandoned channel from the nearest channel floor of the LSinR and the LSR.

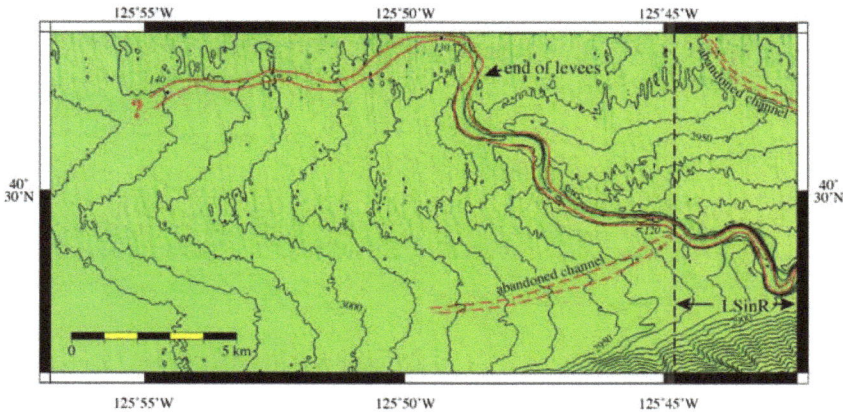

Figure 14. Map view of multibeam bathymetry of the western end of the lower straight reach (LSR) and the western-most lower sinuous reach (LSinR). LSR outlined in red. Ten km black circles mark distances in km in italics from the junction of Mendocino and Mattole Canyons. Red dashed lines outline two abandoned channels. The end of levees on either side of Mendocino Channel indicated with a black arrowhead. Isobath interval 10 m. See Figure 2 for location.

5. Discussion

The new multibeam bathymetry and backscatter data provide an unprecedented quantitative view of the morphology of the entire length of Mendocino Channel. The presence of Mendocino Channel, a sinuous channel on an active margin, begs an explanation for the age of the channel as well as what processes formed the channel. To investigate these questions, several lines of evidence were gathered from the literature and data archives that bear on explanations. The following discussion is focused on these questions.

5.1. The Age of Mendocino Channel

Twelve cores found in archives and the literature are either from within Mendocino Channel or from the northern levees, all located in the upper sinuous reach section (Figure 15). The lithostratigraphies and ^{14}C dates of the cores show a complex pattern that is not straight forward to interpret. Three box cores (BX-1, BX-5 and BX-9) are from Cacchione et al. [5], two piston cores (M9907-47PC and M9907-51PC) are from Goldfinger et al. [30] and one piston core (Y74-1-08) is from www.ngdc.noaa.gov/mgg/curator/ data/melville/avon/avon09mv/051/m9907_051pc_handwritten_corelog.pdf. Each box core sampled the upper 40–50 cm of the channel floor and is composed of numerous thin sandy turbidites interbedded with hemipelagic mud [5]. The sandy turbidites are 2–12 cm thick and conventional ^{14}C dates from BX-1 range from 0.970 kaBP at 5-cm subbottom to 3.595 kaBP at 48-cm subbottom. None of the other box cores have been dated. BX-9 is the most up-channel box core and is located 33.1 km down-channel from the channel convergence point. BX-5 is located 11.4 km down-channel from BX-9 and BX-1 and is located 13.1 km down-channel from BX-5. BX-5 contains the thickest turbidites of the three box cores, more than likely because it was collected on a sharp bend in the reach whereas BX-1 and BX-9 were collected on straight sections of the USR.

The published locations of piston cores M9907-51PC and M9907-47PC are separated by less than 5 m so both piston cores were collected 6.7 km down-channel from BX-1. The top 50 cm of the two piston-core lithostratigraphies do not resemble one another at all and an AMS ^{14}C age at 72 cm subbottom from M9907-51PC is 160 yrBP [30]. No ages have been reported from M9907_47PC. Assuming the turbidites and sands represent events that lasted only days or at most months, then the turbidites can be subtracted from the lithostratigraphies to represent only the hemipelagic mud in the sections. This, of course, assumes that the top of each turbidite is defined by the sand unit and the overlying mud is solely of hemipelagic origin and not deposition from the tail of the turbidity current. Nevertheless, if the 45 cm of sandy turbidites from the top 72 cm of M9907-51PC (the depth of the youngest ^{14}C age) are subtracted from the section, then the linear sedimentation rate of the hemipelagic mud in this section of the core is 169 cm/kyr. Similarly, if the ^{14}C date of 820 yrBP at 400 cm depth in core M9907-51PC is used, and the turbidites in the top 400 cm are subtracted, the hemipelagic sedimentation rate is 241 cm/kyr; both sedimentation rates that appear very fast for hemipelagic mud. For comparison, the nearest Ocean Drilling Program (ODP) site (Leg 167 Site 1020) is in Gorda Basin 80 km northwest of Mendocino Channel and collected an upper sequence of hemipelagic mud and no sandy turbidites with a sedimentation rate of only 11.5 cm/kyr [31]. If the same assumption and calculation is performed on BX-1 (the closest box core to the two M9907-47PC and M9907-51PC piston cores), using the 3.595 ka ^{14}C age at 48 cm, then the linear sedimentation rate of the hemipelagic mud in BX-1 is 13.5 cm/kyr, a hemipelagic sedimentation rate that is comparable to that from the ODP site. Likewise, if the sandy turbidites are subtracted from the top 50 cm of M9907-47PC, and the ^{14}C age at 48 cm in BX-1 is used as the age at 48 cm of M9907-47PC, then the linear sedimentation rate of the top of hemipelagic mud in M9907-47PC is 8.1 cm/yr, a sedimentation rate not too different than that of BX-1 and the nearby ODP site. These order-of-magnitude differences in hemipelagic sedimentation rates in M9907-51PC versus M9907-47PC, BX-1 and the ODP core, and the observation that the ^{14}C age at 70 cm in M9907-51PC is considerably *younger* than the ^{14}C age at 50 cm in BX-1, makes piston core M9907-51PC suspect, either because the published AMS ^{14}C ages are grossly in error or the location of M9907-51PC is in error and the core is not from within Mendocino Channel.

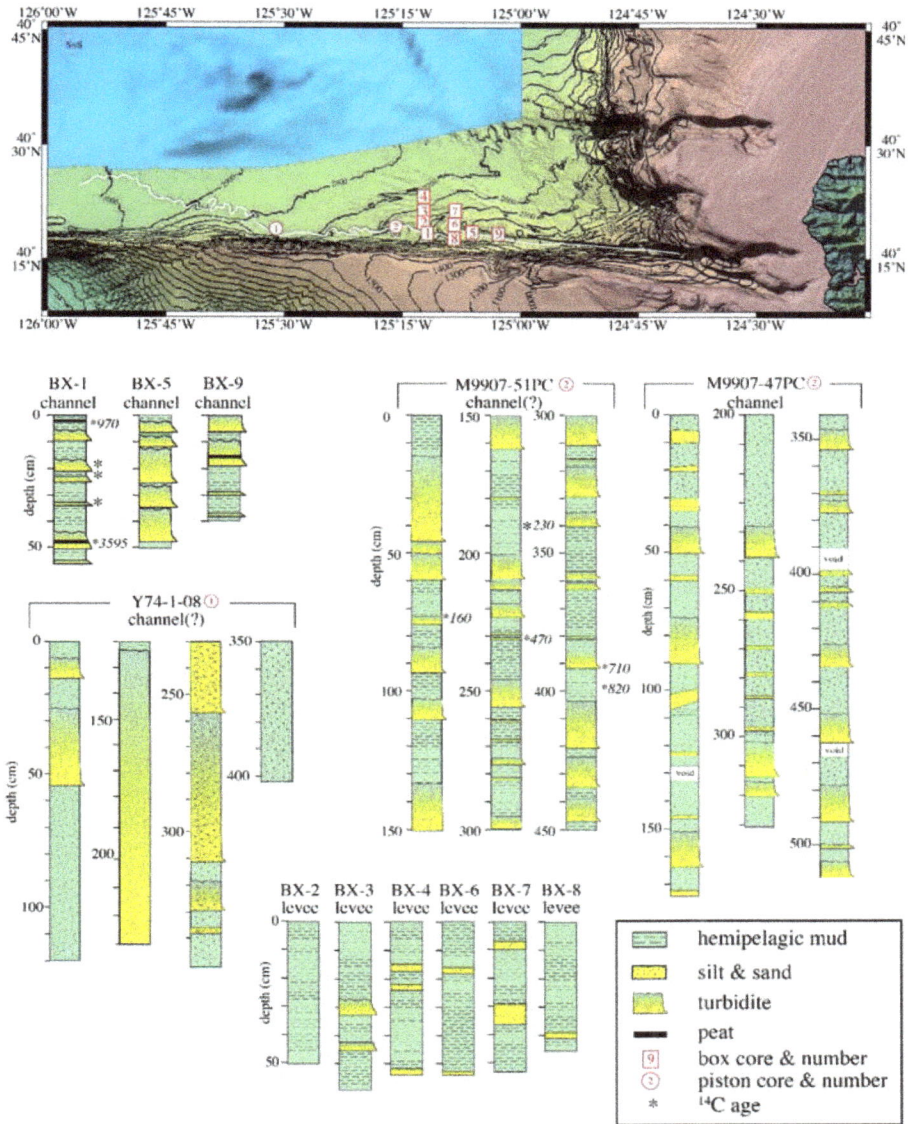

Figure 15. Location map of piston (circles) and box (squares) cores from within the channel and the north levee of Mendocino Channel. Numbers within the symbols indicate the core number in the core descriptions below the map. BX indicates a box core and the other cores are piston cores. Piston core descriptions were downloaded from www.ngdc.noaa.gov/mgg/curator/data/melville/avon/avon09mv/047/. Descriptions of box cores are from [5]. AMS[14]C ages of box cores in italics from [5] and from piston cores from [30].

An undated piston core (Y74-1-08PC) whose location plots on the channel floor contains only two sandy turbidites in the top 50 cm; the top turbidite is 7 cm thick and the next older turbidite is 25 cm thick. When the turbidites are subtracted from the section and the above assumptions are applied, applying the ^{14}C age from BX-1 results in a linear sedimentation rate of 1.3 cm/kyr for the hemipelagic mud in this core, a sedimentation rate that seems too slow for hemipelagic sedimentation. This slow rate suggests the top of Y74-1-08PC is missing, a common occurrence in piston cores. The section below 50 cm subbottom is composed of 67 cm of sandy hemipelagic mud followed by a 119-cm thick sandy turbidite immediately overlying a 56-cm sandy turbidite. Piston core Y74-01-08PC was collected 33 km down-channel from BX-1, the farthest west box core from the channel floor and 91 km down-channel from the confluence point. Piston core Y74-1-08 is only 2.7 km away from the eastern edge of the large landslide from Gorda Escarpment, which may account for the thick sandy turbidite sequence from 50–300+ cm down core.

The lithostratigraphies of the 6 box cores (BX-2, BX-3, BX-4, BX-6, BX-7 and BX-8) collected from the north levee differ from that of the three box cores (BX-1, BX-5, BX-9) and the two piston cores (Y74-01-08PC and M9907-47PC) collected from the channel. The levee cores all have three or fewer thin sandy turbidites from 5–12-cm thick in the upper 50 cm that are interbedded with much thicker hemipelagic mud. Using the ^{14}C age of 3.595 kaBP from BX-1 at 49 cm on the levee box cores, subtracting the thin turbidites using the assumptions above, yields linear sedimentation rates that only range from 11.4–13.9 cm/kyr for the hemipelagic muds on the levee. The sedimentation rates of the hemipelagic muds from the levees, even though there are no age dates, are similar to the calculated sedimentation rates of the hemipelagic muds of BX-1 that has a ^{14}C date as well as to the sedimentation rate from the nearby ODP core.

The box and piston cores demonstrate that that channel has been a conduit for several sandy turbidity currents over at least the past 3.600+ kyrs but there is a conspicuous lack of high-amplitude reflectors in the multichannel seismic profiles that would indicate a significant amount of coarse sediment was deposited in the channel axis over time [4,16,32,33]. Although, the cores do not provide any evidence for how long the channel has existed, the single-channel seismic-reflection profiles from Cacchione et al. [5] and multichannel seismic-reflection profiles from Tréhu et al. [25] (Figure 4) show that the maximum thickness of the Mendocino Channel ranges from ~200–~300 m, using the formula for the conversion of travel time to subbottom depth for the western U.S. margin [34]. If the assumption used above that the turbidites and sands represent events of only days or perhaps months duration, then subtracting the turbidites and sands, and using the occurrences of turbidites in the cores as representative of the turbidites in the seismic section, the hemipelagic sedimentation rate from BX-1 can be used as typical for the entire thickness of the Mendocino Channel (admittedly these assumptions could be challenged, but that is all the data available). If the hemipelagic sedimentation rate of 11.5 cm/kyr from ODP Leg 167 Site 2010 is used as the hemipelagic sedimentation rate for the channels in the multichannel seismic sections, then the initiation of the Mendocino Channel may be as old as 2.6–1.8 Ma (early Quaternary).

5.2. Potential Processes

Mendocino Channel is a rare example of a sinuous channel on an active margin and one that is not associated with a submarine fan. However, Mendocino Channel has many of the characteristics of sinuous channels found on passive margins that are associated with submarine fans. The potential process or processes that formed and maintained Mendocino Channel include (1) frequent seismicity; (2) periodic storm-wave loadings on the narrow shelf that were large enough to re-suspend and transport shelf sediments to the head of Mattole Canyon; (3) periodic low eustatic sea levels throughout the Quaternary that moved the shoreline to the heads of Mattole and Monterey Canyons; (4) tsunamis and (5) hyperpycnal plumes generated by periodic floods of the Mattole River (there is no river that directly feeds to Mendocino Canyon).

5.2.1. Seismicity

Many studies have suggested that nearshore and offshore seismic activity can trigger turbidity currents, (e.g., [30,35,36]; see Pope et al. [37] for a global summary). There is evidence that seismicity has had some impact on Mendocino Channel (e.g., [5,30]). The heads of Mendocino and Mattole Canyons are located at the Mendocino Triple Junction (red star on Figure 1) and the entire area is seismically active with great earthquakes that are suggested to have occurred on this margin with a recurrence of ~500–530 yr and smaller earthquakes along the southern part of this margin with a higher frequency of recurrence [30,38–40]. Although dating the numerous landslides on the California margin adjacent to Mendocino Channel has not yet been attempted, the box cores taken within Mendocino Channel contain a series of thin sandy turbidites interbedded with thicker hemipelagic muds [5]. The top sand in each box core is overlain by 3–5 cm of hemipelagic mud. Conventional ^{14}C ages of organic debris within the sands date the latest turbidite at 970 ± 80 yBP [5], but no turbidite was recovered that would correlate to the 1906 M_w 7.8 San Francisco earthquake. However, the occurrence of numerous large landslides in the immediate area of Mendocino Channel (e.g., [25]), coupled with the local seismicity makes earthquakes prime candidates for the initiation of sediment failures that evolved into turbidity currents that coursed down the margin and onto the basin floor.

5.2.2. Wave-Loading Resuspension

It is widely recognized that some submarine canyons have continued to be conduits for shelf-sediment transport to the deep sea, even during eustatic high stands, e.g., [41]. Wave-loading by periodic storms has been suggested as a mechanism that can re-suspend shelf sediment and generate high concentrations of suspended sediment at canyon heads (e.g., [42] and references within; [43]). This process is more likely in regions with high rainfall with steep terrestrial slopes and extensive mass wasting that generates and transports large volumes of sediment to local coastal rivers. Such rivers would then transport much of the suspended and bed-load sediment to the shelf and form depocenters for eventual resuspension by storm waves (see Nittrouer et al. [44]). If a depocenter is close to a canyon head, then storm-generated re-suspended sediment would have the potential to be transported into and down the canyon as a gravity-driven turbidity current. The coastal mountains of northern California are steep and there is abundant rainfall (annually 1100–2500 mm/yr) and, in addition, major winter storms impact the coast (e.g., 3–10 major storms/yr for the past 50 yr). If major storms occurred during El Niño events, then the major storms would have been intensified [45]. Records of El Niño events have been dated as old as 130 ka [46], so wave-loading resuspension by major storms is a distinctly possible process that could have generated turbidity currents in Mendocino and Mattole Canyons.

5.2.3. Hyperpycnal Plumes

River floods with a high concentration of suspended sediment, with concentrations high enough to make the flood water denser than the sea water it flows into, could form a hyperpycnal plume. An important criterion for the generation of a hyperpycnal flow is the degree of estuarine mixing prior to the flow entering the ocean [47]. Hyperpycnal plumes are generated by typhoons or intense rainfall that couple high river discharge with large suspended-sediment load. The plumes have the potential to flow downslope by gravity and, if the density of the hyperpycnal flow remains denser than the bottom water, sediment-laden plumes have the potential to evolve into a density-driven turbidity current that can flow out onto a basin floor [43,48–51]. Hyperpycnal plumes have been invoked for sediment waves off the California Margin just north of Mendocino Channel [52], in the Atlantic off Morocco [53], for deposits on Var Fan in the Mediterranean Sea [20,54], and off Taiwan [50,55,56]. This process has even been invoked as an alternate explanation for Humboldt Slide off northern California [57], although this interpretation is controversial (see discussion by Nittrouer et al. [51]). A dilute turbidity current was observed by Sumner and Paull [58] on an ROV dive in the head of Mendocino Canyon but only in water depths of less than 400 m. Whether this turbidity current reached the seafloor at water depths of 2000–3000 m is unknown, although Kao et al. [55] observed a hyperpycnal flow off Taiwan that reached water depths of 3000–3700 m. Regardless, this process would only be possible for the Mattole River and Canyon because the Mendocino Canyon has no adjacent river to feed sediment to the head of Mendocino Canyon.

5.2.4. Eustatic Sea Levels

Quaternary episodes of major regressions of eustatic sea levels located river mouths closer to adjacent canyon heads. Marine and ice-core studies have demonstrated that Quaternary eustatic sea levels dropped 120–130 m relative to today's sea level during maximum global glacial conditions ([59]; see De Boer et al. [60] for discussion and references). The shoreline effect of the glacial eustatic low sea levels at the heads of Mendocino and Mattole Canyons is shown in Figure 16. The mouth of the Mattole River is presently 2.5 km from the head of Mattole Canyon. A 120 m drop in eustatic sea level would have fed the Mattole River directly into the canyon head. A 130 m drop in eustatic sea level would have fed the river farther down the canyon head. The head of Mendocino Canyon is presently 3.2 km offshore the shoreline. A 120 m drop in eustatic sea level would have placed the shoreline 370 m from the head of Mendocino Canyon and a 130 m eustatic lowering would have placed the shoreline an additional 75 m closer to the canyon head. In these instances, there was virtually no shelf between the shoreline and the two canyon heads. The last eustatic low stand occurred about 20 ka, but the youngest turbidite in BX-1, collected within Mendocino Channel, dates at about 0.97 ka and the fifth turbidite in the sequence at 50 cm subbottom in BX-1 is dated at 3.59 ka, so eustatic sea levels could not be the dominant forcing process that generated many of the late stage channel turbidites.

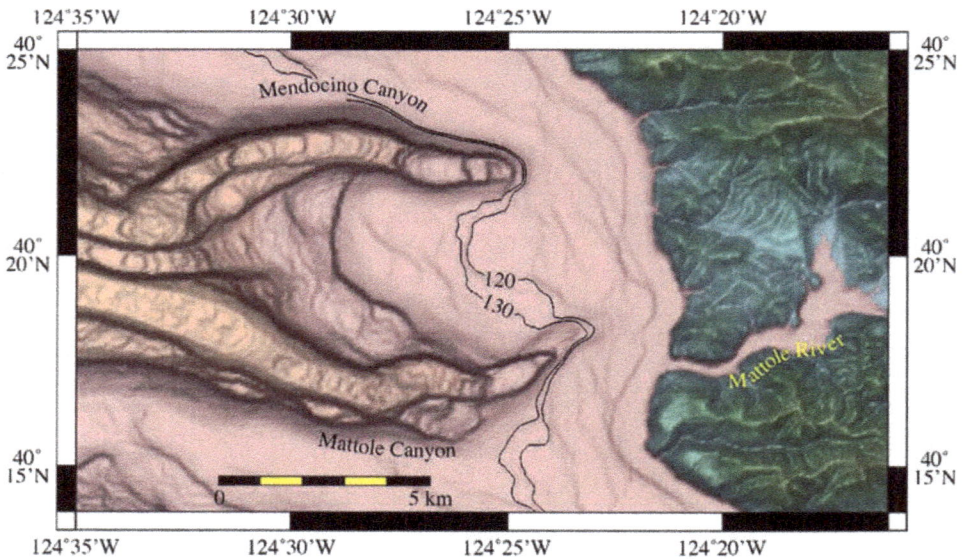

Figure 16. Map view of Coastal Relief Model bathymetry of heads of Mendocino and Mattole Canyons. The 120 and 130 m isobaths represent the range of the shoreline at the glacial maximum isostatic low stands of eustatic sea level. See Figure 2 for location.

5.2.5. Tsunamis

Tsunamis can have a profound effect on a continental shelf, especially on a very narrow shelf ([61,62] and references within; [63–65]). The outgoing tsunami waves, after first eroding the shore, have the capacity to transport mud, sand and even large boulders to the ocean in the backflow. Once in the head of a canyon, gravity and the sediment-laden density of the backflows would transport the sediment down-canyon as gravity-driven sediment flows that potentially could reach a basin floor. Atwater and Hemphill-Haley [66] and Hemphill-Haley [67] reported tsunami deposits on coastal Washington State and Garrison-Laney et al. [68] reported on 5 tsunami deposits within the last 3500 years at the southern end of the Cascadia Subduction Zone, near the shoreline adjacent to the heads of both Mendocino and Mattole Canyons. However, Weiss [69] modeled tsunami effects on non-cohesive sediments from the shelf and slope and calculated that even the 2004 Boxing Day tsunami in the Indian Ocean transported fine sand only 335 m offshore in a water depth of 985 m. This study suggests that if a nearby earthquake generated a submarine landslide that in turn initiated a tsunami, the resuspension and transport of fine sand and silt could have reached the heads of both Mendocino and Mattole Canyons.

6. Conclusions

The new multibeam bathymetry and backscatter data provide the first quantitative description of the entire Mendocino Channel. A more thorough understanding of the immediate area of Mendocino Channel was gained by combining the new analyses of the multibeam data with interpretations of published multichannel seismic and sediment cores from the channel and immediate vicinity. Several aspects of the entire Mendocino Channel system are intriguing. For instance, the concave-down profile of Mendocino Canyon channel suggests it is presently largely inactive and has accumulated sediment in the canyon-channel reaches. Conversely, the concave-up profile of Mattole Canyon channel suggests it is presently active and either has or is in the process of establishing an equilibrium profile. The similar water depths of the flat surfaces of terraces T_g and T_d on either side of Mattole Canyon, together with the sharp 80 m drop of the floor of Mattole Canyon at the capture point by Mendocino Canyon, is strong evidence that Mattole Canyon channel is older than Mendocino Canyon channel and that it was captured by Mendocino Canyon channel. The lack of obvious high-amplitude reflectors in the subsurface on multichannel seismic profiles (Figure 4) suggests Mendocino Channel floor has not been the site of significant deposits of coarse sediments. However, the maximum ~350 m thick north-side levees of the USR suggests a large amount of sediment has transited the channel over the life span of Mendocino Channel.

What changed the nature of the reaches from straight to sinuous and back to straight? Typically, major changes of planform characteristics of fluvial channels are the result of changes of sediment load, an increase in peak discharge or a change in channel gradient or some combination of the three [70]. The gradient of the USR decreases from 0.71° at its beginning to 0.12° at the beginning of the USinR. The gradient at the beginning of the LSinR is 0.18° and varies down-channel less than ±0.2° to the end of the reach. Is this small gradient enough to cause a change from linear to sinuous in channel character? Why did the levee heights suddenly decrease in height at two locations? The large north-side levee heights along the USR suddenly decrease from 200 m above the channel floor at 25 km down-channel to 100 m high, about the location where the channel emerges from the steep continental margin, but the change in channel character from the USR to the USinR occurs 41.8 km down-channel where the north and south levees remain less than 100 m above the channel floor. The sudden decreases in north levee heights might reflect the end of continental margin sediment mixed with the north levee sediments. The overall decrease in north and south levee heights in the USR is fairly smooth until the abrupt decrease in height at 22–25.5 km down-channel (Figure 5a). If this is the case, then the sudden decrease in north and south levee heights in the USinR at 62–65 km down-channel must reflect a reduction in the carrying capacity of the turbidity currents and a subtle decrease in channel gradient. The gradient of the channel floor of the USR just before it changes to the USinR is 0.71° and it flattens to 0.12° at the beginning of the USinR. The remainder of the USinR has a gradient of ~0.33° throughout the reach and both north and south levees suddenly end at 60–65 km down channel, 80+ km before the channel is no longer resolved in the bathymetry data. Nothing in the multichannel seismic subsurface data suggests any basement control for these changes.

It seems that a combination of significant and numerous earthquakes (seismicity) and wave-loading resuspension are the most likely processes that created the turbidity currents that formed and maintained Mendocino Channel. Eustatic lowered sea levels are possible contributors to the initial formation of the channel, but the lack of high-amplitude reflectors and the young ages for the turbidites in BX-1 suggest that eustatic sea levels are not the major forcing for activity within Mendocino Channel. Answers to the intriguing aspects and questions raised by the analyses presented here must await future dedicated seismic and sediment sampling cruises that can provide the necessary data to address these observations.

7. Summary

- Newly acquired multibeam bathymetry and backscatter data, together with published multichannel seismic and sediment core data provides a quantitative geomorphometric analysis of the entire extent of Mendocino Channel and to explore the age and possible causes that may have contributed to the formation and maintenance of the channel.
- Mendocino Channel has evolved from the confluence of Mendocino and Mattole Canyon channels at the point where Mattole Canyon channel was captured by Mendocino Canyon channel.
- The concave-up profile of Mattole Canyon channel suggests it is in the process of reaching, or has reached, an equilibrium profile whereas the concave-down profile of Mendocino Canyon channel suggests it is largely inactive.
- The 148 km length of Mendocino Channel can be subdivided into sinuous and linear reaches based on planform geometry.
- Mendocino Channel trends westward along the base of Gorda Escarpment and descends 365 m along an inclined perch towards basin water depths until the channel is deflected to the northwest by a large landslide deposit.
- Changes from a linear reach to a sinuous reach and back to a linear reach are abrupt with 90° bends.
- There are several 180° meanders and 2 cut-off meanders in the sinuous reaches.
- Both inside-bend and outside-bend terraces are found on the sinuous reaches; some are paired terraces and some are unpaired.
- Although landslides are evident along the north wall of Gorda Escarpment, there is very little evidence of landslide debris within Mendocino Channel.
- Dated box and piston cores provide an estimate of the hemipelagic sedimentation rates that, when applied to the thickness of the channel-levee complex suggests the channel may be as old as 2.6–1.8 Ma (early Quaternary).
- A combination of significant and numerous earthquakes (seismicity) and wave-loading resuspension are the most likely processes that created the turbidity currents that have flowed down Mendocino Channel.

Acknowledgments: The data for this paper were collected on NOAA Ship *Okeanos Explorer* cruise EX0903 funded by the National Oceanic and Atmospheric Administration. The Captain, crew and technical staff aboard the cruise were critical to achieving the success we had. I am grateful for all their help at sea. I am especially indebted to Mashkoor Malik, my co-Chief Scientist on the cruise and to David A. Cacchione who discovered Mendocino Channel on a 1984 USGS GLORIA cruise and sparked my initial interest in Mendocino Channel. I appreciate the constructive reviews of an earlier version of the manuscript by Larry A. Mayer and two anonymous reviewers. I especially appreciate the thorough and thought-provoking review by Jeffrey Peakall and I absolve all of the reviewers of any responsibility for my interpretations. The multibeam echosounder data were collected to support the bathymetry mapping for the U.S. Extended Continental Shelf efforts conducted by the University of New Hampshire's Center for Coastal & Ocean Mapping/Joint Hydrographic Center. The data collection and analyses were funded under NOAA grants NA15NOS4000200 and NA10NOS4000073. The U.S Extended Continental Shelf bathymetry effort is unique among maritime nations in that the data collected are freely available within a month after completion of each cruise for use by the scientific and industrial communities and the public. The data are available at http://ccom.unh.edu/data/mendocino-east-bathymetry as well as from https://maps.ngdc.noaa.gov/viewers/bathymetry/. Any use of the data should be referenced to doi:10.7289/V5B56GRV (bathymetry) and doi:10.7289/V56D5R03 (backscatter).

Conflicts of Interest: The author declares no conflict of interest.

References

1. Damuth, J.E.; Flood, R.D. Morphology, sedimentation processes, and growth pattern of the Amazon deep-sea fan. *Geo-Marine Lett.* **1983**, *3*, 109–117. [CrossRef]
2. Flood, R.D.; Damuth, J.E. Quantitative characteristics of sinuous distributary channels on the Amazon deep-sea fan. *Geol. Soc. Am. Bull.* **1987**, *98*, 728–738. [CrossRef]
3. Shanmugam, G.; Moiola, R.J. Submarine fans: Characteristics, models, classification, and reservoir potential. *Earth Sci. Rev.* **1988**, *24*, 383–428. [CrossRef]
4. Wynn, R.B.; Cronin, B.T.; Peakall, J. Sinuous deep-water channels: Genesis, geometry and architecture. *Mar. Pet. Geol.* **2007**, *24*, 341–387. [CrossRef]
5. Cacchione, D.A.; Drake, D.E.; Gardner, J.V. Recent meandering channel at the base of Mendocino Escarpment. In *Geology of the U.S. Seafloor: The View from GLORIA*; Gardner, J.V., Field, M.E., Twichell, D.C., Eds.; Cambridge University Press: Cambridge, UK, 1996; pp. 181–192.
6. Gardner, J.V. Program EEZ-SCAN: A Reconnaissance View of the Western U.S. Exclusive Economic Zone. Available online: https://books.google.com.hk/books?id=-4vKMaCm2RUC&pg=PA125&lpg=PA125&dq=A+reconnaissance+view+of+the+western+U.S.+Exclusive+Economic+Zone&source=bl&ots=WjSK8mYdC1&sig=XZ7J2vHuNt7rLG1nlrsph_XPfWM&hl=en&sa=X&ved=0ahUKEwjf08CU6q3XAhVDT7wKHTMTBgQQ6AEIOjAH#v=onepage&q=A%20reconnaissance%20view%20of%20the%20western%20U.S.%20Exclusive%20Economic%20Zone&f=false (accessed on 9 September 2017).
7. EEZ-SCAN 84 Scientific Staff. *Atlas of the Exclusive Economic Zone, Western Conterminous United States*: *U.S. Geological Survey Miscellaneous Investigations Series I-1792*; U.S. Geological Survey: Reston, VA, USA, 1986.
8. Damuth, J.E.; Kolla, V.; Flood, R.D.; Kowsmann, R.O.; Monteiro, M.; Gorini, M.A.; Palma, J.J.; Belderson, R.H. Distributary channel meandering and bifurcation patterns on the Amazon deep-sea fan as revealed by long-range side-scan sonar (GLORIA). *Geology* **1983**, *11*, 94–98. [CrossRef]
9. Primez, C.; Flood, R.D. Morphology and Structure of Amazon Channel. Available online: http://www-odp.tamu.edu/publications/155_IR/VOLUME/CHAPTERS/ir155_03.pdf (accessed on 9 September 2017).
10. Curray, J.R.; Moore, D.G. Growth of the Bengal deep-sea fan and denudation in the Himalayas. *Geol. Soc. Am. Bull.* **1971**, *82*, 563–572. [CrossRef]
11. Weber, M.E.; Wiedicke, M.H.; Kudrass, H.R.; Hübscher, C.; Erienkeuser, H. Active growth of the Bengal Fan during sea-level rise and highstand. *Geology* **1997**, *25*, 315–318. [CrossRef]
12. Hubscher, C.; Spieβ, V.; Breitzke, M.; Weber, M.E. The youngest channel-levee system of the Bengal Fan: Results from digital sediment echosounder data. *Mar. Geol.* **1997**, *141*, 125–145. [CrossRef]
13. Curray, J.R.; Emmel, F.J.; Moore, D.G. The Bengal Fan: Morphology, geometry, stratigraphy, history and processes. *Mar. Pet. Geol.* **2003**, *19*, 1191–1223. [CrossRef]
14. Schwenk, T.; Spieβ, V.; Hübscher, C.; Breitzke, M. Frequent channel avulsions within the active channel-levee system of the middle Bengal Fan-an exceptional channel-levee development derived from Parasound and Hydrosweep data. *Deep-Sea Res. II* **2003**, *50*, 1023–1045. [CrossRef]
15. Deptuck, M.E.; Steffens, G.S.; Barton, M.; Pirmez, C. Architecture and evolution of upper fan channel-belts on the Niger Delta slope and in the Arabian Sea. *Mar. Pet. Geol.* **2003**, *20*, 649–676. [CrossRef]
16. Kastens, K.A.; Shor, A.N. Evolution of a channel meander on the Mississippi deep-sea fan. *Mar. Geol.* **1983**, *71*, 165–175. [CrossRef]
17. Twichell, D.C.; Kenyon, N.H.; Parson, L.M.; McGregor, B.A. Depositional patterns of the Mississippi Fan surface: Evidence from GLORIA II and high-resolution seismic profiles. In *Seismic Facies and Sedimentary Processes of Submarine Fans and Turbidite Systems*; Weimer, P., Link, M.H., Eds.; Springer: New York, NY, USA, 1991; pp. 349–363.
18. Droz, L.; Bellaiche, G. Rhone Deep-Sea Fan: Morphostructure and growth pattern. *Am. Assoc. Pet. Geol. Bull.* **1985**, *69*, 460–479.

19. Droz, L.; Marsset, T.; Ondréas, H.; Lopez, M.; Savoye, B.; Spy-Anderson, F.-L. Architecture of an active mud-rich turbidite system: The Zaire Fan (Congo-Angola margin southeast Atlantic): Results from ZaïAngo 1 and 2 cruises. *Am. Assoc. Pet. Geol. Bull.* **2003**, *87*, 1145–1168. [CrossRef]

20. Mulder, T.; Savoye, B.; Piper, D.J.W.; Syvitski, J.P.M. The Var submarine sedimentary system: Understanding Holocene sediment delivery processes and their importance to the geological record. In *Geological Processes on Continental Margins: Sedimentation, Mass-Wasting and Stability*; Stocker, M.S., Evans, D., Cramp, A., Eds.; Special Publication 129; Geological Society of London: London, UK, 1998; pp. 145–166.

21. Lonsdale, P.; Hollister, C.D. Cut-offs at an abyssal meander south of Iceland. *Geology* **1979**, *7*, 597–601. [CrossRef]

22. Lewis, K.B.; Pantin, H.M. Channel-axis, overbank and drift sediment waves in the southern Hikurangi Trough, New Zealand. *Mar. Geol.* **2002**, *192*, 123–151. [CrossRef]

23. Gardner, J.V.; Malik, M.; Walker, S. Plume 1400 meters high discovered at the seafloor off the northern California margin. *EOS* **2009**, *90*, 275. [CrossRef]

24. Peakall, J.; McCaffrey, B.; Kneller, B. A process model for the evolution, morphology; architecture of sinuous submarine channels. *J. Sediment. Res.* **2000**, *70*, 434–448. [CrossRef]

25. Tréhu, A.M.; Stakes, D.S.; Bartlett, C.D.; Chevallier, J.; Duncan, R.A.; Goffredi, S.K.; Potter, S.M.; Salamy, K.A. Seismic and seafloor evidence for free gas, gas hydrates; fluid seeps n the transform margin offshore Cape Mendocino. *J. Geophys. Res.* **2003**, *108*. [CrossRef]

26. Ware, C.; Knight, W.; Wells, D. Memory intensive statistical algorithms for multibeam bathymetric data. *Comput. Geosci.* **1991**, *17*, 985–993. [CrossRef]

27. Schumm, S.A.; Brakenridge, G.R. River responses. In *The Geology of North America, North America and Adjacent Oceans during the Last Deglaciation*; Ruddiman, W.F., Wright, H.E., Jr., Eds.; v. K-3; The Geological Society of America: Boulder, CO, USA, 1987; pp. 221–240.

28. Shepard, F.P.; Dill, R.F. *Submarine Canyons and Other Sea Valleys*; Rand McNally & Company: Chicago, IL, USA, 1966.

29. Nakajima, T.; Peakall, J.; McCaffrey, W.D.; Paton, D.A.; Thompson, P.J.P. Outer-bank bars: A new intra-channel architectural element within sinuous submarine slope channels. *J. Sediment. Res.* **2009**, *79*, 872–886. [CrossRef]

30. Goldfinger, C.; Nelson, C.H.; Morey, A.E.; Johnson, J.E.; Patton, J.R.; Karabanov, E.; Gutiérrez-Pastor, J.; Eriksson, A.T.; Gàrcia, E.; Dunhill, G.; et al. Turbidite Event History–Methods and Implications for Holocene Paleoseismicity of the Cascadia Subduction Zone. Available online: http://www.wou.edu/las/physci/taylor/g473/Goldfinger_etal_2011_excerpt_reading.pdf (accessed on 9 September 2017).

31. Lyle, M.; Koizumi, I.; Richter, C. *Proceedings of the Ocean Drilling Program, Initial Reports*; Site 1020; U.S. Government Printing Office: Washington, DC, USA, 1997.

32. Manley, P.I.; Flood, R.D. Cyclic sediment deposition within Amazon deep-sea fan. *Am. Assoc. Pet. Geol. Bull.* **1988**, *72*, 912–925.

33. Flood, R.D.; Manley, P.L.; Kowsmann, R.O.; Appi, C.J.; Pirmez, C. Seismic facies and late Quaternary growth of Amazon submarine fan. In *Seismic Facies and Sedimentary Processes of Submarine Fans and Turbidite Systems*; Weimer, P., Link, M.H., Eds.; Springer: New York, NY, USA, 1991; pp. 415–433.

34. Carlson, R.L.; Gangi, A.F.; Snow, K.R. Empirical reflection-traveltime/depth and velocity/depth functions for the deep-sea sediment column. *J. Geophys. Res.* **1986**, *91*, 8249–8266. [CrossRef]

35. Gorsline, D.S.; De Diego, T.; Nava-Sanchez, E.H. Seismically triggered turbidites in small margin basins: Alfonso Basin, western Gulf of California and Santa Monica Basin, California Borderland. *Sediment. Geol.* **2000**, *135*, 21–35. [CrossRef]

36. Babonneau, N.; Cattaneo, A.; Ratzov, G.; Déverchère, J.; Yelles-Chaouche, A.; Lateb, T.; Bachir, R. Turbidite chronostratigraphy off Algiers, central Algerian margin: A key for reconstructing Holocene paleo-earthquake cycles. *Mar. Geol.* **2017**, *384*, 63–80. [CrossRef]

37. Pope, E.L.; Talling, P.J.; Carter, L. Which earthquakes trigger damaging submarine mass movements: Insights from a global record of submarine cable breaks? *Mar. Geol.* **2017**, *384*, 131–146. [CrossRef]

38. Adams, J. Paleoseismicity of the Cascadia subduction zone–Evidence from turbidites off the Oregon-Washington margin. *Tectonics* **1990**, *9*, 569–583. [CrossRef]

39. Clarke, S.H., Jr.; Carver, G.A. Late Holocene tectonics and paleoseismicity, southern Cascadia subduction zone. *Science* **1992**, *255*, 188–192. [CrossRef] [PubMed]

40. Smith, S.; Knapp, J.S.; McPherson, R.C. Seismicity of the Gorda plate, structure of the continental margin, and an eastward jump of the Mendocino triple junction. *J. Geophys. Res.* **1993**, *98*, 8153–8171. [CrossRef]

41. Covault, J.A.; Graham, S.A. Submarine fans at all sea-level stands: Tectono-morphologic and climatic controls on terrigenous sediment delivery to the deep sea. *Geology* **2010**, *38*, 939–942. [CrossRef]

42. Puig, P.; Ogston, A.S.; Mullenbach, B.L.; Nittrouer, C.A. Shelf-to-canyon sediment-transport processes on the Eel continental margin (northern California). *Mar. Geol.* **2003**, *193*, 129–149. [CrossRef]

43. Parsons, J.D.; Friedrichs, C.T.; Traykovski, P.A.; Mohrig, D.; Imran, J.; Syvitski, J.P.M.; Parker, G.; Puig, P.; Buttles, J.L.; Garcia, M.H. The mechanics of marine sediment gravity flows. In *Continental Margin Sedimentation: From Sediment Transport to Sequence Stratigraphy*; Nittrouer, C.A., Austin, J.A., Field, M.E., Kravity, J.H., Syvitski, J.P.M., Wiberg, P.L., Eds.; International Association of Sedimentologists Special Publication 37; Wiley-Blackwell: Hoboken, NJ, USA, 2007; pp. 275–337.

44. Nittrouer, C.A.; Austin, J.A.; Field, M.E.; Kravitz, J.H.; Syvitski, J.P.M.; Wiberg, P.L. *Continental Margin Sedimentation: From Sediment Transport to Sequence Stratigraphy*; International Association of Sedimentologists Special Publication 37; John Wiley & Sons, Inc.: Hoboken, NJ, USA, 2007.

45. Jin, F.-F.; Bourharel, J.; Lin, I.-I. Eastern Pacific tropical yclones intensified by El Nino delivery of subsurface ocean heat. *Nature* **2014**, *516*, 82–84. [CrossRef] [PubMed]

46. Cane, M.A. The evolution of El Niño, past and future. *Earth Planet. Sci. Lett.* **2005**, *230*, 227–240. [CrossRef]

47. Felix, M.; Peakall, J.; McCaffrey, W.D. Relative importance of processes that govern the generation of particulate hyperpycnal flows. *J. Sediment. Res.* **2006**, *76*, 382–387. [CrossRef]

48. Mulder, T.; Syvitski, J.P.M. Turbidity currents generated at river mouths during exceptional discharges to the world oceans. *J. Geol.* **1995**, *103*, 285–299. [CrossRef]

49. Imran, J.; Syvitski, J.P.M. Impact of extreme river events on the coastal ocean. *Oceanography* **2000**, *13*, 85–92. [CrossRef]

50. Mulder, T.; Syvitski, J.P.M.; Migeon, S.; Faugères, J.-C.; Savoye, B. Marine hyperpycnal flows: Initiation, behavior and related deposits: A review. *Mar. Pet. Geol.* **2003**, *20*, 861–882. [CrossRef]

51. Dadson, S.J.; Hovius, N.; Chen, H.; Date, W.B.; Lin, J.-C.; Hsu, M.-L.; Lin, C.-W.; Horng, M.J.; Chen, T.-C.; Milliman, J.; et al. Earthquake-triggered increase in sediment delivery from an active mountain belt. *Geology* **2004**, *32*, 733–736. [CrossRef]

52. Nittrouer, C.A.; Austin, J.A.; Field, M.E.; Kravitz, J.H.; Syvitski, J.P.M.; Wiberg, P.L. Writing a Rosetta stone: Insights into continental-margin sedimentary processes and strata. In *Continental Margin Sedimentation: From Sediment Transport to Sequence Stratigraphy*; International Association of Sedimentologists Special Publication 37; Wiley-Blackwell: Hoboken, NJ, USA, 2007.

53. Wynn, R.B.; Weaver, P.P.E.; Ercilla, G.; Stow, D.A.V.; Masson, D.G. Sedimentary processes in the Selvage sediment-wave field, NE Atlantic: New insights into the formation of sediment waves by turbidity currents. *Sedimentology* **2007**, *47*, 1181–1197. [CrossRef]

54. Piper, D.J.W.; Savoye, B. Processes of late Quaternary turbidity current flow and deposition of the Var deep-sea fan, north-west Mediterranean Sea. *Sedimentology* **1993**, *40*, 557–582. [CrossRef]

55. Kao, S.J.; Dai, M.; Selvaraj, K.; Zhai, W.; Cai, P.; Dhen, S.N.; Yang, J.Y.T.; Liu, J.T.; Liu, C.C.; Syvitski, J.P.M. Cyclone-driven deep sea injection of freshwater and heat by hyperpycnal flow in the subtropics. *Geophys. Res. Lett.* **2010**, *37*, 1–5. [CrossRef]

56. Gavey, R.; Carter, L.; Liu, J.T.; Talling, P.J.; Hsu, R.; Pope, E.; Evans, G. Frequent sediment density flows during 2006 to 20154, triggered by competing seismic and weather events: Observations from subsea cable breaks off southern Taiwan. *Mar. Geol.* **2017**, *384*, 147–158. [CrossRef]

57. Lee, H.J.; Locat, J.; Desgagnés, P.; Parsons, J.D.; McAdoo, B.G.; Orange, D.L.; Puig, P.; Wong, F.L.; Dartnell, P.; Boulanger, E. Submarine mass movements on continental margins. In *Continental Margin Sedimentation: From Sediment Transport to Sequence Stratigraphy*; Nittrouer, C.A., Austin, J.A., Field, M.E., Kravitz, J.H., Syvitski, J.P.M., Wiberg, P.L., Eds.; International Association of Sedimentologists Special Publication 37; Wiley-Blackwell: Hoboken, NJ, USA, 2007; pp. 213–274.

58. Sumner, E.J.; Paull, C.K. Swept away by a turbidity current in Mendocino submarine canyon, California. *Geophys. Res. Lett.* **2014**, *41*, 7611–7618. [CrossRef]

59. Shackleton, N.J. Oxygen isotopes, ice volume and sea level. *Quat. Sci. Rev.* **1987**, *6*, 183–190. [CrossRef]

60. De Boer, B.; van de Wal, R.S.W.; Bintanja, R.; Lourens, L.J.; Tuenter, E. Cenozoic global ice-volume and temperature simulations with 1-D ice-sheet models forced by benthic $\delta^{18}O$ records. *Ann. Glaciol.* **2010**, *51*, 23–33. [CrossRef]

61. Shanmugam, G. The tsunamite problem. *J. Sediment. Res.* **2006**, *76*, 718–730. [CrossRef]

62. Bourgeois, J. Geologic effects and records of tsunamis. In *The Sea*; Robinson, A.R., Bernard, E.N., Eds.; v. 15: Tsunamis; Harvard University Press: Cambridge, MA, USA, 2009; pp. 53–91.

63. Paris, R.; Fournier, J.; Poizot, E.; Etienne, S.; Morin, J.; Lavigne, F.; Wassmer, P. Boulder and fine sediment transport and deposition by the 2004 tsunami in Lhok Nga (western Banda Aceh, Sumatra, Indonesia): A coupled offshore-onshore model. *Mar. Geol.* **2010**, *268*, 43–54. [CrossRef]

64. Yamazaki, Y.; Cheung, K.F. Shelf resonance and impact of near-field tsunami generated by the 2010 Chile earthquake. *Geophys. Res. Lett.* **2011**, *38*. [CrossRef]

65. Sakuna, C.; Szczuciński, W.; Feldens, P.; Schwarzer, K.; Khokiattiwong, S. Sedimentary deposits left by the 2004 Indian Ocean tsunami on the inner continental shelf offshore of Khao Lak, Andaman Sea (Thailand). *Earth Planets Space* **2012**, *64*, 931–943. [CrossRef]

66. Atwater, B.F.; Hemphill-Haley, E. *Recurrence Intervals for Great Earthquakes of the Past 3500 Years at Northeastern Willapa Bay*; U.S. Government Printing Office: Washington, DC, USA, 1997; p. 108.

67. Hemphill-Haley, E. Diatom evidence for earthquake-induced subsidence and tsunami 300 yr ago in southern coastal Washington. *Geol. Soc. Am. Bull.* **1995**, *107*, 367–378. [CrossRef]

68. Garrison-Laney, C.E.; Abramson-Ward, H.F.; Carver, G.A. Late Holocene tsunamis near the southern end of the Cascadia subduction zone. *Seismol. Res. Lett.* **2002**, *73*, 248.

69. Weiss, R. Sediment grains moved by passing tsunami waves: Tsunami deposits in deep water. *Mar. Geol.* **2008**, *50*, 251–257. [CrossRef]

70. Schumm, S.A. Patterns of alluvial rivers. *Annu. Rev. Earth Planet. Sci.* **1985**, *13*, 5–27. [CrossRef]

![geosciences logo] *geosciences*

MDPI

Article

Geomorphology and Late Pleistocene–Holocene Sedimentary Processes of the Eastern Gulf of Finland

Daria Ryabchuk [1,2,*], Alexander Sergeev [1], Alexander Krek [3], Maria Kapustina [3], Elena Tkacheva [3,4], Vladimir Zhamoida [1,2], Leonid Budanov [1,5], Alexandr Moskovtsev [1] and Aleksandr Danchenkov [3,4]

[1]   A.P. Karpinsky Russian Geological Research Institute (VSEGEI), 74, Sredny prospect,
      199106 Saint Petersburg, Russia; sergeevau@yandex.ru (A.S.); Vladimir_Zhamoida@vsegei.ru (V.Z.);
      leon_likes@mail.ru (L.B.); aleks_moskovtsev@vsegei.ru (A.M.)
[2]   Institute of Earth Science, St. Petersburg State University, 7–9 Universitetskaya Embankment,
      199034 Saint Petersburg, Russia
[3]   Shirshov Institute of Oceanology, Russian Academy of Sciences, 36, Nahimovskiy prospekt,
      117997 Moscow, Russia; av_krek_ne@mail.ru (A.K.); Kapustina.mariya@ya.ru (M.K.);
      elenatkacheva_kg@mail.ru (E.T.); aldanchenkov@mail.ru (A.D.)
[4]   Institute of Environmental Management, Urban Development and Spatial Planning,
      I. Kant Baltic Federal University, 14, Nevskogo Alexandre street, 236016 Kaliningrad, Russia
[5]   Faculty of Geological Prospecting, Saint-Petersburg Mining University, 21 Line, 2,
      199106 Saint Petersburg, Russia
*   Correspondence: Daria_Ryabchuk@mail.ru; Tel.: +7-921-789-3367

Received: 8 December 2017; Accepted: 14 March 2018; Published: 18 March 2018

Abstract: In 2017, a detailed study of the Eastern Gulf of Finland (the Baltic Sea) seafloor was performed to identify and map submerged glacial and postglacial geomorphologic features and collect data pertinent to the understanding of sedimentation in postglacial basins. Two key areas within the Gulf were investigate using a multibeam echosounder, SeaBat 8111 and an EdgeTech 3300-HM acoustic sub-bottom profiling system. High-resolution multibeam bathymetric data (3-m resolution) were used to calculate aspect, slope, terrain ruggedness and bathymetric position index using ArcGIS Spatial Analyst and the Benthic Terrain Modeler toolbox. These data and resultant thematic maps revealed, for the first time, such features as streamlined till ridges, end-moraine ridges, and De Geer moraines that are being used for the reconstruction of the deglaciation in the Eastern Gulf of Finland. This deglaciation occurred between 13.8 and 13.3 ka BP (Pandivere–Neva stage) and 12.25 ka BP (Salpausselkä I stage). Interpretations of the seismic-reflection profiles and 3D models showing the surfaces of till, and the identification of the Late Pleistocene sediment and modern bottom relief, indicate deep relative water-level fall in the Early Holocene and, most likely, several water-level fluctuations during this time.

Keywords: geomorphology; submerged glacial bedforms; deglaciation; sedimentation; multibeam; acoustic-seismic profiling

---

1. Introduction

The Baltic Sea is an ideal natural laboratory [1] for the study of geologically-induced driving forces of different sea bottom and coastal zone processes (e.g., the intensity of sedimentation and erosion processes and areas of their distribution, sediment flows, geochemical processes, benthic landscape development). Understanding these forces is important for the sustainable use of natural resources and environmental protection [2,3]. In addition, some Baltic Sea basins are crucial for the understanding of postglacial geological history and recent sedimentation processes. These basins are located in subsiding areas within the Southern and Southwestern Baltic Sea [1,4–6] and regions of uplift are located within the Northern Baltic Sea [7]. Several key areas of the Eastern Baltic Sea are very important for modelling historic and present tectonic processes (e.g., Eastern Gulf of Finland and Pärnu Bay), which are characterized by very low (from 0 to +3 mm/year) rates of uplift—a near zero rate of recent sea level change. According to a datum obtained by a Kronshtadt gauge measurement, the rate of sea level rise from 1835 to 2005 was 0.7 mm/year [8]. The combination of tectonic processes, total sea bottom coverage by Quaternary deposits, relatively smooth and shallow relief, and widespread Holocene accretion, both above and below the recent sea level, is useful geological and geomorphological information that can be used to reconstruct palaeoenvironmental changes [9–13].

Between 1986 and 2000, the Eastern Gulf of Finland was mapped by the Russian Research Geological Institute (VSEGEI) [14], resulting in the publication of a State Geological Survey Map (1:200,000 scale). Approximately 8000 km of seismic reflection sub-bottom profiles (SBP) and more than 6000 sampling sites were used in the mapping effort. The SBP lines were oriented mainly parallel to meridians and spaced approximately 2 km apart. Two acoustic sources were synchronously operated, one with a fundamental frequency of 500 Hz (sparker) and another at a frequency of 7.5 kHz (piezoceramic transmitter). More than 6000 sediment samples including 4000 gravity cores were collected and used for SBP data interpretation. In 2009 and 2011, during joint Russian–Finnish field cruises on board the R/V *Aranda* of the Finnish Environment Institute, approximately 800 km of 12-kHz pinger sub-bottom profiles were collected by Meridata Ltd. [15]. Recently, these profiles were digitized and analysed using GIS methods. Assessment of regional deglaciation processes and postglacial sedimentation was undertaken to form 3D models compiled of pre-Quaternary relief, consisting of moraines and Late Pleistocene surfaces, from which the thicknesses of till, glacio-lacustrine, and Holocene sediments were calculated [16].

Despite a relatively high confidence in the geology reported to date for the area of our study, there still remain important unsolved problems, specifically those pertaining to the postglacial geological history including (i) the location of end moraines and glaciofluvial deposits in the Gulf of Finland; (ii) the age of deglaciation and rate of glacial front retreat; and (iii) the number of Holocene sea level fluctuations and the amplitude of relative regressions [13,16]. The best known regional palaeoreconstructions of deglaciation were reported by I. Krasnov (International Map of the Quaternary deposits of Europe, [17], E. Zarrina [18], D. Kvasov [19], A. Raukas [20], D. Subetto [21], and J. Vassiljev [22]. These reconstructions were based on terrestrial investigations where locations of ice-sheet margins within the Gulf of Finland basin are generally shown on maps by dashed lines (inferred) or attributed as unknown [23]. Hypothesized locations of terminal moraines formed during the Pandivere–Neva stage east of Kotlin Island [18] or near the Eastern coast of Narva Bay to Cape Peschany [21] have not been documented as there is no geomorphological or chronological evidence for these glacial forms. The problem regarding Holocene sea level fluctuation, especially the location of regression levels is the subject of much discussion today [9–13].

Recent advances in marine geological methodology and technology using multiple instruments, such as high-resolution geophysical SBP, multibeam echosounder bathymetry and backscatter, and side-scan sonargraphs have provided data that is useful for the 3D geomorphometric visualization of submerged surfaces, which has significantly improved our knowledge of sea bottom relief and geological structure [24–29]. In the Eastern (Russian) part of the Gulf of Finland, the first attempts at using these state-of-the-art geological and geophysical methodologies focused on the study of geological hazards, such as submarine landslides, pockmarks [30,31], and processes associated with Fe–Mn concretion and regeneration, based on exploration [32] undertaken by VSEGEI between 2012 and 2014. The exploration was part of a Russian National project entitled "State Monitoring of Geological Environment of Near-Shore Areas of Russian Baltic, Barents and White Seas" and a Russian–Finnish project entitled "Transboundary Tool for Spatial Planning and Conservation of the Gulf of Finland" (TOPCONs) (a project of the European Neighbourhood and Partnership Instrument

(ENPI) Program) [3]. Unfortunately, the areas covered using high-resolution geophysical equipment in the Eastern Gulf of Finland during these campaigns were limited.

This paper presents the first results of the interpretation of multibeam and SBP data collected during a joint Atlantic Branch of the Institute of Oceanology of Russian Academy of Science (ABIO RAS) and VSEGEI cruise on board R/V Academic Nikolaj Strakhov in the Eastern Gulf of Finland during the summer of 2017. The main goal of this paper is to present new high resolution geophysical data and morphometric analyses from that area of the Gulf of Finland which had previously lacked good data for understanding and mapping the marine geology. A high resolution geophysical survey carried out within two key areas of the Eastern Gulf of Finland imaged, for the first time, submerged glacial and postglacial bottom features and provided new data that can be used to interpret sedimentation within the postglacial basins of the Northeastern Baltic Sea (Figure 1A). The main criteria for choosing the key areas was the possible occurrence of variety in submerged glacial and glaciofluvial landforms and transition zones from relative bathymetric high areas to sediment basins.

Study Area

The Eastern Gulf of Finland is located at the boundary between the Russian Plate and the Southeastern margin of the Baltic crystalline shield. A peneplain surface, now gently sloping south–southeast, was eroded into Paleo–Mesoproterozoic basement prior to the commencement of the Late Vendian–Early Paleozoic terrigenous sedimentation event. Tertiary denudation, including dominant glacial erosion, was instrumental in forming the present bedrock topography, both in the Southern sediment-covered part of the region and in the shield [14].

Quaternary sediment almost completely covers the bottom and coastal areas of the Eastern Gulf of Finland, except locally in Vyborg Bay and on some islands. The Pleistocene geological record locally exhibits till (ground moraines), deposited during the last glaciation, glacio-lacustrine sediments of ice marginal lakes and the Baltic Ice Lake (varved and homogenous clays), and local glaciofluvial sands. The lower part of the Holocene is represented by the Ancylus Lake clays with microlenses of black amorphous Fe–sulphides. The upper part of the sedimentary section was formed during the Littorina and Post-Littorina sea phases of the Baltic Sea's development. These sediments are represented by silty clay mud and sands. Boulders and pebbles form the tops of submarine highs and the upper parts of coastal slopes in the areas of intense submarine erosion. Sands of different grain sizes and geneses (from unsorted relict sands of the Gulf proper to fine-grained, very well sorted wave accretion sands of the near shore zone) are the most widespread type of bottom sediment. Sandy clays and silts are usually connected with areas of non-sedimentation (transitional zones) and weak submarine currents. Clayey mud accumulation occurs within bottom depressions. A unique type of sediment is represented by Fe–Mn concretions [14].

2. Material and Methods

2.1. Broad-Scale Data Analysis

To pinpoint the potential location of submarine end moraine complex data from previous investigations of the Eastern Gulf of Finland, Quaternary deposits, bottom relief, and palaeogeographic development since the last deglaciation were utilized. To obtain new data about regional deglaciation and postglacial development, 3D models of pre-Quaternary relief, moraine and Late Pleistocene surfaces were compiled; the thicknesses of the till, glacio-lacustrine and Holocene sediments were calculated based on interpretations of the seismic reflection profiles collected between 1986 and 2000. A broad scale bathymetric model of the Eastern Gulf of Finland bottom was analysed using traditional GIS instruments, which allowed us to localize submarine landforms, which were further interpreted using all available geological and geophysical information (Figure 1). The results of these interpretations were used to select key areas for field multibeam and acoustic-seismic surveys.

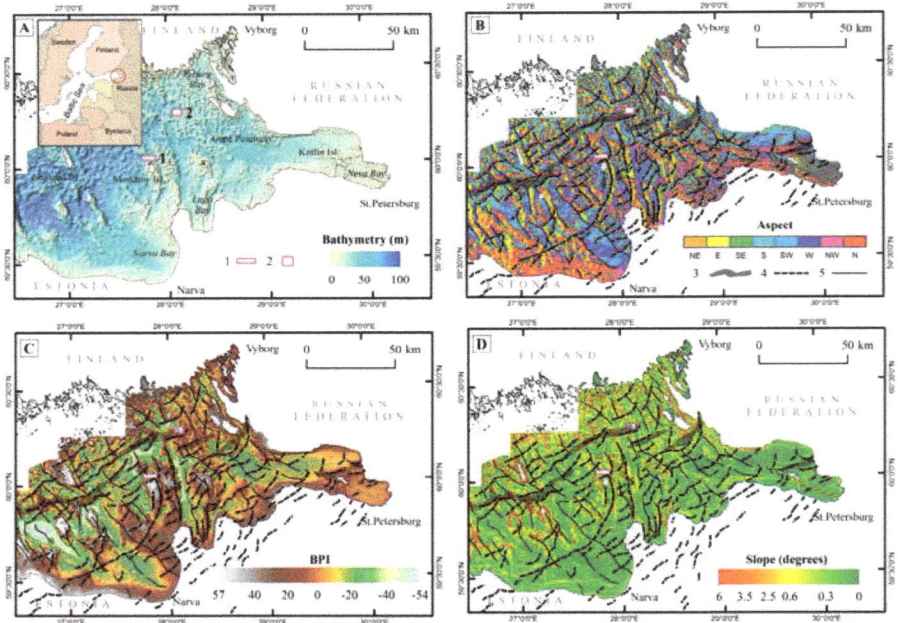

Figure 1. Broad scale models of the Eastern Gulf of Finland bottom relief. (**A**) bathymetry. Bottom relief is relatively shallow. Depths gradually rise from the east (2–5 m in Neva Bay) to west (70–80 m around Gogland Island); (**B**) bathymetry aspect map overlayed upon the till surface. The results of analysis facilitated the interpretation of the most probable location of recessional moraines and streamlined moraine features; (**C**) Bathymetric Position Index (BPI); (**D**) slope. Note: most parts of the Eastern Gulf of Finland bottom relief are very smooth (with a slope angle of less than 1°). Maximum slope angles (up to 6°) are observed in the Northwestern part of the study area (blocks 1, 2, key areas selected for multibeam surveys in 2017; 1 = Moschny Island area; 2 = Vyborg Bay area in (**A**); 3 = predicted end-moraine areas according to sub-bottom profile (SBP) data (1986–2000) interpretations (see **B** for symbol); 4 = assumed till relief features parallel to ice-sheet margin (see **B** for symbol); 5 = assumed streamlined till relief features (see **B** for symbol) [16]. The colour ramp has been selected to highlight the main features of the terrain.

2.2. Multibeam Echosounding and Acoustic-Seismic Profiling Equipment

During the cruise of the R/V Academic Nikolaj Strakhov (19–25 July 2017) two key areas in the Eastern Gulf of Finland were studied using a multibeam echosounder (MBES) and SBP (Figure 2). In the Vyborg Bay key area, thirty-nine lines of MBES and SBP tracks (total length is 164 km) were run to provide a coverage area of 9 km^2 (Figure 2A). Thirty-three lines of MBES and SBP (total length of 119 km) in the Moschny Island key area covered 6.8 km^2 (Figure 2B).

The R/V Academic Nikolaj Strakhov is equipped with a Teledyne RESON Seabat 8111-E208-3F66 Dry MBES system and an EdgeTech 3300-HM sub-bottom profiler (with Discover Sub-Bottom v3.36). Data collection and primary processing were undertaken using PDS2000 v3.7.0.47 processing software.

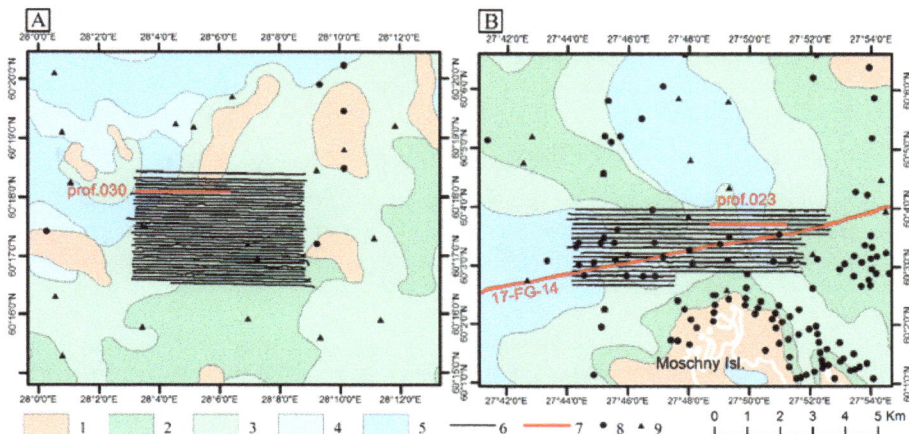

Figure 2. Part of a Quaternary deposits map of key study areas [14] with multibeam echosounder (MBES) and SBP profile tracks (solid black lines) and sampling stations (black dots). (**A**) Vyborg Bay; (**B**) Northern area off Moschny Island. Late Pleistocene sediment shown as (1) till; (2) varved clays; (3) homogenous clays of the Baltic Ice Lake. Holocene sediment shown as (4) Ancylus Lake clays, (5) Littorina—Post-Littorina marine silty-clayey muds. Geophysical data collection lines shown as (6) MBES and SBP (2017); (7) SBP, described in this article. Sample locations: (8) grab-corers and box-corers, (1984–2017); (9) long gravity corers (1984–2000).

The MBES data were collected using an operating frequency of 100 kHz. The area was covered by 101 beams at $1.5° \times 1.5°$, $150°$ perpendicular to the direction of travel and $1.5°$ in the direction of travel. The resolution of each beam, regardless of the distance away from nadir, was 3.7 cm. The EdgeTech 3300-HM fundamental frequency range was 2–10 kHz with a pulse width of 5 to 100 ms at a sampling rate of either 20, 25, 40, or 50 kHz, depending on the pulse of the higher frequency. To calculate depths and positions, data from external GPS and motion sensors fixed to the vessel were used. The position of these sensors and GPS antennas were located at the same reference point, corresponding to the point of the OCTANS sensor arrangement, which was installed at the waterline as close as possible to the x-axis of the vessel. Georeferencing of data received from the sensors was carried was in the WGS-84 coordinate system.

An additional seismic-reflection profile (17-FG-14) was collected in the Moschny Island key area on board the R/V SN 1303 in September 2017 (Figure 2B). The SBP survey was performed using a GEONT-HRP "Spektr-Geophysika" Ltd. (Russian) sparker, operated at a working frequency range of 0.03 to 2 kHz. The vessel continuously recorded positions using the Differential Global Positioning System (DGPS) with ± 5 m accuracy (Furuno GP7000F system in combination with Vector VS330 (Hemisphere GNSS, Scottsdale, AZ, USA).

2.3. Calibration

Before the bathymetric survey, the MBES system was calibrated on a flat seabed and along a slope. The purpose of calibration was to determine the systematic errors in the measured values (depth, coordinates) and to exclude these errors from the good data. The MBES standard calibrations (roll, pitch, yaw) were performed using the PDS2000 automatic program. The corrections received were used during the survey. A Sea&Sun Tech CTD90 multi-parameter probe was used prior to the survey, in the deepest part of the acoustic polygon, to obtain water column velocities which were recorded in the PDS2000 user files.

2.4. Multibeam Bathymetry Data Processing

Three main stages of data processing were designed using the PDS2000 and ArcGIS 10.2 software packages. Subsequent processing and calculation of seafloor attributes led to the construction of geomorphic and substrate maps.

2.4.1. Stage 1. Initial Data Processing Using the Default Software Package, PDS2000

Primary processing of the data was carried out using PDS2000 software. This process consisted of two main steps: (1) data cleaning with filters; and (2) post-processing and manual cleaning. Systematic depth distortions along with edge beam inaccuracies were removed with data filtering. Further manual processing removed non-systematic distortions. After processing, the data were exported in ASCII format (xyz) to a GIS project.

A 10-m cell size was chosen for creating the preliminary bathymetric grid. The grid was used for the visual assessment of systematic errors in the data array (e.g., furrows, scanning lines). If there were no visible artefacts, all of the accepted depths were exported in ASCII (xyz) format for further morphometric analysis.

2.4.2. Stage 2. Data Import into ArcGIS and Generation of the Digital Bathymetric Model (DBM)

Importation of a separate survey was carried out by ArcGIS tool ASCII 3D to Feature Class for converting large data sets into the 3D ASCII XYZ format. The data sets were then coupled into a single sonar point cloud data with extraction of Z-data as an additional text field from the 3D Shp file.

Due to the high density of the initial data sets (1–2 m between the cloud points), the Natural Neighbor method was used to construct the digital terrain model, a method based on Voronoy tessellation for discrete data [33]. A normal digital terrain model constructed using a 3-m-grid (25 m^2), which resulted in striped artefacts associated with the initial reflections (side and central beams).

2.4.3. Stage 3. Reduction of Bathymetric Data Artefacts with the Use of the ArcGIS Software

The reduction of the depth artefacts consisted of two main steps. The first step was noise smoothing on the surface of the digital model using a local linear polynomial approximation (the nearest 15 measurements were used for the approximation). The second step was de-striping the measured data to eliminate distortions within uniform bottom areas.

The local polynomials method has been used for the smoothing and reconstruction of bedforms. This method is part of deterministic methods that assume the existence of analytical dependencies between values in space. The polynomials method of the deterministic group uses a polynomial defined at some point from its coordinates. For example, for a point i with coordinates (x, y) the function F (x, y) = P_n (x, y), where P_n is a polynomial of degree n can be applied. The first-order polynomials are used for the most common two-dimensional cases [34]. The local polynomial interpolation fits many polynomials, each within specified overlapping neighbourhoods, providing an accurate approximation for the points. This is a simple method devoid of bias that allows smoothing of local data microvariations and extreme values (usually noise) that are not common for the neighbourhood. Using this method allowed us to eliminate noise on the digital model surface with a 0.13 m root mean square error.

The System for Automated Geoscientific Analyses (SAGA) GIS was used to remove stripe artefacts by means of convolution filters [35,36], thus allowing preservation of the general angle exposition and reduction of error angles from the data array. Nevertheless, this approach did not provide for the creation of accurate bathymetric terrain models because the depth difference between the initial and resulting data was too large.

It is assumed that the "banding" was caused by incorrectly setting the tilt of the sonar head (HeadTilt) at 1 along with equipment error. There were no significant improvements or noticeable changes in data quality when processing the data using the PDS2000 software when values (e.g., roll 0 and −1) were changed. Temperature may have played a role in the failure as the recorded temperature in the echosounder menu (the temperature of the radiator is fixed at close to absolute zero) is a reliable error. A similar situation was not observed in the following surveys (September–October 2017).

Artefacts can limit the interpretation of morphometric properties when constructing DBMs. Band-shaped distortions form elongated shapes, which are more evident on flat surfaces of the sea bottom, resulting in erroneous characteristics of inclines and orientation of the seabed (Figure 3C). Elimination of such errors using the procedure of de-striping [35,36] allows the true slope to be kept (Figure 3D).

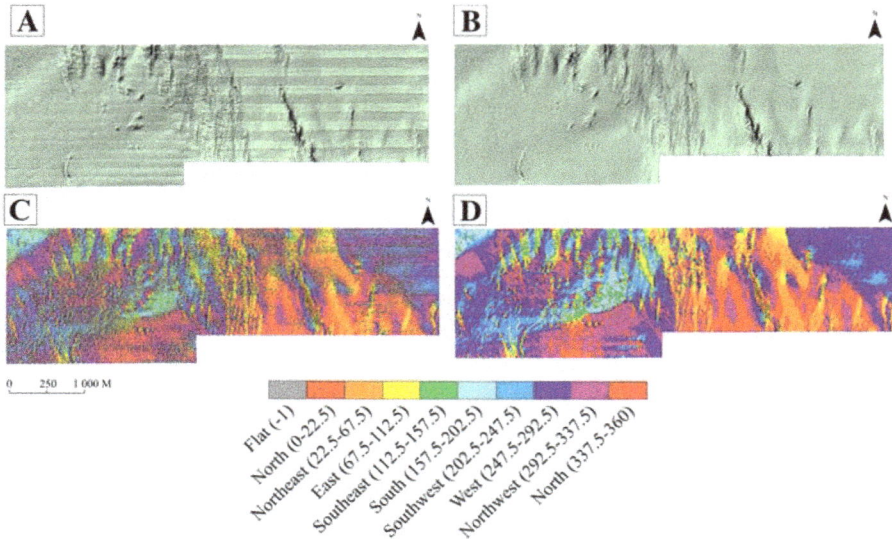

Figure 3. Orientation of the seabed calculated from the models (Moschny Island key area). (**A**) initial shaded relief prior to approximation; (**B**) shaded relied after approximating and de-striping; (**C**) initial seabed slope prior to approximation; (**D**) seabed representation after approximating and de-striping.

This process consisted of two main steps: (1) data cleaning with filters, and (2) post-processing and manual cleaning. Systematic depth distortions along with edge beam inaccuracies were removed by filtering the acquired bathymetric data. Non-systematic distortions were removed by manual processing.

2.5. GIS Analyses of Bottom Relief

GIS analyses of the seafloor bottom relief and the accompanying dataset of geophysical data were carried out using ArcGIS software. MBES data (at 3 m resolution) were used to calculate aspect, slope, and terrain ruggedness using ArcGIS Spatial Analyst and the Benthic Terrain Modeler toolbox (Figures 4–7). The software allowed the identification of specific geomorphological features, such as downslope direction, degrees of slope inclination, terrain ruggedness, and local relative relief of rises and depressions (Table 1) [37,38]. After processing and cleaning, digital terrain models (DTMs) were constructed in GIS, devoid of most artifacts. The regularity and coincidence of the artifacts with nadir made for easy identification of the artifacts. To calculate the Bathymetric Position Index (BPI), we used the Fine scale of the Bathymetric Position Index (BPI). To calculate the neighborhood, the inner radius used was 5 and the outer radius corresponded to 50. When calculating the aspect and slope, the default parameters were used.

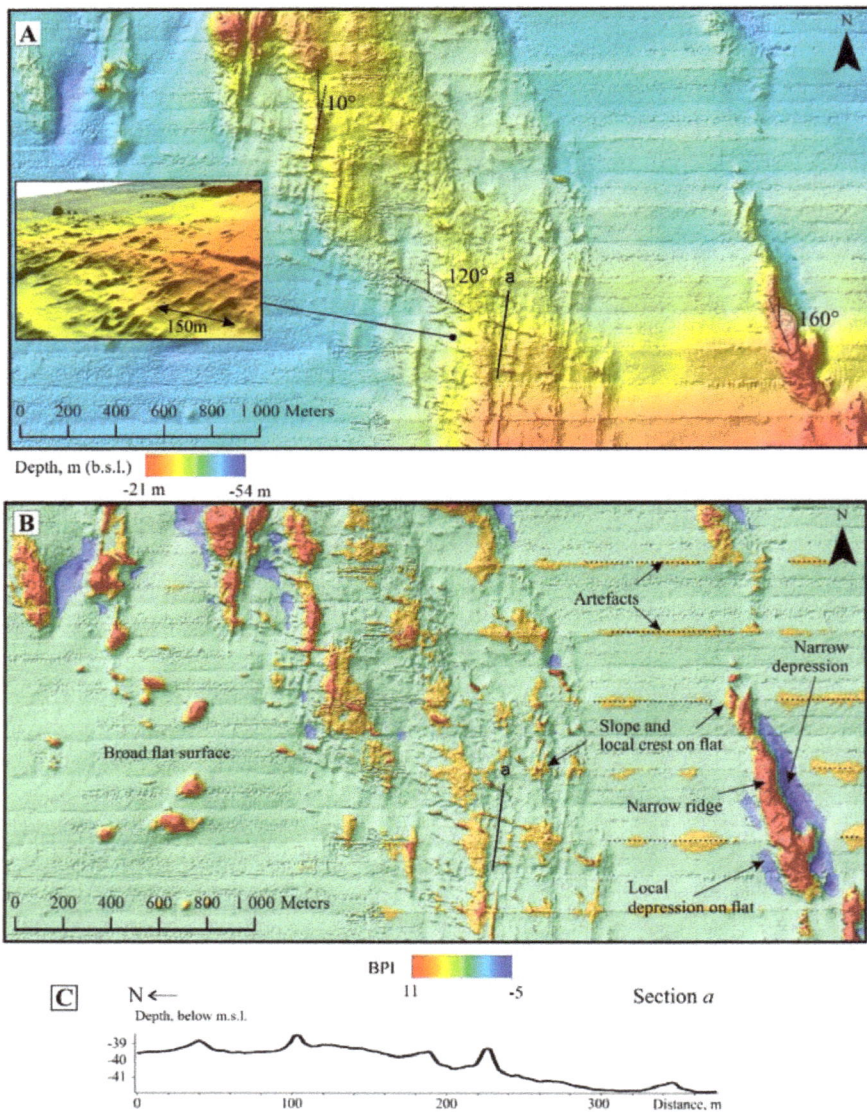

Figure 4. Digital bathymetric model (DBM) of the Moschny Island key area; (**A**) MBES image; (**B**) BPI map; (**C**) N–S oriented bathymetric profile constructed across small ridges.

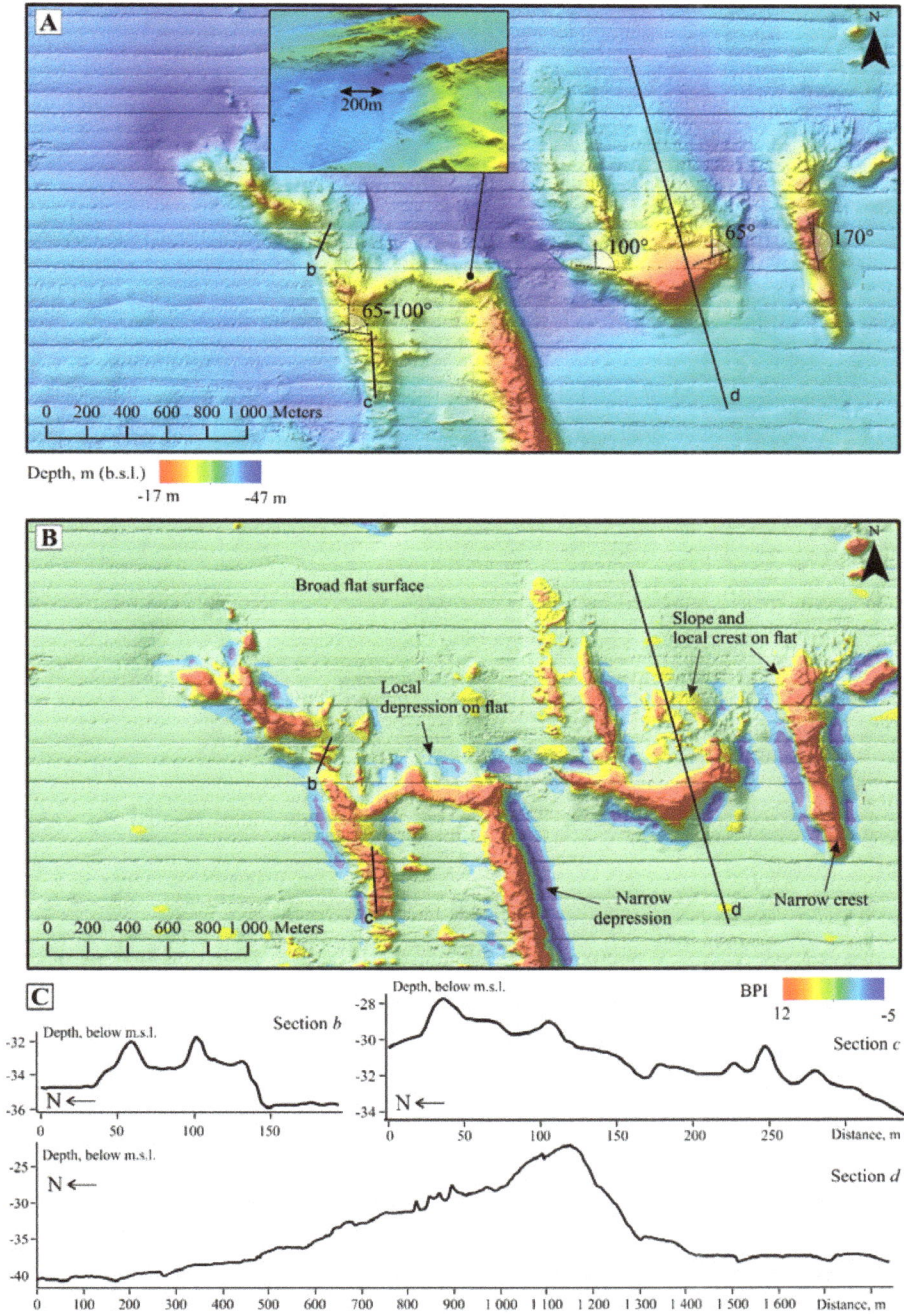

Figure 5. Digital bathymetric model (DBM) of the Vyborg Bay key area; (**A**) MBES image; (**B**) BPI; (**C**) bathymetric profiles across high relief forms.

Table 1. Results of morphometric analyses of MBES and SBP data.

Location	Shape	Direction (Azimuth)	Height Above Around Seafloor, m	Height of Ridge Above Its Base, m	Width of Ridge Base, m	Length of Ridge, m	Slope, Degrees	Crests Interval, m	Geological Interpretation
Isl. Moschny	Linear	NNE–SSW (10°)	0.5–1	-	20–60	1200	1–3°	-	Glacial erosion ridges
	Linear	SSE–NNW (160°)	5–8	15–20	100	1000	5–20°	-	Streamlined moraine ridges
	Linear	SE–NW (120°)	0.5–1.5	1–2	8–10	1300	5–15°	50–150 (av.85)	De Geer moraine
Vyborg Bay	Linear	SSE–NNW (170°)	10–15	15–20	130–170	1000	5–20°	-	Streamlined moraine ridges
	Crescent	NE–SW and SE–NW (65° and 100°)	10–20	10–25	from 70–200 to 300–1000	more than 4300	3–4° N slope 10° S slope	-	End-moraine ridges
	Crescent	NE–SW and SE–NW (65° and 100°)	0.5–1.5	1–2	8–10	300	5–20°	50	De Geer moraine

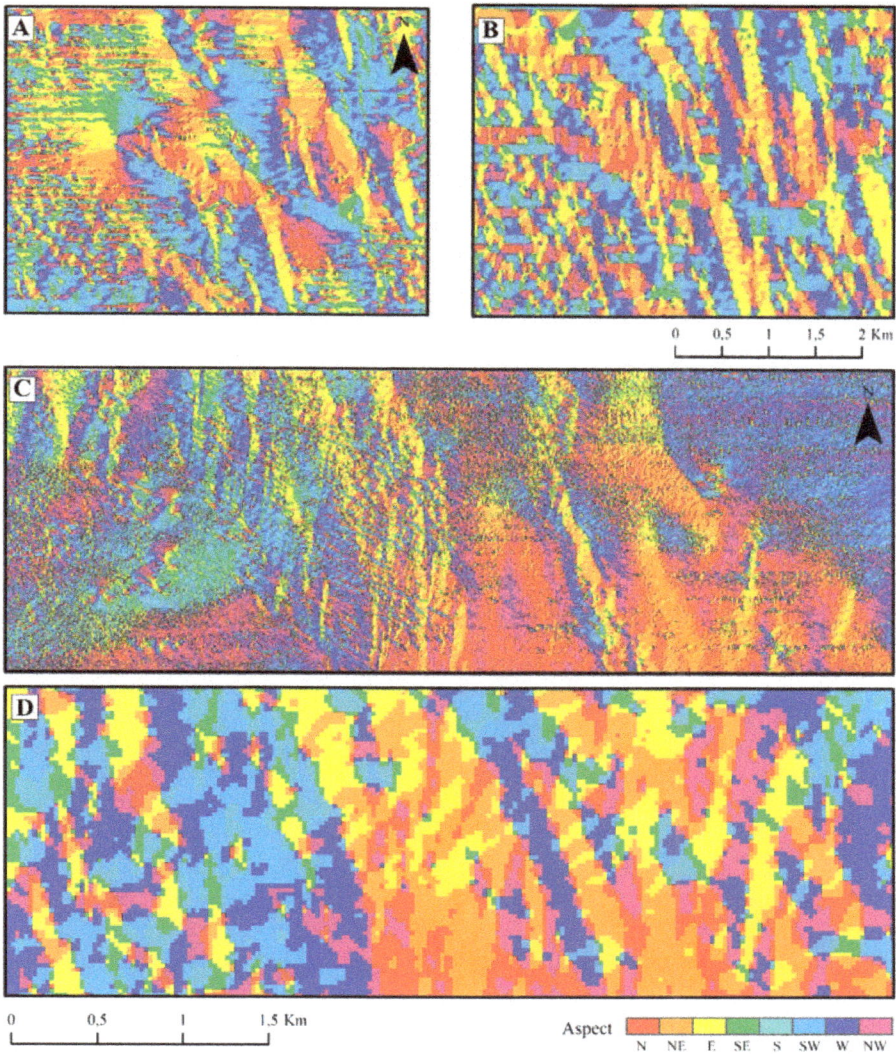

Figure 6. Aspect of slopes based on the MBES DBMs; (**A**) multibeam relief for the Vyborg key area, (**B**) glacial moraine surface of the Vyborg key area, (**C**) multibeam relief of the Moschny Island key area, (**D**) glacial moraine surface of the Moschny Island key area.

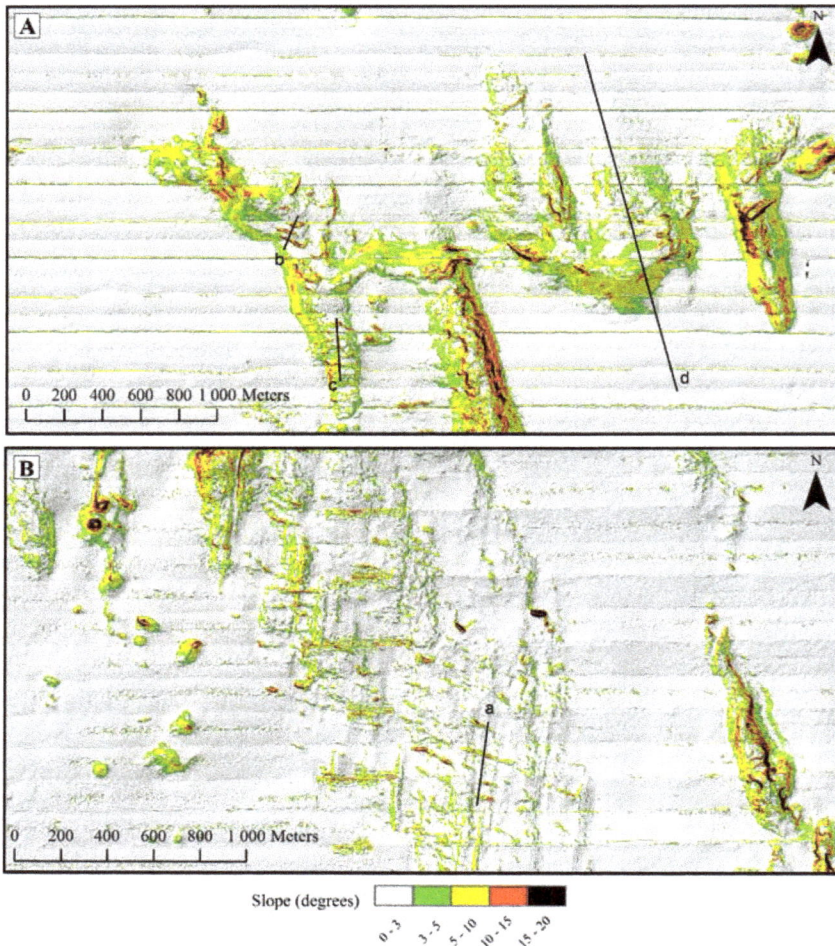

Figure 7. Slope in degrees (**A**) Moschny Island key area; (**B**) Vyborg key area. See Figures 5 and 6 for the locations of Profiles A–D. The colour ramp has been selected to highlight the main features of the terrain.

SBP Data Processing

The SBP data processing was carried out using RadExPro software. This software is commonly used for in-depth High/Ultra-high resolution (HR/UHR) marine seismic processing, real-time marine 2D/3D seismic quality control (QC), and complete processing of the shallow seismic-reflection data. Five distinct acoustic units were identified in the seismic-reflection SBP based on specific acoustic characteristics (Figure 8).

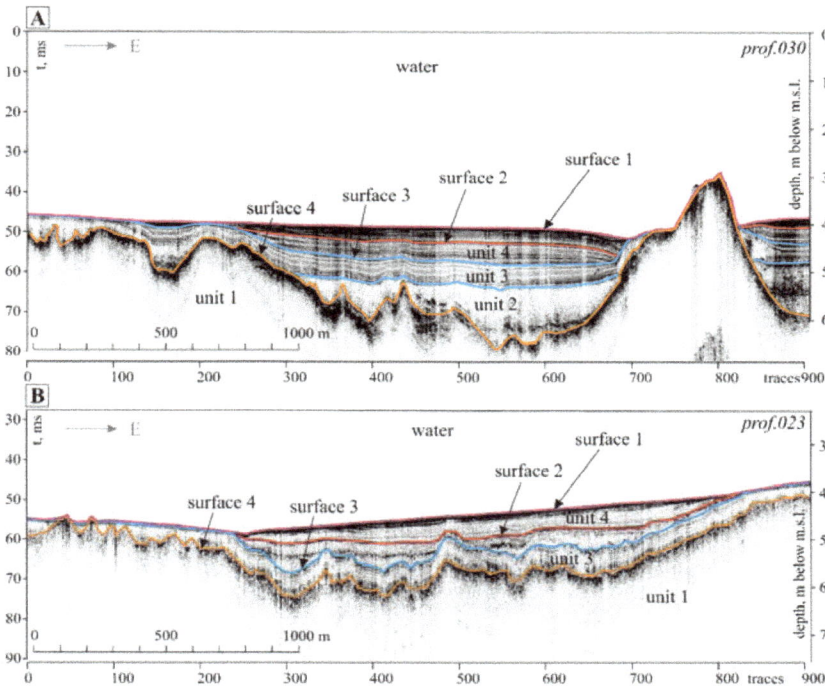

Figure 8. Seismic-reflection profiles: (**A**) Moschny Island key area (profile 023); (**B**) Vyborg Bay key area (profile 030). Interpreted acoustic units consist of Unit 1 = till (moraine); Unit 2 = questionably glaciofluvial sandy deposit; Unit 3 = Late Pleistocene glacio-lacustrine clays; Unit 4 = Holocene lacustrine and marine silty-clayey muds. Surface 1 = sea bottom surface; Surface 2 = inner erosion surface of Holocene mud; Surface 3 = top of Late Pleistocene clays; Surface 4 = top of till.

The upper limits of Unit 1 (Surface 4), Unit 3 (Surface 3), the seafloor (Surface 1), and the lower inner erosion boundary (Surface 2) were easily traced using the intensive axes of in-phase refractions of Unit 4. As a result, spread sheets that contain the geographic coordinates, and the absolute height of each point of the acoustic boundaries (surfaces) were constructed. These data were imported into ArcGIS 10.0 and interpolated with the Natural Neighbour method at 10-m cell size for the erosional boundary in Unit 4 and the top of Unit 3. All other surfaces were interpolated at a 25-m cell size. Border imaging was performed in ArcScene. The difference in window sizes is caused by array variety. Erosion surfaces have sub-horizontal relief, while the other surfaces have a rugged one. The density of the bottom relief data is high enough in all directions, and it allows interpolation, even with a smaller cell size, without any errors. Data about sub-bottom surfaces are redundant along the profile lines, but there is nothing between them to avoid the occurrence of error interpolation with a greater cell size. Using the tools "Aspect" and "Slope", transformations of the surfaces were accomplished to show the azimuth of sloping platforms and steepness of slopes (see Figure 6B,D).

The SBP data processing and interpolation was used to construct a 3D model of the distinct acoustic surfaces (see Figure 9). However, the resolution of this model is much lower than the bottom relief digital bathymetric model (DBM), but it provides some useful information that we feel is important for geological interpretation.

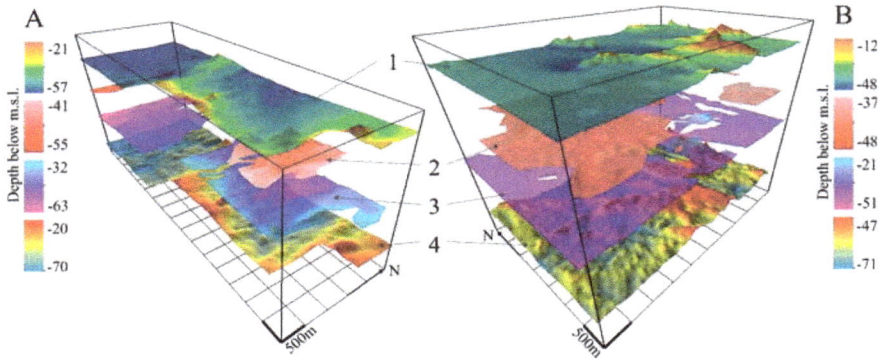

Figure 9. A 3D model of acoustic surfaces based on SBP processing; (**A**) Moschny Island key area, (**B**) Vyborg Bay key area. Geological interpretation of surfaces consists of: 1 = recent bottom surface, 2 = inner erosion surface of Holocene mud, 3 = surface of Late Pleistocene clays, 4 = till surface.

3. Results

The geological interpretation of the SBP and MBES data (see Figures 8 and 10) is based on the analyses of grab samples and gravity cores collected from Quaternary sediment from the geological surveys of the 1990s [13,14]. Although the cores' penetration depths were only 2 to 4 m, till, Late Pleistocene glacio-lacustrine clays, Holocene silty-clay, and mud were sampled from distinct seafloor exposures in the Eastern Gulf of Finland. All of these sedimentary units are characterized by a distinct lithologic composition and sedimentary structures that facilitates correlating them to our interpreted acoustic units with a high degree of confidence.

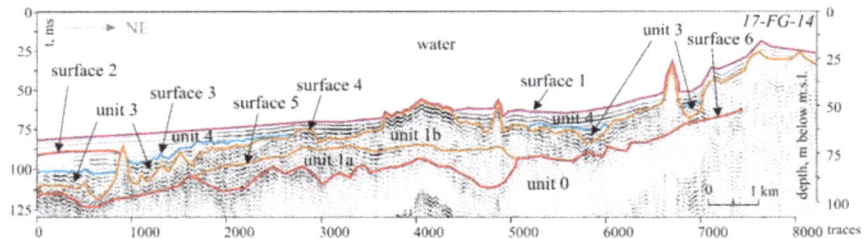

Figure 10. A seismic-reflection profile (17-FG-14) near the Moschny Island key area. Geological interpretation consists of Unit 0 = Vendian deposits, Unit 1a = lower till, Unit 1b = upper till, Unit 3 = Late Pleistocene glacio-lacustrine clays, and Unit 4 = Holocene lacustrine and marine silty-clayey muds. Surface 1 = sea bottom surface; Surface 2 = inner erosional horizon in Holocene mud, Surface 3 = top of Late Pleistocene clays, Surface 4 = top of till, Surface 5 = inner surface between two moraines. Surface 6 = top of Vendian deposits.

In the Vyborg Bay key area (see Figures 1A and 2A), the depth of the seafloor varies from 12 to 15 m along the crests of moraine ridges and between 35 to 40 m within the sedimentary basins. Figure 8B shows interpreted seismic-reflection profiles collected in the key area that exhibit separate and distinct acoustic characteristics. The top (Surface 4) of Unit 1 is characterized by discontinuous high-amplitude reflections lacking an explicit in-phase correlation. An acoustical transparent zone lacking reflections lies below Unit 1, which can be traced over the entire survey area. Unit 1 corresponds to the basal Quaternary sedimentary unit, which overlies the Vyborg complex of rapakivi granite–gabbroic anorthosites and is composed of 20 to 30 m thick tills deposited during the last glaciation. Maximum thicknesses are found along moraine ridges. The results of our study present a much more complicated picture of moraine ridge distribution compared to that previously shown [17–19] (see Figures 2B and 11A).

Using the results from sediment sampling of exposed till (Unit 1) and the presence of numerous boulders on the seafloor, we were able to map the aerial extent of the till. In addition, we mapped specific seafloor depressions with depths <1 m associated with exposed moraine ridges whose encircling depressions result from contourite currents (Figure 11A).

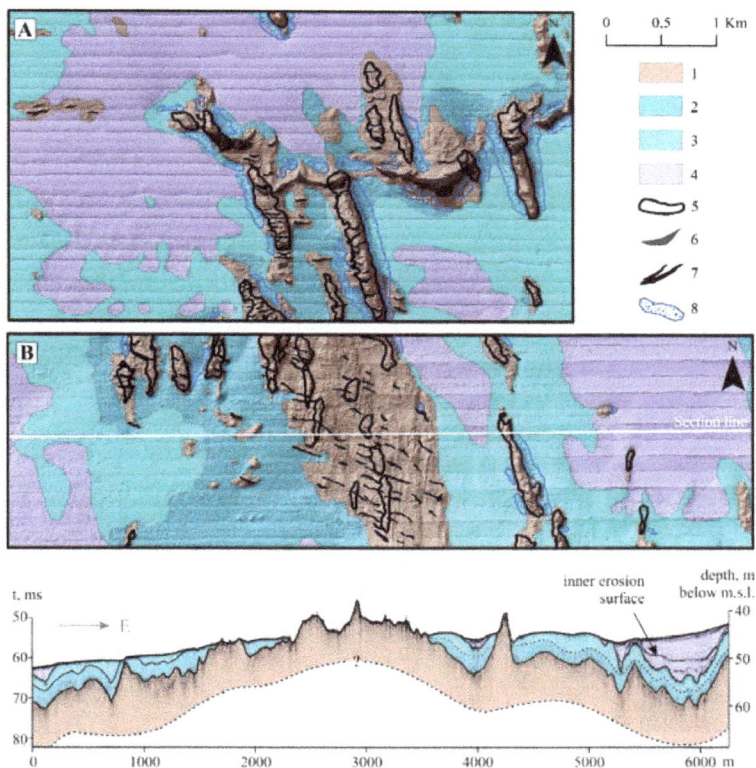

Figure 11. Distribution of Quaternary sedimentary units and seabed morphology based on geological interpretation of SBP and MBES data; (**A**) Vyborg Bay key area; (**B**) Moschny Island key area; (**C**) W–E cross-section. Sedimentary units (see explanation in panel A) consist of 1 = seafloor exposures of Late Pleistocene till with unconformable surface; 2 = seafloor exposures of Late Pleistocene varved glacio-lacustrine clays with unconformable surface; 3 = seafloor exposures of unconformable to disconformable Late Pleistocene homogenous (non-laminated) glacio-lacustrine clays, 4 = sediment basin filled with Holocene marine silty-clayey muds. Interpreted geomorphological features consist of 5 = streamlined moraine ridges; 6 = end-moraine ridges; 7 = De Geer moraine; 8 = depressions formed as a result of contourite current action around exposed moraine ridges.

Unit 3 exhibits parallel rhythmic reflections with in-phase axes of varying amplitudes in the seismic-reflection profiles that are conformable with the surface of the underlying acoustic unit (Figure 8B). Unit 3 is widely distributed in the key areas, and its upper-most reflection (Surface 3) is identified by a bright spot (high amplitude reflection), characteristic of the Late Pleistocene glacio-lacustrine varved and homogenous clays that accumulated in ice marginal lakes and the Baltic

267

Ice Lake. The exposed glacio-lacustrine clays are characterized in the MBES surface roughness images and can be divided into two types (Figure 11A): (1) a rough surface, interpreted as exposed varved clay (lower part of Unit 3) that formed due to selective removal of fine-grained material and enrichment of coarse-grained (gravel, pebbles) sediment, forming a surficial lag deposit; (2) the smoothed surface imaged from the MBES bathymetry is characteristic of homogeneous lacustrine-glacial outcrops. These outcrops are found in areas of little erosion or sediment starved areas. The thickness of the glacio-lacustrine clays ranges from 5 to 10 m.

Unit 4 represents Holocene Ancylus Lake and marine Littorina and Post-Littorina silty-clayey mud in the Vyborg Bay key area (Figure 8B). Here, sediment basins are filled with Holocene muds that dominate the seafloor in this key area (Figure 11A). Locally, mainly in the Northwestern part of the Vyborg Bay key area, acoustic Unit 4 is characterized by sub-horizontal, in-phase axes with variable amplitude reflections. Holocene marine muds >10 m fill local basins located in the Northwestern part of the key area (Figure 11A).

In the Moschny Island key area (Figures 1A, 2B and 11B) the seafloor is characterized by a relatively smooth surface that varies in depth from 54 m within the basins in the Western part, to 21 m on the crests of moraine ridges. In the south, near Moschny Island, depths are markedly shallower. Late Pleistocene varved and homogenous clays deposited in ice marginal lakes and the Baltic Ice Lake cover the seafloor in this key area, as reported from previous investigations [14]. Two local sedimentation basins filled by Holocene silty-clay mud are found in the Eastern and Western parts of the Moschny Island area (Figure 2B). The MBES survey data show a complicated distribution of Quaternary sediment and geomorphic features (Figure 11B).

Pre-Quaternary deposits in the Moschny Island key area are represented by Vendian sediment rocks [14]. The Quaternary basal unit (acoustic Unit 1) is represented by till (<25 m thick). Relative to the seafloor, depths of rare narrow depressions surrounding moraine ridges are up to 2 m deep. Locally above the tills, acoustically transparent zones occur within local depressions (Unit 2 in Figure 8A). The upper boundary of acoustic Unit 2 is sub-horizontal to the lower part of the overlying acoustic Unit 3. Unit 2 is up to 12 m thick. Unit 2 is not exposed in this key area and consequently was not sampled.

Acoustic Unit 3 is characterized by a series of parallel reflections of varying amplitudes, conformable with the reflections of the underlying unit (Figure 8A). This unit represents Late Pleistocene glacio-lacustrine clays that are widely distributed in the Moschny Island key area. As previously described for the Vyborg Bay key area, it is possible to divide outcrops of varved clays and homogenous clays using MBES (Figure 11B). The total thickness of the glacio-lacustrine deposits rarely exceed 6 to 8 m. Acoustic Unit 4 represents the Holocene lacustrine and marine silty-clay muds up to 10 m thick that have accumulated within local sediment basins.

Based on the MBES data, the DBMs of Moschny and Vyborg Bay areas were constructed to illustrate the results of our surveys (Figures 5 and 6). Aspect map analyses revealed several orientations of seafloor geomorphic features within both key areas. The primary orientation for well-defined, elongated, linear ridges up to 1000 m long, 100 to 170 m wide and 15–20 m high (Table 1, Figures 4 and 5) is SSE–NNW (160°–170°), observed on the till surface (Figure 7B,D). These ridges are characterized by the maximum degree of slope in the aspect map (Figure 7).

A secondary orientation is NNE–SSW (10°) for a less well-defined (than that observed for the primary orientation) bottom relief that represents linear elongations up to 1200 m long, 20 to 60 m wide and 0.5–1 m high (Table 1, Figure 4). These features are only observed within the Moschny Island key area (Figure 6C,D).

A tertiary orientation for bottom relief is SE–NW (120°), for small, 0.5 to 1.5 m (some up to 2 m) high, 8 to 10 m wide, and up to 1300 m long, rhythmic, parallel ridges, spaced from 50 to 150 m apart (Table 1, Figure 4). These features are prominent within relatively bathymetric high areas of the Moschny Island key area where they are characterized as interrupted linear features (Figure 6C). In the bathymetric low area, the ridges are subdued or absent. These features are not exhibited in the till surface (Figures 6D and 9), most likely due to information being deleted during the processing of the DBM. This may be the result of interpolation caused by high data resolution along the seismic-reflection survey line and the absence of data between the survey lines [28]. However, very distinct individual ridge crests exhibited in the seismic-reflection profiles (Figure 8B) can be traced for quite a long distance (up to 1300 m) on the surface of a relatively flat bathymetric rise.

In the Vyborg Bay area, the >4300 m long curved bedforms, and those oriented roughly NE–SW to SE–NW (65° to 100°) can be traced on the till surface (Figure 7A,B). Morphologically, this orientation corresponds to relatively high (10–20 m) and wide (70–200 m to 300–1000 m) ridges (Table 1, Figure 5).

Small rhythmically-curved, lined bedforms of similar orientation are observed on the ridges' surfaces. These bedforms are 0.5 to 1.5 m (up to 2 m) high, 8 to 10 m wide and up to 300 m long (Table 1, Figure 5). In depressions, these ridges are not visible; thus they are not traced on the glacial moraine DBM (Figure 7A,B), but like in the Moschny key area, they can be identified on seismic-reflection profiles.

4. Discussion

Reconstruction of the Late Weichselian ice sheet retreat through the Eastern Gulf of Finland basin is still a problem. Morphological and chronological deglaciation markers are primarily located onshore [23,39–46]. A lack of data from the marine environment and existing contradictory hypotheses has led to large uncertainties in the reconstruction of the ice margin retreat within recently submerged areas [23], including the Eastern Gulf of Finland [16]. High resolution MBES and SBP data can provide important geomorphological indicators of the ice sheet retreat, which occurred between 13.8 and 13.3 ka BP (Pandivere–Neva stage in the Southern coast of the gulf) [36,38] and 12.25 ka BP (Salpausselkä I stage, Southeastern Finland) [40,41].

The MBES and SBP data collected in 2017 and reported herein represent the first set of high resolution bathymetric data for the Eastern Gulf of Finland obtained for scientific study. These data provide a unique opportunity to describe and map significant small relief bedforms. The geological interpretations we present here are sound, even though not all artifacts in the MBES data could be removed. Since the seismic-reflection profiles were collected along the nadir of the MBES lines, morphometric analyses were validated with a high degree of confidence.

The dominant seafloor geomorphology in the Eastern Gulf of Finland are the 1000 m long, 100 to 170 m wide and 15 to 20 m high SSE–NNW (160°–170°) oriented elongated linear oval-shaped ridges, interpreted as streamlined moraine ridges and classified as drumlins. These features formed beneath an ice sheet as it moved over the land, parallel to the ice direction and are composed of basal till or lodgement till [45–47].

The 10 to 20 m high, 70 to 1000 m wide, approximately 4300 m long crescent ridge, oriented NE–SW to SE–NW (65°–100°), mapped in Vyborg Bay, is interpreted as an end-moraine, parallel to the terminal ice margin. Its asymmetrical shape (Figure 5C, section d) with the steeper (10° dipping) Southern (lee-side) and gentler (3–4°) Northern (stoss-side) slopes is characteristic of typical end-moraine morphology [45–47]. Although the general location of the end-moraine was predicted based on broad-scale bottom relief in previous GIS analyses [16], the precise location and morphological evidence was not known until our study. The end-moraine ridges that we have mapped are smaller in comparison to the Pandivere–Neva and Salpausselkä I glacial forms. The Pandivere–Neva terminal moraine zone consists of a curved and interrupted belt of push end-moraines, glaciofluvial deltas and eskers [48]. The heights of these features range from only a few metres up to 20 m [49], because Holocene wave erosion has substantially levelled them. In contrast, the Salpausselkä I moraine is 0.5 to 4 km wide and is approximately 20 to 80 m high [50,51].

Among the most interesting submerged glacial relief features found in our study are series of rhythmic parallel ridges, 1 to 1.5 m high, elongated at a slightly different direction within our two study areas. These ridges are overlaid with both drumlin and end-moraines. The small parallel ridges can be identified as De Geer moraines, which are typically described as relatively low till ridges, with rhythmic parallel linear to curved positive relief features in plain view. Such clusters of de Geer moraines, with dimensions of <5 m high, 10 to 50 m wide, and >100 m long, are exemplified in the Replot and Björkö areas of the Kvarken Archipelago in Finland [45–47]. On the Northern slope of Slupsk Bank in the Southwestern Baltic, De Geer moraines exhibit ridges from 2 to 8 m high with a distance of 200 to 500 m between crests [52]. On the German Bank of the Southern Scotian Shelf of Atlantic Canada [53], the De Geer moraine is represented by sub-parallel ridges, varying from 1.5 m high and 40 m wide to 5 to 8 m high and 100 to 130 m wide. Individual ridges can be traced horizontally for a few hundred metres up to almost 10 km [53]. Numerous closely spaced linear and curvilinear ridges occur east of Shetland. The range of the ridges varies from 1 to 20 km in length, with distances from 700 to 2000 m. Most ridges have a vertical seafloor expression of between 10 and 20 m. These straight, sub-parallel, sharp-crested glacial features are interpreted as De Geer-type moraines [54]. Recent hypotheses on the origin of De Geer moraines are presented here. A study on the Kvarken Archipelago [45,47] proposed that the moraine there was formed in crevasses running parallel to the ice margin and deposited under sub-aquatic conditions. A study of the De Geer moraine on the German Bank indicated that its formation was at the grounding line of a tidewater glacier, the result of deposition and/or pushing action during minor re-advances, and and stillstands [53]. Therefore, a De Geer moraine reflects the probable position of a retreating ice margin. The MBES data collected in Vyborg Bay support this hypothesis as the direction of De Geer moraine coincides with elongation of the end-moraine ridge crest in Vyborg Bay.

A slight difference in orientations of the De Geer moraines mapped near Moschny Island and in Vyborg Bay indicates a rotation of the ice margin front during deglaciation, validating an earlier hypothesis [16,55]. However, this MBES study identified and comprehensively mapped De Geer moraines in the Eastern Gulf of Finland for the first time, even though an extensive SBP coverage had previously been done.

The low linear ridges located north of Moschny Island and oriented NNE–SSW (10°) were difficult to interpret. Acoustic units, interpreted as glacial moraine in seismic-reflection profiles (see profile 17-FG-14, Figure 10 for example), are different from the acoustically chaotic reflections absent of continuous reflections that generally characterize moraines. In profile 17-FG-14, well-defined, but irregular, reflections (e.g., Surface 5 of Figure 10) are interpreted to represent a reflection within the moraine units. The lower section (Unit 1a) of the moraine deposits appears to have been eroded by an overlying advancing glacier, followed by the deposition of another (upper) moraine deposit (Unit 1b).

Acoustic Surface 4 (Figure 8) is interpreted as the upper surface of till. Till, identified within both key areas, is mostly overlaid by a 5 to 7 m thick layer of varved clays, a rhythmic pair of brown clay layers and grey silty-microlayers [56]. The varved clays are very well identified in the seismic-reflection profiles (lower part of Unit 3, Figures 8 and 10) due to their laminated stratigraphy and "clothing" bedding. Homogeneous glacio-lacustrine clays, as determined from sediment core analyses [14,56], form the upper part of acoustic Unit 3 (Figures 8 and 10). Generally, in the Eastern Gulf of Finland, the boundary between typical varved clays and homogeneous clay is a facies transition. Locally, in the troughs of the moraine surface, between the till and the glacio-lacustrine deposits, there is a layer of sediment that appears to be glacio-fluvial sand (unit 2, Figure 8).

Another sharp erosive contact is exhibited as Surface 3 (Figure 8), which is proposed to lie between Late Pleistocene and Holocene sediments. Locally, erosion was so intense that up to 10 m of the Late Pleistocene glacio-lacustrine clays were eroded. This interpretation supports the interpretations made from other acoustic-seismic profiles [13,56], which support the hypothesis of a deep and relatively long period of low water level in the Eastern Gulf of Finland during the Early Holocene [57,58]. The general trend of relief transformation between the final stage of the ice sheet retreat and the beginning of the Holocene is relief smoothing as a result of glacio-fluvial and glacio-lacustrine sedimentation in the Late Pleistocene and intense submarine erosion after the final draining of the Baltic Ice Lake (Figures 8 and 9).

The distribution and thickness of Holocene silty-clay mud within the study area indicates a drastic change in the sedimentation processes since the beginning of the Ancylus Lake stage. Since that time, accumulation has occurred within local sedimentary basins, while large areas in the Gulf of Finland were sediment-starved or experiencing erosion. Recent silty-clayey mud accumulation in the Eastern Gulf of Finland occurs in local sedimentary basins staggered along a ladder-shape bottom profile that steps down from a depth of 5 to 6 m in Neva Bay, to 80 m south of Gogland Island and is separated by vast sediment-starved or eroding areas (Figure 12). Three-dimensional geological models of our key areas and SBP data demonstrate that Holocene marine mud sedimentation occurred within local depressions at depths from 40 to 45 m in the Eastern part of the Moschny Island key area, and from 55 to 60 m in Vyborg Bay and the Western Moschny Island area. The thickness of the Holocene clayey mud varies from tens of cm on the slopes of ridges to 8 to 10 m in the deepest part of the sedimentary basins.

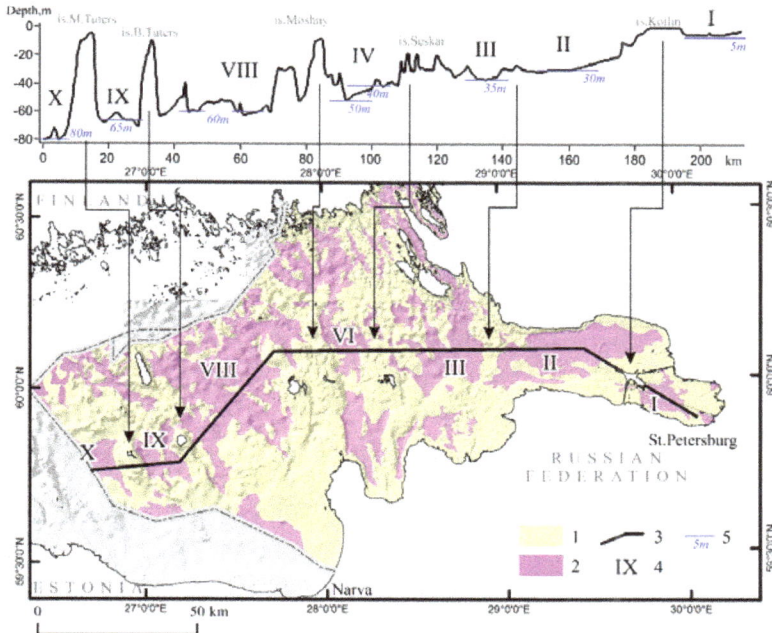

Figure 12. Schematic representation of the sedimentary basins in the Eastern Gulf of Finland: 1 = areas of submarine erosion and non-sedimentation, 2 = sediment basins with silty-clayey mud accumulation, 3 = cross-section profile line, 4 = sedimentary basin number, 5 = mean level of the modern clayey mud surface. Sediment basins are located along a ladder-shape profile, which steps down from depths of 5 to 6 m in the Neva Bay to 80 m south of Gogland Island and is separated by vast areas of non-sedimentation and erosion.

Our interpretation of the seismic-reflection profiles fixes one to two erosional surfaces within the Holocene mud. Construction of a 3D geological model allowed us to identify an erosional horizon within the muds that could be traced through all sedimentary basins within our study area at a depth of 45 to 60 m (Figure 9). We interpret the erosion to have taken place during a time of decreased water level. A lower water level in pre-Littorina time is supported by high-resolution sediment core investigations [15], modelling of submarine terrace formation [57], and onshore geoarcheological research [59]. More detailed study of Littorina sea-level fluctuations requires the continuation of high-resolution sediment coring.

5. Conclusions

1. Geophysical research in key areas of Vyborg Bay and north of Moschny Island using high-resolution MBES and SBP data revealed, for the first time, well-defined streamlined moraine ridges, De Geer moraines, and end-moraine ridges. Mapping of these glacial features established the location and orientation of the ice sheet margins during different stages of deglaciation in the Eastern (Russian) Gulf of Finland.

2. Our interpretations of high-resolution seismic-reflection profiles and 3D models of the till surfaces, Late Pleistocene sediments, and the mapping of the modern seafloor relief indicate that a significant fall in the water level in the Early Holocene and that, most likely, several water-level fluctuations occurred, documented by erosion surfaces (acoustic unconformable horizons) in silty-clay mud.

3. We conclude that the distribution and thickness of Holocene silty-clay mud mapped within our study areas indicates a drastic change in sedimentation since the beginning of the Ancylus Lake stage. Since the Ancylus, accumulation occurred within local sedimentary basins, while conditions of large bottom areas of the Gulf of Finland remained starved of sediment or suffered from erosion. Modern silty-clay accumulation in the Eastern Gulf of Finland occurs in local sedimentary basins, located at depths from 5 to 6 m in Neva Bay, to 80 m south of Gogland Island, which are separated by vast sediment starved and/or erosional areas.

Acknowledgments: Multibeam data processing was supported by the state assignment No. 0149-2018-0012. Data analyses and interpretation was carried out under project No. 17-77-20041 of Russian Science Foundation. The authors thank H. Gary Greene for helpful and important comments and language revision.

Author Contributions: The drafting of the research program, the primary processing of multibeam echosounder data in the software package PDS2000 and a description of the methods for obtaining and processing primary data was undertaken by Krek Alexander. The registration of primary MBES data and primary processing of the data array in the software package PDS2000 was accomplished by Kapustina Maria. The registration of the primary data for the upper sediment layer characterized in the SBP and the primary processing of the bathymetric data array in the PDS2000 software package was done by Tkacheva Elena. Danchenkov Aleksandr completed the final processing of bathymetric survey data in ArcGIS and SAGAGIS, including the elimination of artifacts in the digital models. He also contributed text for the section on research methods. Processing of seismic-reflection profiles was carried out by Budanov Leonid and Moskovtsev Alexandr. Geological interpretation was undertaken by Ryabchuk Daria, Sergeev Alexander and Zhamoida Vladimir.

Conflicts of Interest: The authors declare no conflict of interest.

References

1. Harff, J.; Björck, S.; Hoth, P. *The Baltic Sea Basin: Introduction. The Baltic Sea Basin Central and Eastern European Development Studies (CEEDES)*; Springer: Berlin/Heidelberg, Germany, 2011; pp. 3–9.

2. Uścinowicz, S. *Geochemistry of the Baltic Sea Surface Sediments*; Polish Geological Institute—National Research Institute: Warsaw, Poland, 2011; p. 356. ISBN 978-83-7538-814-5.

3. Kaskela, A.; Rousi, H.; Ronkainen, M.; Orlova, M.; Babin, A.; Gogoberidze, G.; Kostamo, K.; Kotilainen, A.; Neevin, I.; Ryabchuk, D.; et al. Linkages between benthic assemblages and physical environmental factors: The role of geodiversity in Eastern Gulf of Finland ecosystems. *Cont. Shelf Res.* **2017**, *142*, 1–13. [CrossRef]

4. Uścinowicz, S. *Relative Sea Level Changes, Glacio-Isostatic Rebound and Shoreline Displacement in the Southern Baltic*; Polish Geological Institute: Warsaw, Poland, 2003; Volume 10, p. 79.

5. Lampe, R.; Meter, H.; Ziekur, R.; Janke, W.; Endtmann, E. Holocene Evolution of the Irregularly Sinking Baltic Sea Coast and the Interactions of the Sea-Level Rise, Accumulation Space and Sediment Supply. In *SINCOS–Sinking Coasts. Geosphere, Ecosphere and Anthroposhere of the Holocene Southern Baltic Sea*; Harff, J., Luth, F., Eds.; Bericht der Romisch-Germanischen Kommission; Philipp von Zabern: Darmstadt, Germany, 2007; pp. 15–46.

6. Harff, J.; Meyer, M. *Coastlines of the Baltic Sea–Zones of Competition Between Geological Processes and a Changing Climate: Examples from the Southern Baltic*; The Baltic Sea Basin Central and Eastern European Development Studies (CEEDES) Book Series; Springer: Berlin/Heidelberg, Germany, 2011; pp. 149–164.

7. Hyttinen, O.; Kotilainen, A.T.; Virtasalo, J.J.; Kekäläinen, P.; Snowball, I.; Obrochta, S.; Andrén, T. Holocene stratigraphy of the Ångermanälven River estuary, Bothnian Sea. *Geo-Mar. Lett.* **2016**, *37*, 273–288. [CrossRef]

8. Gordeeva, S.; Malinin, V. *Gulf of Finland Sea Level Variability*; RSHU Publishers: St. Petersburg, Russia, 2014; p. 179. ISBN 978-5-86813-403-6. (In Russian)

9. Rosentau, A.; Veski, S.; Kriiska, A.; Aunap, R.; Vassiljev, J.; Saarse, L.; Hang, T.; Heinsalu, A.; Oja, T. *Palaeogeographic Model for the SW Estonian Coastal Zone of the Baltic Sea*; The Baltic Sea Basin Central and Eastern European Development Studies (CEEDES) book series; Springer: Berlin/Heidelberg, Germany, 2011; pp. 165–188.

10. Rosentau, A.; Muru, M.; Kriiska, A.; Subetto, D.A.; Vassiljev, J.; Hang, T.; Gerasimov, D.; Nordqvist, K.; Ludikova, A.; Lõugas, L.; et al. Stone Age settlement and Holocene shore displacement in the Narva-Luga Klint Bay area, eastern Gulf of Finland. *Boreas* **2013**, *42*, 912–931. [CrossRef]

11. Sandgren, P.; Subetto, D.A.; Berglund, B.E.; Davydova, N.N.; Savelieva, L.A. Mid-Holocene Littorina Sea transgressions based on stratigraphic studies in coastal lakes of NW Russia. *GFF* **2004**, *126*, 363–380. [CrossRef]

12. Miettinen, A.; Savelieva, L.; Subetto, D.; Dzhinoridze, R.; Arslanov, K.; Hyvärinen, H. Palaeoenvironment of the Karelian Isthmus, the easternmost part of the Gulf of Finland, during the Litorina Sea stage of the Baltic Sea history. *Boreas* **2007**, *36*, 441–458. [CrossRef]

13. Ryabchuk, D.; Zhamoida, V.; Amantov, A.; Sergeev, A.; Gusentsova, T.; Sorokin, P.; Kulkova, M.; Gerasimov, D. Development of the coastal systems of the easternmost Gulf of Finland, and their links with Neolithic–Bronze and Iron Age settlements. *Geol. Soc. Lond. Spec. Publ.* **2016**, *411*, 51–76. [CrossRef]

14. Petrov, O.V. *Atlas of Geological and Geoecological Maps of the Russian Part of the Baltic Sea*; VSEGEI: St. Petersburg, Russia, 2010; 78p.

15. Virtasalo, J.J.; Ryabchuk, D.; Kotilainen, A.T.; Zhamoida, V.; Grigoriev, A.; Sivkov, V.; Dorokhova, E. Middle Holocene to present sedimentary environment in the easternmost Gulf of Finland (Baltic Sea) and the birth of the Neva River. *Mar. Geol.* **2014**, *350*, 84–96. [CrossRef]

16. Ryabchuk, D.; Hyttinen, O.; Zhamoida, V.; Kotilainen, A.; Sergeev, A.; Amantov, A.; Budanov, L.; Moskovtsev, A. Deglaciation of the Eastern Gulf of Finland basin. Unpublished manuscript. 2017.

17. Krasnov, I.I.; Duphorn, K.; Voges, A. (Eds.) *International Quaternary Map of Europe 1:2,500,000*; Sheet 3, Nordkapp; UNESCO: Hanover, Germany, 1971.

18. Zarina, E.P. Geochronology and Paleogeography of Late Pleistocene of the North-West of Russia. In *Periodization and Paleogeography of Late Pleistocene*; VSEGEI: Leningrad, Soviet Union, 1970; pp. 27–33. (In Russian)

19. Kvasov, D.D. *Late Quaternary History of the Large Lakes and Inner Seas of the Eastern Europe*; Nauka: Leningrad, Russia, 1975; p. 278. (In Russian)

20. Raukas, A.; Hyvärinen, H. (Eds.) *Geology of the Gulf of Finland*; Valgus: Tallinn, Estonia, 1991; p. 422.

21. Subetto, D.A. *Lake Sediments: Paleolimnological Reconstructions*; Herzen Russian State Pedagogical University: St. Petersburg, Russia, 2009; p. 339. ISBN 978-5-8064-1444-2. (In Russian, resume in English).

22. Vassiljev, J.C.B.C.; Saarse, L.; Rosentau, A. *Palaeoreconstruction of the Baltic Ice Lake in the Eastern Baltic*; The Baltic Sea Basin Central and Eastern European Development Studies (CEEDES) Book Series; Springer: Berlin/Heidelberg, Germany, 2011; pp. 189–202.

23. Hughes, A.L.C.; Gyllencreutz, R.; Lohne, Ø.S.; Mangerud, J.; Svendsen, J.I. The last Eurasian ice sheets—A chronological database and time-Slice reconstruction, DATED-1. *Boreas* **2016**, *45*, 1–45. [CrossRef]

24. Dorschel, B.; Wheeler, A.J.; Monteys, X.; Verbruggen, K. *On the Irish Seabed. Atlas of the Deep-Water Seabed*; Springer: New York, NY, USA, 2010; p. 164.

25. Harris, P.T.; Baker, E.K. *GeoHab Atlas of Seafloor Geomorphic Features and Benthic Habitats. Seafloor Geomorphology as Benthic Habitat*; Elsevier: Amsterdam, Netherlands, 2012; pp. 871–890.

26. Diesing, M.; Green, S.L.; Stephens, D.; Lark, R.M.; Stewart, H.A.; Dove, D. Mapping seabed sediments: Comparison of manual, geostatistical, object-based image analysis and machine learning approaches. *Cont. Shelf Res.* **2014**, *84*, 107–119. [CrossRef]

27. Dove, D.; Arosio, R.; Finlayson, A.; Bradwell, T.; Howe, J.A. Submarine glacial landforms record Late Pleistocene ice-Sheet dynamics, Inner Hebrides, Scotland. *Quat. Sci. Rev.* **2015**, *123*, 76–90. [CrossRef]

28. Lecours, V.; Dolan, M.F.J.; Micallef, A.; Lucieer, V.L. A review of marine geomorphometry, the quantitative study of the seafloor. *Hydrol. Earth Syst. Sci.* **2016**, *20*, 3207–3244. [CrossRef]

29. Flemming, N.C.; Harff, J.; Moura, D.; Burgess, A.; Bailey, G.N. (Eds.) *Submerged Landscapes of the European Continental Shelf*; John Wiley & Sons, Inc.: Hoboken, NJ, USA, 2017; p. 533. ISBN 9781118922132.

30. Petrov, O.V.; Lygin, A.M. (Eds.) *Information Bulletin on Geological Environment Assessment of the Near-Shore Areas of the Barents, White and Baltic Seas*; VSEGEI: St. Petersburg, Russia, 2014; p. 136. (In Russian)

31. Raateoja, M.; Setälä, O. *The Gulf of Finland Assessment*; Reports of the Finnish Institute Series; Finnish Environment Institute: Helsinki, Finland, 2016; Volume 27, p. 363.

32. Zhamoida, V.; Grigoriev, A.; Ryabchuk, D.; Evdokimenko, A.; Kotilainen, A.T.; Vallius, H.; Kaskela, A.M. Ferromanganese concretions of the eastern Gulf of Finland–Environmental role and effects of submarine mining. *J. Mar. Syst.* **2017**, *172*, 178–187. [CrossRef]

33. Sibson, R. Chapter 2: A Brief Description of Natural Neighbor Interpolation. In *Interpolating Multivariate Data*; John Wiley & Sons: New York, NY, USA, 1981; pp. 21–36.

34. Demyanov, V.; Savelyeva, E. *Geostatistics: Theory and Practice*; Nauka: Moscow, Russia, 2010; p. 327. ISBN 978-5-02-037478-2.

35. Oimoen, M.J. An Effective Filter for Removal of Production Artifacts in U.S. Geological Survey 7.5-Minute Digital Elevation Models. In Proceedings of the Fourteenth International Conference on Applied Geologic Remote Sensing, Las Vegas, NV, USA, 6–8 November 2000; pp. 311–319.

36. Peregro, A. SRTM DEM Destriping with SAGA GIS: Consequences on Drainage Network Extraction. Available online: http://www.webalice.it/alper78/saga_mod/destriping/destriping.html (accessed on 7 December 2017).

37. Wright, D.J.; Lundblad, E.R.; Larkin, E.M.; Rinehart, R.W.; Murphy, J.; Cary-Kothera, L.; Draganov, K. ArcGIS Benthic Terrain Modeler. Available online: http://www.csc.noaa.gov/products/btm/ (accessed on 20 January 2017).

38. Lundblad, E.; Wright, D.J.; Miller, J.; Larkin, E.M.; Rinehart, R.; Battista, T.; Anderson, S.M.; Naar, D.F.; Donahue, B.T. A benthic terrain classification scheme for American Samoa. *Mar. Geod.* **2006**, *29*, 89–111. [CrossRef]

39. Bulletin de la Commission Pour L'etude du Quaternaire. Available online: http://ginras.ru/library/pdf/2_1930_bull_quatern_comission.pdf (accessed on 10 November 2017).

40. Donner, J. *The Quaternary History of Scandinavia*; Cambridge University Press: London, UK, 1995; p. 200. ISBN 978-0-521-01831-9.

41. Saarnisto, M.; Saarinen, T. Deglaciation chronology of the Scandinavian Ice Sheet from the Lake Onega Basin to the Salpausselkä End Moraines. *Glob. Planet. Chang.* **2001**, *31*, 387–405. [CrossRef]

42. Hang, T. A local clay-Varve chronology and proglacial sedimentary environment in glacial Lake Peipsi, eastern Estonia. *Boreas* **2003**, *32*, 416–426. [CrossRef]

43. Kalm, V. Pleistocene chronostratigraphy in Estonia, southeastern sector of the Scandinavian glaciation. *Quat. Sci. Rev.* **2006**, *25*, 960–975. [CrossRef]

44. Vassiljev, J.; Saarse, L. Timing of the Baltic Ice Lake in the eastern Baltic. *Bull. Geol. Soc. Finl.* **2013**, *85*, 9–18. [CrossRef]

45. Breilin, O.; Kotilainen, A.; Nenonen, K.; Virransalo, P.; Ojalainen, J.; Stén, C.-G. Geology of the Kvarken Archipelago. Appendix 1 to the application for nomination of the Kvarken Archipelago to the World Heritage list. In *Metsähallitus Western Finland Natural Heritage Services, West Finland Regional Environment Centre*; Regional Council of Ostrobothnia: Vaasa, Finland, 2004; p. 12.

46. Breilin, O.; Kotilainen, A.; Nenonen, K.; Räsänen, M. The unique moraine morphology, stratotypes and ongoing geological processes at the Kvarken Archipelago on the land uplift area in the western coast of Finland. In *Quaternary Studies in the Northern and Arctic Regions of Finland, Proceedings of the Workshop Organized within the Finnish National Committee for Quaternary Research (INQUA), Kilpisjärvi Biological Station, Finland, 13–14 January 2005*; Ojala, A.E.K., Ed.; Special Paper 40; Geological Survey of Finland: Espoo, Finland, 2005; pp. 97–111.

47. Kotilainen, A.T.; Kaskela, A.M. Comparison of airborne LiDAR and shipboard acoustic data in complex shallow water environments: Filling in the white ribbon zone. *Mar. Geol.* **2017**, *385*, 250–259. [CrossRef]

48. Karukäpp, R.; Raukas, A. Deglaciation History. In *Geology and Mineral Resources of Estonia*; Raukas, A., Teedumäe, A., Eds.; Estonian Academy Publishers: Tallinn, Estonia, 1997; pp. 263–267. ISBN 9985-50-185-3.

49. Talviste, P.; Hang, T.; Kohv, M. Glacial varves at the distal slope of Pandivere-Neva ice-Recessional formations in western Estonia. *Bull. Geol. Soc. Finl.* **2012**, *84*, 7–19. [CrossRef]

50. Glückert, G. The First Salpausselkä at Lohja, southern Finland. *Bull. Geol. Soc. Finl.* **1986**, *58*, 45–55. [CrossRef]

51. Donner, J. The Younger Dryas age of the Salpausselkä moraines in Finland. *Bull. Geol. Soc. Finl.* **2010**, *82*, 69–80. [CrossRef]

52. Uścinowicz, S. Moreny De Geera na Ławicy Slupskiej-nowe dowody na subakwalną deglacjację obszaru południowego Bałtyku. In Proceedings of the XVIII Konferencja Naukowo-Szkoleniowa. Stratigrafia Plejstocenu Polski, Stara Kiszewa, Poland, 5–9 January 2011. (In Polish)

53. Todd, B.J. De Geer moraines on German Bank, southern Scotian Shelf of Atlantic Canada. Geological Society, London, UK. *Memoirs* **2016**, *46*, 259–260. [CrossRef]

54. Bradwell, T.; Stoker, M.S.; Golledge, N.R.; Wilson, C.K.; Merritt, J.W.; Long, D.; Everest, J.D.; Hestvik, O.B.; Stevenson, A.G.; Hubbard, A.L.; et al. The northern sector of the last British Ice Sheet: Maximum extent and demise. *Earth-Sci. Rev.* **2008**, *88*, 207–226. [CrossRef]

55. Amantov, A.V.; Amantova, M.V. Modeling of postglacial development of Lake Ladoga and eastern part of the Gulf of Finland. *Reg. Geol. Metallog.* **2017**, *69*, 5–14. (In Russian)

56. Spiridonov, M.; Ryabchuk, D.; Kotilainen, A.; Vallius, H.; Nesterova, E.; Zhamoida, V. The Quaternary deposits of the Eastern Gulf of Finland. In *Holocene Sedimentary Environment and Sediment Geochemistry of the Eastern Gulf of Finland, Baltic Sea*; Vallius, H., Ed.; Geological Survey of Finland: Espoo, Finland, 2007; pp. 5–18. ISBN 978-951-69-09-953.

57. Amantov, A.V.; Zhamoida, V.A.; Ryabchuk, D.V.; Spiridonov, M.A.; Sapelko, T.V. Geological structure of submarine terraces of the eastern Gulf of Finland and modeling of their development during postglacial time. *Reg. Geol. Metallog.* **2012**, *50*, 5–27. (In Russian)

58. Amantov, A.; Ryabchuk, D.; Fjeldskaar, W.; Zhamoida, V.; Amantova, M. Possible role of Hydroisostasy in Peculiarities of Late Glacial-Postglacial Sedimentation of the Eastern part of the Gulf of Finland and Lake Ladoga. In *EGU General Assembly, Geophysical Research Abstracts*; EGU General Assembly: Vienna, Austria, 2013; Volume 15.

59. Sergeev, A.; Ryabchuk, D.; Gusentsova, T.; Sorokin, P.; Nesterova, E.; Zhamoida, V.; Spiridonov, M.; Kulkova, M.; Glukhov, V. Reconstruction of the paleorelief of the Littorina Sea coastal zone within St. Petersburg based on a study of the archaeological site Okhta 1.59. *Ger. Anz. Romis.-Ger. Kom. Deutsch. Archaologischen Inst.* **2014**, *92*, 33–59.

geosciences

MDPI

Article

Shape and Size Complexity of Deep Seafloor Mounds on the Canary Basin (West to Canary Islands, Eastern Atlantic): A DEM-Based Geomorphometric Analysis of Domes and Volcanoes

Olga Sánchez-Guillamón [1,*] , Luis Miguel Fernández-Salas [2] , Juan-Tomás Vázquez [1] ,
Desirée Palomino [1] , Teresa Medialdea [3] , Nieves López-González [1] , Luis Somoza [3]
and Ricardo León [3]

[1] Spanish Institute of Oceanography (IEO), Oceanographic Center of Málaga, Puerto Pesquero, S/N,
    29640 Málaga, Spain; juantomas.vazquez@ieo.es (J.-T.V.); desiree.palomino@gmail.com (D.P.);
    nieves.lopez@ieo.es (N.L.-G.)
[2] Spanish Institute of Oceanography (IEO), Oceanographic Center of Cádiz, Muelle de Levante, S/N,
    11006 Cádiz, Spain; luismi.fernandez@ieo.es
[3] Geological Survey of Spain (IGME), Ríos Rosas 23, 28003 Madrid, Spain; t.medialdea@igme.es (T.M.);
    l.somoza@igme.es (L.S.); r.leon@igme.es (R.L.)
*   Correspondence: osanchezguillamon@gmail.com; Tel.: +34-678-898-622

Received: 20 November 2017; Accepted: 17 January 2018; Published: 23 January 2018

Abstract: Derived digital elevation models (DEMs) are high-resolution acoustic technology that has proven to be a crucial morphometric data source for research into submarine environments. We present a morphometric analysis of forty deep seafloor edifices located to the west of Canary Islands, using a 150 m resolution bathymetric DEM. These seafloor structures are characterized as hydrothermal domes and volcanic edifices, based on a previous study, and they are also morphostructurally categorized into five types of edifice following an earlier classification. Edifice outline contours were manually delineated and the morphometric variables quantifying slope, size and shape of the edifices were then calculated using ArcGIS Analyst tools. In addition, we performed a principal component analysis (PCA) where ten morphometric variables explain 84% of the total variance in edifice morphology. Most variables show a large spread and some overlap, with clear separations between the types of mounds. Based on these analyses, a morphometric growth model is proposed for both the hydrothermal domes and volcanic edifices. The model takes into account both the size and shape complexity of these seafloor structures. Grow occurs via two distinct pathways: the volcanoes predominantly grow upwards, becoming large cones, while the domes preferentially increase in volume through enlargement of the basal area.

Keywords: seafloor geomorphometry; domes; volcanoes; digital elevation models (DEMs); Canary Basin; Atlantic Ocean

---

1. Introduction

The interdisciplinary science of geomorphometry, the quantitative representation of topography and terrain modeling [1], has entered a new era taking advantage of the constant advancements in elevation data acquisition and the spatial resolvability of digital elevation models (DEMs). DEM-based applications in underwater exploration were previously limited by the dimensionality and low resolution of the available bathymetric data [2]. However, the improvement in acoustic technology over the last two decades has resulted in increasing use of digital elevation models [3]. The advent of new echo-sounding techniques is improving traditional geomorphometric techniques, including quantitative measurements derived from marine DEMs, despite the fact that submarine environments are inherently more difficult to sample. Most DEM-based studies highlight particular submarine patterns and model different geomorphic feature values in the seascapes (i.e., they analyze morphometric attributes). These involve general geomorphometric techniques for characterizing ocean floor physiographic domains [4–6], mapping habitats [7,8], analyzing textures [9,10] and studying hydrodynamics [11,12].

For characterizing and classifying specific seascapes, DEM-based studies are still relatively nascent and less numerous in the marine environment than terrestrial landscapes. Other authors have focused on modeling submarine mass movements [13,14], canyons [15,16] and pockmarks [17–19]. In vulcanology, this type of study is used for determining the key processes involved in the construction and modification of seamounts [20–22], calculating lava volumes [23] and looking at the evolution of volcanic areas such as the mid ocean ridges [24]. DEMs are combined with similar acoustic datasets, such as backscatter and sonar imagery, in order to increase the potential for understanding volcanic and tectonic processes [25]. In this sense, significant volcano-related studies, in which geomorphology and the statistical analyses of volcanic features have become essential for morphometric studies, have been carried out over the past 20 years. Some of the most notable of these studies are those that geomorphometrically determined the evolution and development of cones in ridge areas such as the Mid-Atlantic Ridge and the Azores Plateau ([25–27], and references therein). Moreover, abundant links have been found between the geological and geomorphic characteristics of volcanic landforms (e.g., [28,29]) and biological habitats [30,31] with geomorphic characteristics such as size, shape, and degree of isolation being key.

Landform characterizations based on specific morphometric attributes are of fundamental importance in the theory of geomorphometry [32]. Standardized, comparable and systematic methods for extracting DEM-based morphometric information on seafloor elevations are still scarce. Even so, several approaches can be found in literature that have been successfully applied in sub-aerial environments for mapping and quantifying volcanic landforms [33,34]. In particular, the authors of [35,36] developed a quantitative well-formalized method of systematically studying the morphometry of volcanic edifices by delimiting the basal contour and extracting morphometric attributes (i.e., size, slope, and shape descriptors). This method has been widely applied to arc volcanoes [35] and cinder cone fields [37] in Central America; to glacio-volcanoes and shield volcanoes in Iceland [38]; as well as in a global dataset of composite volcanoes [39]. There are no comparable studies in the submarine environment, and novel applications of morphometric analyses to submarine structures are needed. Research in this direction will contribute to comprehensively classifying and morphometrically characterizing different positive seascapes namely seamounts, cold-water coral mounds, mud volcanoes, and other features of the seabed such as pockmarks, bedforms, and landslides.

This paper aims to further this objective by focusing on a set of seabed features located west of the Canary Islands (Eastern Atlantic) at a depth of 4800–5200 m, for which a new detailed morphometric method based on DEM analysis (Figure 1A) has been used. The study analyses and characterizes submerged volcanic structures and their driving processes and proposes evolutionary trends. These seafloor features are both circular and elongated in shape, with diameters ranging from 2 to 24 km, heights of up to 250 m, and flank slopes ranging from 2 to 24° [40]. In earlier studies, they were characterized by [41] as various types of structures including both hydrothermal domes and different types of volcanoes related to recent volcanic and intrusive activity. The authors of [42] also classified them into five morphostructural types of edifices (MT1 to MT5), intimately linked to specific origins (Figure 1B). These forty structures were selected with the aim of analyzing edifices with different sizes, shapes, and morphostructural settings, in order to explore the efficacy of the current and well-tested systematic method of [36], rigorously characterizing the morphometric footprint of these submarine edifices. We have applied statistical analysis methods to determine which morphometric variables correlate most strongly in the characterization of the complex morphology of these edifices.

Finally, we have used these quantitative characteristics to establish a morphometric growth model that includes the complexity of size and shape of the different seafloor edifices.

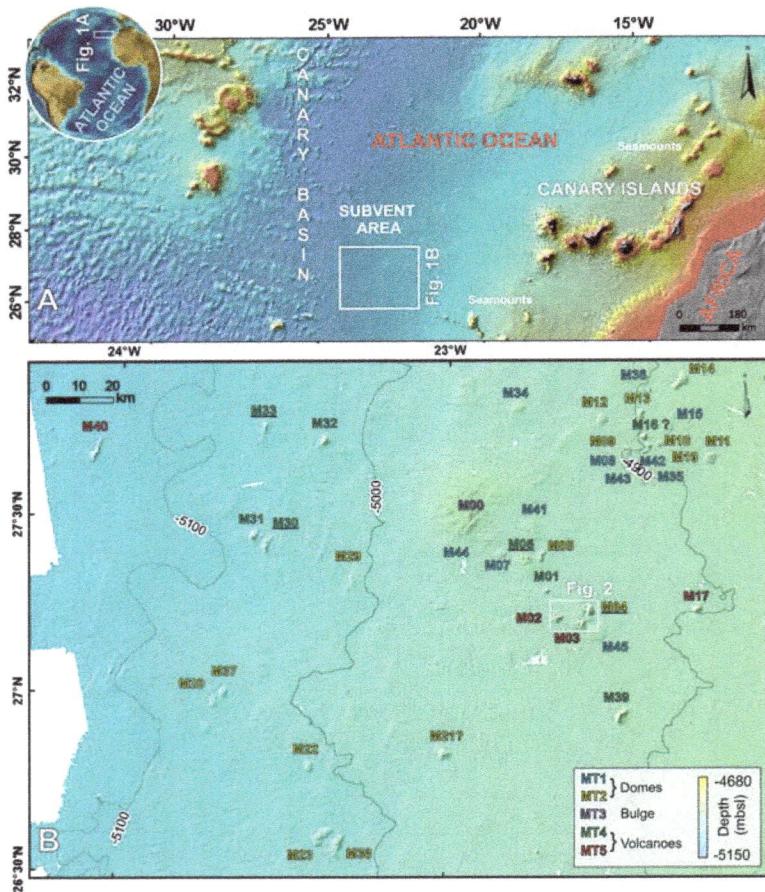

Figure 1. Location of the study area. (**A**) Overview map of the western Canary lower continental slope and location of the study area in the Canary Basin (Central Eastern Atlantic Ocean); (**B**) Subvent Area bathymetric base map where the forty mounds are morphostructurally classified into five types of edifices [42]. The highlighted mounds are categorized according to their origin following [41].

2. Study Area

The study area is situated in the central Canary Basin, approximately 500 km west of the volcanic archipelago of the Canary Islands (Figure 1). The Canary Basin is located in an intraplate setting over Jurassic to present oceanic crust [43]. This basin has been characterized as having a heterogeneous distribution of various volcanic elevations including seamounts, hills, and seafloor mounds [42,44]. Nevertheless, in the central area of this basin, known as the Subvent Area, these seafloor mounds are hydrothermal domes and scattered volcanoes related to Quaternary intrusive activity that gave rise to a huge magmatic sill complex together with volcanic activity [41]. Indeed, different morphostructural types (MT) of seafloor mounds have been differentiated according to the height, slope and basal area of these seafloor elevations [42] in agreement with various linked sill intrusions and volcanoes, as reported by [41].

The different genetic processes occurring in the Subvent Area have resulted in a striking seabed landscape including: hydrothermal domes formed above inclined but mainly saucer-shaped intrusions (MT1 and MT2); a huge bulge with a central depression associated with inclined sills at the top of a buried seamount (MT3); volcanic mounds developed at the top of basement highs and hydrothermal volcanic complexes associated with stratified sills (MT4 and MT5) [41,42] (Figure 1B). The classification according to size and slope variables has proved successful in discriminating hydrothermal domes from volcanic edifices, and just a few mismatches in the classification have been identified (i.e., M04, M06, M30, and M33) [42]. M06, M30 and M33 were morphostructurally classified as hydrothermal domes (MT2) despite the fact they have been confirmed as volcanic edifices, whereas M04 was regarded as volcanic mounds (MT4) even though it had been characterized as a peripheral dome of a volcanic/sill complex [41]. All the mounds have been labeled in accordance with the aforementioned classification, but the highlighted mounds are reclassified in regard to their confirmed origin (Figure 1B).

3. Method

Geomorphometric methods are commonly implemented in several steps, beginning with the acquisition, processing, and analyses of the bathymetric data, followed by Geographic Information System (GIS) applications (Figure 2). The methodological approach to computing features extracted from the DEM is based on multibeam data and its subsequent analysis using the ArcGIS © desktop v. 10.4.1 software packages. This contains various applications for the nearly automated extraction of variables from DEMs and further analyses based on these.

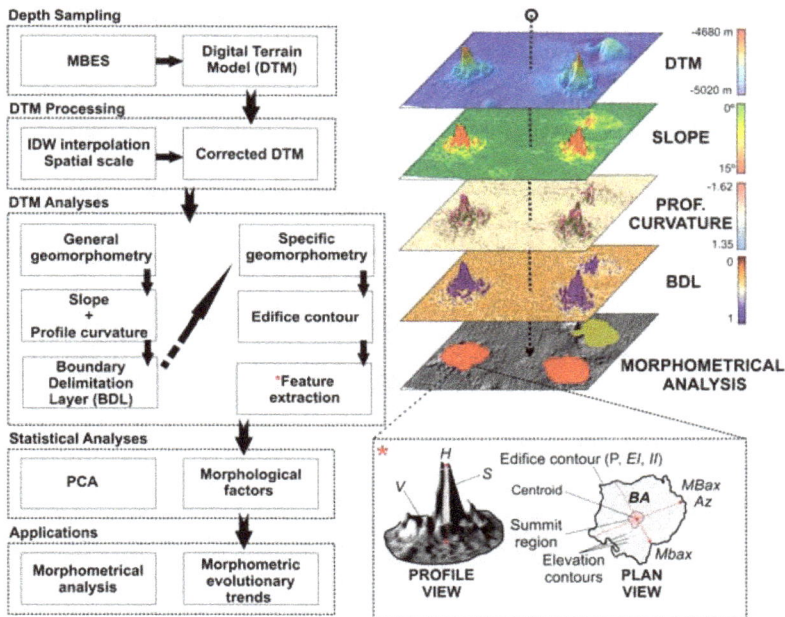

Figure 2. Methodological workflow for the acquisition, processing, analyses, and applications of DEM in this study. The main features extracted are listed as follows: H (m) height; S (degrees) slope; V (km^3) volume; BA (km^2) basal area; P (km) perimeter; EI—ellipticity index; II—irregularity index; MBax and mBax (km) major and minor basal axes and Az (degrees) azimuth.

3.1. Dataset Sampling and Processing

To record seafloor depth, four oceanographic cruises (GAROÉ-2010, GAIRE-2011, AMULEY-ZEEE-2012 and MAEC-SUBVENT-2013) were undertaken, from 2010 to 2013, by the Spanish R/V *Hespérides* and R/V *Sarmiento de Gamboa*. The datasets obtained cover the lower and middle continental slope offshore the Canary Islands up to 5400 m below sea level (mbsl) (Figure 1). Full coverage and high-resolution bathymetric datasets were acquired using Kongsberg-Simrad EM-120 and Atlas Hydrosweep DS-3 multibeam echosounder systems (MBES). The sonar frequency of the Simrad EM-120 deep system was 13 kHz, with an angular coverage sector of up to 150 degrees and 191 beams. The transducer opening was as narrow as 1 degree. The achievable swath width was normally up to 3.5 times the water depth. The angular coverage sector and beam pointing angles were set to vary automatically with depth according to achievable coverage. The beam spacing was normally equidistant with available equiangle. Atlas Hydrosweep DS system was operated at a frequency of 14 to 16 kHz with a maximum of 345 beams spread over 140 degrees. The bathymetric across-track coverage was 3.5 times the water depth. The acoustic footprints were arranged in either an "equal-angle" or "equal-distant" pattern. In the Subvent Area (5000 m water depth), the vertical resolution and was 10 m and the footprint range was between approximately 50 and 90 m. Obtaining the DEMs required the sampled depths to be processed using CARIS HIPS and SIPS © software (i.e., sound velocity correction, noise filtering, cleaning of the data, interpolation, and choice of optimal spatial scale) in order to generate a corrected surface model yielding a DEM with 150 m of spatial resolution (Figure 1B). This resolution was chosen following the full homogenization of the data by hydrographers from the Spanish Navy's Hydrographic Institute.

3.2. DEM Analyses: Delimiting the Contour and Morphometric Variables

From the corrected DEM, we undertook topographic modeling based on the aforementioned method [36], which blends both primary and secondary DEM-derived products; slope and profile curvature; into a single boundary delimitation layer (BDL), using ArcGIS Analysis tools (Figure 2). This final product was used for the next step where each edifice analyzed was spatially delimited based on concave breaks in slope around it. All outlined edifices had distinct positive topographies, with a relief of more than 10 m, basal diameters of greater than 1 km, and areas not smaller than 2 km^2 (i.e., 90 pixels) according to the available spatial resolution. Based on these criteria, we analyzed forty mounds consisting of topographically recognized shapes.

The edifice boundary was used to directly compute the following size variables: Perimeter (P), which is the outlining edifice contour; Basal and summit areas (Ba and Sa), and edifice Height and Volume (H, V) were measured as the absolute difference between the summit and basal relief points or surfaces of the selected edifice outline, respectively, using ArcGIS Analysis tools. Major and minor basal and summit axis (MBax, mBax, MSax and mSax), so called diameters or widths, were manually measured to calculate the aspect ratio of the edifices. Other morphological size ratios were calculated including Flatness (FL = MBax/MSax and Sa/Ba) [45,46], defined as the ratio of the summit diameter or area to the basal diameter or area; and Sigma Value (SV = 2 × H/(MBax−MSax)), normally inversely proportional to FL (1-FL) [46]. Aspect ratios such as Eccentricity (mBax/MBax) and the H/MBax index were also calculated [47]. Slope (S) was calculated as the first derivative of the DEM [48], also using also ArcGIS Analysis tools. This corresponds to the maximum elevation change over a given distance and is indicated in degrees. The plan shape of the edifices was characterized using the Ellipticity Index (EI = Π × (MBax)^2/Ba), which quantifies edifice elongation, and the Irregularity Index (II = (P/2 × Ba) × ($\sqrt{}$Ba/Π)-1), which quantifies edifice complexity (for details see [36]). Orientation variables were represented by the azimuth of the MBax (Az), which was calculated from the most northerly to the most southerly point, indicating the edifice elongation orientation.

The morphological variables used in this study should be highly correlated because they derive from a unique source (DEM). All these variables contain relevant information on the geometry of the edifices since they have previously proven to be good indicators for correlating seafloor features with constructive- and destructive-related processes. They are used in correlation and statistical analyses to examine whether these edifices display similar morphometries (i.e., maintaining a uniform H/MBax) as observed in other morphometric analyses [27,45,47].

3.3. Principal Components Analysis

Once all the aforementioned variables and ratios had been calculated for each edifice, we applied a principal components analysis (PCA [49]) method to the data using XLSTAT software v.7.5. This is a common statistical procedure used to convert a set of observations into uncorrelated groups of correlated variables (i.e., principal components (PCs) or factors (F)) (e.g., [16]). The use of PCA is a standard practice in current exploratory data analysis [50], used for extracting the dominant patterns from the datasets in terms of loading plots and reducing the non-relevant information. Each PC explains a percentage of the total variability of the original dataset. The first PC accounts for the highest variance, the second accounts for the next most important variance, and so on. A decision criterion based on the eigenvalues is often used to determine how many PCs are relevant in the analysis [51]. In this study, only factors with eigenvalues higher than 1 were considered, following the Kaiser-Harris criterion, because this metric represents the most useful factors. PCAs and their key applications in oceanography can be considered diagnostic tools for multivariate analysis since they enable the simplification of the complex variance patterns found in many physical and geological systems.

4. Results

The forty mounds analyzed comprise a wide spectrum of seafloor forms and sizes. The integrated combination of automatic morphological attributes mapping and feature-based quantitative representation was useful for extracting the relevant morphometric information. We assessed which were the most highly correlated and well-determined variables (Figure 3) in the morphometric relationships among the mounds and related variables. These relationships were plotted to evaluate whether there are any natural clusters in the data (Figures 4 and 5), as well as the size and shape growth patterns of the various types of mounds (Figures 6 and 7).

4.1. Main Morphometric Variables

The PCA (Figure 3) was conducted using ten of the seventeen variables listed in Section 3.2. The variables driving the PCA were height, slope, sigma value, flatness, basal area, volume, diameter, irregularity and ellipticity indexes, and azimuth. These variables display loading of more than ±0.5 and contribute to defining the main factors explaining 84% of the total variance (in geometric variability (size and shape) of the mounds), with eigenvalues larger than one. H, S, and SV are positively loaded in Factor 1, while FL is negatively loaded; together these explain 39% of the variance in these seafloor edifices. This factor has been considered an indicator of the overall vertical growth and inclination of the edifice and we have denominated it Profile Shape Factor (F1). BA, V, and MBax contribute to defining Factor 2, which we have called Size Factor (F2). This explains 27% of the variability in the seafloor edifices. Finally, Factor 3 explains 18% of the variance and is considered the Plan Shape Factor (F3) since it is positively determined by the shape descriptors EI and II, and negatively loaded by the azimuth of the edifices.

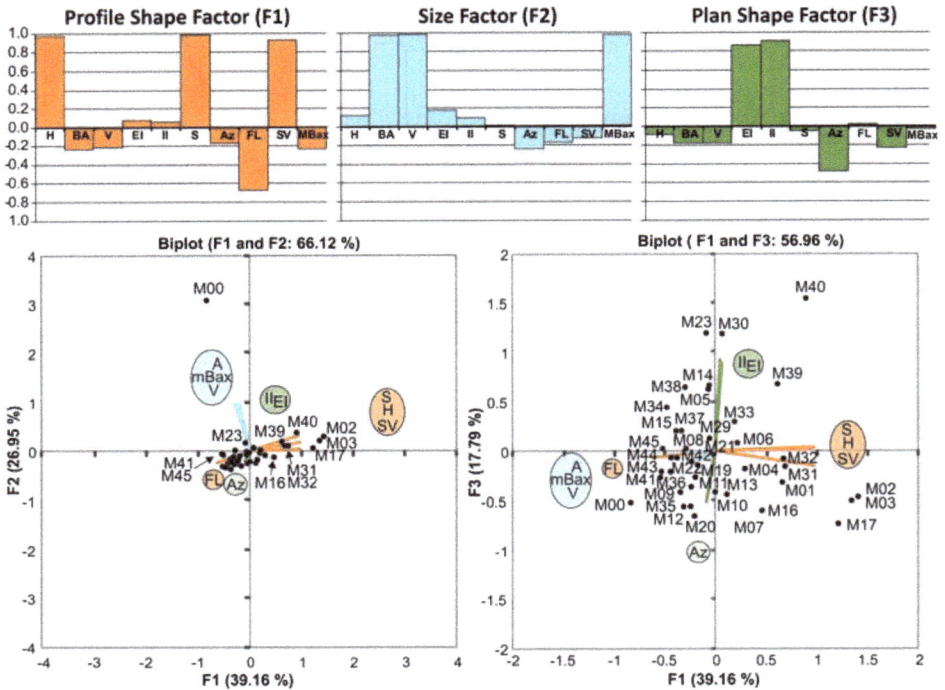

Figure 3. Factor loading plots of the three main principal components that account for 84% of the total variance of the edifices. Factor loads explain the correlations between the PCs and the original variables. Bar plots illustrate which variables are driving the PCs. We have named each factor according to the most significant variables of the analysis. The PCA biplots allow visualization of both the observations (edifices) and the variables among F1-F2 and F1-F3 relationships. Edifices are displayed as points while variables are displayed as linear axes. Standardized values are used to represent the factors.

The relationships among the edifices and the variables are plotted in the F1 vs. F2 and F3 biplots (Figure 3) where each dot represents one of the forty edifices and the ten variables are represented by linear vectors. Axis X (F1) relates to the height and slope measurements, with edifices on the left having both lesser heights and slopes (i.e., M41 and M45), while the edifices on the right have larger values (i.e., M02 and M03). Axis Y (F2) is a measure of the basal size, with most of the edifices on the bottom being small, while edifices on the top are large (i.e., M23) to extreme in size (i.e., M00). Axis Y (F3) is a measure of the shape, with edifices on the bottom being circular (i.e., M07 and M17), while edifices on the top show varied elongation and shape complexity (i.e., M30). Dots that are close together correspond to edifices that have similar values (i.e., M31 and M32, or M15 and M37) (Figure 3). Dots that are relatively far apart correspond to edifices that are different to the others, for example: M02, M03, and M17 have high Profile Shape Factor (F1) values, M00 has the highest Size factor (F2) value and M40 has the highest value for the Basal Shape Factor (F3).

4.2. Relationships between the Main Morphometric Variables

In order to investigate the relationships between the different types of mounds and related variables, several XY scatter plots and their linear correlation coefficients (R^2) are shown among the morphometric variables that have determined morphometric factors F1, F2 and F3. The highest correlations measure the strength of the linear relationship between two of the plotted variables and are determined by the size and slope variables. The data is discriminated by the types of edifices in different color-coded tiles (Figure 4A–E).

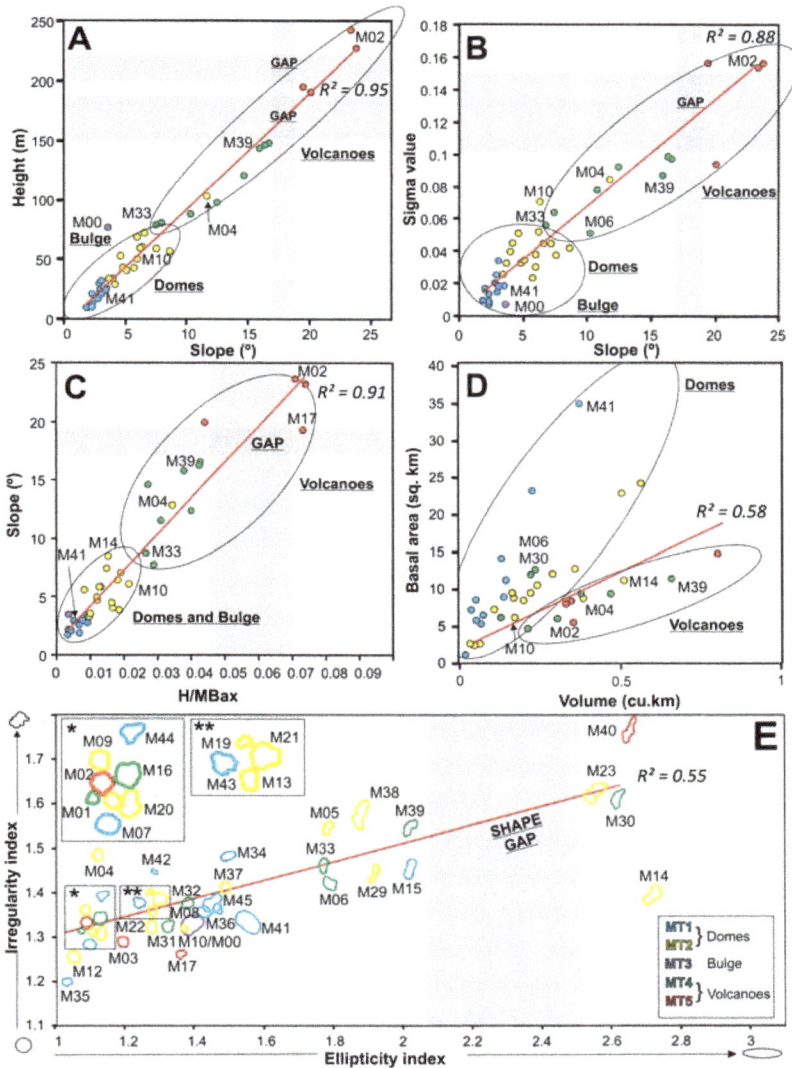

Figure 4. XY scatter plots of correlated variables for the forty mounds (n: 40). (**A,B**) Slope (X) vs. Height (Y) and Sigma Value, respectively; (**C**) Height/Basal length (X) vs. Slope (Y); (**D**) Volume (X) vs. Area (Y) where M00 is excluded being considered an outlier for Size Factor (F2); and (**E**) Ellipticity index (X) vs. Irregularity index (Y). The basal contour of each of the forty mounds is shown. All the mounds conserve their scale with the exception of M00. The red lines indicate the trend of each correlation and the R^2 coefficient is displayed next to them. The ellipses indicate the natural tendency of the edifices types to cluster, and gaps in the distribution are highlighted by shaded regions.

4.2.1. Profile Shape Factor (F1)

Slope correlates very positively with height, sigma value and height/basal length ratio, having R^2 coefficients of 0.95, 0.88 and 0.91, respectively. They have the strongest linear relationship and a large scatter between the mounds generating gaps in the data. Slopes range from 1.8 to 24°, with a mean of 7.7°, while height ranges from 10 to 245 m, with a mean value of 70.5 m (Figure 4A). The sigma value has an inverse relationship with flatness and ranges from 0.005 to 0.16, having a mean value of 0.05 (Figure 4B). In this way, high values of SV= 0.1 indicate conical or multipeaked shapes with high height and slope values. Basal lengths range from 1 to 6.5 km with the exception of M00, which is up to 24 km long, and the H/MBax ratio reaches 0.07, with a mean value of 0.021. A gap in the H/MBax ratio exists between 0.04 and 0.07 (Figure 4C). The data plotted on these graphs shows some overlaps and gaps in the distribution although an important agglomeration is observed at values of less than 80 m in height, 10° of slope, 0.06 SV, and 0.025 H/MBax ratio. This agglomeration of observations includes all the domes apart from M04, whereas the volcanoes are distributed into in several skewed groups at levels of more than from 75 m high, 12° slope, 0.06 SV, and 0.023 H/MBax ratio, with the exception of M30 and M33 (Figure 4A–C). This suggests that processes controlling the size and slope of the domes are not similar to those controlling the volcano building.

4.2.2. Size Factor (F2)

The correlation between basal area and volume is positively proportional and linear, with a high R^2 coefficient of 0.58 (Figure 4D). They have a highly linear relationship but there is a large degree of scatter between the mounds and diffuse limits. Basal area range from 2 to 35 km^2 and volumes from 0.005 to 0.7 km^3, with the exception of M00, which has maximum values of 440 km^2 and 11 km^3, respectively. The mean values are 20.66 km^2 and 0.5 km^3. M00 has atypical values in this distribution and is considered an outlier that is 3xIQR (inter-quartile range) above the third quartile and for this reason it has not been included. On this plot, domes and volcanoes are clearly separated into two distinct groups that follow different size increase trends, with the exception of a few edifices (M04, M06, M14 and M30) located on the contrary cluster. This suggests that the smaller volcanoes are morphometrically similar to the domes, and larger domes attain the same size values as volcanoes. Several mounds do not fit the best linear regressions (i.e., M41 in Figure 4D) but most plot fairly close to the rest of the mounds.

4.2.3. Basal Shape Factor (F3)

The ellipticity (EI) and irregularity (II) ratios range from 1 to 2.7 and 1.2 to 1.8, respectively (Figure 4E), with means of 1.5 and 1.4. They have a positive asymmetric correlation with $R^2 = 0.55$. The II distribution is more roughly symmetrical, whereas the EI distribution is largely skewed and unevenly spread with a gap between 2 and 2.6 (Figure 4E). There are no significant relationships between MTs among these descriptors although three slight groupings can be identified in the data: near circular edifices (low II and EI values), such as M03 and M12; more complex forms (from irregular edifices that are not very elongated (i.e., M05) to elongated edifices that are less irregular (i.e., M14); and elongated complex edifices (high EI and II values) such as M40.

4.2.4. Other Morphometric Relationships

The above-mentioned variables do not correlate significantly with others, although natural clusters classifying domes and volcanoes are observed when we plot them against one another (Figure 5). Flatness ranges from 0.05 to 0.47 with a mean value of 0.23. There is a weak correlation between height and flatness, with $R^2 = 0.3$ (Figure 5A). As height increases, flatness decreases; however, there are two further rough trends: at heights above approximately 90 m, flatness invariably ranges from 0.05 to 0.15 in the majority of the volcanoes, except M33. At heights below 90 m, flatness is highly variable and corresponds to the domes (Figure 5A). The greatest range in flatness is observed for mounds with diameters< 5 km and heights of lower than 77 m (Figure 5A,B). This indicates that mean flatness may decrease with diameter, and the most typical shape of larger edifices is pointed or with multipoint geometry (i.e., M02 and M40). Other edifices, whose diameters encompass and extend beyond the sizes of the widest mounds, tend to be flatter (i.e., M41).

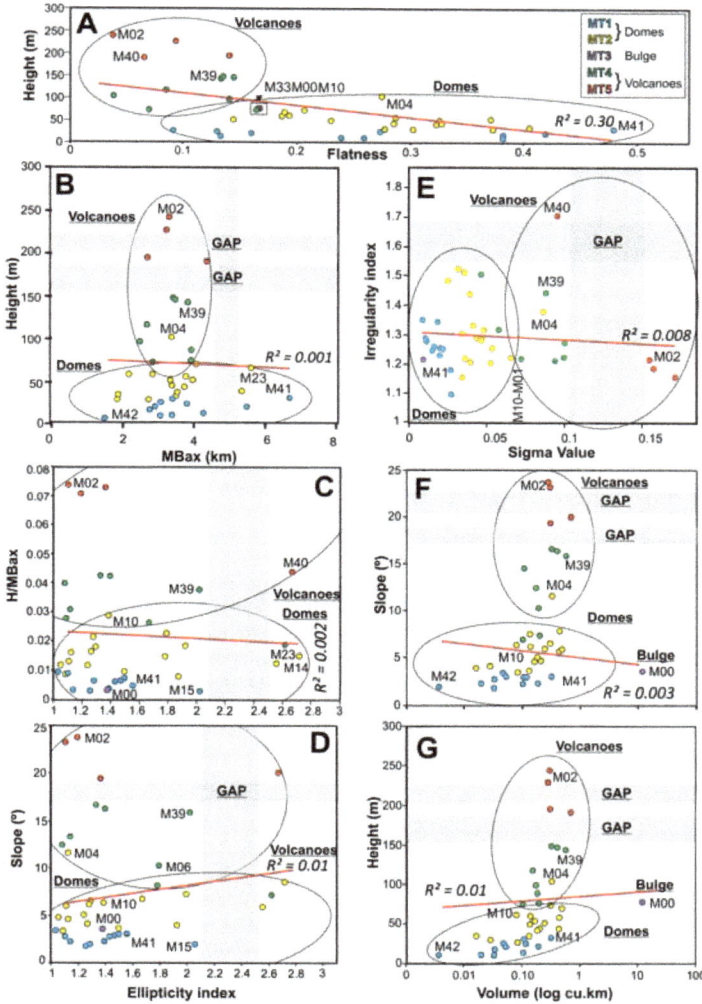

Figure 5. XY scatter plots of the main variables for the forty edifices (n: 40). (**A**) Flatness (X) vs. Height (Y); (**B**) Basal length (X) vs. Height (Y) excluding M00 that is considered an outlier; (**C** and **D**) Ellipticity index (X) vs. Height/Basal length (Y) and Slope (Y), respectively; (**E**) Sigma value (X) vs. Irregularity index (Y); and (**F** and **G**) Logarithmic (log 10) Volume (X) vs. Slope (Y) Height and (Y), respectively. The five morphometric edifice types are shown and some mismatched edifices are highlighted. The red lines indicate the trend line of each correlation and the R^2 coefficient is displayed next to them. The ellipses indicate the natural tendency of the edifice types to cluster, and gaps in the distribution are highlighted by shaded regions.

The height vs. basal length graph shows no correlation with R^2 near 0 showing gaps in both distributions even though there are not many differences in the basal length compared to the height increment (Figure 5B). The height increases roughly with decreasing basal width for large mounds (from 140 m to 240 m and 4.5 km to 3 km), corresponding to volcanoes, whereas domes show a wider range of variability (1 km to 6.5 km). The relationship between H and the basal length (MBax) (H/MBax ratio) suggests a distinct mechanism of growth and, in turn, is indicative of the slope angle and flatness of the mounds (Figures 4 and 5). This suggests that the maximum steepness is not constant, but rather decreases with increasing size. Mounds with basal lengths of less than 3.3 km have higher H/MBax ratios, up to 0.07, while mounds with basal lengths of more than 4 km have smaller H/MBax ratios, lower than 0.03 (Figures 4 and 5).

Similar to the height vs. flatness relationship, the H/MBax ratio shows a rough negative correlation with both shape indices, although not as strong. This suggests that lower H/MBax ratios are related to increasing complexity (i.e., M14 and M23 in Figure 5C). Furthermore, these indices (EI and II) show a lack of correlation with slope and size (i.e., SV) variables (R^2 = 0.01) (Figure 5D,E). Therefore, larger edifices tend to be more circular (i.e., M02 in Figure 5C), and steep edifices tend to have higher values for both indexes, even though circular shapes are also seen in intermediate to small edifices (i.e., M10) (Figure 5D). Despite the fact that the slope vs. EI plot shows no correlation and an asymmetrical distribution, domes and volcanic agglomerations are easily distinguished, as the largest and steepest mounds (i.e., M02 and M03) have an antithetic relationship with the elongation (EI) (Figures 4E and 5D). The sigma value is inverse to the flatness and ranges from 0.005 to 0.16, having a mean value of 0.05. Hence, high values of SV= 0.1 indicate conical or multipeaked geometries with high height and slope values. These observations correspond to the majority of the volcanoes with different basal shapes (circular to elongate) (Figure 5E). A medium to low range of SV indicates high degree of flatness with intermediate to low heights and slope angles, corresponding to the majority of the domes. Domes range from circular to irregular and have an II of up to 1.5 (Figure 5E).

Conversely, height and slope seem to correlate positively with volume and basal area (Figure 5F,G), although they do not correlate with R^2 near 0. The edifices seem to increase H/V and S/V ratios in two different manners: domes appear to follow the same morphological trend, increasing their volume yet not their height or slope (i.e., M42, M10, and M00); while the development of large volcanic edifices seems to be either through increased height and slope, making them the highest and steepest edifices (i.e., M02), or an enlarged volume and length, where they become the most elongated edifices (i.e., M40) (Figure 5F,G).

4.3. Geomorphometric Observations on the Morphological Types

A comparison of the variables indicates a clear separation between the five morphostructural types. However, there are certain overlaps in MTs and a diffuse spread, together with gaps in the distribution and mismatches in the different type of edifices. In other words, some edifices classified as domes have H/MBax and slope values that are much higher than expected (i.e., M04) and some volcanoes have lower H/MBax and slopes than expected (i.e., M06, M30 and M33) (Figures 4 and 5). Nevertheless, there is a striking difference between domes and volcanoes of increasing size (BA, V and MBax), decreasing slope and aspect, (S, H/MBax) and complexity (EI, II and summit shape (SV, F).

The five types of edifice morphometry are illustrated in Figure 6. MT1has an H/MBax ratio of lower than 0.01 and shows the lowest slope angle of ~2.6°, as well as a high degree of flatness, from 0.1 up to 0.5 (Figure 6A,F). The typical profile shape is a small, smooth, wide dome that gently breaks the surrounding seafloor slope, with a maximum width of 6.8 km (Figure 6F). MT2 has an intermediate H/MBax ratio of 0.016 (from 0.01 to 0.027), a variable slope angle of >2.75 up to 11°, and a mean flatness of 0.25 (varying significantly from 0.15 to 0.4) (Figure 6B,F). This MT has the greatest variety, and ranges from fairly small to intermediate edifices in terms of volume (0.005 to 0.3 km^3) and basal area (2 to 13 km^2), according to the great variety of edifice shapes (Figures 4–6). Its profile ranges from steeply domes to irregular double and crater-topped edifices. MT3 has the lowest H/MBax ratio, 0.003, together with a low slope value of ~3.5° and a low degree of flatness of 0.17 (Figure 6C,F). It is represented by a single elevation (M00), which is remarkably different from the others and should be regarded as an exception in the Subvent Area. Its profile shape is quite irregular, with a small summit area of 5 km length compared with its 25 km basal diameter (Figure 6C,F). MT4 has a variable intermediate to high H/MBax ratio (from 0.02 to 0.044) with a high slope angle of up to 15°, with flatness varying from low to intermediate, i.e., ~0.06 to 0.15 (Figure 6D,F). Some of these edifices have similar elongation azimuths, of between 22.5° and 40° and large profile shapes, up to 6 km in length (Figure 6D,F). Finally, MT5 has an H/MBax ratio of ~0.065, including a maximum value of 0.07, and shows the highest slope angle, up to 24°, and the lowest flatness value, ~0.04 (Figure 6E,F). Some of these structures have an approximately conical geometry with profile shapes of less than 5 km in length (Figure 6F).

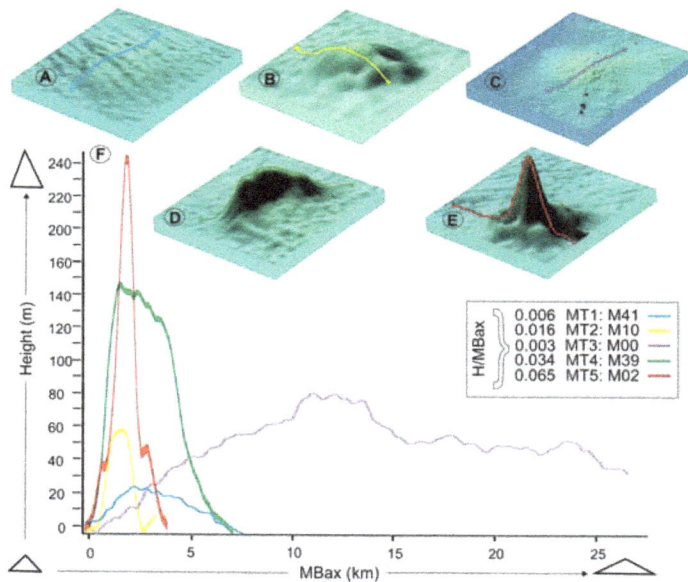

Figure 6. Representative geometry and bathymetric profiles for each type of edifice illustrating the variation in profile shapes. (**A**) M41 (MT1); (**B**) M10 (MT2); (**C**) M00 (MT3); (**D**) M39 (MT4); (**E**) M02 (MT5) and (**F**) Bathymetric profiles of five types of morphometries. The mean aspect H/MBax ratio of each MT indicates that domes are considerably lower than volcanoes.

5. Discussion

The results of this study illustrate how a morphometric sub-aerial method [35,36] can be successfully adapted to the submarine environment. This should contribute to further quantitative investigations of the various seafloor elevations. Morphometric relationships based on size, slope and shape variables distinguish volcanic edifices from hydrothermal domes for quantitative purposes. Nonetheless, quantifying the limits between edifice types is difficult due to some mismatching and overlap in form, as identified in several edifices. This morphometric analysis is also used for assessing if there are any evolutionary trends among the different types of edifices analyzed (Figure 7).

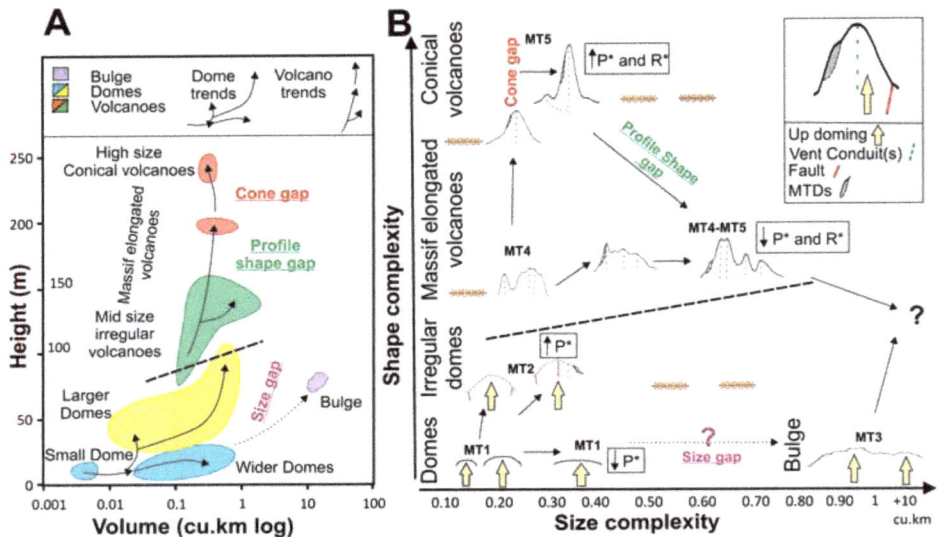

Figure 7. Morphometric model that incorporates different types of seafloor edifices. (**A**) Height versus volume diagram showing best-fit fields for the five types of seafloor edifice and possible evolutionary trends. The straight dashed line is the threshold used to separate domes from volcanoes; (**B**) Possible evolutionary growth paths of domes and volcanoes are highlighted. P* is pressure balance and R* is resistance balance, according to the model proposed by [35]. Arrows represent edifice growth trends and orange crosses represent size and shape morphometries that do not occur in the Subvent Area.

5.1. Relevance of Morphological Variables in the Geomorphometry of the Mounds

The three morphological factors are objectively related to the complex interaction between both primary constructive and secondary destructive, geological and geomorphic processes that could have occurred in these edifices [36,39]. Each morphological variable is relevant in regard to the seafloor edifice evolution, and analyzing these through PCA overcomes the problem of the subjectively selecting the variables [16,52]. This analysis reveals that there is a kinship between the geometry of these edifices and the morphological variables used to describe them.

Height and slope provide insight into how all these edifices grow vertically and could be related to gravitational processes, since gravitational force controls the instability of bottom surfaces [53]. In the case of volcanoes, height accounts for how explosive the eruptive event was [54,55] as well as the type of eruption (single point versus fissure-type of opening) [56]. Both height and slope, and the average values, have been strongly related to the evolution of edifices towards steady state or equilibrium concave-up profiles controlled by gravity-driven mass transfer [39]. The height/basal width ratio has also been widely used in previous studies of sub-glacial and post-glacial aerial and sub-aerial volcanic environments [56,57] as well as arc volcanoes [35], and has been considered a useful profile estimator due to the fact it correlates positively with the slope, which tends to decrease with size [39]. This relationship also is evidenced in this deep submarine environment (Figure 4C). Furthermore, this measurement contributes to better understanding the geometry and mechanics in the up-doming processes related to intrusions [40], as it is considered in theoretical and laboratory models [58].

According to these assumptions, the edifices described have experienced different growth patterns that reflect the various mechanisms causing their increased vertical growth and steepness, as domes range from heights of 10 to 90 m and volcanoes can be up to 250 m high (i.e., M42 and M02), with mean slope values varying from <11° to >11°, up to 25°. The volcanoes seem also seem to have experienced different extrusion rates, and they display both spire and multiple-peaked summits (Figure 6).

Basal area and volume provide an indicative measure of the geometry (size and shape) of the plumbing system and magmatism (eruption size) [22]. MT1, MT2 and MT3 domes are related to sill intrusion geometry, whereas MT4 and MT5 are regarded as volcanic systems. Major size differences suggest that more than one variable (e.g., magma emplacement, magma volume at depth, effusion rate, and pre-existing topography) could distinctly govern the size and shape formation of these edifices, from gentle domes to volcanic cones (Figure 6). In volcanoes, gravitational processes could mask the real eruptive volume of the edifice affected by flank instabilities (i.e., M06 and M39). This could imply that the original volumes of MT4 and MT5 edifices, preferentially affected by mass transports deposits [40,42], could be diminished.

Finally, the shape ratios of these edifices are quite diverse, as evidenced by the spread in the plan shape plot (Figure 4E). This suggests that the differing morphological evolution could be related to instability (erosion) processes. Variations in the irregularity are influenced by the local tectonic setting, as well as the thickness and cohesion of the sediment cover [36,58]. These indices do not correlate with size and aspect variables (Figure 5), although the predominant shape of both domes and volcanoes is circular, whereas elongated edifices tend to be volcanoes. Volcanic edifices with a low degree of elongation are interpreted to be central-vent controlled edifices (i.e., M02) while high EI values are related to fissure-controlled edifices (i.e., M40), as suggested in previous morphometric studies of volcanoes [39]. According to [59], discontinuous eruptive events may be related to multiple point sources and be suitable models for larger, elongated edifices, as observed in MT4 (i.e., M40).

5.2. Morphometry of the Domes

As expected, the clearest difference between edifice types is that domes are generally lower and smoother than volcanic edifices, with lower H/MBax ratios and slopes, high degree of flatness (Figures 5 and 6). The H/MBax distributions in domes are positively more uniform (around 40 m high in height and 3.5 km in length) than in volcanic edifices. They increase in gradient toward the base (MBax up to 90 times their heights, therefore having a ratio of 1:90). This implies that domes primarily grow by increasing their diameters and not significantly altering their height. In turn, domes exhibit no trend in flatness with respect to doming size, suggesting that doming processes and the basal diameters of the resulting seafloor domes do not have a linear relationship. Domes span a wide range of sizes and shapes, and this heterogeneity complicates size parameterization in other small seafloor morphologies such as pockmarks [17,60].

Small domes (i.e., M35 in MT1) that do not exceed 30 m in height, are quite circular (mean II ~1.33) with an elongation of up to 1.5 (Figure 4E). However, intermediate and larger sized domes (up to 90 m high) (i.e., M37 and M21 in MT2) domes are steeper and have a lower degree of flatness than the former, yet are quite irregularly shaped (mean value of II and EI greater than 1.5). Therefore, a 1.5 EI value seems to be a threshold in differentiating dome type, while II values do not appear to be diagnostic variables for domes (Figures 4 and 5). Domes such as M14 and M15 show the greatest elongations (up to 2.6) related to the occurrence of two merged domes and mass transport deposits extending westwards [42]. This EI range (1.5 to 2.6) could be considered a threshold value for detecting merging in domes.

The primary circular structure of larger domes could have been modified by the formation of major collapse/deformation features, erosion, and mass wasting processes [46]. It has been observed that domes need to reach a height up to 60 m high and mean slope of at least 6.5° before being possibly affected by sector collapses [42] (i.e., M04 and M13). In this way, M10 is regarded as a transitional edifice between the two types of domed edifices (MT1 and MT2) with threshold morphometric characteristics of 60 m height, 6.5° of slope, and 0.08 km³ of volume, and around 1.35 for plan shape (Figure 6). This morphometric factor may be an indicator representing the morphologic trigger of possible lateral instabilities in domes in the Subvent Area. These results demonstrate that most of the domes share a common profile and size, regardless of their more or less complex basal shape, and indicate that their formation has common controls. Nevertheless, two types of domes are present, with morphometric distinctions seeming to occur from heights of 60 m.

The flat or unrecognizable summits of domes must be related to constructional processes (i.e., folding and faulting). Hydrothermal domes are often small structures but their heterogeneity depends on the distinct geometry and size of the intrusions at depth (i.e., inclined or saucer-shaped sills) [41,58] that cause the different uplifting, forced folding, and faulting of the overburden. Intermediate to larger domes manifest boundary faults vertically linked to the tips of the underlying sills that induce hot fluid flow transport to surficial units, as in the case of M12, M13 and M14 (MT2) [41]. Differences in doming linked to faulting have played a key role in dome formation: faulting during magmatic intrusions is likely to be an important process in the northeastern part of the Subvent Area (Figure 1B), where larger domes are ubiquitously distributed, whereas in the central and southern parts peripheral domes are scattered and show lower and more irregular shapes. One of these peripheral domes is M04, which does not match in the existing morphometric dome variability, reaching up to 100 m in height and having a slope of 11.6° (Figures 4 and 5). This exception reflects drawbacks in dome morphometrics, likely related to incompleteness of the delimitation of summit and basal contours due to the presence of two larger collapse depressions at the base that could have altered its real shape.

The above observations agree well with the models proposed by [58] in laboratory experiments. These authors associated gradual slope (up to 4.5°) and H/MBax ratio (<0.03) to variations in seafloor (surface) deformation linked to both types of sheet intrusions and cohesion of the overburden. In this sense, it can be assumed that the aspect H/MBax ratio is a good estimator of the shapes of the intrusions, generating similar morphometric trends in the two types of domes in the Subvent Area (MT1 and MT2) related to saucer-shaped sill intrusions [41].

5.3. Morphometry of the Volcanoes

In contrast to domes, volcanic edifices have varied summit morphologies and dimensions, including single and multi-peaked as well as spire mounds and summit craters [40,42]. Summits may reflect the width of the conduit(s) feeding the volcano [61] yielding either a near-flat surface if lava has cooled at the top (i.e., M31 and M32), a spire cone (i.e., M02 and M03), multi-peaked edifices if the edifice continues growing to a critical height when it starts building outwards (i.e., M06, M30, M39 and M40), or a summit crater if there is drain-back of the lava occurs (i.e., M16). The profile of these volcanoes also varies widely, from smooth shapes (i.e., M16) to more complex geometries (i.e., M39) (Figure 6), mostly with sharp slopes.

The height of the volcanic mounds increases with decreasing or holding diameters, presenting shapes similar to seamounts. Their height and MBax profile distributions are lower than in domes and quite variable, with mean values of around 140 m height and 3.5 km length (MBax about 25 times the height, thus 1:25), which normally increase in gradient toward the summit (Figures 4–6). Previous studies have determined similar aspect ratios for intraplate seamounts, such as 1:6 [46]; higher, 1:10 to 1:15 [62], for Indian seamounts; and 1:8 being observed in Pacific Seamounts [63]. Flatness does not have a positive relationship with respect to eruption size (volume), suggesting that the height of the edifices follows a variable and unremarkable relationship with formation type (central or fissure controlled) [46], basal shape, and eruptive size of the volcanic edifices [22] (Figure 7). The volumes of the seafloor volcanic structures suggest a predominance of central vent formation (3.1 km^3) compared to the fissure ridges (2.8 km^3). However, this takes into account neither secondary mass transport deposits (MTD) dropping down the slopes nor the material intruded at depth.

Nonetheless, two different trends can be observed in this relationship: firstly, most of the MT5 edifices increase positively towards the summit (from 190 to 250 m high, holding diameters of around 3.3 km wide) (i.e., 1:15) while MT4 edifices, although being high (up to 200 m high), are also more elongated (1:32), usually following a NE trend, and they consequently have an increase volume

(Figures 5–7). This fact implies that volcanoes grow primarily by increasing their heights, probably connected to a central-controlled source, and later they also start to grow also laterally, in relation to multiple-point sources. This suggests a different style of volcanic growth [64]. In this sense, volcanoes grow vertically up to a critical height, becoming unstable with respect to their base, and secondary destructive processes may affect the edifice geometry. Instability processes play an important role in shaping the seafloor, and calculating the dimensions of submarine volcanoes is essential for understanding their consequent morphological changes [44].

At lower to intermediate sizes (from 75 to 150 m high, 11° of mean slope, and volumes of greater than 0.20 km^3 up to 0.60 km^3), volcanoes tend to be either complex-massif elongated edifices or irregularly circular volcanoes (i.e., M30 and M31 in MT4) with II values ~1.4 and EI up to 2. The smallest volcano is at least 75 m high, with this height value being regarded as a lower threshold in volcano growth in the Subvent Area. The morphometry of some of the small volcanoes (M06, M30 and M33) is similar to that of domes in terms of height, slope, and volume [42], but clearly differs in complexity, with 1.8 being the threshold value in volcanoes (Figure 4E). In this sense, basal shape descriptors are regarded as key contributors to distinguish edifice type. Mid-sized volcanoes that continue growing up to 200 m in height, maintaining their plan shape as elongated ridge volcanoes (II and EI further than 1.8 and 2.6, respectively), evolve towards more complex shapes (i.e., M39 and M40 in MT4 and MT5). This suggests a migration in the location of the volcanic vent and a sort of linear fissure, from single to multiple conduits [38,53], as observed in the neighboring seamounts of Drago and The Paps, in the Canary Islands Volcanic Province [44]. Multipeak elongated edifices, such as M06, M30, M39 and M40, seem to grow from independent branches of the main eruptive conduit, channelizing multiple but small paths towards the seafloor, so their summits do not reach to similar heights.

At larger sizes (up to 250 m in height, 24° of slope, and 0.3 km^3), volcanoes tend to have the smallest flatness values and therefore more pointed top(s) (highest H/MBax ratios) (i.e., M02 in MT5), suggesting this shape becomes preferential as edification progresses, regardless of origin [63]. In addition, it seems that the tallest and most conical mounds (MT5) do not require a larger base for support, as they are formed by intermediate and lower volcanoes, with irregularly circular and elongated shapes (MT4). Hence, the height and summit radius ratio shows little correlation, also finds in H/MBax ratio, the biggest one (up to 0.07, thus around 1:15 for M02 and M03).

The complex shapes of these volcanic edifices could also be the result of later modifications related to flank instabilities together with MTD processes (i.e., M30 and M40) [42] that typically affect seamounts morphology (e.g., [44,65,66]) and which are widely recorded in complex-shaped edifices in the Atlantic Ocean [24]. Favored by the steep slopes (Figure 4C), erosion participates in the dismantling of these elongated volcanoes (Figures 6 and 7). An EI threshold of greater than 1.8 is used to distinguish fissure-controlled volcanic edifices from other types of eruptive edifices (Figure 4E). This limit is very similar to that proposed by [38] to differentiate tindars from the other three types of glacio-volcanic edifices.

The volcanoes of the Subvent Area are related to both buried oceanic basement highs (i.e., M32 and M33) and a deep Quaternary volcanic hydrothermal system related to an intraplate hotspot (M01, M02 and M03) [41]. The increase in size and shape of these volcanoes is favored by repeated magmatic extrusions along the summit that would work as a siphon conduit, as observed in other seamounts affected by hydrothermal circulation through ridge flanks [67]. In this way, M02 and M03 may represent faster or higher extrusion events than MT4 volcanoes, meaning a significant variation in height, summit morphology (FL and SV values), and others morphological features, such as hydrothermal rings [41,42], where height and slope gaps point to a distinct mode of origin (Figures 6 and 7).

5.4. A Morphometric Evolution Model for the Seafloor Edifices

The wide spectrum of profile and plan shapes and sizes seems to represent specific and recognizable different growth trends. Previous studies have suggested a striking, positive relationship in their morphological, generalized evolutionary model of arc volcanoes shapes [35]. These authors related the Pressure balance (P* = PM/PL; the balance between magma (or forced folds) pressure (PM) or (PF), in volcanoes or domes, respectively and lithostatic pressure (PL)), and the Resistance balance (R* = RE/RC; established among conduit resistance (RC) and edifice resistance (RE)), with the size and shape complexity of the edifices [35]. Pressure factors are commonly used to explain maximum edifice heights while resistance factors depend on the predominant edifice material and the degree of faulting, which are related to structural conditions [35], especially in volcanoes. These relationships have been applied in other vulcanology studies (e.g., [54,68]), and are here used to assess the different morphometric trends in volcanoes and hydrothermal domes that depend on the predominant genetic processes and which may be controlled by the pressure and resistance balances (P* and R*) (Figure 7).

From the smallest morphometries represented by MT1 domes (i.e., <0.004 km^3 and 10 m height, thus M42) the most recognizable trends are: (1) the smallest irregularly circular domes (MT1) maintain their height but not their volume, suggesting low PF that is possibly enabled by the high RE attributed to wider domes (i.e., M41 in MT1); and (2) the near circular domes increase in height and volume becoming larger domes (MT2), suggesting high PF maintaining a high pressure balance and continued upward grow (i.e., M22) (Figure 7A). It also suggests that a low RE could be attributed to the bending of the sedimentary units [42]. The threshold limits for this change are 60 m in height and a 6.5° of slope (i.e., M10). This evolution both in height and volume is continuous until a height of 100 m high and slopes of 10° of slope is reached (i.e., M04 in MT2) (Figure 7A). This coincides with an interval of abundant mounds where both the domes and volcanoes overlap and there are still morphometric mismatches between edifices (M04, M06, M30 and M33), which can be differentiated by their shape complexities (M04 has lower II-EI ratios). Pressure balance (P*) and conduit resistance (RC) in these domes seem to be high enough, possibly related to a combination of PM and PF in sub-shallow depths [35]. At this morphometric point (domes of more than 60 m in height, 10° of slope, and II and EI higher than 1.5) (i.e., M04 and M13), the domes are characterized by irregular basal perimeters related to the shape of the intrusive systems [41] (doming and faulting) and the MTD processes [42] (Figure 7B).

The evolutionary trend of domes is therefore characterized by significant volume increases with minor height increments, enlargement of the basal and summit area (Sa/Ba increases) and low complexity growth (Figure 7B). There is the suggestion of possible evolution from dome to a bulge (MT3) (Figure 7A). It would maintain a height of 77 m. This type of seafloor edifice is related to intrusive systems and major tectonic structures, such as basement highs [41], is here considered an outlier with a unique morphometry [40].

There is a morphometric evolution from mid-sized, irregularly sub-conical volcanoes towards both elongated and conical volcanic edifices, through either preferentially increased volume (MT4) or height (MT5), respectively. This may reflect a critical height range, where two distinct evolutionary paths are possible; irregular sub-cones either continue growing upwards and become large cones (MT5), or grow sideways and become large massif edifices (MT4), resulting in a scarcity of cones in this height range (150–240 m high) (Profile Shape gap) (Figure 7A). The change in eruptive characteristics and features, and consequently volcanic edifice size (Figure 7B), supports an evolution towards these complex shapes with large, high summit areas. MT4 volcanoes may have more than one main vent, thus P* and R* are lower, and evolve from initially smaller, sub-conical volcanoes by vent-up migration. These volcanoes usually have a smooth conical profile in the mBax direction but another that is multi-peaked in the MBax direction (Figures 6 and 7B), as also observed in the morphometric evolutionary trends of American arc volcanoes [35].

Elongated massif volcanoes are characterized by volume increases with minor height increases (H/MBax is similar or slightly decreased compared to circular volcanoes, from 1:32 to 1:50); there is enlargement of the basal and summit areas (Sa/Ba increases), and major growth complexity (II and EI increase). Once these mid-sized massif volcanic edifices have formed, they can continue growing, becoming the largest and most elongated volcanic edifices with increasing complexity (e.g., M39 and M40). Even so, the largest elongated massif volcanoes may also have evolved from larger conical MT5 volcanoes, producing a massif trend reaching the threshold of 200 m in height and a volume of 0.7 km^3 (Figure 7B). In this sense, the high PM and subsequent P* in these edifices seem to be related to shallow magma reservoirs where low RE or high RC promote multiple new volcanic vent sites as suggested by [35].

Occasionally, sub-conical shapes evolve greater heights, and another height gap is observed where the volume is maintained (up to a height of 243 m high and a volume of 0.3 km^3) (i.e., M02 and M03 in MT5, Figures 6 and 7). This gap in the morphometric trend of the volcanic cones could reflect a critical height from which a second process may take over, allowing the top of the cones to continue growing; it is possible that this process is linked to hydrothermal circulation after the eruption. This evolutionary trend of cones is characterized by height increases with a conserved volume (H/MBax is greatly increased from 1:32 to 1:15); there is decreased conicality of the summit area (Sa/Ba), and reduced complexity (the lowest II and EI), suggesting a predominantly centrally controlled activity. Once these largest edifices are formed, they can continue growing toward the summit, either increasing the H/MBax ratio and reaching typical seamount ratios (around 1:6) [46,62], or the very large elongated volcanic edifices could develop increasing complexity, becoming even larger than M39 and M40 (Figure 7B). Eventually, the P* in these edifices is expected to be the highest, with open magmatic fluid-filled conduits, meaning they are dominated by hydrothermal migration. The cone height gap may be the point where P* and R* reduce to a critical threshold [35].

Finally, it is important to mention the influence of flank instabilities on these evolutionary trends. Elongated volcanoes are most prone to flank collapses and these rapidly reduce the height and regularity of the structures (e.g., M39 and M40). On the other hand, sub-conical and higher cones (e.g., M02) tend to maintain their conical shape after suffering these erosive processes, although they may become shorter and wider [69].

6. Conclusions and Outlook

The application of a GIS-based method for delimiting sub-aerial edifice boundaries and automatically extracting the morphometry has proven to be effective regardless of edifice type or origin. Correlation and principal component analyses have enabled the identification of ten main variables (accounting for 84% of the total variance) that define edifice profile shape, basal size and basal shape, including flatness, elongation and irregularity. Several threshold limits aid in the identification of domes and volcanoes in the complex Subvent Area. Evolutionary trends between contrasting morphologies are proposed in a new scheme using the concepts of size and shape complexity. We have found that these evolutionary trends can be characterized, for volume and height increases, into two distinct pathways; volcanoes mainly grow upward and become large cones, while domes preferentially increase in volume, enlarging their basal area.

The evolutionary model presented may have implications for other works that rely on the understanding volcanoes and other seafloor edifices through the analysis of bathymetric datasets acquired at this spatial resolutions or great depths. Our morphometric model links morphological trends to tectonic and magmatic (intrusive and extrusive) phenomena, thus providing a framework for future marine morphometric studies integrating geophysical and geochemical data.

Acknowledgments: This work is a contribution to SUBVENT (CGL2012-39524-C02-01), EXPLOSEA (CTM2016-75947-R) and EXARCAN (CTM2010-09496-E) projects (Spanish MINECO) and the Marine and Coastal Geophysical and Geology PAIDI Group (RNM328). Olga Sánchez Guillamón is benefited by a MINECO doctoral grant (BES-867 2013-062657). We thank the crews as well as the technical staff of the UTM and the scientific party of SUBVENT and EXARCAN projects. Thanks are extended to the two anonymous reviewers and academic editors who have provide very helpful and constructive comments that greatly helped to improve the manuscript.

Author Contributions: Olga Sánchez Guillamón and Luis Miguel Fernández Salas conceived the idea of the study and wrote the paper. Juan Tomás Vázquez was the main researcher of the SUBVENT project and Luis Somoza was the main researcher of the EXARCAN project and the coordinator of the work for the Submission of Data and Information on the Limits of the Continental Shelf of Spain to the West of the Canary Islands. Desiree Palomino and Ricardo León were responsible for the bathymetric data processing. Olga Sánchez Guillamón, Juan Tomás Vazquez and Luis Miguel Fernández-Salas interpreted the bathymetric data to carry out the morphometric analyses. Nieves Lopez-Gonzalez performed and interpreted the statistical analyses. Ricardo León, Teresa Medialdea and Luis Somoza had substantively revised the final manuscript. All authors provided guidance for the analysis and assist with the realization of the figures. They all also have collaborated in the data acquisition during the oceanographic cruises.

Conflicts of Interest: All authors have approved the manuscript and they agree with submission to the Special issue in "Marine Geomorphometry" in Geosciences journal. There are no conflicts of interest to declare among authors. The founding sponsors had no role in the design of the study; in the collection, analyses, or interpretation of data; in the writing of the manuscript, and in the decision to publish the results.

References

1. Pike, R.J. Geomorphometry—Diversity in quantitative surface analysis. *Prog. Phys. Geogr.* **2000**, *24*, 1–20.
2. Krause, D.C.; Menard, H.W. Depth distribution and bathymetric classification of some seafloor profiles. *Mar. Geol.* **1965**, *3*, 169–193. [CrossRef]
3. Lecours, V.; Dolan, M.F.J.; Micallef, A.; Lucieer, V.L. A review of marine geomorphometry, the quantitative study of the seafloor. *Hydrol. Earth Syst. Sci.* **2016**, *20*, 3207–3244. [CrossRef]
4. Fox, C.G.; Hayes, D.E. Quantitative methods for analyzing the roughness of the seafloor. *Rev. Geophys.* **1985**, *23*, 1–48. [CrossRef]
5. Gorini, M.A.V. Physiographic classification of the ocean floor: A multiscale geomorphometric approach. In Proceedings of the geomorphometry, Zurich, Switzerland, 31 August–2 September 2009.
6. Harris, P.T.; Macmillan-Lawler, M.; Rupp, J.; Baker, E.K. Geomorphology of the oceans. *Mar. Geol.* **2014**, *352*, 4–24. [CrossRef]
7. Wilson, M.F.J.; O'Connell, B.; Brown, C.; Guinan, J.C.; Grehan, A.J. Multiscale Terrain Analysis of Multibeam Bathymetry Data for Habitat Mapping on the Continental Slope. *Mar. Geodesy* **2007**, *30*, 3–35. [CrossRef]
8. Micallef, A.; Berndt, C.; Masson, D.G.; Stow, D.A.V. A technique for the morphological characterization of submarine landscapes as exemplified by debris flows of the Storegga Slide. *J. Geophys. Res.* **2007**, *112*, F02001. [CrossRef]
9. Orpin, A.R.; Kostylev, V.E. Towards a statistically valid method of textural sea floor characterization of benthic habitat. *Mar. Geol.* **2006**, *225*, 209–222. [CrossRef]
10. Lucieer, V.; Lucieer, A. Fuzzy clustering for seafloor classification. *Mar. Geol.* **2009**, *264*, 230–241. [CrossRef]
11. Sandwell, D.T.; Gille, S.T.; Smith, W.H.F. *Bathymetry from Space: Oceanography, Geophysics, and Climate*; Geoscience Professional Service: Bethesda, MD, USA, 2002.
12. Gille, S.T.; Metzger, E.J.; Tokmakian, R. Seafloor topography and ocean circulation. *Oceanography* **2004**, *17*, 47–54. [CrossRef]
13. Micallef, A.; Le Bas, T.P.; Huvenne, V.A.I.; Blondel, P.; Hühnerbach, V.; Deidun, A. A multi-method approach for benthic habitat mapping of shallow coastal areas with high resolution multibeam data. *Cont. Shelf Res.* **2012**, *39*, 14–26. [CrossRef]
14. Rovere, M.; Gamberi, F.; Mercorella, A.; Leidi, E. Geomorphometry of a submarine mass-transport complex and relationships with active faults in a rapidly uplifting margin (Gioia Basin, NE Sicily margin). *Mar. Geol.* **2014**, *356*, 31–43. [CrossRef]
15. Porter-Smith, R.; Lyne, V.D.; Kloser, R.J.; Lucieer, V.L. Catchment-based classification of Australia's continental slope canyons. *Mar. Geol.* **2012**, *303*, 183–192. [CrossRef]
16. Ismail, K.; Huvenne, V.A.I.; Masson, D.G. Objective automated classification technique for marine landscape mapping in submarine canyons. *Mar. Geol.* **2015**, *362*, 17–32. [CrossRef]
17. Andrews, B.D.; Brothers, L.L.; Barnhardt, W.A. Automated feature extraction and spatial organization of seafloor pockmark, Belfast Bay, ME, USA. *Geomorphology* **2010**, *124*, 55–64. [CrossRef]
18. Harrison, R.; Bellec, V.K.; Mann, D.; Wang, W. A new approach to the automated mapping of pockmarks in multi-beam bathymetry. *IEEE Image Proc.* **2011**, *18*, 2777–2780.

19. León, R.; Somoza, L.; Medialdea, T.; González, F.J.; Giménez-Moreno, C.J.; Pérez-López, R. Pockmarks on either side of the Strait of Gibraltar: Formation from overpressured shallow contourite gas reservoirs and internal wave action during the last glacial sea—Level lowstand? *Geo-Mar. Lett.* **2014**, *34*, 131–151. [CrossRef]
20. Mitchell, N.C. Susceptibility of mid-ocean ridge volcanic islands and seamounts to large-scale landsliding. *J. Geophys. Res.* **2003**, *108*, 2397–2419. [CrossRef]
21. Passaro, S.; Milano, G.; D'Istanto, C.; Ruggieri, S.; Tonielli, R.; Bruno, P.P.; Sprovieri, M.; Marsella, E. DTM-based morphometry of the Palinuro seamount (Eastern Tyrrhenian Sea): Geomorphological and volcanological implications. *Geomorphology* **2010**, *115*, 129–140. [CrossRef]
22. Wormald, S.C.; Wright, I.C.; Bull, J.M.; Lamarche, G.; Sanderson, D.J. Morphometric analysis of the submarine arc volcano Monowai (Tofua-Kermadec Arc) to decipher tectono-magmatic interactions. *J. Volcanol. Geotherm. Res.* **2012**, *239*, 69–82. [CrossRef]
23. Caress, D.W.; Clague, D.A.; Paduan, J.B.; Martin, J.F.; Dreyer, B.M.; Chadwick, W.W.; Denny, A.; Kelley, D.S. Repeat bathymetric surveys at 1-metre resolution of lava flows erupted at Axial Seamount in April 2011. *Nat. Geosci.* **2012**, *5*, 483–488. [CrossRef]
24. Mitchell, N.C.; Tivey, M.A.; Gente, P. Seafloor slopes at mid ocean ridges from submersible observations and implications for interpreting geology from seafloor topography. *Earth Planet. Sci. Lett.* **2000**, *183*, 543–555. [CrossRef]
25. Mitchell, N.C.; Livermore, R.A. Speiss Ridge: An axial high on the slow-spreading Southwest Indian Ridge. *J. Geophys. Res.* **1998**, *103*, 15457–15471. [CrossRef]
26. Head, J.W.; Wilson, L.; Smith, D.K. Mid-ocean ridge eruptive vent morphology and substructure: Evidence for the dike widths, eruption rates, and axial volcanic ridges. *J. Geophys. Res.* **1996**, *101*, 28265–28280. [CrossRef]
27. Stretch, R.C.; Mitchell, N.C.; Portaro, R.A. A morphometric analysis of the submarine volcanic ridge south-east of Pico Island, Azores. *J. Volcanol. Geotherm. Res.* **2006**, *156*, 35–54. [CrossRef]
28. Macdonald, G.A. *Volcanoes*; Prentice-Hall: Upper Saddle River, NJ, USA, 1972; p. 510.
29. Francis, P.W. *Volcanoes. A planetary Perspective*; Oxford University Press: Oxford, UK, 1993; p. 443.
30. Rowden, A.A.; Clark, M.R.; Wright, I.C. Physical characterization and a biologically focused classification of "seamounts" in the New Zealand region. *N. Z. J. Mar. Freshw. Res.* **2005**, *39*, 1039–1059. [CrossRef]
31. Clark, M.R.; Watling, L.; Rowden, A.A.; Guinotte, J.M.; Smith, C.R. A global seamount classification to aid the scientific design of marine protected area networks. *Ocean Coast Manag.* **2011**, *54*, 19–36. [CrossRef]
32. Florinsky, I.V. *Digital Terrain Analysis in Soil Science and Geology*; Elsevier: London, UK, 2012; pp. 7–30.
33. Camiz, S.; Poscolieri, M.; Roverato, M. Geomorphometric comparative analysis of Latin-American volcanoes. *J. S. Am. Earth Sci.* **2017**, *76*, 47–62. [CrossRef]
34. Favalli, M.; Fornaciai, A. Visualization and comparison of DEM-derived parameters. Application to volcanic areas. *Geomorphology* **2017**, *290*, 69–84. [CrossRef]
35. Grosse, P.; van Wyk de Vries, B.; Petrinovic, I.A.; Euillades, P.A.; Alvarado, G. Morphometry and evolution of arc volcanoes. *Geology* **2009**, *37*, 651–654. [CrossRef]
36. Grosse, P.; van Wyk de Vries, B.; Euillades, P.A.; Kervyn, M.; Petrinovic, I.A. Systematic morphometric characterization of volcanic edifices using digital elevation models. *Geomorphology* **2012**, *136*, 114–131. [CrossRef]
37. Di Traglia, F.; Morelli, S.; Casagli, N.; Garduño-Monroy, V. Semi-automatic delimitation of volcanic edifice boundaries: Validation and application to the cinder cones of the Tancitaro Nueva Italia region (Michoacán-Guanajuato Volcanic Field, Mexico). *Geomorphology* **2014**, *219*, 152–160. [CrossRef]
38. Pedersen, G.B.M.; Grosse, P. Morphometry of subaerial shield volcanoes and glaciovolcanoes from Reykjanes Peninsula, Icealnd: Effects of eruption environment. *J. Volcanol. Geotherm. Res.* **2014**, *282*, 115–133. [CrossRef]
39. Grosse, P.; Euillades, P.A.; Euillades, L.D.; de Vries, B.V.W. A global database of composite volcano morphometry. *Bull. Volcanol.* **2014**, *76*, 784. [CrossRef]
40. Sanchez-Guillamón, O.; Vázquez, J.T.; Somoza, L.; Palomino, D.; Fernández-Salas, L.M.; Medialdea, T.; León, R.; López-Gonzalez, N.; y González, F.J. Morphological characteristics and superficial structure of submarine mounds in the lower slope of the Canary continental margin (W of Canary Islands). In *Volumen de Comunicaciones Presentadas en el VIII Simposio Sobre el Margen Ibérico Atlántico*; del Río, V.D., Barcenas, P., Fernández-Salas, L.M., López-Gonzalez, N., Palomino, D., Rueda, J., Sánchez-Guillamón, O., Vázquez, J.T., Eds.; Ediciones Sia Graf: Málaga, Spain, 2015; pp. 177–180.

41. Medialdea, T.; Somoza, L.; González, F.J.; Vázquez, J.T.; de Ignacio, C.; Sumino, H.; Sánchez-Guillamón, O.; Orihashi, Y.; León, R.; Palomino, D. Evidence of a modern deep-water magmatic hydrothermal system in the Canary Basin (Eastern Central Atlantic Ocean). *Geochem. Geophys. Geosyst.* **2017**, *18*. [CrossRef]

42. Sánchez-Guillamón, O.; Vázquez, J.T.; Palomino, D.; Medialdea, T.; Fernández-Salas, L.M.; León, R.; Somoza, L. Morphology and shallow structure of seafloor mounds in the Canary Basin (Central Eastern Atlantic Ocean). *Geomorphology* **2018**, under revision.

43. Ranero, C.R.; Banda, E. The crustal structure of the Canary Basin: Accretion processes 1064 at slow spreading centers. *J. Geophys. Res.* **1997**, *102*, 10185–10201. [CrossRef]

44. Palomino, D.; Vázquez, J.T.; Somoza, L.; León, R.; López-González, N.; Medialdea, T.; Fernández-Salas, L.M.; González, F.J.; Rengel, J.A. Geomorphological features in the southern Canary Island Volcanic Province: The importance of volcanic processes and massive slope instabilities associated with seamounts. *Geomorphology* **2016**, *255*, 125–139. [CrossRef]

45. Smith, D.K. Shape analysis of Pacific seamounts. *Earth Planet. Sci. Lett.* **1988**, *90*, 457–466. [CrossRef]

46. Das, P.; Iyer, S.D.; Kodagali, V.N. Morphological characteristics and emplacement mechanism of the seamounts in the Central Indian Ocean Basin. *Tectonophysics* **2007**, *443*, 1–18. [CrossRef]

47. Clague, D.A.; Moore, J.G.; Reynolds, J.R. Formation of flat topped volcanic cones in Hawaii. *Bull. Volcanol.* **2000**, *62*, 214–233. [CrossRef]

48. Wood, J.D. The Geomorphological Characterisation of Digital Elevation Models. Ph.D. Thesis, University of Leicester, Leicester, UK, 1996.

49. Jolliffe, I.T. *Principal Component Analysis*, 2nd ed.; Springer Series in Statistics: New York, NY, USA, 2002.

50. Lebart, L.; Piron, M.; Morineau, A. *Statistique Exploratoire Multidimensionnelle, Visualisation et Inférence en Fouille de Données*; Dunod: Paris, France, 2006; 464p.

51. Kabacoff, R.I. *R in Action*, 2nd ed.; Manning Publication: Shelter Island, NY, USA, 2013; pp. 378–380.

52. Al-Hamdani, Z.; Reker, J. Towards Marine Landscapes in the Baltic Sea. BALANCE Interim Report #10, 2007. Available online: http://balance-eu.org/ (accessed on 10 November 2017).

53. Pain, C.F. Size does matter: Relationships between image pixel size and landscape process scales. In *MODSIM 2005 International Congress on Modelling and Simulation Modelling and Simulation Society of Australia and New Zealand*; Zerger, A., Argent, R.M., Eds.; MSSANZ: Perth, Australia, 2005; pp. 1430–1436.

54. Eaton, J.P.; Murata, K.J. How volcanoes grow? *Science* **1960**, *132*, 925–938. [CrossRef] [PubMed]

55. Bolongaro-Crevenna, A.; Torres-Rodríguez, V.; Sorani, V.; Frame, D.; Ortiz, M.A. Geomorphometric analysis for characterizing landforms in Morelos State, Mexico. *Geomorphology* **2005**, *67*, 407–422. [CrossRef]

56. Rossi, M.J. Morphology and mechanism of eruption of postglacial shields in Iceland. *Bull. Volcanol.* **1996**, *57*, 530–540. [CrossRef]

57. Smellie, J.L. Quaternary volcanism: Subglacial landforms. In *Encyclopedia of Quaternary Sciences*; Elias, S.A., Ed.; Elsevier: Amsterdam, The Netherlands, 2007; pp. 784–798.

58. Schmiedel, T.; Galland, O.; Breitkreuz, C. Dynamics of sill and laccolith emplacement in the brittle crust: Role of host rock strength and deformation mode. *J. Geophys. Res. Solid Earth* **2017**, *122*. [CrossRef]

59. Smith, D.K. Comparison of the shapes and sizes of seafloor volcanoes on Earth and "pancake" domes on Venus. *J. Volcanol. Geotherm. Res.* **1996**, *73*, 47–64. [CrossRef]

60. Judd, A.G.; Hovland, M. *Seabed Fluid Flow: The Impact of Geology, Biology and the 1182 Marine Environment*; Cambridge University Press: Cambridge, UK, 2007; 475p.

61. Fornari, D.J.; Ryan, W.B.F.; Fox, P.J. The evolution of craters and calderas on young seamounts: Insights from Sea MARC I and Sea Beam sonar surveys of a small seamount group near the axis of the East Pacific Rise at ~10°N. *J. Geophys. Res.* **1984**, *89*, 11069–11083. [CrossRef]

62. Mukhopadhyay, R.; Iyer, S.D.; Ghosh, A.K. The Indian Ocean nodule field: Petrotectonic evolution and ferromangenese deposits. *Earth-Sci. Rev.* **2002**, *60*, 67–130. [CrossRef]

63. Smith, D.K.; Jordan, T.H. Seamount statistics in the Pacific Ocean. *J. Geophys. Res.* **1988**, *93*, 2899–2918. [CrossRef]

64. Chaytor, J.D.; Keller, R.A.; Duncan, R.A.; Dziak, R.P. Seamount morphology in the Bowie and Cobb hot spot trails, Gulf of Alaska. *Geochem. Geophys. Geosyst.* **2007**, *8*, 9. [CrossRef]

65. McGuire, W.J. Volcano instability: A review of contemporary themes. In *Volcano Instability on the Earth 200 and Other Planets*; McGuire, W.J., Jones, A.P., Neuberg, J., Eds.; Geological Society of London Special Publications: London, UK, 1996; pp. 1–23.

66. Tempera, F.; Hipolito, A.; Madeira, J.; Vieira, S.; Campos, A.S.; Mitchell, N.C. Condor seamount (Azores, NE Atlantic): A morpho-tectonic interpretation. *Deep-Sea Res.* **2013**, *98*, 7–23. [CrossRef]

67. Fisher, A.T.; Wheat, C.G. Seamounts as conduits for massive fluid, heat, and solute fluxes on ridge flanks. *Oceanography* **2010**, *23*, 74–87. [CrossRef]

68. Davison, J.; De Silva, S. Composite volcanoes. In *Encyclopedia of Volcanoes*; Sigurdsson, H., Ed.; Academic Press: New York, NY, USA, 2000; pp. 663–681.

69. Vezzoli, L.; Tibaldi, A.; Renzulli, A.; Menna, M.; Flude, S. Faulting-assisted lateral collapses and influence on shallow magma feeding system at Ollagüe volcano (Central Volcanic Zone, Chile-Bolivia Andes). *J. Volcanol. Geotherm. Res.* **2008**, *171*, 137–159. [CrossRef]

geosciences

MDPI

Article

Origin of High Density Seabed Pockmark Fields and Their Use in Inferring Bottom Currents

Kim Picard [1,*], Lynda C. Radke [1], David K. Williams [2], William A. Nicholas [3],
P. Justy Siwabessy [1], Floyd J. F. Howard [1], Joana Gafeira [4], Rachel Przeslawski [1],
Zhi Huang [1] and Scott Nichol [1]

[1] Geoscience Australia, Crn Jerrabomberra and Hindmarsh Avenue, Symonston 2607, Australia;
    lyndacradke@gmail.com (L.C.R.); justy.siwabessy@ga.gov.au (P.J.S.); floyd.j.f.howard@gmail.com (F.J.F.H.);
    rachel.przeslawski@ga.gov.au (R.P.); zhi.huang@ga.gov.au (Z.H.); scott.nichol@ga.gov.au (S.N.)
[2] Australian Institute of Marine Science, Arafura Timor Research Facility, University Ave.,
    North Darwin 0811, Australia; dk.williams@aims.gov.au
[3] GeoQuEST Research Centre, School of Earth & Environmental Sciences, University of Wollongong,
    Wollongong 2522, Australia; tony1nicholas@gmail.com
[4] British Geological Survey, The Lyell Centre, Research Avenue South, Edinburgh EH14 4AP, UK;
    jdlg@bgs.ac.uk
* Correspondence: kim.picard@ga.gov.au; Tel.: +61-262-499-548

Received: 19 March 2018; Accepted: 23 May 2018; Published: 30 May 2018

Abstract: Some of the highest density pockmark fields in the world have been observed on the northwest Australian continental shelf ($>700/km^2$) where they occur in muddy, organic-rich sediment around carbonate banks and paleochannels. Here we developed a semi-automated method to map and quantify the form and density of these pockmark fields (~220,000 pockmarks) and characterise their geochemical, sedimentological and biological properties to provide insight into their formative processes. These data indicate that pockmarks formed due to the release of gas derived from the breakdown of near-surface organic material, with gas accumulation aided by the sealing properties of the sediments. Sources of organic matter include adjacent carbonate banks and buried paleochannels. Polychaetes biodiversity appears to be affected negatively by the conditions surrounding dense pockmark fields since higher biodiversity is associated with low density fields. While regional bi-directionality of pockmark scours corresponds to modelled tidal flow, localised scattering around banks suggests turbulence. This multi-scale information therefore suggests that pockmark scours can act as proxy for bottom currents, which could help to inform modelling of benthic biodiversity patterns.

Keywords: multibeam sonar; carbonate banks; semi-automated mapping; polychaete; Northwestern Australia; Oceanic Shoals Australian Marine Park; Bonaparte Basin; Timor Sea

1. Introduction

Pockmarks are seafloor depressions that form in soft sediment by fluid-flow processes. They have been reported since the 1970s [1] and are present worldwide [2]. Pockmarks are most commonly found in areas of thick Holocene sediment accumulation along continental shelves, in tropical to glacial settings. The expression of fluid-flow via pockmarks at the seafloor is important for several reasons. These include the need to identify potential hydrocarbon resources and areas where carbon sequestration is not possible due to the potential leakage of the targeted reservoir [3]; areas of potential geohazards (e.g., gas venting, slope failure, basement faulting) [4,5]; and areas of significant biodiversity associated with seep-related habitats, such as bioherms and reefs [1,6,7].

Pockmarks have been observed aggregated in low to very high density pockmark fields, such as the ~900/km^2 observed in the Barents Sea, Norway [8], and in both a random and non-random distribution due to the presence of underlying faults or buried paleo-channels [2]. Pockmarks range in size over almost four orders of magnitude (decimetres to km diameter), with mega-pockmarks (>3 km diameter) observed in the South China Sea [9], but are most commonly observed with diameters of the order of in the tens to hundreds meters [2]. Their vertical relief ranges from decimeters to about 30 m. Pockmarks tend to have distinct v- to u-shaped profile, a variable shape in plan view, and outline that can be circular or elongated in shape due to modification by seabed currents [10].

Pockmarks can be maintained through the semi-continuous fluid or gas escape, a lack of sedimentation to infill them or self-maintenance and scouring caused by local turbulence created when currents flow over them [11,12]. Pockmarks may affect faunal abundance or biotic composition through differences in environmental parameters (sedimentology, geochemistry, boundary layer dynamics, oxygen saturation) between pockmarks and surrounding areas [13–16]. Research on the infaunal communities of the Bonaparte Basin (offshore northern Australia) have shown distinct polychaete assemblages on sediment plains compared to other geomorphic features, but there was no investigation of the possible relationships to pockmarks [17].

Analysis of pockmark morphology and spatial distribution in relation to seabed geological and oceanographic processes can provide insight into the processes driving their development, the sub-seabed plumbing system and their potential influence on infaunal distribution [2]. With the growing use of multibeam sonar to image the seafloor at very high-resolution, small pockmarks clustered in large concentrations are more commonly observed. Characterising these consistently and efficiently can be challenging, thus several semi-automated digital mapping methods have been recently developed [18–20].

This paper has the following objectives:

(1) Identify and map pockmarks on the shelf of the western Timor Sea, Australia using a new semi-automated method.
(2) Document the geochemical and sedimentological properties of pockmark sediments, and of associated infaunal communities.
(3) Present a conceptual model for pockmark formation that links pockmark form and density to local environmental conditions, including hydrodynamics and sedimentary processes.

1.1. Regional Setting

1.1.1. Geological and Physiographic Settings

The study area is located in the tropical Oceanic Shoals Australian Marine Park (AMP; 71,740 km^2) within the Timor Sea on the continental shelf of northwestern Australia. The area lies within the sedimentary Bonaparte Basin, which is expressed at the seabed by the Van Diemen Rise, Sahul Shelf, Malita Graben, Sahul Syncline, Petrel Sub-basin and the Londonderry Rise (Figure 1). The Sahul Shelf, Van Diemen Rise and Londonderry Rise form carbonate-dominated platforms characterised by a complex suite of submerged flat-topped carbonate banks and terraces, separated by valleys, channels and plains [21–25]. Water depths range from 10 m on the shallowest banks to over 200 m in shelf-incising valleys. The central basin is about 270 km long with water depths of 90 to 140 m. The seabed of the greater Bonaparte Basin, including the Oceanic Shoals AMP region represents a drowned landscape consisting of coastal lowlands, estuarine valleys and nearshore environments. These are the product of periodic and repeated formation of a semi-enclosed sea during the Quaternary and most recently inundated during the Late Pleistocene to Holocene postglacial rise in sea level [26–28].

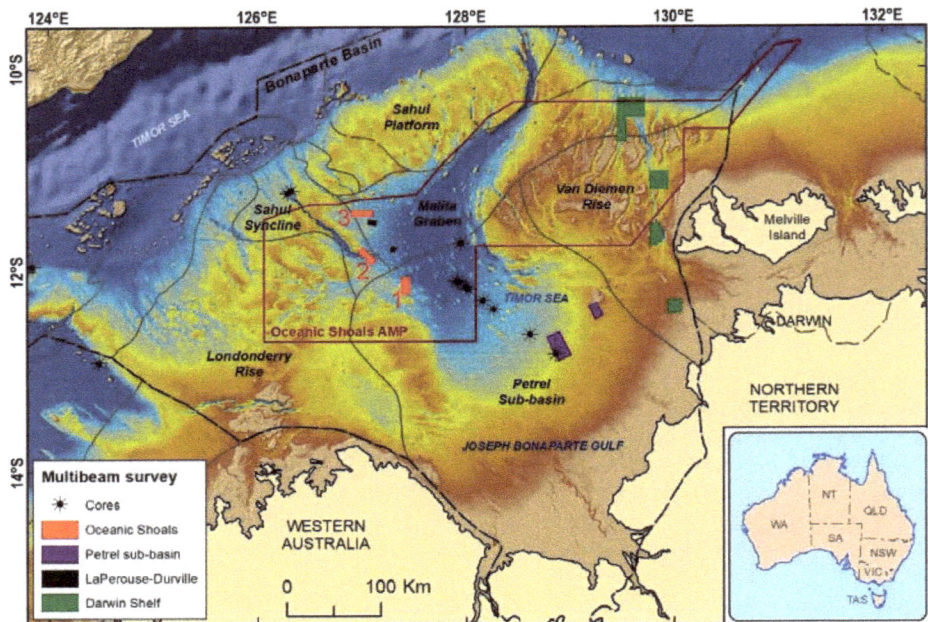

Figure 1. Location map showing the study area (Oceanic Shoals survey) within the Oceanic Shoals Australian Marine Park (brown outline) in the Timor Sea, Northern Australia. Geological setting is also included with the encompassing Bonaparte Basin as a bold dashed line and its subdivisions as continuous line. Post-survey reports that are associated with the survey outlined in this figure are as followed: Oceanic Shoals [21], Petrel Sub-basin [29], LaPerouse-Durville [29], Darwin Shelf [24,25], cores [27]. Figure modified from [21]

1.1.2. Environmental and Oceanographic Settings

The Oceanic Shoals study area is within a tropical region, with a pronounced southern hemisphere summer monsoonal season that commonly includes cyclones. The Indonesian Through-Flow, passing between Timor and northern Australia delivers relatively warm (~26 °C to 31 °C) low salinity water from the north. Across the shelf, macrotidal conditions drive strong tidal flows and wind-driven waves and seasonal cyclones come from the east. These conditions result in an anticlockwise ocean circulation across the shelf strongly influencing sediment transport [30]. The re-suspended seabed sediments coupled with sediment discharged from coastal river systems, contribute to a turbid shelf environment that is likely to suppress photosynthetic activity at the seabed. This therefore may be an important determinant of the composition of benthic assemblages across the region [31].

The study area encompasses three areas located along the north and west portion of the central basin (Figure 1). Of these, Area 1 is located within the Petrel Sub-basin, Area 2 straddles the southern portion of the cross-shelf valley within the Sahul Syncline (Figure 1), and Area 3 is located across the Sahul Syncline to Malita Graben transition. The extent of this general area spans a cross-shelf transect of approximately 110 km (Figure 1). Each area was positioned to encompass a variety of seabed geomorphic features and water depths, as identified in the national geomorphic features database [21,23].

On most of the Australian continental shelf, pockmarks are seldom found because of the dominance of coarse sediments (sand and gravel) in most regions. There is also a limited input of terrigenous organic–rich sediments, and a moderate to high percentage of carbonate sediments on the shelves [27,30]. However, high-density fields of pockmarks are present within the broad continental shelf of northern Australia, including the extensive Bonaparte Basin where seabed sediments do incorporate terrigenous fine-grained material. During the Quaternary, the present seabed of the Bonaparte Basin has been repeatedly exposed either in part, or completely, from eustatic sea level changes [27,28,32]. During low sea-levels, banks, shoals and terraces became exposed, resulting in Joseph Bonaparte Gulf forming a semi-enclosed basin [33]. Within the Bonaparte Basin, the Petrel Sub-basin, Malita Graben, Sahul Syncline and adjacent structural elements have acted as a depocentre for marine and riverine sediment since the Paleozoic [26,27,33–36]. This resulted in hydrocarbon generation occurring at depths in older sedimentary strata. Recently, seabed sediment with a particularly high organic matter content derived from the preservation of mangroves and other plant matter has been reported within paleo-river channels mapped on the Petrel Sub-basin mid-shelf and presently covered by a thin sediment veneer [33]. Additionally, the reworking of buried terrestrial sediments upwards to the seabed via pockmark formation has been identified within the Petrel Sub-basin using rare earth element chemistry [37]. Previous studies suggested that banks of the greater Oceanic Shoal region could result from one of these two processes: (1) hydrocarbon release from depth and its movement upward in faults providing a food source for the colonising reef-builders, or (2) thinner sediment cover over organics resulting in faster release of the product of organic breakdown [29].

2. Materials and Methods

2.1. Survey Summary

The data presented herein were collected during the Oceanic Shoals environmental marine survey (SOL5650/GA0339) undertaken in September 2012 [21]. High-resolution multibeam echosounder (MBES) bathymetry and backscatter data were collected over three areas covering a total of about 425 km^2 of seabed (Figure 1). MBES bathymetric data was processed using Caris Hips and SIPS v.7.1. Seabed backscatter data was processed with CMST-GA MB Process toolbox software co-developed by the Centre for Marine Science and Technology at Curtin University of Technology and Geoscience Australia [38]. Both data types were gridded at 1 m horizontal resolution for analysis and interpretation purposes.

Shallow sub-bottom profiles (SBP) were collected over 140 line km using a sparker source and processed with a C-view processing unit using 1 sec firing rate and 500 ms record length, then filtered using band-pass filter and converted to SGY file. Seabed samples were collected in duplicates at 61 stations using a Smith-McIntyre grab or box corer (Figure 2). Each sample was analysed for sedimentological characteristics (sediment texture as Mud, Sand and Gravel %, grain-size distribution, CaCO$_3$ content) and geochemical composition (incl. porosity, CO$_2$ production rates (TCO$_2$), and Total Organic Content (TOC)) as described in [21,39] method, and polychaetes biodiversity (57 stations only—identified to species level according to [17]).

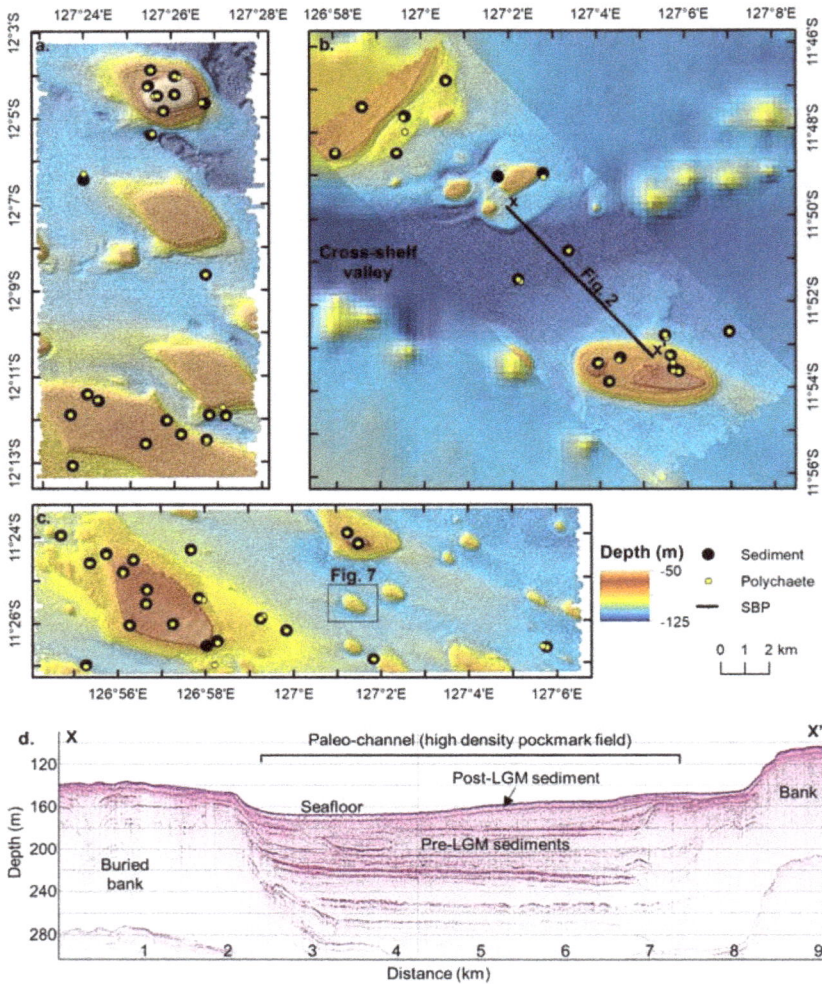

Figure 2. Hillshaded bathymetry data of the three survey areas in the Oceanic Shoals Australian Marine Park (AMP) and location of sediment samples used for the various analyses. (**a**) Area 1, (**b**) Area 2 overlaid on the 250 m bathymetry grid [38] and showing the area in relation to the cross-shelf valley existing within the Sahul Syncline, and (**c**) Area 3 also showing the location of Figure 7. (**d**) Sub-bottom profile across Area 2 cross-shelf valley (**b**).

2.2. Geomorphic Analysis and Automated Pockmarks Characterisation

Geomorphic feature mapping was completed manually using MBES data and derivative datasets such as slope and contour to identify discrete features (Figure 3) [21]. Feature definitions were based on [23]. For this study, bank morphometric profiles were calculated using ArcGIS 3D (v. 10.1, Esri, Redlands, CA, USA) profiling across the narrowest axis of the bank (Figure 3b).

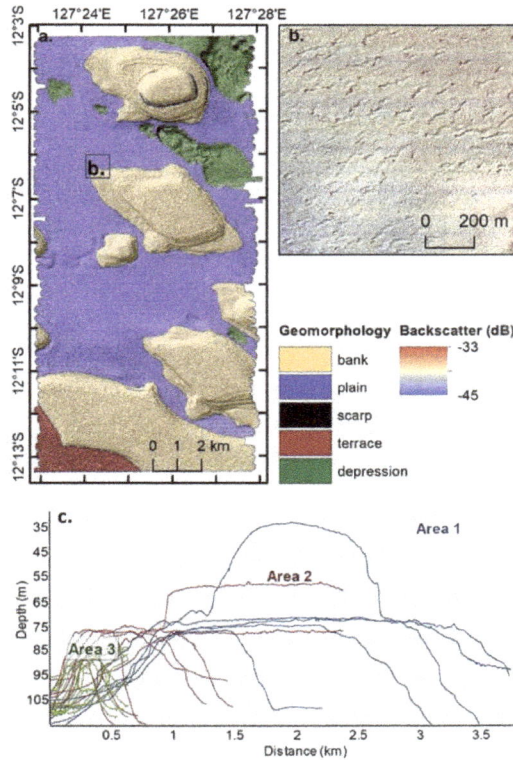

Figure 3. (**a**) Geomorphic map of Area 1. (**b**) Hillshaded backscatter inset showing scoured pockmarks with high backscatter intensity centers (red). (**c**) Bank profiles taken along the narrow axis of all banks in the study area. Colours are associated with the area in which the bank was profiled.

Due to their large number, manual mapping of individual pockmarks was deemed unsuitable. Therefore, an automated method (GA1) was developed to identify both pockmarks and associated scour marks. The overall approach was similar to other published automated and semi-automated methods [10,18], in which a multi-stage process was undertaken. The main stages included (1) identifying pockmarks and differentiating scoured from non-scoured pockmarks, (2) determining scour orientation, (3) calculating field density and (4) assessing confidence in the results (Figure 4). Each step of the method used an existing suite of ESRI spatial analytical tools, such as the hydrological and surface toolboxes. Details of the method can be found in the supplementary material. Density classes were used as a common unit for comparative analyses with other survey datasets.

Stage 1: Pockmark identification and classification
ArcGIS Spatial Analyst using Bathymetric Position Index (BPI)

1. Identify pockmarks in plain areas
Fine and Broad-scale BPI (inner and outer radius: 15 and 500 respectively)
Pockmarks (BPI ≤ -0.07 and area between 25 and 2000 m²)
Plains (BPI between -3 and 3 and area > 100 m²)

2. Calculate pockmark geometry (centroid, length, width, and scour false bearing)
Minimum bounding geometry – Rectangle by area

3. Differentiate pockmarks with and without scours
Polygon length < 30 m = without scour and > 30 m = with scours

Stage 2: Scour orientation
ArcGIS 3D Analyst (Hydrology tools: accumulation and stream)

4. Eliminate pockmarks with scours resulting from data noise
Bearing between 135° and 150° (directions associated to sonar outer beam overlap)

5. Determine scouring direction
Hydrology tools – Flow accumulation, Stream and
ETGeowizard measures bearings

Stage 3: Density calculation
ArcGIS Spatial Analyst (Kernel density and polygon conversion)

6. Create pockmark density polygons
Kernel density tools (radius = 250 m and cell size = 10 m)
Convert raster to polygon using 4 classes (Separating values = 25, 200, 700)
Add polygons representing Banks produced during geomorphic analysis [18]

7. Manual edit polygons
Merge intermediate density polygons representing "transition zones" due to the use
of a large radius (250 m) with adjacent density polygons

Stage 4: Confidence assessment:
Comparative analysis with BGS & GA2 methods

8. QA GA1 picks with BGS picks
Compare total counts and coverage area

9. Large sample size
180,645 without scour (~80%); 36,863 with scours (~20%)

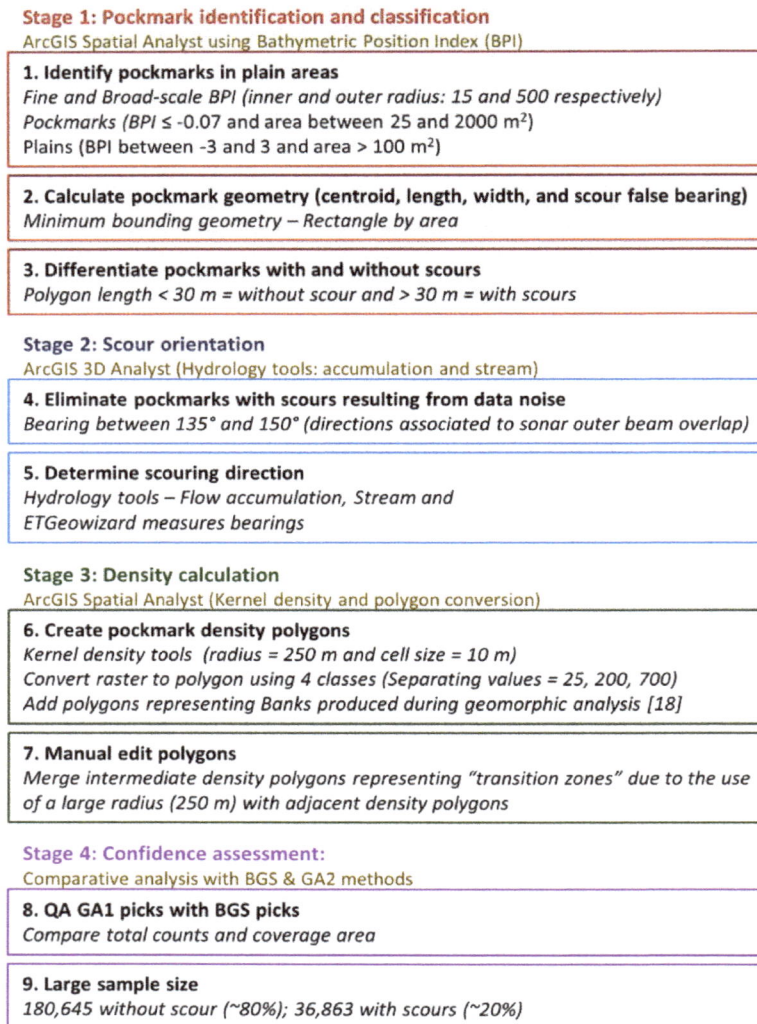

Figure 4. Summary steps involved in the extraction and characterisation of pockmarks. See supplementary material for details.

The pockmarks identified using the GA1 method were compared with results from two other methods using a small subset area of Area 3. These methods were a combination of the British Geological Survey (BGS) Seabed Mapping toolbox [18,20] and the Geomorphometry and Gradient Metrics (GGM) toolbox [40], and another tool tested by Geoscience Australia (GA2). The subset area included a carbonate bank and the associated clustered pockmarks surrounding its base, as well as other scoured and non-scoured pockmarks distributed in the plain area.

For the BGS combined approach, the landform tool in the GGM toolbox was used to create the surface curvature landform index (Bolstad's variant) using a square neighborhood analysis window of 12 by 12 cells. The derived raster was used as input for the Feature delineation [Derived] tool within the BGS Seabed Mapping Toolbox [20]. The mapped features corresponded to the areas with a surface curvature landform index lower than −0.04, that are wider than 5 m with a minimum size ratio higher than 0.2 and that include cells with values lower than −0.05. The final outline was achieved by adding a 2 m buffer.

The GA2 method shared similarities with BGS approach, in which Hydrology toolset in ArcGIS™ was used to identified sinks and pits from the bathymetry data. However, GA2 used a fuzzy features tool in landserf (based on topographic terrain analysis) to calculate multiscale pitness [41] and the openness tool (based on the viewshed principle) to calculate positive openness [42]. Both calculations were based on a search radius of 21 cells. Where pitness was greater than the mean plus one standard deviation, pits were identified. From these, only the ones associated with a positive openness values of less than the mean minus one standard deviation were selected and converted into polygons (simplify boundary function ON). Selection was further constrained by only selecting polygons that intersected either sinks, or pits or both. Finally, polygons with area less than 10 m² were deleted and resultant polygons were regarded as pockmarks. The key issue related to these semi-automated methods is the choices of value thresholds, which is a somewhat subjective process.

2.3. Infaunal Data Analyses on Grab Samples

Multivariate analyses were performed only on samples that returned polychaetes identified to species level, while univariate analyses of species richness and abundance included samples from which no polychaetes were identified. All samples from banks were excluded from statistical analyses because these did not have an associated value for pockmark density. Assemblage data and species richness data were square-root transformed to reduce the influence of dominant groups [43] and meet ANOVA assumptions, respectively. A PERMANOVA using unrestricted permutation of raw data and type III SS [44] was undertaken to determine if polychaete assemblages varied among pockmark density classes, while a 1-way ANOVA was undertaken to determine if species richness varied among pockmark density classes.

2.4. Hydrodynamic Model

To understand the processes contributing to the presence of pockmarks and associated scouring, a coastal shelf model was constructed to evaluate bed shear stress at the three different survey areas (Figure 1). A finite element model utilising an unstructured grid was constructed. The bathymetry used in the model was Australian Bathymetry and Topography Model [45] gridded at 250 m and the multibeam dataset used here gridded at 10 m resolution.

The boundary of the model was set beyond the shelf break in order for the currents to stabilise by the time they reach the study area. The northern boundary conditions were set as open water boundaries between islands of the Indonesian archipelago and Papua New Guinea. Both the eastern and western boundaries were set as open water. Two types of ocean boundary conditions were trialed and both included the semi-diurnal tides. (1) Tidal elevations derived from the TOPEX/Poseidon satellite altimetry global tidal model driver (TMD) toolbox version 2.05 with all constituents implemented to create hourly tidal elevation files, and (2) tidal currents derived from the same toolbox using all harmonic constituents. These currents were multiplied by the width and depth of the boundary to calculate the hourly flow rate. Wind stresses from the Bureau of Meteorology gridded datasets were also applied in both of the seasonally dominant directions of the North West and South East. The wind forcing, while changing the surface currents, did not change the currents near the sea bed and the bed shear stress calculations. The model was refined near the study area and, with salinity kept constant, run for a three-month period to represent mean tidal conditions. From these model runs, the bed shear stress time series and exceedance of erosion was also calculated.

3. Results

3.1. Geomorphic Analysis

High-resolution seafloor mapping revealed multiple carbonate banks rising tens of meters above otherwise flat pockmarked plains [21].

3.1.1. Banks

Relatively steep-sided (slopes up to 35 degrees) and flat-topped banks are located in water depths of 45 m to123 m (Figure 2). Along their flanks, banks have terraces of various widths, which are regarded as possible indicators of lower sea levels. Banks and their associated features are characterised by the highest backscatter intensity values measured from the multibeam sonar data and coarsest sediment composition sampled [21]. Sub-bottom profiles reveal a mostly acoustically transparent seismic facies across the banks (Figure 2). The dimensions (length and height) of the banks reduces northward towards the shelf edge (Figure 3).

3.1.2. Plains

Plains dominate the seascape, occur between 94m and 146 m water depth and deepen northward. They record gradients averaging 2.5 degrees and are characterised by low backscatter intensity values, representative of the fine-grained sediment observed in the samples [21]. Their sub-surface is mostly characterised by a thick sequence (>150 ms TWT) of acoustically transparent seismic facies closest to the seafloor, underlain by a semi-stratified facies, which in places is replaced by cut and fill facies (Figure 2d). The thickest section (>350 ms TWT) is recorded in the deepest region of Area 2, which corresponds to the start of a deep cross-shelf valley (Figures 1 and 2).

3.1.3. Pockmarks: Density, Scour Direction Results

Semi-automated mapping identified 220,508 pockmarks throughout the plain areas, including at the foot of the banks. Some bank tops also hosted pockmarks, but these were not included in this analysis. Pockmarks displayed wide ranging geometries, varying from circular to elongate. Their dimensions varied from 10 to 30 m in diameter and 0.5 to 2 m in depth. In most cases, the centers of the pockmarks recorded higher backscatter intensity values than their surroundings [21].

Elongated pockmarks represented mainly pockmarks accompanied by a semi-linear depression shallowing in a particular direction (Figure 3b). The depression, commonly but not always, formed a v-shape on the seabed, with the open end of the 'v' on the downstream side of the pockmark. These depressions extended for up to ~200 m, but on average less than 80 m, and are interpreted as a scour mark (Figure 5). For the purpose of this study, pockmarks are divided in two main types: with and without scour. Pockmark types are intermixed throughout the areas they are found within and pockmarks without scour are at most 4.5 times more common (see supplementary tables). Scoured pockmarks show strong bi-directionality, which is associated with an overall east-west quadrant ratio of ~1 (Figure 5). The scour directions in Areas 1 and 2 are slightly more biased towards the eastern quadrant, whereas the directions recorded in Area 3 dominate the western quadrant. More specifically, the bi-directionality is strongest in Area 3, with over 75% of the scour directions divided relatively equally between W-NW and E-SE quadrants, while Area 1 is the most scattered (44% within the W-NW and E-SE quadrant; Figure 5 and supplementary tables).

Pockmarks form fields of dense clusters, with densities greater than $700/km^2$ for non-scoured pockmarks (Figure 6). High-density areas are mostly concentrated around the foot of the banks and represent over 17% of the total surveyed area (supplementary Figure S2). The most common density for both types of pockmarks is between 25 and $200/km^2$. Area 1 contains the highest coverage of banks and low density pockmark fields ($<25/km^2$; Figure 6 and supplementary Figure S2). Area 2 contains a large field of high density pockmarks within the cross-shelf valley (Figures 2 and 6) and the largest area of intermediate-density ($200–700$ pockmark/km^2) fields of non-scoured pockmarks (Figure 6 and supplementary material). Many of the scoured pockmarks in the central plain of Area 2, which extends across the southern end of the Sahul Valley (Figure 1), are oriented so that a flow from north to south is indicated for the Sahul Valley. This is consistent with the general anticlockwise circulation suggested for the basin [15]. In Area 3, dense non-scoured pockmarks fields surround the banks and similarly to Area 2, many of the pockmark scours are oriented such that a flow from northwest to southeast is implicated, also consistent with the anticlockwise circulation [30].

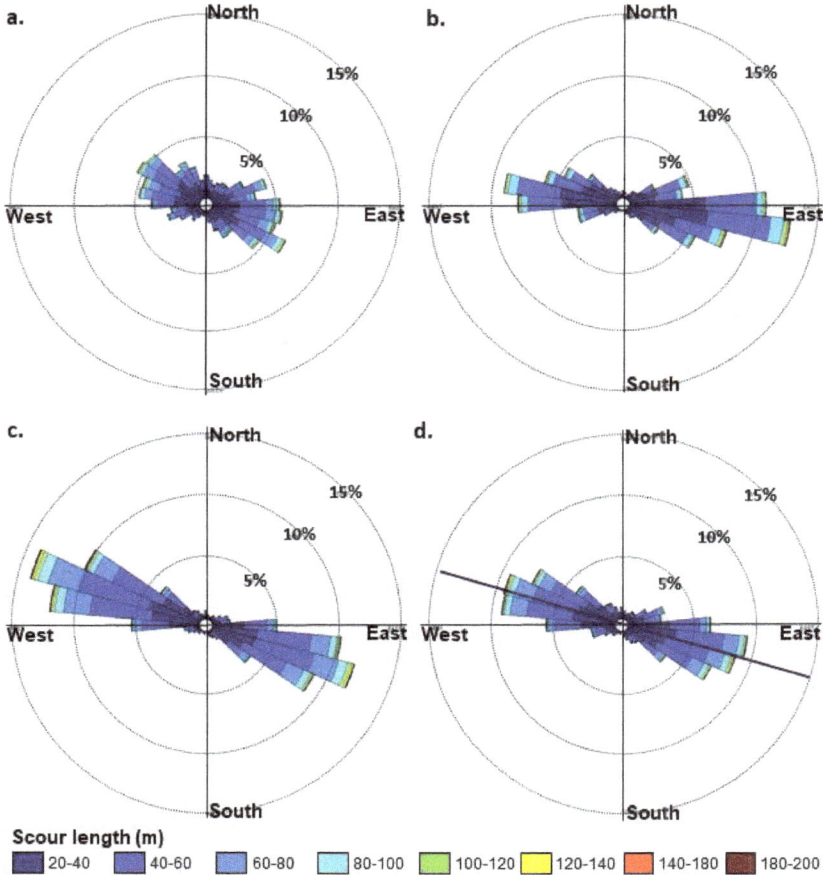

Figure 5. Rose diagrams of observed scour direction for each of the area analyzed. (**a**) Area 1; (**b**) Area 2; (**c**) Area 3; and (**d**) combined areas.

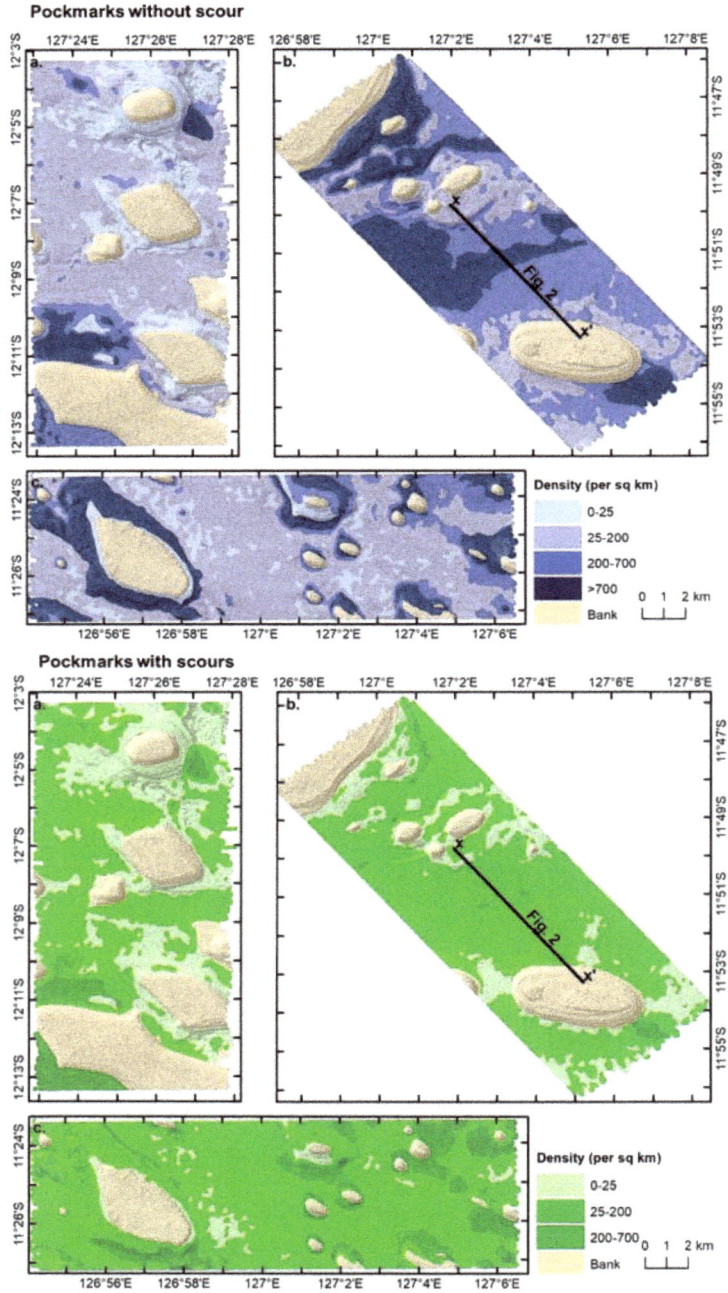

Figure 6. Density distribution of pockmark without scour (to; blue shades) and with scours (bottom; green shades) within the three survey areas ((**a**) Area 1, (**b**) Area 2, and (**c**) Area 3). The location of the sub-bottom profile presented in Figure 2d is also displayed on b. to indicate the location of the buried paleochannel within the cross-shelf valley of Area 2.

3.1.4. Comparison Analysis and Confidence Assessment

Comparison analysis between the results of GA1, GA2 and BGS methods shows that all three methods identified most pockmarks present, within 10% of each other (Figure 7). However, discrepancies exist between their pockmark total count and coverage area (Figure 7i). For example, BGS method identifies 60% more pockmarks than the GA1 method (Figure 7i), but their total coverage areas are similar. These discrepancies are attributed to GA methods often combining closely spaced pockmarks or identifying pockmark scours. For GA1 method, this resulted respectively in reducing the overall pockmark count while increasing the total coverage area (Figure 7i).

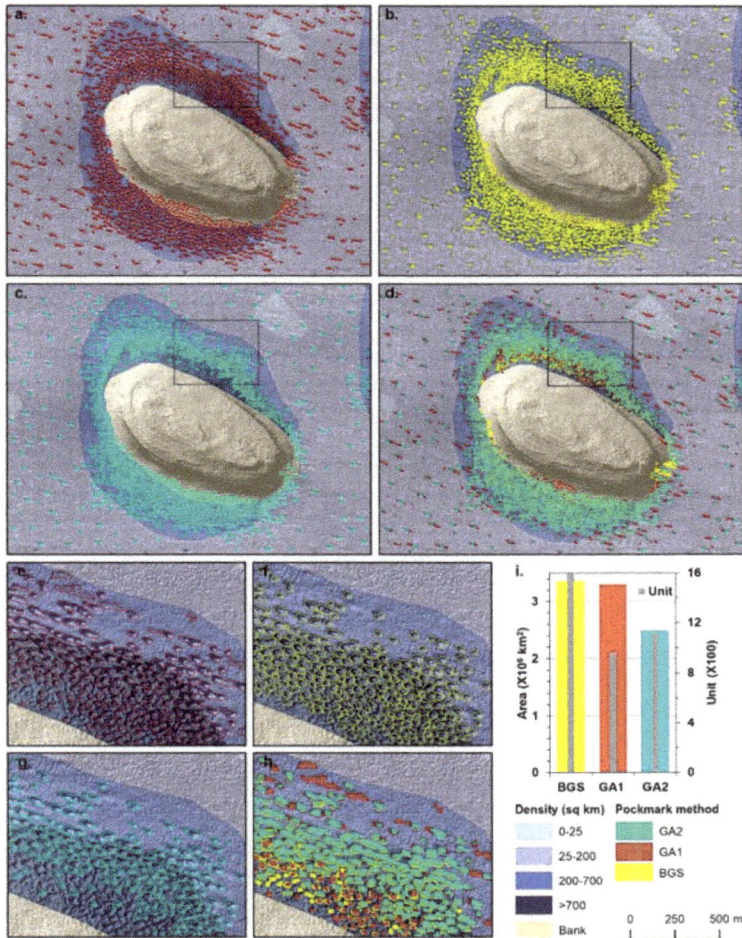

Figure 7. Examination of pockmarks identified using the three methods: GA1 (red), British Geological Survey (BGS) (yellow), GA2 (blue). Panels (**a–d**) show the picks using the various methods for the entire test area, while panels (**e–h**) show an inset of the test area in order to present more details. (**i**) Histogram comparing the total pockmark count (grey bars) and coverage area (coloured bars) between the three automated methods used to identify pockmarks.

3.2. Seabed sediment sample analyses

3.2.1. Sediment Texture Analyses

Sediment textures provide some insight into pockmark development. Seabed sediments in the mapped areas have mixed textures, including carbonate-dominated sand (90–95% CaCO$_3$) and gravel (~95% CaCO$_3$), and calcareous mud (CaCO$_3$ values around 25–35%). Some of these sediment samples have the highest mud concentrations measured from the Australian continental shelf [40]. Most of this mud is composed of silt-sized particles, with clay content up to ~25% (Figure 8). The clay is expected to be about 50% smectite, based on data from seabed sediments in the Eastern Joseph Bonaparte Gulf [46]. Low-density pockmark fields have the lowest mud content and highest sand content (Figure 9). This relationship reverses for higher density fields but mud content is greatest in the plain areas where pockmark densities are less than 200/km^2. Gravelly mud and muddy gravel generally occur where the highest pockmark field densities are found (i.e., foot of the banks), with the mud to sand ratio less pronounced than for the plains (Figure 9). These textural relationships are also directly correlated with the average seabed backscatter intensities, in which the highest backscatter intensities are associated with the banks and where the sand to mud ratio is greater than 1. The opposite relationship is observed for the areas of lowest backscatter intensities (Figure 9).

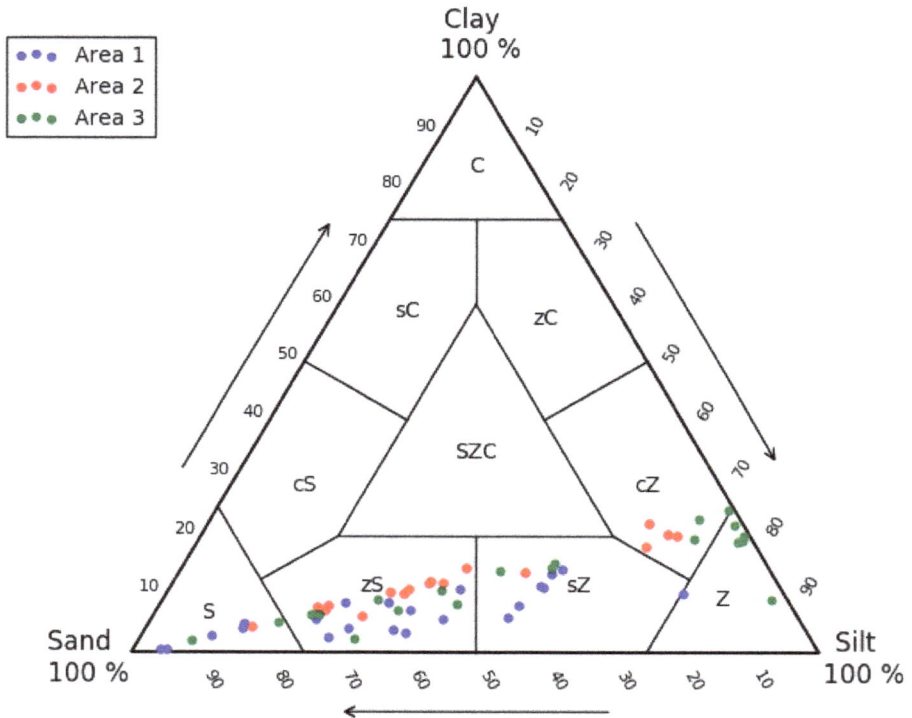

Figure 8. Ternary diagram illustrating the relative proportions of Sand, Silt, and Clay (after [47]) indicating the dominance of silt over clays in the muddy sediments of the Oceanic Shoals.

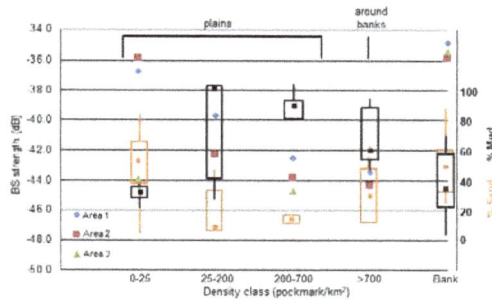

Figure 9. Sediment characteristics (box plots) and averaged seabed backscatter strength (coloured symbols) for each non-scoured pockmark field density class per area. Similar results were observed for the scoured pockmark.

3.2.2. Geochemistry Analyses

Oceanic Shoals sediments have the highest Total Organic Carbon (TOC) concentrations of sediments collected so far on the Australian continental shelf, which average 0.8% (Figure 10a, [39]). Moreover, there is a positive linear relationship between the mud content, surface area (SSA) and TOC as common for Australian Shelf sediments [48]. While most Oceanic Shoals sediments have surface area-normalised TOC concentrations that are in the typical range for the continental shelf [39], the sediments from the dense pockmark fields have less TOC per unit surface area than sediments from the low-density pockmark fields (Figure 10b).

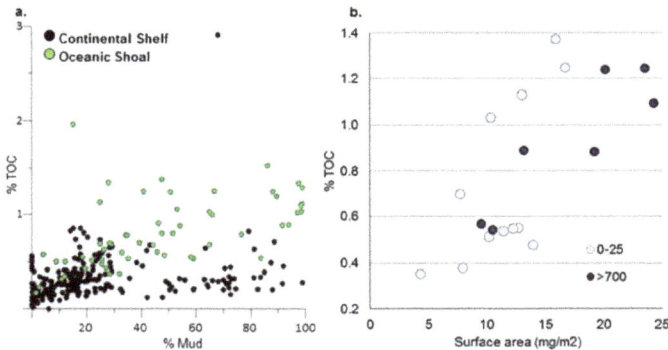

Figure 10. (**a**) Total organic carbon content (%) of the samples collected in the study area compared to other samples collected on the Australian Shelf highlighting the high Total Organic Carbon (TOC) concentrations at a given grain-size. The data shown here is from samples collected from the 0–2 cm sediment horizon during surveys in the period from October 2007 to May 2013 (see Table 1 in [39]). (**b**) TOC versus specific surface area of sediments highlighting that sediments from the dense pockmark fields have less TOC per square of sediment than sediments from the low-density pockmark fields.

When organic matter breaks down in marine sediments, dissolved inorganic carbon (DIC) is released to porewaters where it may build up in concentration. Porewater DIC concentrations of seabed sediments on Australia's continental margin (as TCO_2 pool) increase approximately exponentially with porosity, and the slope of this relationship is about two-fold higher for pockmarked sediment of the Oceanic Shoals area (Figure 11).

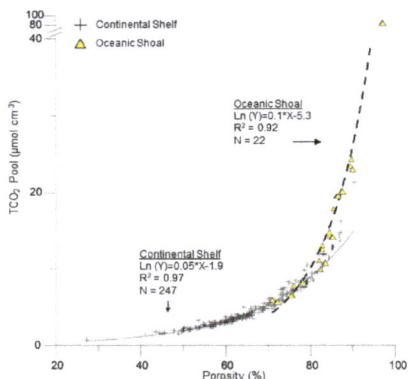

Figure 11. Total carbon dioxide porewater pool concentration (0–2 cm) versus porosity (0–2cm) of the samples collected on the Australian Shelf between October 2007 to May 2013 (see Table 1 in [39] for survey details). The yellow triangles pertain to pockmarked sediment in the Oceanic Shoals AMP survey areas.

3.2.3. Polychaete Analysis

A total of 38 grab and box core samples were considered for assemblage data. There were three outliers (one for each area) in assemblage data that were identified by their isolation from the main cluster. Statistical analyses were thus performed with and without these (see supplement Figure S5 for details on the distribution). There were no significant differences in polychaete assemblages for pockmark density without scours (including outliers: Pseudo F = 1.2583, p = 0.051; excluding outliers: Pseudo F = 1.1909, p = 0.107). In contrast, there were significant differences in assemblages among pockmark density classes with scours (including outliers: Pseudo F = 1.4036, p = 0.023; excluding outliers: Pseudo F = 1.14548, p = 0.018). Pairwise comparisons revealed significant assemblage (p = 0.005) differences between only the lowest densities classes of scoured pockmarks (<25/km^2 and 25 to 200/km^2).

A total of 51 samples were considered for species richness data. Among pockmark density classes without scours, species richness was significantly different (df = 3, F = 8.7349, p = 0.0001). Tukeys HSD pairwise tests showed significantly more species in the lowest density class (<25 pockmarks/km^2) than in the other three higher density classes (Figure 12a). Similarly, species richness was significantly different among pockmark density classes with scours (df = 2, F = 8.6768, p = 0.0006), with more species collected in the lowest density class than the higher density classes (Figure 12b).

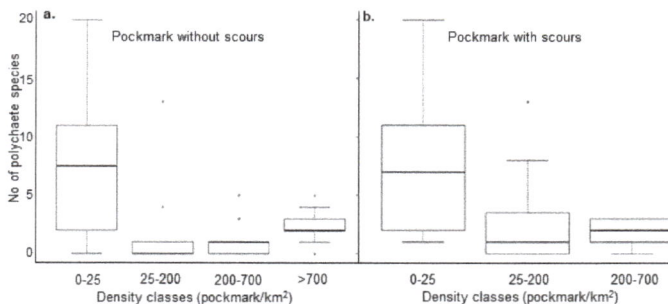

Figure 12. Boxplots of polychaete species richness among pockmark density classes (**a**) without scours and (**b**) with scours.

3.2.4. Hydrodynamic Model Analysis

The performance of the model developed here was considered good as the model replicated the growth of the tidal amplitude from the edge of the shelf to the shoreline. The model results were compared to the Timor Sea mooring array, part of the Integrated Marine Observation System (IMOS), that runs to the shelf edge and records a tidal amplitude that grows from plus 3 m at the shelf edge to 6 m on the inner shelf. Results were also correlated with other mooring sites and tide gauges, including Darwin Harbour, Melville and Milner Bay, and inshore the Gulf of Carpentaria.

There is no data available on the bed shear stress erosion threshold for this area above which sediment transport would occur. However, a value of 0.1 N/m^2 is considered an appropriate value for reasonably cohesive sediments, such as the ones found in this study area [49] Over the three months period modelled for the study area, bed shear stress exceeded 0.1 N/m^2 for over 33% of the time and 0.2 N/m^2 for over 8% of the period (Figure 13). Area 1 is exposed to the longest period of bed shear stress exceeding the threshold (49%), while Area 2 the least (12%). Over periods of many years, even the periods of high bed shear stress would be sufficient to provide scour energy on the seabed. Analyses of the modelled current velocities (directions) for all areas suggest a strong bi-directionality with a slight dominance of the current in the NW direction (Figure 14).

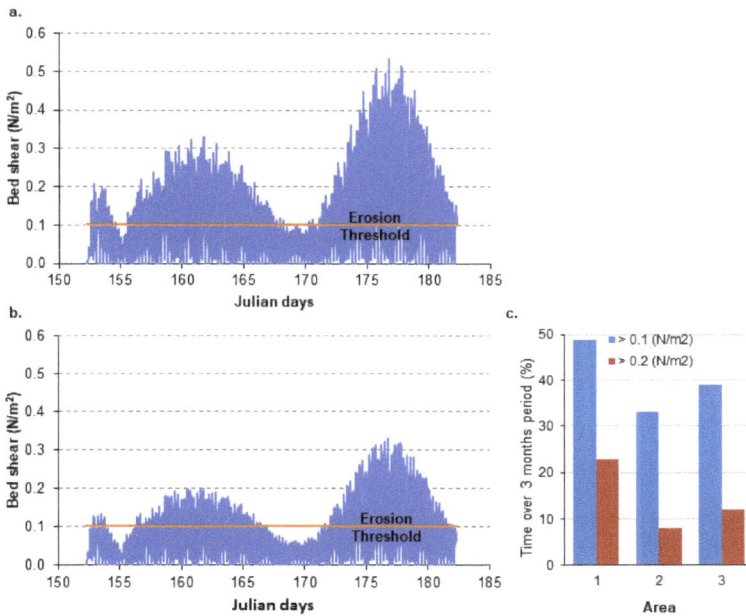

Figure 13. Modelled bed shear stress results for the three months period in (**a**) Area 1 and (**b**) Area 3. (**c**) Comparison of the percent time in the three months modelled period where the bed shear stress exceeded the erosion threshold of 0.1 N/m^2 considered for the type of sediments found in the study area [49] and 0.2 N/m^2.

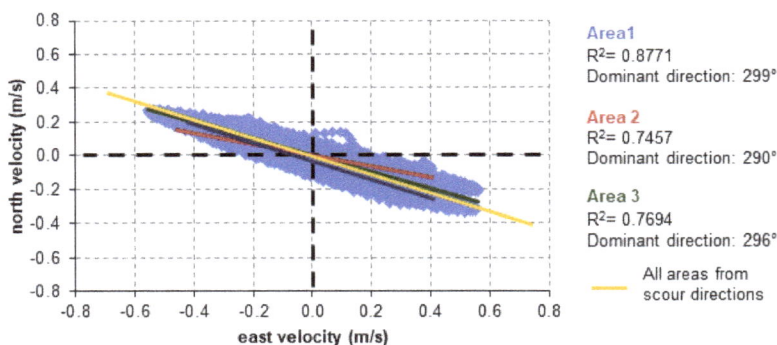

Figure 14. Near seabed velocities showing the directionality of current for Area 3. Trends for Areas 1 (blue) and 2 (red), as well as the summary trend (yellow) for all the observed data (scours) are also plotted. Note the dominant direction indicated is towards the bearing value indicated.

4. Discussion

4.1. Confidence Assessment of the Semi-Automated Method

Method GA1 used to identify pockmarks and accompanying scour mark was compared to the BGS combined delineation approach and a second unpublished method developed by Geoscience Australia (GA2). The comparative analysis shows that while all pockmarks in the area were identified within 10% uncertainty and is consistent with other semi-automated methods [10], the BGS method identified 60% more individual pockmarks than GA1. This suggests that there are likely considerably more individual pockmarks in the overall study area, than the 220,508 pockmarks identified. Uncertainties associated with pockmark scour direction, although not quantified here, were considered low due to the large sample number (~40,000; supplementary material). However, for the 'slice' of directions associated with bearings having similar bearings to the survey lines (i.e., between 135° and 150°, and its opposite 315° to 330°), this 'slice' may be misrepresented because step 4 of GA1 method eliminated these polygons (Figure 5).

Considering that all three methods were developed separately and with different objectives in mind, the results are promising in terms of semi-automatically identifying pockmarks from a range of pockmark fields of varying density (Figure 7). The analysis suggests that while the objective of the study may drive the choice of a preferential method, e.g., scoured pockmarks may be better mapped using a method based on BPI algorithm (GA1), using a combined approach may be ideal to fully characterise any type of pockmarks. Moreover, the comparative analysis provides confidence in using the results of GA1 method over the whole study area to address hypotheses about pockmark formative processes.

4.2. Pockmark Formation and Ecosystem Contribution

Examination of shallow sub-bottom profiles across the survey areas suggests an absence of deeply-sourced gas with no evidence of gas migration from the sub-surface (Figure 2). However, in the sediment samples examined, the concentration of CO_2 in sediment porewater (as TCO_2) is particularly high, approximately twice as high in these sediments compared to elsewhere on the Australian continental shelf (Figure 11). Without evidence for gas sourced from depth, this suggests that the CO_2 is derived from recent (i.e., shallow) sediments, and in turn must imply that organic matter (OM) trapped in the shallow sediment is breaking down rapidly. Our hypothesis for the origin of high pockmark densities in the Oceanic Shoals involves the concept of OM priming (or cometabolism), which pertains to the enhanced remineralisation of refractory sediment-bound OM through the breakdown of more labile OM (see conceptual model in Figure 15) [50]. Carbonate banks in the study area rise tens of meters above soft sediment plain where seabed sediments are often in the euphotic zone (~<60 m) and conditions thus are suitable for the growth of benthic algae as the local source of OM [37]. The shedding of labile organic matter from banks is a contemporary mechanism and likely occurs due to the relatively high current velocities in the region. This is supported by the hydrodynamic current model results (Figures 13 and 14) and the observed scouring of the pockmarks (Figure 3). The lower TOC concentrations per unit surface area of sediment in the high-density pockmark fields may provide evidence for OM priming, on the assumption that some particle bound OM had been mineralised into dissolved CO_2 (Figure 10b). In our conceptual model (Figure 15), the high porewater CO_2 concentrations observed reflect a combination of factors including elevated sedimentary TOC concentrations, the break-down of labile OM shed from the banks, the cometabolism of particle-bound OM and restricted permeability imposed by fine sediments (including the marl-like $CaCO_3$ content in the mud fraction). Pockmarks could arise if this combination of factors creates saturation with respect to CO_2 followed by gas production.

The data from the Oceanic Shoals region are similar to data from the Petrel Sub-basin (Figure 1) where high TCO_2 in shallow sediments and pockmark formation was attributed to the breakdown of OM and $CaCO_3$ in the shallow subsurface [29,33]. There, pockmarks are abundant and their distribution correlates strongly with the known distribution of multiple shallowly buried paleochannels, similar to Area 2 here (Figure 2). In the Petrel Sub-basin study, mangrove-derived OM was abundant. Whilst in this study the OM is understood to be sourced from marine biota that inhabit the banks, the co-location and similarity in shape of the high-density field with the cross-shelf valley suggest that OM may also be sourced from the buried paleochannel. Although the OM source generally appears different for the study area (mainly banks), there is a broad similarity in pockmark formation in these two separate seabed regions of the Bonaparte Basin. This hypothesis, in turn, suggests that it likely can be applied elsewhere in this immediate region, including the pockmark fields identified on the Darwin Shelf [24]. It has been suggested that pockmark formation in this region may be due to the presence of banks focusing fluid flow from the shallow sub-surface [29]. The evidence from this study neither supports nor negates the latter hypothesis.

The biogeochemical and physical environment of the pockmark fields appear to drive some of the polychaetes biodiversity in this area. Infauna analysis showed distinct polychaete assemblages and higher richness associated with low density pockmark fields (<25/km). These fields occupy 15% of the study area, are located randomly (Figure 6) and are associated with coarser grain-size (large proportion of shelly sand and gravel texture, as per high % $CaCO_3$) where TCO_2 concentrations are at their lowest (Figure 15). In some environments, sediment grain size can regulate infaunal communities, with sand content increasing oxygen permeability and associated habitat availability beneath the seafloor surface [51]. However, sediment grain-size is unlikely to be the primary determinant of infaunal species distributions, with other factors such as hydrodynamics, nutrient availability, and biological interactions being key drivers [52]. This is supported by previous analyses of polychaete assemblages in northern Australia in which sediment grain-size was shown to be weakly related to polychaete assemblages and richness [53,54]. Alternatively, as per the intermediate disturbance hypothesis [15], polychaetes may be responding to seabed disturbances associated with active pockmark fields, with successional communities being more prevalent in pockmarks as disturbances such as gas expulsion occur (Figure 15). For example, two groups of pockmarks in the Bay of Fundy were each associated with discrete megafaunal communities, one category supporting pre-equilibrium communities and the other supporting equilibrium communities [53]. Further work is needed on this relationship, including targeted sampling within and outside individual pockmarks.

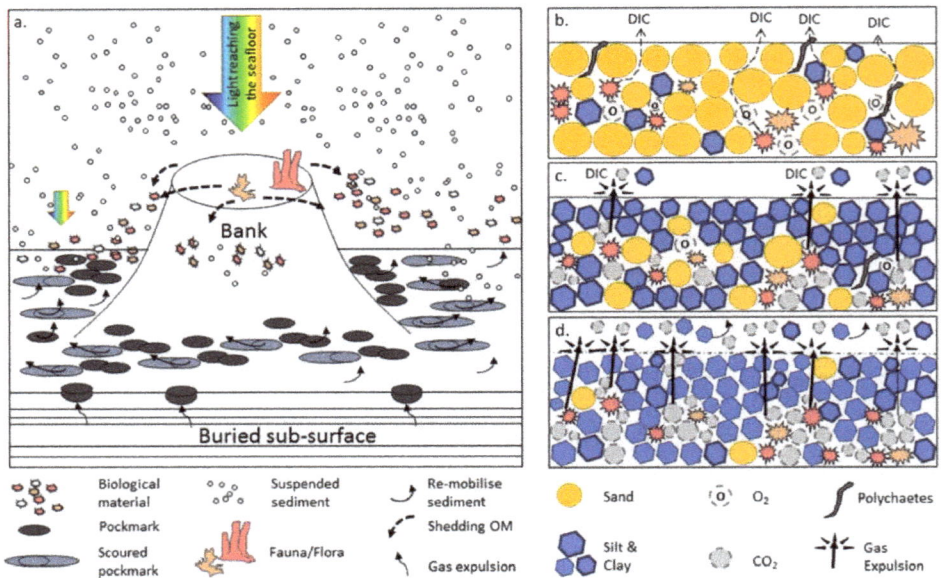

Figure 15. Conceptual model summarising the hypothesis about the formative processes of the pockmark fields in the study area. (**a**) Overview of the bank-plain system; (**b**) Low pockmark density scenario where more elevated sand content results in lower porosity compared to finer grain-sizes (shown here as clay and silt particles) and presumably higher permeability (interconnected pore space). Therefore, when OM is mineralised into DIC it does not build up to saturation levels for CO_2. Polychaete biodiversity is relatively high because of the increase gas exchange (O_2) and more stable conditions. (**c**) Medium pockmark density scenario where porosity is higher than (**a**). Permeability may become occluded within sediment micro-niches due to the occurrence of fine sediments (silt and clays) and the swelling of smectite. Limited pockmark formation occurs due to gas build up. Polychaete biodiversity is likely low due to the low O_2 concentrations and increasing CO_2 concentrations caused by restriction in permeability and the physical disruption to the sediments due to pockmark formation. (**d**) High pockmark density scenario in which sediments contain more silts and clays creating more micro-niches for gas formation. The gas is expulsed and the seafloor becomes "rugged". Clay particles have less sediment-bound OM due to OM priming. Sediments are mobilised from pockmarks due to strong near-seabed currents. Low polychaetes biodiversity is presumably due to physical disturbance.

4.3. Pockmark Maintenance and Contribution to Hydrodynamic Knowledge

Correlation of the observed pockmark scours (Figure 5) with the hydrodynamic model results (Figure 14) suggest that at a regional scale, seabed currents are dominated by the SE-NW tidal flow. Results also indicate that small differences in the predominant modelled and observed directions between each mapped area exist (Figures 5 and 14). Given this strong regional agreement, the fine-scale variations observed within an area can give valuable insight into seabed and oceanographic processes at a scale finer than model resolution (Figures 3 and 7). For example, the scattering of the scour directions observed in Area 1 compared to the more dominant bi-directionality observed in Areas 2 and 3 (Figure 5) suggest that the presence of larger and shallower banks generates more turbulence.

Though the modern seabed within the Bonaparte Basin is in places sediment-starved [21,28], turbulence within pockmarks enhances sediment suspension [11,12]. Given the strong tides here, the presence of open pockmarks within the organic-rich seabed sediments of the Bonaparte shelf suggests that they are active and that there is insufficient sedimentation to bury them faster than they form. Cores collected within the central Bonaparte Basin to the southeast of this study area suggest that post-LGM sediment thickness in the center of the basin is less than few meters [26]. This may explain the small size and depth of the pockmarks with smaller pockmarks typically forming in thinner sedimentary strata [55]. Furthermore, even though high turbidity in the water column was observed [20], the overall bed shear stress is enough to support erosion and mobilisation, while preventing settling of the sediments (Figure 14).

The wide spectrum of shapes exhibited by the pockmarks indicates that they provide an integrated record of the various stages of their formation (Figure 7). It is however, difficult to attribute whether scouring is associated with the formation of the pockmark at a specific time within a tidal cycle or that it reflects a long-term scouring process attributed to the integrated tidal regime. Some studies show that eccentricity of pockmarks is made at the time of the pockmark formation [18] and that it is unlikely that modern currents are reshaping inactive pockmarks, while others suggest that pockmark-scale turbulence plays a role in the excavation and scouring of pockmarks [11,12]. The intermixed variability in the shapes and scour directions observed over a small area here indicates that currents at the time of formation drive the eccentricity of the pockmark, but that scouring is ongoing thereafter.

5. Conclusions

On the northern Australian continental shelf, pockmark formation appears related to the decomposition of organic material which is richly preserved within the shallow sub-surface. Where carbonate banks are the primary point source for organic detritus, pockmarks become concentrated around the banks and are modified by tidal currents that are steered by the banks. As such, the morphometric data extracted here for large numbers of pockmarks provides a robust proxy to support modelling of local to regional hydrodynamics. Local variability in sediment texture and geochemistry around carbonate banks also reflects local organic input and appears to influence polychaetes biodiversity within sediments; but it also relates to seabed disturbance as driven by pockmark formation. In summary, there are multiple interactions that influence pockmark form and distribution. As more pockmarks fields are unveiled by the increased availability of high-resolution seabed data, semi-automated characterisation of pockmarks will become essential and better understanding of the seabed processes will be achieved.

Supplementary Materials: Supplementary materials can be found at http://www.mdpi.com/2076-3263/8/6/195/s1.

Author Contributions: K.P. performed the GIS analysis of the pockmarks and their scours and led this study, L.C.R. performed the geochemical experiments and data analyses, D.K.W. designed and performed the hydrodynamic model, W.A.N. analysed the grain-size data and contributed expertise in pockmark formation for the region, J.P.S. analysed the multibeam backscatter, F.H. mapped the geomorphology of the study area, J.G. and Z.H. designed and performed the pockmark delineation with BGS and GA2 methods respectively, and R.P. performed polychaetes analyses. S.N. led the field work. All co-authors wrote the sections of the manuscript corresponding to their expertise.

Acknowledgments: This work was partly undertaken for the Marine Biodiversity Hub, a collaborative partnership supported through funding from the Australian Government's National Environmental Science Programme (NESP). Thank you to Chris Glasby and Charlotte Watson who identified the polychaetes. We also acknowledge Johnathan Kool and Chris Battershill for sampling infauna onboard the vessel, and Maggie Tran and Andrew Carroll for sorting infauna in the laboratory. Thank you also to Todd Nowack for helping with the figures. We would also like to thank the two anonymous reviewers for their constructive review. Finally, thank you to the crew of the R. V. Solander, Australian Institute of Marine Science for support during the Oceanic Shoals survey in 2012. Joana Gafeira publishes with the permission of the Executive Director of the British Geological Survey. This manuscript is published with permission of the CEO, Geoscience Australia.

Conflicts of Interest: The authors declare no conflict of interest. The founding sponsors had no role in the design of the study; in the collection, analyses, or interpretation of data; in the writing of the manuscript, and in the decision to publish the results.

References

1. King, L.H.; MacLean, B. Pockmarks on the Scotian shelf. *Geol. Soc. Am. Bull.* **1970**, *81*, 3141–3148. [CrossRef]
2. Judd, A.G.; Hovland, M. *Seabed Fluid Flow: The Impact on Geology, Biology and the Marine Environment*; Cambridge University Press: Cambridge, UK, 2007; p. 475.
3. Nicholas, W.A.; Carroll, A.; Picard, K.; Radke, C.L.; Chen, J.; Howard, F.; Siwabessy, J.P.; Dulfer, H.; Tran, M.; Consoli, C.; et al. *Seabed Environments, Shallow Sub-Surface Geology and Connectivity, Petrel Sub-Basin, Bonaparte Basin, Timor Sea*; Geoscience Australia: Canberra, Australia, 2014; p. 124.
4. Hovland, M.; Gardner, J.V.; Judd, A.G. The significance of pockmarks to understanding fluid flow processes and geohazards. *Geofluids* **2002**, *2*, 127–136. [CrossRef]
5. Tjelta, T.I.; Svanø, G.; Strout, J.M.; Forsberg, C.F.; Planke, S.; Johansen, H. Gas seepage and pressure build-up at a North Sea platform location: Gas origin, transportation, and potential hazards. In Proceedings of the Offshore Technology Conference (OTC), Houston, TX, USA, 30 April–4 May 2007; p. 11, OTC paper no. 18699.
6. Harris, P.T.; Baker, E.K. *Seafloor Geomorphology as Benthic Habitat*, 1st ed.; Elsevier: London, UK, 2012; p. 900.
7. Decker, C.; Olu, K. Does macrofaunal nutrition vary among habitats at the Håkon Mosby mud volcano? *Cahiers Biol. Mar.* **2010**, *51*, 361–367.
8. Rise, L.; Bellec, V.K.; Chand, S.; Bøe, R. Pockmarks in the southwestern Barents Sea and Finnmark fjords. *Nor. J. Geol. Norsk Geol. Foren.* **2015**, *94*, 263–281. [CrossRef]
9. Sun, Q.; Wu, S.; Hovland, M.; Luo, P.; Lu, Y.; Qu, T. The morphologies and genesis of mega-pockmarks near the Xisha Uplift, South China Sea. *Mar. Petroleum Geol.* **2011**, *28*, 1146–1156. [CrossRef]
10. Andrews, B.D.; Brothers, L.L.; Barnhardt, W.A. Automated feature extraction and spatial organization of seafloor pockmarks, Belfast Bay, Maine, USA. *Geomorphology* **2010**, *124*, 55–64. [CrossRef]
11. Brothers, L.L.; Kelley, J.T.; Belknap, D.F.; Barnhardt, W.A.; Koons, P.O. Pockmarks: Self-scouring seep features? In Proceedings of the 7th International Conference on Gas Hydrates (ICGH 2011), Edinburgh, Scotland, UK, 17–21 July, 2011; p. 10, Paper no. 326.
12. Brothers, L.L.; Kelley, J.T.; Belknap, D.F.; Barnhardt, W.A.; Andrews, B.A.; Landon Maynard, M. Over a century of bathymetric observations and shallow sediment characterization in Belfast Bay, Maine USA: Implications for pockmark field longevity. *Geo-Mar. Lett.* **2011**, *31*, 237–248. [CrossRef]
13. Szpak, M.T.; Monteys, X.; O'Reilly, S.S.; Lilley, M.K.S.; Scott, G.A.; Hart, K.M.; McCarron, S.G.; Kelleher, B.P. Occurrence, characteristics and formation mechanisms of methane generated micro-pockmarks in Dunmanus Bay, Ireland. *Cont. Shelf Res.* **2015**, *103*, 45–59. [CrossRef]
14. Webb, K.E.; Barnes, D.K.A.; Plankea, S. Pockmarks: Refuges for marine benthic biodiversity. *Limnol. Oceanogr.* **2009**, *54*, 1776–1788. [CrossRef]
15. Webb, K.E.; Barnes, D.K.A.; Gray, J.S. Benthic ecology of pockmarks in the Inner Oslofjord, Norway. *Mar. Ecol. Prog. Ser.* **2009**, *387*, 15–25. [CrossRef]
16. Wildish, D.J.; Akagi, H.M.; McKeown, D.L.; Pohle, G.W. Pockmarks influence benthic communities in Passamaquoddy Bay, Bay of Fundy, Canada. *Mar. Ecol. Prog. Ser.* **2008**, *357*, 51–66. [CrossRef]
17. Przeslawksi, R.; Glasby, C.; Nichol, S. Polychaetes (Annelida) of the Oceanic Shoals region, northern Australia: Considering small macrofauna in marine management. *Mar. Freshw. Res.* **2018**. submitted.
18. Gafeira, J.; Long, D.; Diaz-Doce, D. Semi-automated characterisation of seabed pockmarks in the central North Sea. *Near Surf. Geophys.* **2012**, *10*, 303–314. [CrossRef]
19. Weiss, A.D. Topographic Position and Landforms Analysis. In Proceedings of the ESRI International User Conference, San Diego, CA, USA, 9–13 July 2001.
20. Gafeira, J.; Dolan, M.F.J.; Monteys, X. Geomorphometric Characterization of Pockmarks by using a GIS-based Semi-automated Toolbox. *Geosciences* **2018**, *8*, 154. [CrossRef]
21. Nichol, S.L.; Howard, F.J.F.; Kool, J.; Stowar, M.; Bouchet, P.; Radke, L.; Siwabessy, J.; Przeslawski, R.; Picard, K.; Alvarez de Glasby, B.; et al. *Oceanic Shoals Commonwealth Marine Reserve (Timor Sea) Biodiversity Survey: GA0339/SOL5650 Post-Survey Report*; Geoscience Australia: Canberra, Australia, 2013; p. 112.
22. Van Andel, T.H.; Veevers, J.J. Morphology and sediments of the Timor Sea. Department of National Development, Bureau of Mineral Resources. *Geol. Geophys. Bull.* **1967**, *83*, 186.
23. Heap, A.D.; Harris, P.T. Geomorphology of the Australian margin and adjacent seafloor. *Aust. J. Earth Sci.* **2008**, *55*, 555–585. [CrossRef]

24. Przeslawski, R.; Daniell, J.; Anderson, T.; Barrie, V.; Heap, A.; Hughes, M.; Li, J.; Potter, A.; Radke, L.; Siwabessy, J.; et al. *Seabed Habitats and Hazards of the 54 Oceanic Shoals Marine Biodiversity Survey: GA0339/SOL5650—Post Survey Report Joseph Bonaparte Gulf and Timor Sea, Northern Australia*; Geoscience Australia Record: Canberra, Australia, 2011; p. 156.

25. Heap, A.D.; Przeslawski, R.; Radke, L.; Trafford, J.; Battershill, C.; Shipboard Party. *Seabed Environments of the Eastern Joseph Bonaparte Gulf, Northern Australia: SOL4934 Post Survey Report*; Geoscience Australia Record: Canberra, Australia, 2010; p. 81.

26. West, B.G.; Conolly, J.R.; Blevin, J.E.; Miyazaki, S.; Vuckovic, V. *Petroleum Prospectivity of the East Malita Graben area, Bonaparte Basin*; Bureau of Mineral Resources, Geology and Geophysics Record: Canberra, Australia, 1992.

27. Yokoyama, Y.; De Deckker, P.; Lambeck, K.; Johnston, P.; Fifield, L.K. Sea-level at the Last Glacial Maximum: Evidence from northwestern Australia to constrain ice volumes for oxygen isotope stage 2. *Palaeogeogr. Palaeoclimatol. Palaeoecol.* **2001**, *165*, 281–297. [CrossRef]

28. Yokoyama, Y.; Lambeck, K.; De Deckker, P.; Johnston, P.; Fifield, K. Timing of the Last Glacial Maximum from observed sea-level minima. *Nature* **2000**, *406*, 713–716. [CrossRef] [PubMed]

29. George, T.; Cauquil, E. A multi-disciplinary site investigation for the assessment of drilling geohazards and environmental impact within the northern Bonaparte Basin. *Preview* **2010**, *148*, 41–44.

30. Lees, B.G. Recent terrigenous sedimentation in Joseph Bonaparte Gulf, Northwestern Australia. *Mar. Geol.* **1992**, *103*, 199–213. [CrossRef]

31. Taupp, T.; Hellmann, C.; Gergs, R.; Winkelmann, C.; Wetzel, M.A. Life Under Exceptional Conditions—Isotopic Niches of Benthic Invertebrates in the Estuarine Maximum Turbidity Zone. *Estuar. Coasts* **2017**, *40*, 502. [CrossRef]

32. Yokoyama, Y.; Purcell, A.; Lambeck, K.; Johnston, P. Shore-line reconstruction around Australia during the Last Glacial Maximum and Late Glacial Stage. *Quat. Int.* **2001**, *83–85*, 9–18. [CrossRef]

33. Nicholas, W.A.; Nichol, S.L.; Howard, F.J.F.; Picard, K.; Dulfer, H.; Radke, L.C.; Carroll, A.G.; Tran, M.; Siwabessy, P.J.W. Pockmark development in the Petrel Sub-basin, Timor Sea, Northern Australia: Seabed habitat mapping in support of CO_2 storage assessments. *Cont. Shelf Res.* **2014**, *83*, 129–142. [CrossRef]

34. Gunn, P.J. Bonaparte Basin: Evolution and structural framework. In Proceedings of the Petroleum Exploration Society of Australia, Symposium the North West Shelf, Perth, Australia, 10–12 August 1988.

35. Bourget, J.; Nanson, R.; Ainsworth, R.B.; Courgeon, S.; Jorry, S.J.; Al-anzi, H. Seismic stratigraphy of a Plio-Quaternary intra-shelf basin (Bonaparte Shelf, NW Australia). In Proceedings of the Petroleum Exploration Society of Australia Symposium, the Sedimentary Basins of Western Australia IV, Perth, Australia, 18–21 August 2013.

36. De Deckker, P.; Yokoyama, Y. Micropalaeontological evidence for Late Quaternary sea-level changes in Bonaparte Gulf, Australia. *Glob. Planet. Chang.* **2009**, *66*, 85–92. [CrossRef]

37. Radke, L.C.; Li, J.; Douglas, G.; Przeslawski, R.; Nicholl, S.; Siwabessy, J.; Huang, Z.; Trafford, J.; Watson, T.; Whiteway, T. Characterising sediments for a tropical sediment-starved shelf using cluster analysis of physical and geochemical variables. *Environ. Chem.* **2015**, *12*, 204–226. [CrossRef]

38. Gavrilov, A.N.; Siwabessy, P.J.W.; Parnum, I.M. *Multibeam Echo Sounder Backscatter Analysis*; Centre for Marine Science and Technology: Perth, Australia, 2005.

39. Radke, L.; Nicholas, T.; Thompson, P.; Li, J.; Raes, E.; Carey, M.; Atkinson, I.; Huang, Z.; Trafford, J.; Nichol, S. Baseline biogeochemical data from Australia's continental margin links seabed sediments to water column characteristics. *Mar. Freshw. Res.* **2017**. [CrossRef]

40. Evans, J.S.; Oakleaf, J.; Cushman, S.A.; Theobald, D. An ArcGIS Toolbox for Surface Gradient and Geomorphometric Modeling, version 2.0-0. 2014. Available online: http://evansmurphy.wix.com/evansspatial (accessed on 19 March 2018).

41. Fisher, P.; Wood, J.; Cheng, T. Where is Helvellyn? Fuzziness of multi-scale landscape morphometry. *Trans. Inst. Br. Geogr.* **2004**, *29*, 106–128. [CrossRef]

42. Yokoyama, R.; Shirasawa, M.; Pike, R.J. Visualizing topography by openness: A new application of image processing to digital elevation models. *Photogramm. Eng. Remote Sens.* **2002**, *68*, 257–265.

43. Clarke, K.R.; Warwick, R.M. *Change in Marine Communities: An Approach to Statistical Analysis and Interpretation*, 2nd ed.; PRIMER-E: Plymouth, Australia, 2001; p. 172.

44. Anderson, M.J.; Gorley, R.N.; Clarke, K.R. *PERMANOVAþ for PRIMER: Guide to Software and Statistical Methods*; PRIMER-E: Plymouth, UK, 2008.

45. Geoscience Australia. The Australian Bathymetry and Topography 2009 250 m. 2009. Available online: http://www.ga.gov.au/metadata-gateway/metadata/record/89581/ (accessed on 19 March 2018).

46. Radke, L.C.; Webber, E. *Seafloor Environments of the eastern Timor Sea, Northern Australia: Mineralogy of Seabed Sediments*; Geoscience Australia: Canberra, Australia, 2014.

47. Shepard, F.P. Nomenclature based on sand-silt-clay ratios. *J. Sediment. Petrol.* **1954**, *24*, 151–158.

48. Keil, R.G.; Mayer, L.M.; Quay, P.D.; Richey, J.E.; Hedges, J.I. Loss of organic matter from riverine particles in deltas. *Geochim. Cosmochim. Acta* **1997**, *61*, 1507–1511. [CrossRef]

49. Van Rijn, L.C. Principles of Sediment Transport in Rivers, Estuaries and Coastal Seas. 1993. Available online: www.aquapublications.nl (accessed on 19 March 2018).

50. Van Nugteran, P.; Moodley, L.; Brummer, G.-J.; Heip, C.H.R.; Herman, P.M.; Middleburg, J.J. Seafloor ecosystem functioning: The importance of organic matter priming. *Mar. Biol.* **2009**, *156*, 2277–2287. [CrossRef] [PubMed]

51. Snelgrove, P.V.R.; Butman, C.A. Animal-sediment relationships revised: Cause versus effect. *Oceanogr. Mar. Biol. Annu. Rev.* **1994**, *32*, 111–177.

52. McArthur, M.; Brooke, B.; Przeslawski, R.; Ryan, D.A.; Lucieer, V.; Nichol, S.; McCallum, A.W.; Mellin, C.; Cresswell, I.D.; Radke, L.C. On the use of abiotic surrogates to describe marine benthic biodiversity. *Estuar. Coast. Shelf Sci.* **2010**, *88*, 21–32. [CrossRef]

53. Przeslawski, R.; McArthur, M.A.; Anderson, T.J. Infaunal biodiversity patterns from Carnarvon Shelf (Ningaloo Reef), Western Australia. *Mar. Freshw. Res.* **2013**, *64*, 573–583. [CrossRef]

54. Connell, J.H. Diversity in tropical rain forests and coral reefs. *Science* **1978**, *199*, 1302–1310. [CrossRef] [PubMed]

55. Brothers, L.L.; Kelley, J.T.; Belknap, D.F.; Barnhardt, W.A.; Andrews, B.D.; Legere, C.; Hughes Clarke, J.E. Shallow stratigraphic control on pockmark distribution in north temperate estuaries. *Mar. Geol.* **2012**, *329–331*, 34–45. [CrossRef]

geosciences

MDPI

Article

Seabed Morphology and Sedimentary Regimes defining Fishing Grounds along the Eastern Brazilian Shelf

Silvia N. Bourguignon [1], Alex C. Bastos [2,*], Valéria S. Quaresma [2], Fernanda V. Vieira [1], Hudson Pinheiro [3], Gilberto Menezes Amado-Filho [4], Rodrigo Leão de Moura [5] and João Batista Teixeira [1]

[1] Programa de Pós-Graduação em Oceanografia Ambiental, Universidade Federal do Espírito Santo, Vitória 29075-910, Brazil; silviabourg@hotmail.com (S.N.B.); fernanda.vedoato@gmail.com (F.V.V.); jboceano@gmail.com (J.B.T.)
[2] Departamento de Oceanografia, Universidade Federal do Espírito Santo, Vitória 29075-910, Brazil; valeria.quaresma@ufes.br
[3] Department of Ecology and Evolutionary Biology, University of California Santa Cruz, Santa Cruz, CA 95060, USA; htpinheiro@gmail.com
[4] Instituto de Pesquisas Jardim Botânico do Rio de Janeiro, Rua Pacheco Leão 915, Rio de Janeiro 22460-030, Brazil; gilbertoamadofilho@gmail.com
[5] Instituto de Biologia and SAGE/COPPE, Universidade Federal do Rio de Janeiro, Ilha do Fundão, Rio de Janeiro 21944-970, Brazil; moura.uesc@gmail.com
* Correspondence: alex.bastos@ufes.br; Tel.: +55-027-4009-2878

Received: 6 November 2017; Accepted: 3 March 2018; Published: 9 March 2018

Abstract: Shelf morphology and sedimentary regimes are influenced by processes operating at different temporal and spatial scales and are important records of sea level changes and sediment supply and/or carbonate production. The northern continental shelf of Espírito Santo (Brazil) contains evidence of different sedimentary regimes that distribute diverse and complex marine habitats. Herein, seabed morphology, acoustic images of the seafloor (side scan sonar and sub-bottom profiler), and sediment samples were used to investigate the influence of sedimentary regimes on physical marine habitat distribution. Seabed mapping is also integrated with available data on fisheries to understand the influence of shelf morphology and sedimentology in the usage of distinct fishing gears. The results indicate five morpho-sedimentary facies: terrigenous mud, terrigenous sand, rhodolith beds, carbonate gravel with rhodoliths, and hardground. Through an integrated analysis of the geomorphology and sedimentary distribution, two morpho-sedimentary domains were identified: a sediment-fed shelf adjacent to the Doce River associated with a major mud depocenter and a delta front morphology characterized by gentle slopes and low terrain ruggedness, and a sediment-starved shelf dominated by carbonate sedimentation showing an irregular morphology associated with higher slopes and terrain ruggedness. These contrasting morpho-sedimentary domains are a result of sedimentary responses to sea level fluctuation during Late Quaternary, specially, during the deglaciation processes after the Last Glacial Maximum. The morphological and sedimentary contrasts along the area define the physical habitat distribution. The sediment supply regime area is associated with a terrigenous fine/muddy sedimentation bed, which control the local morphology and favors coastal and delta front progradation. This physical habitat is a well-known shrimp-fishing ground where intense trawling takes place, as well as gillnet fisheries targeting weakfish and croakers. The accommodation regime or low sediment influx area is characterized by carbonate sedimentation associated with hardgrounds and rhodolith beds. In contrast, this physical habitat with scarce sediment supply, facilitates extensive benthic colonization by crustose coralline algae (CCA), which is primarily associated to line fisheries, longlines, and spearfishing. Rhodoliths show a high diversity of CCA and the occurrence of an endemic kelp species. Long-term processes such as relative sea level fluctuations and sediment supply are a legacy for the distribution of benthic habitats, and their resulting morphology can be a surrogate for predicting fishing activities or a first-base analysis for marine spatial planning. Available low-resolution bathymetric datasets can be a powerful tool, if applied with caution and in a regional scale approach. Here, terrain variables (terrain slope and ruggedness) derived from an extensive available (low-resolution and interpolated) bathymetric dataset distinguished two contrasting morphological domains characterized by rugged and smooth/flat seabeds.

Keywords: benthic habitats; shelf morphology; eastern Brazilian shelf

1. Introduction

Sedimentation patterns and seabed morphology are controlled by different drivers operating at different spatial and temporal scales. Physical processes, tectonics, climate, sediment input, and relative sea level changes define the sedimentary regimes at continental shelves [1–3].

Morpho-sedimentary features of modern shelves are responses to long-term geological processes. For instance, coastal regions with high sediment supply (depositional regimes) are characterized as prograding coasts and are likely characterized by deltas, coastal plains, and tidal flats. Coastal areas with no sediment supply, or a limited supply (regimes dominated by accommodation), exhibit a geomorphology characteristic of retrograding and starving coasts, defined by erosive processes and features such as cliffs, escarpments, and unfilled river valleys that became estuaries or lagoons [1,4,5].

The role of sedimentary regimes also directly influences the substrate formation and composition. According to [4], many studies have shown that the local geological characteristics are an efficient way to identify physical areas (habitats) in the seabed inhabited by specific biological communities. Important information on seafloor geology and biology can be rapidly acquired by mapping the distribution of benthic habitats. Nichol and Brooke [6] have also pointed out the influence of Late Quaternary transgression on the establishment of distinct seabed habitats.

Herein, we investigate the influence of distinct sedimentary regimes and associated geomorphology on physical habitat and local fisheries. The study area is the central part of the Espírito Santo Continental Shelf (ESCS), East Brazil. This is a multiple-user shelf with distinct stakeholders. The shelf includes a multiple-use Marine Protected Area (MPA), a Marine Wildlife Refuge, a Biological Reserve for nesting marine turtles (both no-take zones), important fishing grounds and artisanal fishing communities, and borders oil and gas rigs along the outer shelf and slope.

2. Materials and Methods

The investigation was carried out using a compilation of bathymetric data, and new acoustic and surface sediment data. The fishing ground information was obtained from workshops performed with five local communities and follows the same procedure adopted in [7]. The data were integrated and interpreted using a Geographic Information System (GIS) platform.

2.1. Study Area

The Espírito Santo continental shelf is located along the Eastern coast of Brazil. The shelf width varies from 50 to 200 km. Its widest part is known as the Abrolhos Shelf or the Abrolhos Bank [8–10]. To the south of the Abrolhos Bank, the shelf width averages 50 km and the shelf break occurs at 60 to 80 m water depth. The Doce river is the main sediment source to the shelf, with an average load of 133×10^6 t during the rainy season (November to April) [11]. Shelf morphology varies significantly with four distinct regions: Abrolhos shelf, Doce river shelf, Paleovalley shelf, and the wide-southern inner shelf [10]. Here, we focus on contrasting the Doce river shelf with the Paleovalley shelf (Figures 1 and 2).

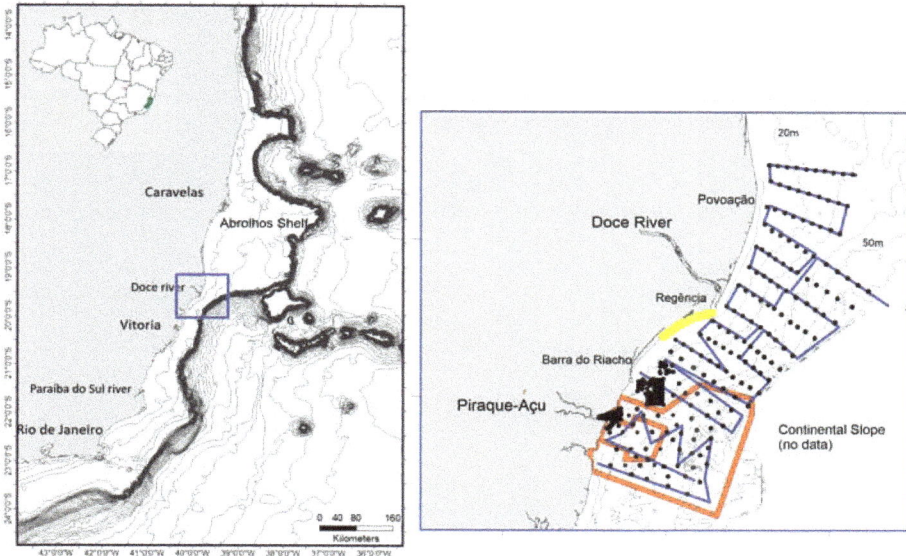

Figure 1. Study area location along the Southeast Brazilian Shelf. Black dots are sampling sites and blue lines are side scan sonar and sub-bottom profiler transects. Orange polygons are the Marine Protected Area—Costas das Algas MPA and the Wildlife Refuge Santa Cruz (inner polygon). Yellow line represents the Comboios Biological Reserve.

2.2. Data Acquisition

2.2.1. Acoustic Imaging

Data collection was conducted using a Side Scan Sonar Edgetech 4100 with a 560p Digital Acquisition System, which included a 272 TD Towfish with Deep-Tow and a Dual-Frequency (100 and 500 Khz) Transducer coupled to a Garmin GPS and the Acquisition Software Discover 4100 (Figure 1). Side-Scan Range was set in 200 or 300 m full swath.

Approximately 1600 km of side scan data were collected. Acoustic images were processed using the software SonarWiz 5 (Chesapeak Technology, Mountain View, CA, USA). Sonograms were corrected for bottom track and signal attenuation. Processed sonograms were exported as geotiff files (1m pixel resolution) and interpreted in a GIS platform.

2.2.2. Sub-Bottom Profiling

Sub-bottom profiles were collected to observe the morphology variations along the shelf using the surface reflector only and not for sub-surface or stratigraphic interpretations. Profiling data were acquired using a sub-bottom profiler, SyQuest StrataBox model, operating at 3.5 KHz. Data were received and recorded by the software StrataBox 3.0.4.1 (SyQuest Technology, San Jose, CA, USA) and processed using the software SonarWiz 5. A total of 1400 km of sub-bottom profile data were acquired.

2.2.3. Sediment Sampling

Sediment samples were collected using a Van Veen grab (n = 187 surficial sediment samples) (Figure 1) and processed for grain size analysis and calcium carbonate content. A laser granulometer (Malvern Mastersizer 2000 (Malvern Panalytical, Nottingham, UK)) was used for particle size analysis of the mud fraction and the sieving method was used for the sand fraction. Calcium carbonate content was determined through dissolution with 10% hydrochloric acid.

The samples were described using the classification proposed by [12], based on carbonate content and grain texture, including: lithoclastic/terrigenous ($CaCO_3$ < 30%), lithobioclastic (30 to 50% $CaCO_3$), biolithoclastic (50 to 70% $CaCO_3$), and bioclastic (>70% $CaCO_3$). Lithobioclastic and biolithoclastic are described as mixed sediments.

Seabed images were obtained with a dropcamera consisting of a high-resolution camera (Go Pro Silver 3) attached to a steel frame, and looking vertically down. The base of the frame is a 60×60 cm^2.

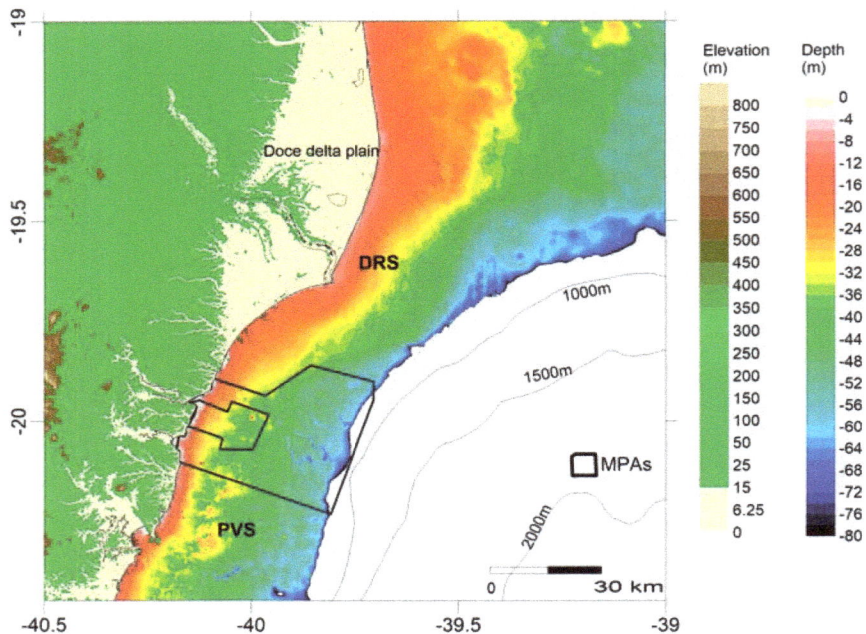

Figure 2. Digital Terrain model of the study region. Continental data was obtained from Topodata database (http://www.dsr.inpe.br/topodata/), and bathymetrica data was digitized from sounding sheets provided by the Brazilian Hydrographic Office. The seabed morphology has a spatial resolution of 200m along the continental shelf. PVS: Paleovalley Shelf; DRS: Doce River Shelf. Datum is WGS 84.

2.3. Shelf Morphology Analysis

The bathymetric map is the result of a compilation of available data (printed sounding sheets and digital files) from the Brazilian Hydrographic Office. Ten sounding sheets were scanned, digitized, and gridded using Surfer 10 (Golden Software, Golden, CO, USA) and ArcGIS. A total of more than 150,000 points were digitized. Sounding sheets represent data collected by single beam echosounders from the 60's to the 80's. Positioning accuracy may vary due to changes in coordinate acquisition systems and methodologies (for more details see [10]). Original datum in the sounding sheets was Corrego Alegre that was transformed and corrected to WGS84, following the available digital data. Original sounding sheets scales varied from 1:10,000 to 1:160,000. All the bathymetric data obtained by the Brazilian Navy follows a protocol, so by not compiling data from other sources (such as satellite derived data), no mismatch was observed. To avoid mismatch, we limited our map to the 100 m isobaths (available sounding data). A data quality analysis was performed qualitatively considering the regional scale of the study. The only observed suspicious features were broad-scale circular lines to the north of the study area. These features seem to be an artifact from the original distribution of the sounding data.

Terrain analysis was carried out using an ArcGIS toolbox, the Benthic Terrain Modeler-BTM [13]. The bathymetric map used for the morphometric calculations has a resolution of 200 m, following the dataset minimum resolution. Slope and Terrain Ruggedness were the input variables used in BTM. Slope represents the inclination of the seabed expressed in degrees and Terrain Ruggedness was calculated using the Vector Ruggedness measure—VRM. This variable represents terrain ruggedness as the variation in three-dimensional orientation of grid cells within a neighborhood, effectively capturing slope and aspect variability into a single measure [14,15]. Slope and VRM calculations is 3 × 3 pixels, following the toolbox default. The morphometric analysis was restricted to seafloor morphology, slope, and terrain ruggedness.

3. Results

3.1. Shelf Geomorphology

The continental shelf presents a contrasting morphology characterized by a flat deltaic lobe with gentle slopes and an irregular relief marked by paleovalleys with greater slopes (Figures 2 and 3). A deltaic lobe is observed associated to the Doce River mouth. In this area, the shelf shows a flatbed (gradients < 0.2°) with a minor step at 30 m depth. Offshore, below 50 m water depth, an increase in seabed gradient (>0.8°) is observed where the shelf breaks. To the south, the inner shelf still presents a gently sloping seabed to Barra do Riacho, but the 30 m isobath gets closer to the coastline. In contrast, adjacent to the Piraque Açu River mouth, the seabed morphology is characterized by an irregular bottom with a wide variation in slope (up to 5° associated with paleovalleys' wall). The shelf morphology in this latter region is characterized by paleovalleys and marine abrasion terraces.

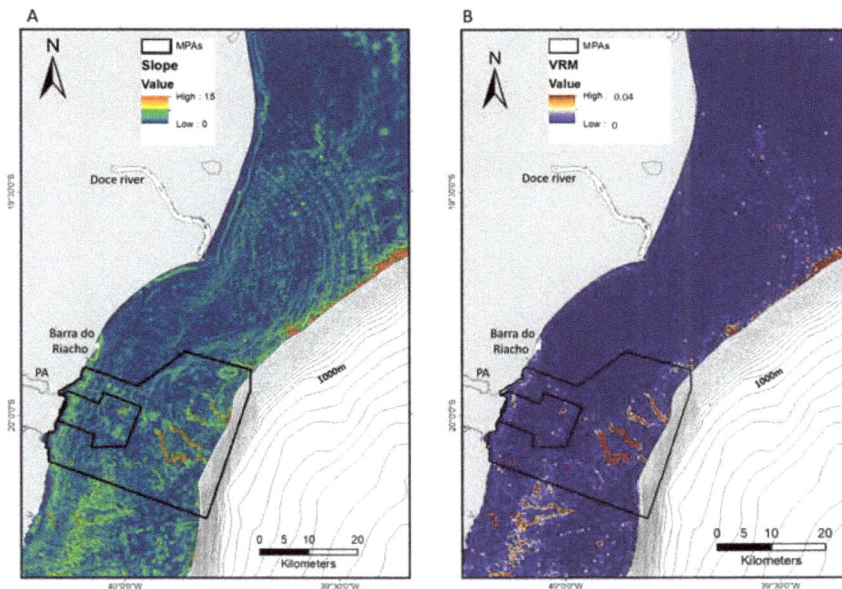

Figure 3. Terrain variables Slope (in degrees) (**A**) and Terrain Ruggedness; (**B**) calculated with the Vector Ruggedness Measure using ArcGIS tool Benthic Terrain Modeler (BTM). Datum is WGS 84. Depth contour intervals: 25 m from 0 to 100 m, 50 m from 100 to 1000 m, and 100 m from 1000 to 2000 m.

Analysis of the terrain slope and ruggedness indicate two contrasting areas. The deltaic lobe shelf has a lower variability in slopes with low values of terrain ruggedness. Higher slopes are related to deltaic lobe fronts, dipping seaward. In contrast, the shelf with paleovalleys shows a higher value of terrain ruggedness, with higher slopes.

3.2. Morpho-Sedimentary Facies—Physical Habitats

The shelf presents a heterogeneous substrate in morphologically distinct seabeds. Five morpho-sedimentary facies (MSF) were identified (Figure 4) based on shelf morphology and sedimentary facies: Terrigenous Mud, Terrigenous Sand, Bioclastic Gravel with Rhodoliths, Rhodoliths, and Coastal Hardground. These MSF are treated herein as physical marine habitats and are described as follows.

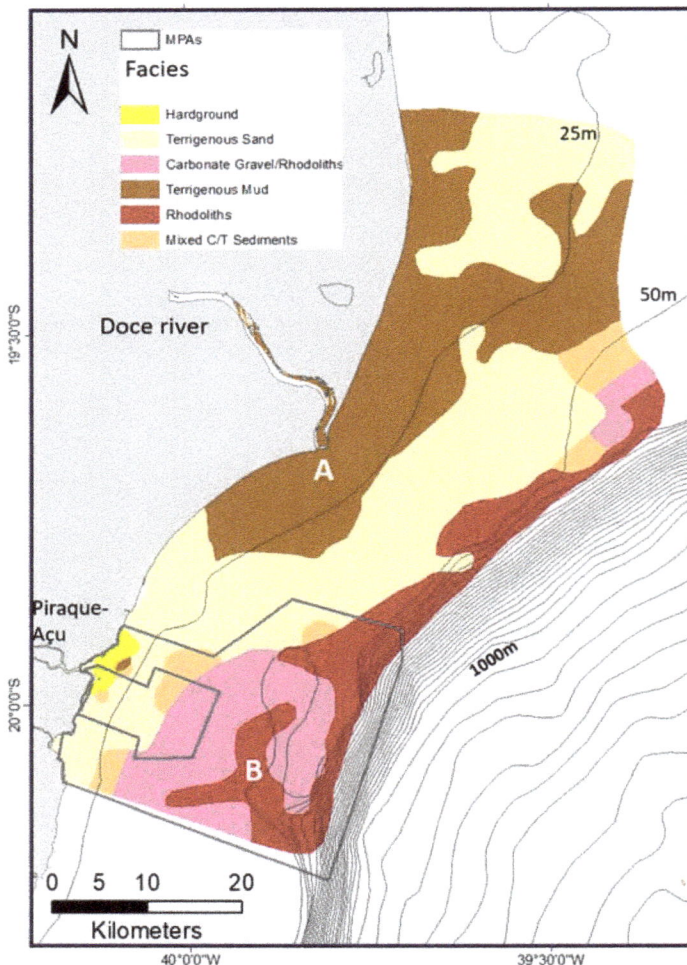

Figure 4. Morpho-sedimentary Facies distribution. (**A**) and (**B**) are sonographic and seabed images locations shown in Figure 5. The dataset used to produce this map is shown in Figure 1, including the full side scan sonar and sub-bottom profiler coverage. Depth contour intervals: 25 m from 0 to 100 m, 50 m from 100 to 1000 m, and 100 m from 1000 to 2000 m.

3.2.1. Terrigenous Mud

This domain is primarily observed adjacent to the Doce River mouth, extending throughout the inner shelf; it was also observed in a small stretch of the middle and outer shelf (Figure 4). This habitat comprised 29% of the study region. The Terrigenous Mud MSF represents the deltaic lobe morphology, with a gentle slope and extending up to 25–30 m water depth (Figure 3). It is composed basically of terrigenous mud and sandy mud. The lithoclastic composition along the coastal region near the Doce River mouth reflects the continental inflow, representing riverine sediments.

3.2.2. Terrigenous Sand

The terrigenous sandy bottom MSF comprises the largest habitat of the studied region (41%). This physical habitat is observed in the inner and mid shelf, but shows a quite irregular distribution (Figure 5). Along the shelf adjacent to the Doce River and to its north, the sandy domain only occurs below 30 m water depth. Southwards, away from riverine mud influence, sandy bottoms occur along the inner shelf, in shallower waters. In morpho-sedimentary terms, this habitat is again characterized by lithoclastic sands, with grain sizes ranging from coarse to fine, but predominantly medium and fine sands.

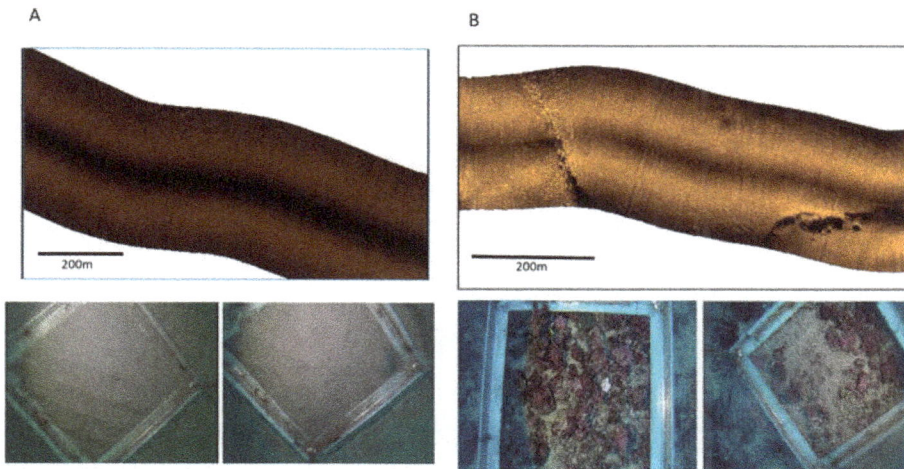

Figure 5. Side scan sonar images and dropcamera seabed frames: (**A**) Terrigenous fine sediment facies characterized by a homogeneous and low backscatter signal; (**B**) rhodoliths characterized by a high backscatter signal with acoustic shadows. For images location, see Figure 4.

The occurrence of these sandy bottoms is also associated with a gentle slope and smooth flatbed along the shelf adjacent to the Doce River, but with a more irregular morphology in the inner shelf adjacent to the Piraque Açu river mouth. Remarkably, this domain is related to a second lobe-like morphology feature along the mid-outer shelf adjacent to the Doce River mouth. In this area, this MSF occurs offshore from the mud deposit (Figure 4).

3.2.3. Rhodolith Beds

The Rhodolith Beds MSF were consistently observed in the outer shelf up to the shelf break (Figure 4), composing 14% of the study area. This facies is represented by a gravelly carbonate fraction, composed from granules to rhodoliths. The extensive rhodolith area indicates that terrigenous sedimentation influence in this region is negligible, since the CCA that form rhodoliths are more likely to grow in areas with low sedimentary inflow [16].

The occurrence of a rhodolith habitat is related to water depth and is adjacent to an irregular bottom morphology. Rhodolith beds are associated to a strong acoustic signature in side scan sonar data. This habitat predominates in depths >40–45 m, being associated with paleochannels and terraces, mainly along the southern part of the shelf. Locally, the formation of hardground or crusts by coalescence of CCA is also observed. Individual rhodoliths exhibited a typical nodular shape, with surface irregularities and diameters ranging between 1 and 10 cm.

3.2.4. Bioclastic Gravel with Rhodoliths

The bioclastic gravel with rhodoliths MSF composes 16% of the study area and is located along much of the shelf adjacent to the Piraque Açu river mouth, within the Costa das Algas Marine Protected Area (Figure 1). This domain is characterized by coarse fragments of biogenic material, including CCA and carbonate skeletons (including shell fragments). Rhodoliths occur along this domain, intermingled with bioclastic gravel. The seafloor is dominated by gravelly sediments and rhodoliths, forming a very irregular morphology, and is associated to paleochannels and hardground terraces.

3.2.5. Coastal Hardground

Coastal hardground is composed mainly by ferrigenous crusts (lateritic crusts) incrusted by organisms. It was only observed at the mouth of the Piraquê-Açu river, although it is a common feature in the intertidal zone along the southern coast of the area. These reefs are formed initially by sediment laterization, followed by marine abrasion during different cycles of sea level changes. Once a hardground, it becomes a substrate for incrustation.

3.3. Fishing Activities

Distribution of fishing grounds was obtained from workshops performed with five local communities. Five major fishery types were described (Table 1).

Table 1. Fishery types and ground areas. Dive includes spearfishing and shellfishing, and Gillnet includes bottom fishing net and fishing net.

	Area (ha) for Each Fishery Type				
	Trawlling	Longline	Angling	Dive	Gillnet
Fishing Ground	40,844	47,003	190,287	10,845	92,055

The relationship between fisheries with distinct MSF and the percentage of each fishery activity over each habitat domain is shown in Figure 6. Most trawling (96%) and all bottom-associated gillnet fisheries are carried out in the terrigenous sediment flatbed domains. Over the carbonate domain with higher terrain ruggedness and slope, fisheries activities are related to line and longline fishing, as well as to spearfishing. Across the Doce river inner shelf, for example, the areas identified as muddy bottom habitats are associated with shrimp trawling artisanal fisheries [17], as well as bottom-associated gillnet fisheries targeting carnivorous fishes such as sciaenids, carangids, and elasmobranchs [18]. According to [19], from Barra do Riacho to the mouth of the Doce River, trawling is performed by both artisanal and industrial boats, mostly targeting shrimp. In a more recent survey from workshops with local fishermen, [20] also highlighted that in this southern region adjacent to the mouth of the Doce River, shrimp fishing is widespread due to the concentration of sandy-muddy bottoms, whereas on seafloors with a muddy composition, whiting fishing is more frequent. Along this stretch, the fishing area is continuous and is used from the coastline to the 40 m isobaths.

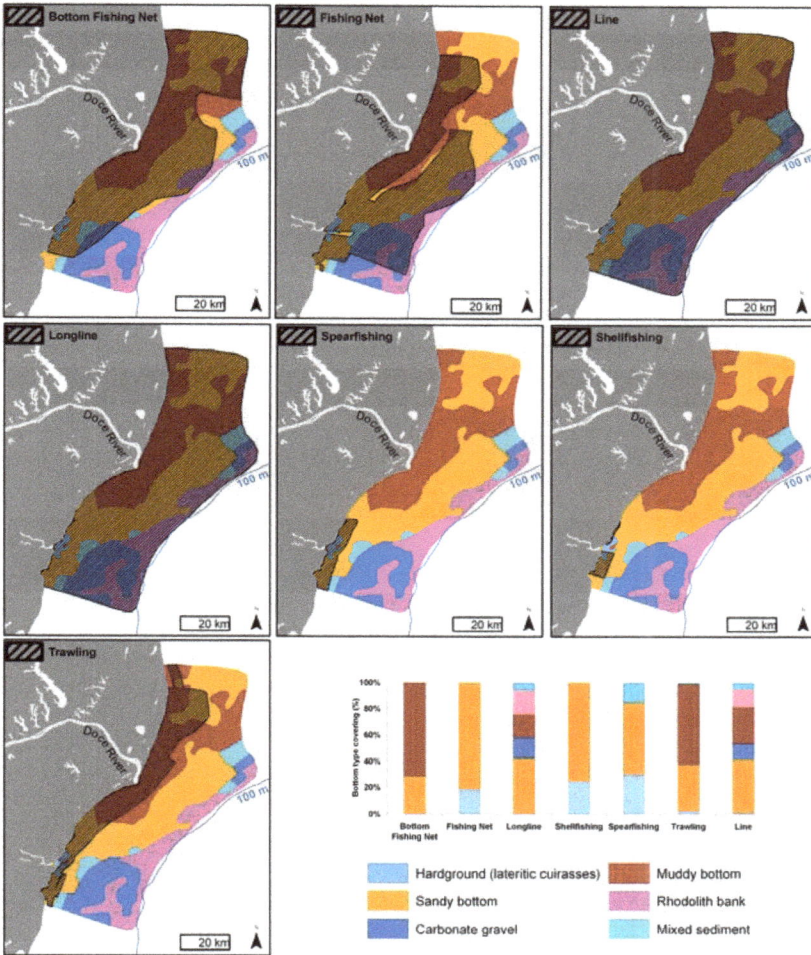

Figure 6. Relationship between different fisheries types and sedimentary facies. The graph shows the percent cover of each fishery in the distinct sedimentary domains.

4. Discussion

The five MSF indicate the occurrence of two main sedimentary regimes: terrigenous supply regime and accommodation/carbonate sedimentation regime. The sedimentation pattern along the inner shelf adjacent to the Doce River is characterized by modern terrigenous sediment input, forming a submarine muddy delta lobe, representing a supply regime area. To the south and offshore from the terrigenous mud deposits, very low or no significant sediment input takes place. Shelf morphology is rather irregular due to erosion or resembles a relict morphological feature, characterizing an accommodation regime [5]. This area is characterized by relict/palimsestic sediments (terrigenous sands) and mixed sediments. Along the outer shelf and parts of the mid shelf, the sedimentary regime is predominantly carbonate, with development of bioclastic gravels and rhodolith beds. It is noteworthy that the carbonate sedimentation develops over an accommodation area characterized by low sediment input and relict morphological features, such as unfilled paleochannels, formed during the last lowstand [10]. Thus, carbonate sedimentation is defined by biologically-controlled precipitation (e.g., crustose coralline algae–rhodoliths), but also by bioclastic particles transport and deposition.

The distribution of these regimes is influenced by processes that operate at different time and spatial scales. Typically, sedimentation is controlled by a combination of autogenic and allogenic processes, which determine the distribution of elements and geological variables in a depositional system [1,5]. Shelf morphology, at this scale, is a product of long-term processes, such as sea level changes.

The alteration of such controls introduces an imbalance between sediment supply and modifications within the basin, characterizing local sedimentary regimes and leading to depositional processes of aggradation or erosion, as well as coastal displacement [1,5,21]. Typically, if the rates of sediment supply are higher than sea level fall, the sedimentary regime is dominated by supply and accompanied by progradation or shoreline regression [1]. Conversely, under no, or minimal sediment supply, the trend is erosion, modification of the shoreline by wave action and, therefore, potential shoreline transgression. This process establishes a sedimentary regime dominated by accommodation accompanied by retrogradation of the coast and marine facies overlying the terrigenous facies [1,5,22].

The sedimentary facies distribution is characterized by lithoclastic (terrigenous), mixed, and bioclastic (carbonate) sediments, and is closely associated to seabed morphology (Figure 6). Terrigenous mud and sand (primarily sandy mud and medium to fine sands) are observed along the deltaic gentle slope flatbeds, with low terrain ruggedness. A muddy patch is found adjacent to the river mouth. Areas with mixed, biolithoclastic, or lithobioclastic sedimentation are also identified; deposits were less common and might represent a transition from terrigenous to carbonate sedimentation. These sediments show a wide range of grain sizes, from sandy mud to gravelly sands. Bioclastic sediments comprise a higher concentration of gravelly and sandy fractions, which dominate the mid-outer shelf in the southern portion of the study area and the outer shelf in its northern part. Rhodolith beds occur along the outer shelf. These carbonate-dominated sediments are associated with higher terrain ruggedness, which is expressed by an irregular shelf morphology with marine abrasion terraces and paleovalleys.

Thus, combining spatial changes in bed slope, which represents the contrasting shelf geomorphology, with sedimentation regimes (terrigenous and carbonate), we can define two major Morpho-Sedimentary Domains (MSD): Terrigenous (terrigenous mud + terrigenous sand facies) and Carbonate (rhodoliths + bioclastic gravel with rhodoliths facies). The Carbonate MSD is related to higher slope, terrain ruggedness, and deeper depths, with carbonate-dominated facies, whereas the Terrigenous MSD is associated with gently inclined bottom, low ruggedness, and shallower areas with siliciclastic sediments. The two MSD are clearly observed in Figure 7, where long and cross-shelf bathymetric profiles show changes in morpho-sedimentary characteristics. MSD distribution in relation to a terrain variable (slope) is shown in Figure 8.

In the Doce River area, the sedimentary input is dominated by supply, characterized by the progradation of the coastline. Typically, the deposits in the region are composed of fine to very fine terrigenous sediments. Such deposits are described as regressive deposits associated to high sediment input, promoting rapid accumulation. Other parts of the Brazilian shelf also exhibit a similar sedimentary regime and morphological characteristics, being dominated by deltas in between starved areas [23]. Four main deltas occur (from south to north): the Paraíba do Sul river delta, Doce river delta, Jequitinhonha river delta, and São Francisco river delta. Sedimentation patterns in the adjacent shelf of these deltas are very similar. Along the narrow eastern Brazilian shelf, [24] identified fine terrigenous sediment facies associated to river mouths or shelf incised valleys and carbonate dominated facies. For the São Francisco River shelf (AL), [25] described a progradation zone in the shelf, characterized by mud and muddy sands.

The presence of muddy bottom and terrigenous sand habitats in the study region suggests that the geological characteristics of the substrate drive demersal and benthic communities structured by scavengers. Substrates with unconsolidated marine deposits are typically associated with flat bottoms and a thick sedimentary cover [26], which was identified in both domains, particularly in the submarine delta lobes of the Doce River. Such morpho-sedimentary properties attract scavenging communities (e.g., shrimps, crabs, catfishes), since the thickness of the terrigenous sediments is directly associated with the levels of organic carbon supplied. In the coastal zone, this carbon is partially recycled and partially deposited with the terrigenous sediments [26]. Such sediments with high concentration of organic carbon provide an important food supply for high-biomass demersal and benthic communities.

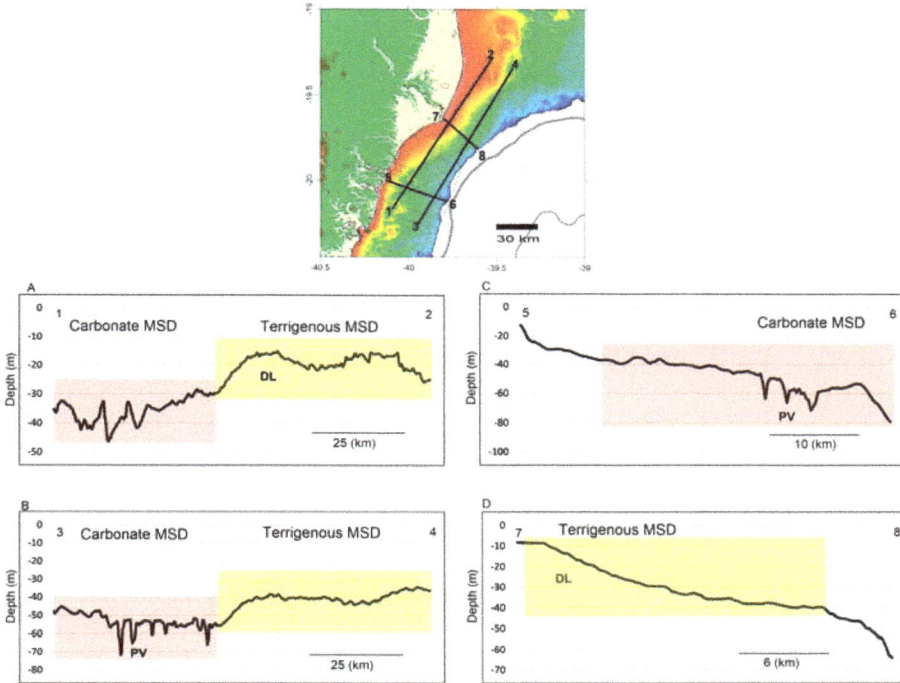

Figure 7. Long-shore and cross-shelf bathymetric profiles and associated MSD. PV—paleovalley; DL—delta front; (**A**) long-shore, inner shelf bathymetric profile showing the morphological and faciologic transition between MSDs, (**B**) long-shore, outer shelf bathymetric profile showing the morphological and faciologic transition between MSDs, (**C**) cross-shelf bathymetric profile along the Carbonate MSD, (**D**) cross-shelf bathymetric profile along the Terrigenous MSD.

Unlike the Doce River region, the southern section is characterized by a contrasting sedimentary regime (Figure 4). The continental shelf is deficient in sediment supply, with very low or negligible inflow, and is characterized by a sedimentary accommodation-dominated regime. The seabed habitats comprise coarse sediments, predominantly coarse sand, bioclastic gravel fractions, and extensive occurrence of rhodolith beds. These aspects support the transgressive characteristics in the area [10].

With the rapid rise in sea level, waves and tides eroded the coastline and shoreface substrates, which generated coarse particle deposits that were later modified and produced low sedimentation rates [5,22]. The low volume of accumulated and deposited sediments characterizes the inner and middle shelf of the region as a "starved" shelf with thinner deposits, coarse relict sediments, and carbonate sedimentation related to the observed habitats (rhodolith beds) [10].

Thus, the irregular morphology and the presence of paleovalleys are a result of marine transgressions under negligible sediment input. These unfilled paleochannels control the habitat morphology and likely act as sediment pathways from the shelf to the slope. A similar situation was recorded in the central continental shelf in Bahia, Northeastern Brazil [27], which was classified as a starving shelf. This classification was especially evident for the middle and outer portions, where several valleys dissecting the shelf were observed. In south Espírito Santo, based on surveys by [7], the accommodation system also dominates this portion of the shelf, which comprises low terrigenous mud and sand contents, while carbonate gravel and rhodolith dominate the shelf.

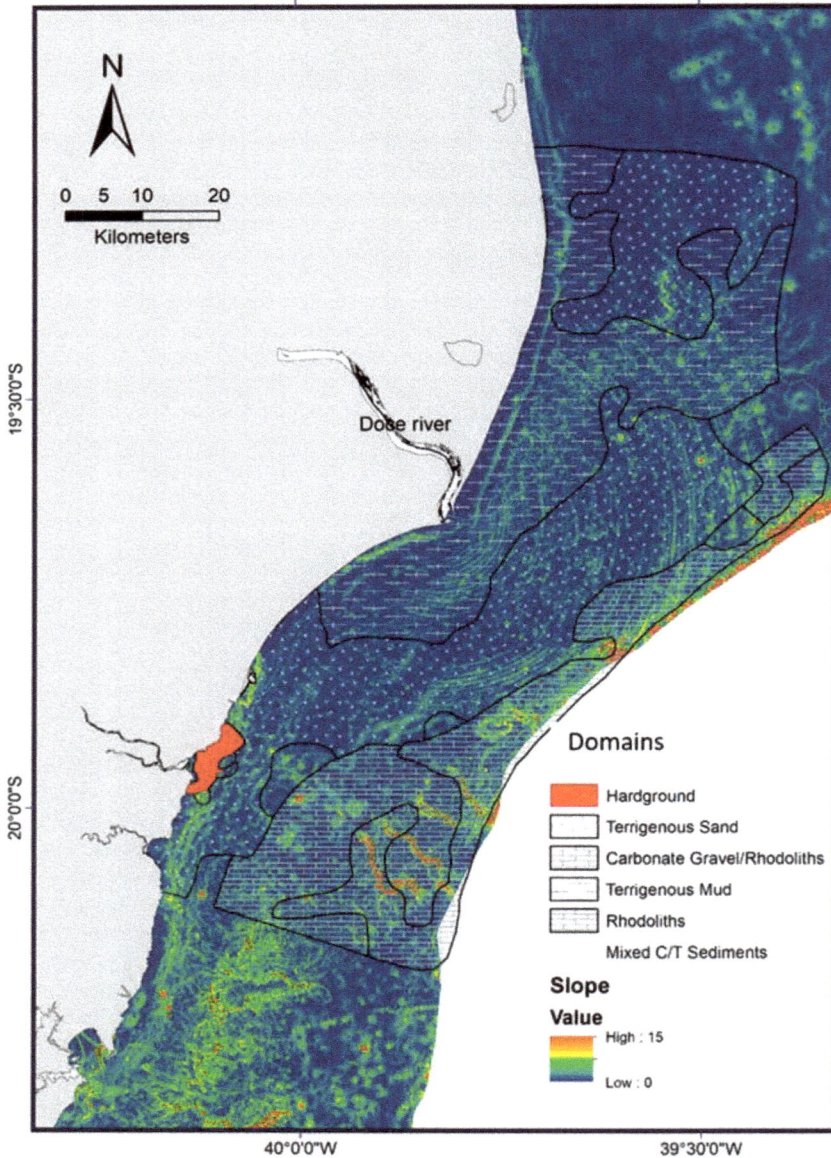

Figure 8. Map combining carbonate and terrigenous domains with a terrain variable (slope).

The outer shelf of the studied area was dominated by an extensive rhodolith bed. The development of rhodolith beds along the eastern Brazilian shelf results from the rising sea level over the last 20 thousand years [28]. The development of the extensive accumulations of CCA on the tropical Brazilian shelf is attributed to ecological conditions that include low terrigenous input, good light penetration, efficient water circulation, and a relatively stable substrate [28,29]. Furthermore, the structural pattern of these rhodolith beds along depth gradients may also be related to the combined extension and inclination of the continental shelf [30,31]. In Abrolhos, for example, [32,33] the geological characteristics of the shelf, combined with meteoceanographic constraints, promote favorable conditions for rhodolith growth.

Rhodoliths highly influence the associated organisms and increase biodiversity, primarily by increasing structural complexity [30,34], being therefore considered ecosystem engineers [16]. The rhodolith structure provides a hard surface and three-dimensional substrate that serves as a microhabitat for a wide variety of associated invertebrates, algae and fish, many of which are economically and ecologically important [35–37]. Moreover, rhodoliths represent one of the world's largest deposits of calcium carbonate ($CaCO_3$), with an estimated 2×10^{11} t [28,38].

The Terrigenous Mud and Terrigenous Sand shelter less diverse benthic assemblages. In the region near the coastal plain of the Doce River, the benthic habitat is not amenable to benthic marine macroalgae fixation, [37] due to the large continental sediment contribution that prevents settlement and development. In contrast, the region further south, adjacent to the Piraque-Açu River, is considered the richest Brazilian region in algal species, with CCA species endemism [39,40] in southern Espírito Santo. Bioclastic Gravel with Rhodoliths and Rhodoliths beds represent an important habitat for several commercially important and threatened fish species [41]. These ecosystems also provide an important habitat for benthic communities and maintain 25% of the known macroalgae species on the Brazilian coast [30]. Thus, it appears that the low sedimentary inflow in the Aracruz and adjacent shelves facilitates the colonization of vast rhodolith beds, conversely to the inner shelf adjacent to the Doce River. Recent studies highlight the presence of 74 fish species in mesophotic rhodoliths beds, including an endemic fish genus and many endangered species [42]. At least 18 commercially important species are associated with this habitat, most of them captured by lines and longlines, such as porgies (*Pagrus pagrus* and *Calamus* sp.), snappers (*Lutjanus buccanella*, *L. cyanopterus* and *Ocyurus chrysurus*), groupers (six species), and amberjacks (*Seriola* spp.) [42].

The shelf morphology and sedimentary distribution over the study area is strongly controlled by long-term processes, such as sea level changes and sediment input. Thus, fishing grounds are also controlled or related to this long-term response, i.e., a legacy of the Late Pleistocene-Holocene sedimentary evolution. Short-term and seasonal oceanographic processes are important, especially in terms of water column variability (temperature, salinity, turbidity, dissolved oxygen, primary production, etc.) and bed disturbance. However, the physical setting controlling the benthic habitat is the seabed morphology and sediment distribution, which is related to the geological background.

Seabed characterization is a major source of environmental information [7], such as environmentally relevant areas and fishing grounds. For instance, trawling activities can be less impacting in the northern areas where muddy bottoms are more widespread than in the southern area. Angling and gillnetting in the south area can help shift the currently high fishing effort over the shallow lateritic cuirasses towards less fished rhodolith bed and sand bottom. In this context, understanding the sedimentation regime is important because it is a key factor that directly influences resources' distribution. Moreover, shelf morphology and terrain analysis reflect sedimentation regimes in the study area and can be used as a primary surrogate by managers and decision-makers in a first step for marine spatial planning.

Finally, it is also relevant to highlight the importance of using available bathymetric datasets, even if they are old or interpolated, but reach minimum data quality. High-resolution multibeam bathymetric models are not commonly available for regional seabed analysis. In this case, available sounding sheets or regional digital single-beam data are the only sources for a morphological or geomorphological analysis of the continental margin. Here, we have used an extensive dataset with a minimum quality and a reasonable accuracy to conduct a terrain analysis from an interpolated digital terrain or bathymetric model. The available data and the derived map are powerful tools for a first or basic analysis for marine management, so the use of these sounding sheets became a must in Brazil and elsewhere. Results can be severely impacted by eventual inaccuracies of the data, such as, errors in data positioning due to changes in technology, datum transformation, or distortion from the printed sounding sheets, if caution is not applied. The results, either terrain variables or the digital terrain model, must be used with caution as limitations apply for seabed morphometric analysis. Usually, interpolated digital terrain models will be able to separate distinct morpho-sedimentary domains in a regional scale [43], but can be very inaccurate if one decides to undertake a high-resolution analysis. Herein, we applied the available dataset with caution and carried out a regional analysis distinguishing end-terms morpho-sedimentary domains; rugged from flat seabeds. The regional contrasting regimes and seabed morphology allowed this interpretation, showing that, in a regional scale, available bathymetric dataset can be very useful.

5. Conclusions

Morphological and sedimentological analysis of a sedimentary regime-contrasting continental shelf revealed the role played by long-term processes in marine physical habitat distribution. Areas with sediment supply regime exhibited physical habitats associated with terrigenous fine sedimentary facies with local prograding morphology. Terrigenous sedimentation was associated with gently inclined beds with low terrain ruggedness. Areas with very low sediment input represent the accommodation regime, in this case, a carbonate dominated shelf. Physical habitats are characterized by bioclastic gravels and living rhodoliths, occurring in deeper areas with irregular morphology, higher slopes, and terrain ruggedness. These distinct habitats explain fishery systems and exploited resources. Understanding the prevailing oceanographic and biological conditions is essential for the management of marine resources, but the study of the geological processes that drive sedimentation regimes allows for a broad understanding of the origin and distribution of marine habitats and fishery resources. In the studied region, the contrasting geomorphology is a reliable surrogate for fishing activities. Terrain analysis involving spatial variability in slope and ruggedness may be more widely used by managers and planners, not only as indicatives of sedimentary regimes, but also for mapping fishing grounds and for designing more effective biodiversity assessments. Moreover, if used with caution, available low-resolution bathymetric datasets can be applied for geomorphometric analysis in a regional scale approach. Here, terrain variables (terrain slope and ruggedness) derived from an extensive available (low-resolution and interpolated) bathymetric dataset distinguished two contrasting morphological domains characterized by rugged and smooth/flat seabeds.

Acknowledgments: Authors want to thank FAPES (Universal and PPE Gerenciamento Costeiro) for two Research Grants that funded data compilation and acquisition and CAPES (Ministry of Education, Brazilian Government) for providing a scholarship for the first Author (SNB). This manuscript is a contribution to the Long-term Ecological Program (PELD ABROLHOS, CNPq/FAPES/CAPES). Financial support for part of the data acquisition was from PD&I ANP/BRASOIL (proc. 48610.011015/2014-55). We thank the reviewers and the Academic editor for their valuable comments that improved the manuscript.

Author Contributions: Silvia N. Bourguignon, Alex C. Bastos, Valéria S. Quaresma, Fernanda V. Vieira, Hudson Pinheiro and João Batista Teixeira have contributed equally to the research and manuscript writing. Rodrigo Leão de Moura and Gilberto Menezes Amado-Filho contributed with final data analysis and manuscript discussion and revision.

References

1. Catuneanu, O. *Principles of Sequence Stratigraphy*; Elsevier Science: Amsterdam, The Netherlands, 2006; p. 386.

2. Cowell, P.J.; Thom, B.G. Morphodynamics of coastal evolution. In *Coastal Evolution: Late Quaternary Shoreline Morphodynamics*; Carter, R.W.G., Woodroffe, C.D., Eds.; University Press: Cambridge, UK, 1994; p. 540.

3. Sternberg, R.; Nowell, A.R.M. Continental shelf sedimentology: Scales of investigation define future research opportunities. *J. Sea Res.* **1999**, *41*, 55–71. [CrossRef]

4. Harris, P.T.; Baker, E.K. 1—Why Map Benthic Habitats? In *Seafloor Geomorphology as Benthic Habitat*; Elsevier: London, UK, 2012; pp. 3–22.

5. Swift, D.J.P.; Phillips, S.; Thorne, J.A. Sedimentation on Continental Margins, IV: Lithofacies and depositional systems. In *Shel and Sandstone Bodies: Geometry, Facies and Sequence Stratigraphy*; Swift, D.J.P., Oertel, G.F., Tillman, R.W., Thorne, J.A., Eds.; Wiley: Hoboken, NJ, USA, 1991; pp. 89–152.

6. Nichol, S.L.; Brooke, B.P. Shelf habitat distribution as a legacy of Late Quaternary marine transgressions: A case study from a tropical carbonate province. *Cont. Shelf Res.* **2011**, *31*, 1845–1857. [CrossRef]

7. Teixeira, J.B.; Martins, A.S.; Pinheiro, H.T.; Secchin, N.A.; Leão de Moura, R.; Bastos, A.C. Traditional ecological knowledge and the mapping of benthic marine habitats. *J. Environ. Manag.* **2013**, *115*, 241–250. [CrossRef] [PubMed]

8. D'agostini, D.P.; Bastos, A.C.; Dos Reis, A.T. The modern mixed carbonate–siliciclastic abrolhos shelf: Implications for a mixed depositional model. *J. Sediment. Res.* **2015**, *85*, 124–139. [CrossRef]

9. Moura, R.L.; Secchin, N.A.; Amado-Filho, G.M.; Francini-Filho, R.B.; Freitas, M.O.; Minte-Vera, C.V.; Teixeira, J.B.; Thompson, F.L.; Dutra, G.F.; Sumida, P.Y.G.; et al. Spatial patterns of benthic megahabitats and conservation planning in the Abrolhos Bank. *Cont. Shelf Res.* **2013**, *70*, 109–117. [CrossRef]

10. Bastos, A.C.; Quaresma, V.S.; Marangoni, M.B.; D'Agostini, D.P.; Bourguignon, S.N.; Cetto, P.H.; Silva, A.E.; Filho, G.M.A.; Moura, R.L.; Collins, M. Shelf morphology as an indicator of sedimentary regimes: A synthesis from a mixed siliciclasticcarbonate shelf on the eastern Brazilian margin. *J. S. Am. Earth Sci.* **2015**, 125–136. [CrossRef]

11. Oliveira, K.S.S.; Quaresma, V.S. Temporal variability in the suspended sediment load and streamflow of the Doce River. *J. S. Am. Earth Sci.* **2017**, *78*, 101–115. [CrossRef]

12. Larsonneur, C. La cartographie de's dépots meubles sur le plateau continental français: Méthode mise du points et utilisée em Manche. *J. Recherche Oceanogr.* **1977**, *2*, 34–39.

13. Wright, D.J.; Pendleton, M.; Boulware, J.; Walbridge, S.; Gerlt, B.; Eslinger, D.; Sampson, D.; Huntley, E. *ArcGISBenthic Terrain Modeler (BTM), 3.0*; NOAA Coastal Services Center, Massachusetts Office of Coastal Zone Management: Boston, MA, USA, 2012.

14. Sappington, J.M.; Longshore, K.M.; Thompson, D.B. Quantifying landscape ruggedness for animal habitat analysis: A case study using bighorn sheep in the Mojave Desert. *J. Wildl. Manag.* **2007**, *71*, 1419–1426. [CrossRef]

15. Jerosch, K.; Kuhn, G.; Krajnik, I.; Scharf, F.K.; Dorschel, B. A geomorphological seabed classification for the Weddell Sea, Antarctica. *Mar. Geophys. Res.* **2016**, *37*, 127–141. [CrossRef]

16. Foster, M.S. Rhodoliths: Between Rocks and Soft Places. *J. Phycol.* **2001**, *37*, 659–667. [CrossRef]

17. Pinheiro, H.T.; Martins, A.S. Estudo comparativo da captura artesanal do camarão sete-barbas e sua fauna acompanhante em duas áreas de pesca do litoral do estado do Espírito Santo, Brasil. *Bol. Inst. Pesca São Paulo* **2009**, *35*, 215–225. (In Portuguese)

18. Pinheiro, H.T.; Joyeux, J.C. Pescarias multi-específicas na região da Foz do Rio Doce, ES, Brasil: Características, problemas e opções para um futuro sustentável. *Braz. J. Aquat. Sci. Technol.* **2007**, *11*, 15–23. (In Portuguese) [CrossRef]

19. Netto, R.F.; Beneditto, A.P.M.D. Diversidade de artefatos da pesca artesanal marinha do Espírito Santo. *Biotemas* **2007**, *20*, 107–119. (In Portuguese)

20. Pinheiro, H.T.; Cordeiro Madureira, J.M.; Joyeux, J.-C.; Martins, A.S. Fish diversity of a southwestern Atlantic coastal island: Aspects of distribution and conservation in a marine zoogeographical boundary. *Check List* **2015**, *11*, 1–17. [CrossRef]

21. Posamentier, H.W.; Jervey, M.T.; Vail, P.R. Eustatic Controls on Clastic Deposition I—Conceptual Framework. *Soc. Econ. Paleontol. Mineral. Spec. Publ.* **1987**. [CrossRef]

22. Posamentier, H.W.; Allen, G.P. Variability of the sequence stratigraphic model: Effects of local basin factors. *Sediment. Geol.* **1993**, *86*, 91–109. [CrossRef]

23. Dominguez, J.M.L. The coastal zone of Brazil. In *Geology and Geomorphology of Holocene Coastal Barriers of Brazil*; Lecture Notes in Earth Sciences; Sergio, R.D., Hesp, P.A., Eds.; Springer-Verlag: Berlin/Heidelberg, Germany, 2009; pp. 17–51.

24. Dominguez, J.M.L.; da Silva, R.P.; Nunes, A.S.; Freire, A.F.M. The narrow, shallow, low-accommodation shelf of central Brazil: Sedimentology, evolution, and human uses. *Geomorphology* **2013**, *203*, 46–59. [CrossRef]

25. Araújo, T.C.M.D.; Santos, R.C.D.A.L.; Seoane, J.C.S.; Seoane, J.C.S. Erosão e Progradação do litoral de Alagoas. In *Erosão e Progradação do Litoral do Brasil*; Muehe, D., Ed.; Minstério de Meio Ambiente: Brasilia, Brazil, 2006; Volume 1, pp. 197–212. (In Portuguese)

26. Baker, E.K.; Harris, P.T. 2—Habitat Mapping and Marine Management. In *Seafloor Geomorphology as Benthic Habitat*; Elsevier: London, UK, 2012; pp. 23–38.

27. Freire, A.F.M.; Dominguez, J.M.L. A seqüência holocênica da plataforma continental central do Estado da Bahia. *Bol. Geociênc. Petrobras* **2006**, *14*, 247–267. (In Portuguese)

28. Amado-Filho, G.M.; Moura, R.L.; Bastos, A.C.; Salgado, L.T.; Sumida, P.Y.; Guth, A.Z.; Francini-Filho, R.B.; Pereira-Filho, G.H.; Abrantes, D.P.; Brasileiro, P.S.; et al. Rhodolith beds are major $CaCO_3$ bio-factories in the tropical South West Atlantic. *PLoS ONE* **2012**, *7*, 41–45. [CrossRef] [PubMed]

29. Milliman, J.D. Role of Calcareous Algae in Atlantic Continental Margin Sedimentation. In *Fossil Algae: Recent Results and Developments*; Flügel, E., Ed.; Springer: Berlin/Heidelberg, Germany, 1977; pp. 232–247.

30. Amado-Filho, G.M.; Bahia, R.G.; Pereira-Filho, G.H.; Longo, L.L. South Atlantic rhodolith beds: Latitudinal distribution, species composition, structure and ecosystem functions, threats and conservation status. In *Rhodolith/Maërl Beds: A Global Perspective*; Riosmena-Rodríguez, R., Nelson, W., Aguirre, J., Eds.; Springer: Boca Raton, FL, USA, 2017; pp. 299–317.

31. Bahia, R.G.; Abrantes, D.P.; Brasileiro, P.S.; Pereira-Filho, G.H.; Amado-Filho, G.M. Rhodolith bed structure along a depth gradient on the northern coast of bahia state, Brazil. *Braz. J. Oceanogr.* **2010**, *58*, 323–337. [CrossRef]

32. Brasileiro, P.S.; Pereira-Filho, G.H.; Bahia, R.G.; Abrantes, D.P.; Guimarães, S.M.P.B.; Moura, R.L.; Francini-Filho, R.B.; Bastos, A.C.; Amado-Filho, G.M. Macroalgal composition and community structure of the largest rhodolith beds in the world. *Mar. Biodivers.* **2016**, *46*, 407–420. [CrossRef]

33. Amado-Filho, G.; Maneveldt, G.; Pereira-Filho, G.; Manso, R.C.C.; Bahia, R.; Barros-Barreto, M.B.; Guimarães, S. Seaweed diversity associated with a Brazilian tropical rhodolith bed. *Cienc. Mar.* **2010**, *36*, 371–391. [CrossRef]

34. Steller, D.L.; Foster, M.S. Environmental factors influencing distribution and morphology of rhodoliths in Bahía Concepción, BCS, México. *J. Exp. Mar. Biol. Ecol.* **1995**, *194*, 201–212. [CrossRef]

35. Foster, M.S.; Riosmena-Rodriguez, R.; Steller, D.L.; Woelkerling, W.J. Living rhodolith beds in the Gulf of California and their implications for paleoenvironmental interpretation. In *Pliocene Carbonates and Related Facies Flanking the Gulf of California, Baja California, Mexico*; Johnson, M.E., Ledesma-Vázquez, J., Eds.; Geological Society of America: Boulder, CO, USA, 1997; pp. 127–139.

36. Steller, D.L.; Riosmena-Rodríguez, R.; Foster, M.S.; Roberts, C.A. Rhodolith bed diversity in the Gulf of California: The importance of rhodolith structure and consequences of disturbance. *Aquat. Conserv. Mar. Freshw. Ecosyst.* **2003**, *13*, S5–S20. [CrossRef]

37. Littler, M.M.; Littler, D.S.; Dennis Hanisak, M. Deep-water rhodolith distribution, productivity, and growth history at sites of formation and subsequent degradation. *J. Exp. Mar. Biol. Ecol.* **1991**, *150*, 163–182. [CrossRef]

38. Amado-Filho, G.M.; Pereira-Filho, G.H. Rhodolith beds in Brazil: A new potential habitat for marine bioprospection. *Rev. Brasil. Farm.* **2012**, *22*, 782–788. [CrossRef]

39. Guimarães, S.M.P.B. A revised checklist of benthic marine Rhodophyta from the state of Espírito Santo, Brazil. *Bol. Inst. Bot.* **2006**, *17*, 145–196.

40. Amado-Filho, G.M.; Maneveldt, G.W.; Manso, R.C.C.; Marins-Rosa, B.V.; Pacheco, M.R.; Guimarães, S.M.P.B. Estructura de los mantos de rodolitos de 4 a 55 metros de profundidad en la costa sur del estado de Espírito Santo, Brasil. *Cienc. Mar.* **2007**, *33*, 399–410. (In Portuguese) [CrossRef]

41. Pimentel, C.R.; Joyeux, J.C. Diet and food partitioning between juveniles of mutton *Lutjanus analis*, dog Lutjanus jocu and lane *Lutjanus synagris* snappers (Perciformes: Lutjanidae) in a mangrove-fringed estuarine environment. *J. Fish Biol.* **2010**, *76*, 2299–2317. [CrossRef] [PubMed]

42. Simon, T.; Pinheiro, H.T.; Moura, R.L.; Carvalho-Filho, A.; Rocha, L.A.; Martins, A.S.; Mazzei, E.; Francini-Filho, R.B.; Amado-Filho, G.M.; Joyeux, J.C. Mesophotic fishes of the Abrolhos Shelf, the largest reef ecosystem in the South Atlantic. *J. Fish Biol.* **2016**, *89*, 990–1001. [CrossRef] [PubMed]

43. Erikstad, L.; Bakkestuen, V.; Bekkby, T.; Halvorsen, R. Impact of scale and quality of digital terrain models on predictability of seabed terrain types. *Mar. Geodesy* **2013**, *36*, 2–21. [CrossRef]

Article

Characteristics and Dynamics of a Large Sub-Tidal Sand Wave Field—Habitat for Pacific Sand Lance (*Ammodytes personatus*), Salish Sea, Washington, USA

H. Gary Greene [1,2,*], David A. Cacchione [3] and Monty A. Hampton [3]

1   SeaDoc Society, Tombolo Mapping Lab, 942 Deer Harbor Road, Eastsound, WA 98245, USA
2   Friday Harbor Labs, University of Washington, WA 98245, USA
3   U.S. Geological Survey, 345 Middlefield Road, Menlo Park, CA 94025, USA;
    dcacchione@comcast.net (D.A.C.); mkhamp@comcast.net (M.A.H.)
*   Correspondence: greene@mlml.calstate.edu

Received: 15 August 2017; Accepted: 19 October 2017; Published: 23 October 2017

Abstract: Deep-water sand wave fields in the San Juan Archipelago of the Salish Sea and Pacific Northwest Washington, USA, have been found to harbor Pacific sand lance (PSL, *Ammodytes personatus*), a critical forage fish of the region. Little is known of the dynamics of these sand waves and the stability of the PSL sub-tidal habitats. Therefore, we have undertaken an initial investigation to determine the dynamic conditions of a well-known PSL habitat in the San Juan Channel within the Archipelago using bottom sediment sampling, an acoustical doppler current profiling (ADCP) system, and multi-beam echo sounder (MBES) bathymetry. Our study indicates that the San Juan Channel sand wave field maintained its shape and bedforms geometry throughout the years it has been studied. Based on bed phase diagrams for channelized bedforms, the sand waves appear to be in a dynamic equilibrium condition. Sea level rise may change the current regime within the Archipelago and may alter some of the deep-water or sub-tidal PSL habitats mapped there. Our findings have global significance in that these dynamic bedforms that harbor PSL and sand-eels elsewhere along the west coast of North America and in the North Sea may also be in a marginally dynamic equilibrium condition and may be prone to alteration by sea level rise, indicating an urgency in locating and investigating these habitats in order to sustain the forage fish.

Keywords: bedforms; forage fish; Pacific sand lance; sediment habitats; bathymetry; currents

1. Introduction

This study investigates the characteristics and mobility of large sand waves located in the central part of the San Juan Channel of the San Juan Archipelago, Washington, USA (Figure 1), here called "The San Juan Channel sand wave field". These sand waves have heights of 1 to 4 m and wavelengths from 10 to 100 m. This work is part of a larger multi-disciplinary investigation of deep-water sandy habitats for Pacific sand lance (*Ammodytes personatus*, formerly identified as *A. hexapterus*), an important forage fish for larger fish including salmon, mammals, and birds in the Pacific Northwest. Sand wave fields of all sizes are potential sub-tidal habitats for this species within the Salish Sea as well as on the continental shelves of northwestern North America.

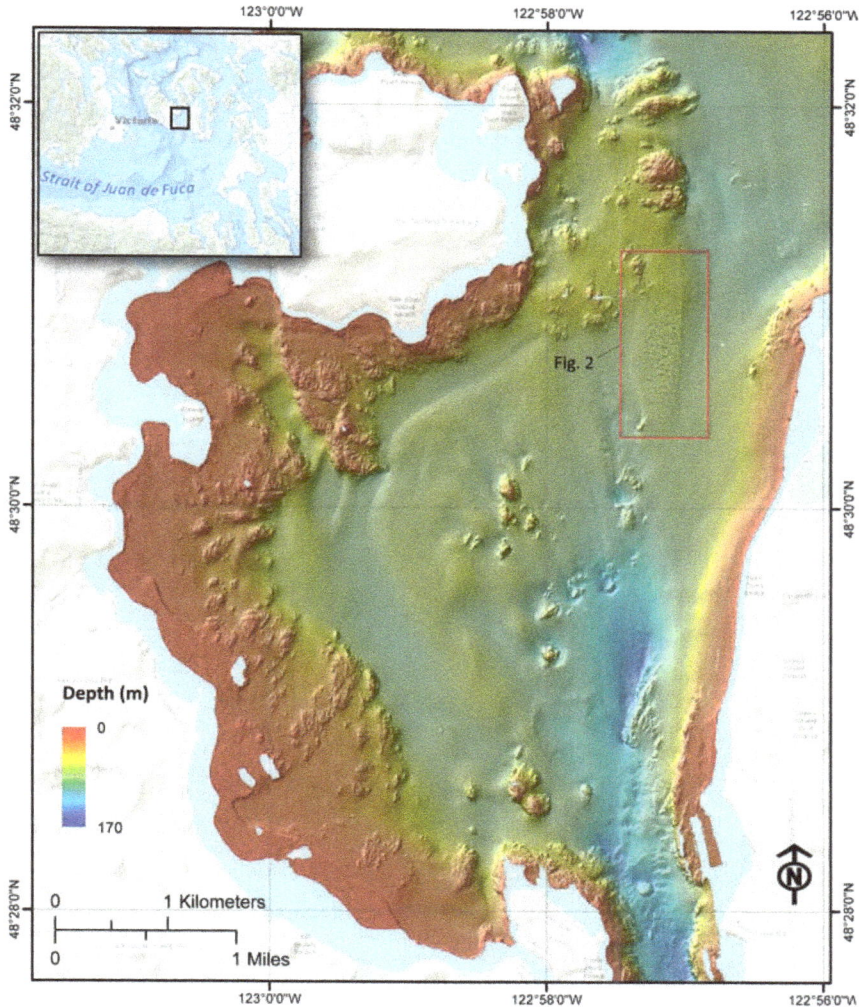

Figure 1. Multibeam echosounder bathymetric map showing the location of the San Juan Channel sediment wave field within the San Juan Archipelago of the Pacific NW, Washington State, USA. The bedform investigated for this study is prominently displayed near the top of the bathymetric image.

The Pacific sand lance (PSL) is found along the coastal North Pacific Ocean from northwestern California to northern Japan, and is one of six species in the genus *Ammodytes* [1,2]. Although PSL are a key component in the marine food web, little is known of this species' sub-tidal habitats. Their burrowing behavior, recruitment rates and conditions, relative abundance and distribution, population structure, local spawning habits, and spawning and burial substrates have been studied [1,3,4]. Previous work on PSL biology and habitat associations has focused on the near-shore and shallow sub-tidal areas [5–7]. It has recently been shown that predominant and important habitats exist in deeper sub-tidal and deep-water environments [8].

1.1. Previous Work

Studies of the lesser sand-eel (*A. marinus*) in the North Sea show that like PSL sand-eels are closely associated with sandy substrate [9,10] where they were reportedly found to inhabit turbulent sandy areas along the edges of sand banks. In their study of the sand-eel fisheries in the North Sea, Wright and others [11] confirmed from their modeling that sand-eels are found at water depths between 30 m and 70 m, consistent with the recorded depth distribution based on the sand-eel fisheries and that the fish preferred sandy substrate. From their studies Wright and others [11] concluded that inferences about the distribution and quality of *A. marinus* habitats in the North Sea could be obtained from knowledge of the sediment fraction and depth, factors they believed to limit their distribution. These authors recommended that further studies of sediment permeability and the relation between ripple geometry and water percolation might help better characterize sand-eel habitat preferences.

Freeman and others [12] used QTC VIEW™ with a single beam sonar system to determine acoustic changes in the seabed where sand-eel were reported to concentrate in the North Sea. These authors, following up on the work of Wright [13], who noted that differences in bottom current velocities contributed to sand-eel concentrations with high densities occurring at sites affected by high tidal velocities, found that current formed features were attractive to sand-eels. They suggested that current direction may be an environmental mechanism influencing the dynamics of sand-eel patches and suggested buried sand-eel form patches in the sediment that lie in the same direction as prevailing tidal currents [12].

Holland and others [14] used the RoxAnn© classification using a single beam sonar to develop a stratified random grab sampling survey in the North Sea based on a sediment map constructed from RoxAnn. Their study was undertaken because they had determined that grab sample stations must be specifically targeted so that adequate numbers of fish are collected, and that based on the work of Greenstreet and others [15] field data did not define sand-eel habitat in sufficient detail to provide an adequate basis for a useful grab survey design with habitat mapping. Therefore, Holland and others [14] concluded that when combined with seabed habitat mapping grab sampling can be directed to those areas where sand-eel habitat is located.

1.2. Objectives

Although much work has been done in relation to deep-water habitat characterization of sand-eel and shallow water habitat characterization of PSL, a better understanding of habitat parameters and controlling processes are needed to better constrain potential habitats. Our focus, therefore, is on deep-water (sub-tidal) habitats of dynamic sediment wave fields with bedforms that harbor dense concentrations of PSL. Since these bedforms appear to be essential habitats for the PSL, their presence and stability is critical to understanding PSL distribution and abundance patterns. The goal of this study is to investigate the physical seafloor processes that form and maintain sand wave fields as a precursor to future habitat association studies on PSL. We address the concerns and recommendations of previous investigators in the study of tidal velocities, direction and relationship to grain sizes.

Deposits of clean sand at water depths where PSL reside in sub-tidal environments (typically <100 m) are common where relatively strong currents sweep the sea floor. To maintain such clean deposits a plentiful sand supply and strong near-bottom currents appear necessary as observed by Wright and others [11] in their study of sand ripples in the North Sea, which are associated with maximum current flows of 1 m s^{-1} [16]. Finer sediment might transit through the area, while coarser sediment might be present as a lag. Bedform fields consisting of ripples, waves, and dunes are common in deep channel areas of the Salish Sea, and several fields have been mapped within the San Juan Archipelago [8,17]. The Sand Juan Channel sand wave field (Figure 2) was studied extensively by Greene and others [8] and was found to be a productive PSL habitat [18]. The sand–wave field is delimited by distinct boundaries where the sand waves are in sharp contact with a relatively featureless surrounding sea floor. Such abrupt seabed transitions have been reported in other nearby sand wave fields [17].

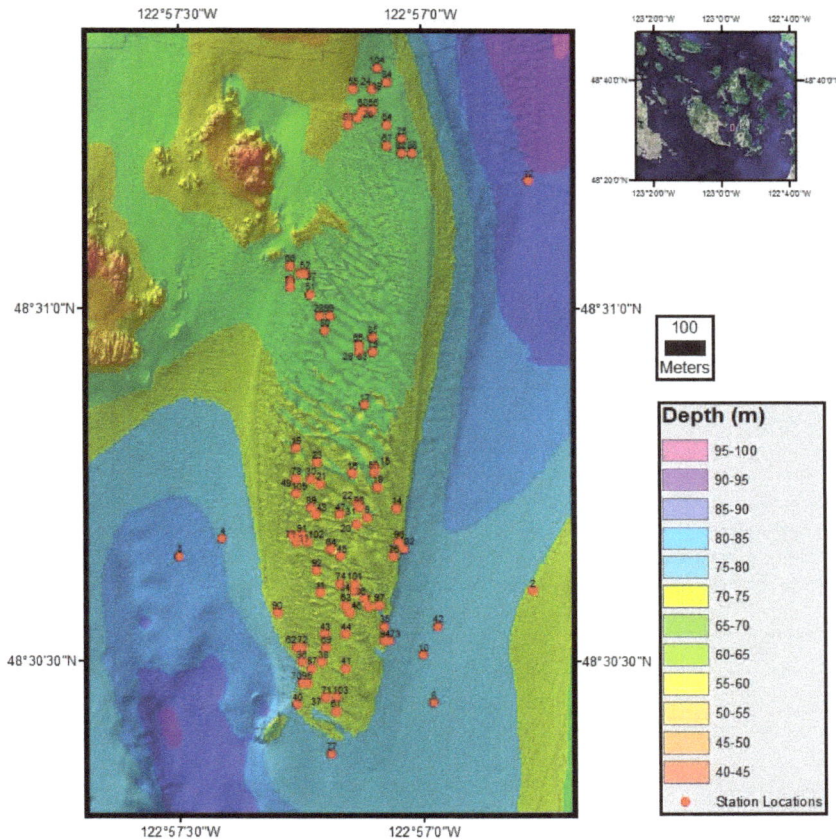

Figure 2. Expanded view of multi-beam echo sounder (MBES) bathymetric image of the San Juan Channel sediment wave field exhibited in Figure 1 with depth illustrated in color. Small dots within circles represent sediment sample locations (diameter of dot equals ~23 m), station numbers are located next to open circles.

The San Juan Channel sand wave field covers an area of approximately 600,000 m^2 at water depths of 20 to 80 m (Figure 2). During 2004 a Washington Department of Fish and Wildlife (WDFG) remotely operated vehicle (ROV) video survey at this location found that PSL were burrowing and emerging from the sediment [8]. In October 2006 and 2008, PSL were again discovered in the sand waves, and the fish and bottom sediment were sampled with a small Peterson grab sampler. The significance of the sediment wave field as a PSL habitat was shown by the work of Blain [18], who collected over 42 Van Veen sediment grab samples, which contained an average of 10.0 individuals/grab. Blain [18] estimated that the local PSL population consisted of over 63 million fish. Every year since 2008 students of Friday Harbor Laboratories, University of Washington, have sampled PSL in this sand wave field (e.g., [19–21]).

Assuming that the San Juan Channel sediment wave field, a tidally shaped field of dynamic bedforms, is a typical PSL habitat, we set out to investigate the processes that shape it to better identify potential habitats. Therefore, our objectives are to determine the seafloor processes that form the sand wave field, its study ripple orientations in relation to sediment permeability, to document that PSL, like sand stability, and source of sand. Our intent is to address the recommendations of Wright and others [11] to -eel, concentrate in high tidal flow areas as proposed by Freeman and others [12], and to improve the grab sampling success of potential habitats as discussed by Holland and others [14], therefore, to determine the physical characteristics of PSL deep-water habitats.

2. Background

During the past decade, studies of marine sand waves in coastal and estuarine environments have been aided by detailed MBES bathymetric and backscatter data. Barnard and others [22] mapped bedforms in San Francisco Bay, California using repeated MBES surveys, and estimated sand wave migration rates and bed-load transport patterns. Barnard and others [22] summarized previous mapping and analytical studies of sand wave dynamics in other coastal regions, and provided a comprehensive analysis of sand wave patterns and movement in San Francisco Bay based on over 3000 sand waves. They applied a technique to estimate sand wave migration rates using bedform asymmetry that was formulated for sand waves off the Dutch coast by Knappen [23]. This same technique is used in this study to estimate movement of the larger sand waves in the San Juan Channel sand wave field.

It is widely accepted that bed form size and type (ripples, megaripples, dunes, sand waves; e.g., [24] are controlled largely by sediment grain size, current velocity, and water depth [25,26]. The practical issues of mixed bed grain size distributions, spatially variable physical bottom roughness, and time-dependent current velocities (e.g., tidal flows) complicate these linkages. The sand waves considered here are described using bed phase diagrams developed for tidally controlled bedforms in San Francisco Bay where water depths are similar [27].

Geologic Setting

The San Juan Archipelago-Georgia Basin region of the central Salish Sea is an active tectonic province whose physiography and geomorphology reflect both Mesozoic to Cenozoic convergence (subduction/accretion) plate tectonic processes and Pleistocene glaciation (glacial scouring/deposition). These processes have juxtaposed and deformed Jurassic–Cretaceous metamorphic rocks with Tertiary–Quaternary sedimentary rocks producing a complex of fjords, grooved and polished bedrock outcrops, erratic boulder concentrations, and moraines [28]. Banks of till and glacial advance outwash deposits also have formed and contribute to the variety of relief and substrate within the region [29]. Present day tidal action has fashioned much of the relic glacial–marine sediments into dynamic bedforms within interisland channels, consisting of sand and gravel waves and dune fields. Locally modern day sedimentary deposits overprint the relic deposits as sand and mud banks and represent materials being supplied to the region by the Fraser River of British Columbia, Canada.

The stability and persistence of the larger sandy bedforms are important factors for the preservation of the PSL deep-water habitats. The surfaces of these sand deposits contain multiple scales of bedforms from ripples to large sand waves and are thought to have originally been formed and shaped during the Holocene, subsequent to the last glacial retreat from the region [29]. The complex action of surface waves and strong tidal currents, particularly during lower sea level, transported mobile sand from coastal and shallow island margins into adjacent deeper channel floors, forming the discrete sand bodies. Presently, the isolated sand deposits are situated in water depths from about 20 to 100 m in many of the inter-island channels where current patterns are complex. The sand deposits rest unconformably on a rugged surface composed of coarse glacial till, often in the current shadow of seafloor rock outcrops, and the boundaries between the till and sand bodies have sharply defined, abrupt contacts [29].

3. Methods

In this study, geomorphic and textural characteristics of the San Juan Channel sand wave field were investigated using detailed multi-beam echo sounder (MBES) bathymetric data, bottom samples, bottom camera and underwater video surveys, and a shipboard acoustic doppler current profiling (ADCP) system.

A sequence of four one-day research cruises was undertaken specifically for this study during 2010 to 2013. Regional MBES surveys that included San Juan Channel were conducted earlier [8,30]. Bottom sediment samples, video-transect images, and underwater sediment camera photographs were obtained during the various cruises within the San Juan Channel on two research vessels (R/V *Centennial* and *Tombolo*). Current profiling transects within the sand wave field were carried out with an ADCP mounted on University of Washington Friday Harbor Labs' R/V *Centennial*. The ADCP current profiles were collected only during flood tidal flow, but measurements during ebb currents within the San Juan Channel sand wave field had been collected previously by Ewing [21], whose data were used in our analyses.

3.1. Marine Geophysical Data

Marine geophysical survey data, primarily wide swath MBES bathymetry and backscatter acquired in cooperation with the Geological Survey of Canada, Canadian Hydrographic Service, Center for Habitat Studies, Moss Landing Marine Labs, and Tombolo/Sea Doc Society, and collected in the Northwest Straits region of the Salish Sea (southern Gulf Islands and the San Juan Archipelago) were used to produce seafloor images of the sediment wave field for this investigation. These data, along with side-scan sonar mosaics and 3.5 kHz sub-bottom seismic-reflection profiles were used to produce habitat types after Greene [31], which were published in a marine benthic habitat map series [30]. Interpretation of these data was used to characterize our study area, and to identify and map dynamic bedforms and other sand wave fields in the region for future sampling.

The MBES surveys were conducted in the San Juan Channel sand wave field originally imaged in 2004 with repeat surveys in 2006 and 2007 with the Canadian Coast Guard launch Otter Bay using a Simrad EM 3000–3002™ (300 kHz) system The objective of these multiple surveys were to estimate changes in the sand wave field bedforms morphologies and to determine if sediment transport across the field is active.

3.2. Seafloor Photographs

We used the USGS sediment grain or "eyeball" camera aboard the *RV Tombolo* to take close-up high-resolution photographs of the bottom sediment. These images were used to calculate median grain size based on the work of Rubin [32]. One advantage of the eyeball camera is that extensive coverage can be obtained across the seafloor (approximately 6 images/minute). These camera transects were useful for observing the bottom sediment characteristics, and for making numerous estimates of sediment grain size and its areal variation. The locations of the eyeball camera photo shots obtained along the camera transect and sediment sample stations are shown in Figure 3. No layback calculations were made as transects were undertaken during slack water and the camera tether was near vertical for most of the photographic runs. Therefore, positioning of the vessel was used to plot the locations within a range of ±23 m (~2.5 times boat length). Precise locations of the photographs were made by identifying the sand wave morphology (crest or trough) in the photos with the position of the boat. Two eyeball photographs and their locations, one in a sand-wave trough and the other close to the adjacent crest, are shown in Figure 4a–c. Additional data on sediment type and morphology were obtained using a USGS bottom video camera system. The mean sediment size obtained by analysis of the image in Figure 4b near a sand wave crest was 0.5 mm. Much of the surficial bottom sediment in the sand wave troughs consists of large, poorly sorted fragments of coarse shell hash (Figure 4c).

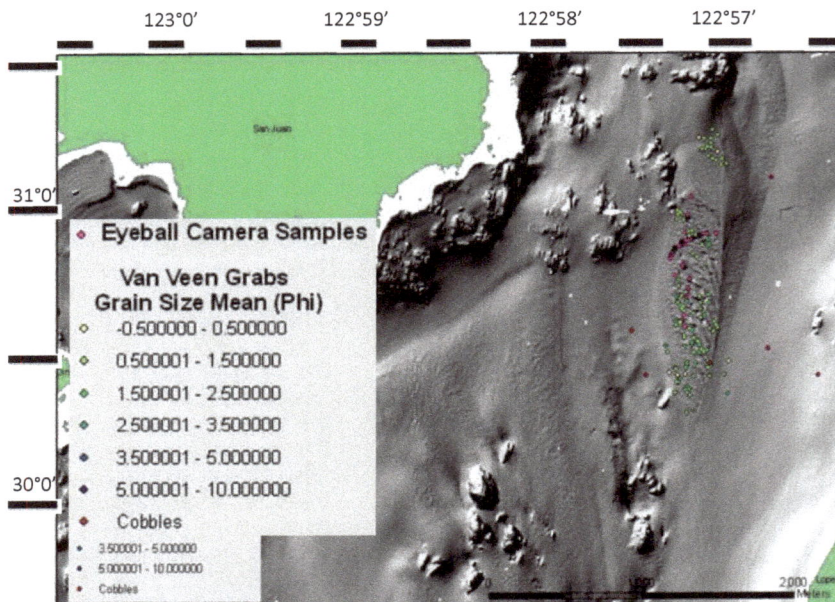

Figure 3. Locations of eyeball photo shots (purple dots) taken along transects of the San Juan Channel sand wave field using the USGS eyeball camera and sediment sample stations showing mean grain sizes (phi scale) in various colors. The objective of this figure is to show the density of sampling points within the sand wave field compared to samples taken outside of the field.

Figure 4. Examples of images collected using the USGS eyeball camera: (**a**) MBES bathymetric image with locations of samples collected in the sand wave field; (**b**) photo of sediment taken in trough; (**c**) photo of sediment taken near crest of a sand wave.

3.3. Sediment Sampling and Laboratory Analysis

Grain-size distribution was determined for 79 sediment samples from the San Juan Channel sand wave field, a bedforms feature we consider typical of PSL deep-water habitat, in water depths of 20 to 80 m (Figures 2 and 3). The samples were collected with a Van Veen grab during seven cruises in 2010 and 2011. All but 9 samples were collected during daylight hours. After drying samples in an oven, grain size was measured with a series of sieves 2, 1, 0.5, 0.25, 0.125, 0.063 mm in sizes; phi [φ] scale sizes from −1 to 4; Wentworth [33] using a Ro-Tap™ machine operated for a period of 15 minutes per sample. Material finer than 0.063 mm (4φ) was retained and included collectively as the "fine fraction", and material resting on the 2.0 mm (−1φ) screen, which could include grains up to a few centimeters diameter, as the "coarse fraction."

The location, water depth, number of fish captured, mean grain size, and statistical parameters at each sampling station are shown in Table 1 and the sample locations are shown in map view (Figure 3). For brevity we report grain size in mm in the remainder of this paper.

Our sediment analyses are compared with that of Blaine [18] to determine if any significant change, both in grain sizes and fish population, has occurred in the sediment wave field since the earlier study of Blaine [18]. Samples from both studies were collected throughout the sediment wave field, as well as outside of the field in identical manner using the same bottom sampler with grain size analyses made using a Rho–Tap measuring machine including identical screens and calculations for the measurements.

Table 1. Sediment Sample Information.

Date	Cruise ID	Sample ID	Latitude	Longitude	Water Depth (Meters)	Fish Count	Mean Grain Size (mm)	Standard Deviation (mm)	Skewness	Kurtosis
10 June 2010 SAN JUAN FIELD	C1-10	C1-10-1	48°30.579′ N	122°57.113′ W	72.0	0	0.6641	0.708419	−0.674404	4.044548
		C1-10-2	48°30.601′ N	122°56.775′ W	73.5	0	1.782087	1.102982	1.469865	3.936629
		C1-10-8	48°30.443′ N	122°56.980′ W	77.6	0	1.168059	1.394146	0.560409	1.904311
		C1-10-9	48°30.706′ N	122°57.115′ W	70.0	0	0.661948	0.571287	−0.428435	4.788353
		C1-10-10	48°30.510′ N	122°57.000′ W	72.1	0	1.050705	1.305547	0.057556	1.367169
		C1-10-11	48°30.598′ N	122°57.211′ W	72.2	0	0.632264	0.666296	−0.686152	4.167434
		C1-10-12	48°31.181′ N	122°56.781′ W	89.6	0	0.803207	1.366305	−0.251335	1.613957
		C1-10-13	48°32.302′ N	122°56.630′ W	24.8	0	0.546959	1.261079	−0.892102	2.727475
7 October 2010 SAN JUAN FIELD	C2-10	C2-10-2	48°30.775′ N	122°57.097′ W	70.0	16	1.004968	1.132628	−0.069622	1.679441
		C2-10-3	48°30.768′ N	122°57.144′ W	70.0	10	0.572152	0.727196	−0.994183	4.054206
		C2-10-4	48°30.865′ N	122°57.119′ W	70.0	13	0.650857	0.759763	−0.738298	3.498874
		C2-10-5	48°30.749′ N	122°57.092′ W	71.2	10	0.877883	1.009105	−0.38097	2.072258
		C2-10-6	48°30.804′ N	122°57.260′ W	70.9	25	0.844679	0.86165	−0.464404	2.655142
		C2-10-7	48°30.696′ N	122°57.136′ W	70.5	43	0.808321	0.734541	−0.632845	3.490027
		C2-10-8	48°30.753′ N	122°57.212′ W	70.3	16	0.791851	0.550303	−0.942019	5.131413
18 November 2010 SAN JUAN FIELD	C3-10	C3-10-1	48 30.726′ N	122°57.134′ W	71.9	13	0.68329	0.811266	−0.575055	3.01503
		C3-10-2	48 30.784′ N	122° 7.217′ W	71.7	0	0.751196	0.695075	−0.344332	3.743417
22 January 2011 SAN JUAN FIELD	C1-11	C1-11-1	48° 31.311′	122° 57.108′ W	60.8	20	1.094493	1.081723	0.059713	1.812171
		C1-11-2	48° 31.242′	122° 57.043′ W	70.4	62	0.809117	1.045999	−0.54649	2.129322
		C1-11-3	48° 31.274′	122° 57.138′ W	61.3	35	1.034332	0.996408	−0.208243	1.898367
		C1-11-4	48° 31.057′	122° 57.257′ W	60.6	33	1.010241	0.916506	−0.329582	2.1571
		C1-11-5	48° 30.991′	122° 57.211′ W	64.6	50	0.742111	0.732595	−0.602172	3.538112
		C1-11-6	48° 30.947′	122° 57.139′ W	65.7	12	0.828904	0.911726	−0.533595	2.555615
		C1-11-7	48° 30.769′	122° 57.238′ W	70.1	11	0.679105	0.653803	−0.557637	4.032605
		C1-11-8	48° 30.727′	122° 57.135′ W	69.8	31	0.741641	0.630577	−0.594211	4.270618
		C1-11-9	48° 30.666′	122° 57.043′ W	73.6	9	0.908589	1.049034	−0.341861	1.92245
		C1-11-10	48° 30.681′	122° 57.271′ W	70.3	20	0.754138	0.677989	−0.69112	4.104119
		C1-11-11	48° 30.605′	122° 57.143′ W	72.2	0	0.770197	0.91318	−0.560431	2.660286
		C1-11-12	48° 30.554′	122° 57.088′ W	73.3	68	0.926035	0.954556	−0.367446	2.192787

Table 1. Cont.

Date	Cruise ID	Sample ID	Latitude	Longitutude	Water Depth (Meters)	Fish Count	Mean Grain Size (mm)	Standard Deviation (mm)	Skewness	Kurtosis
		C1-11-13	48° 30.503'	122° 57.252' W	74.7	19	0.78769	0.869861	−0.54983	2.942144
		C1-11-14	48° 30.458'	122° 57.207' W	73.2	32	0.657471	0.717478	−0.825559	4.281199
		C1-11-15	48° 30.505'	122° 57.210' W	72.8	34	0.86808	0.87255	−0.527988	2.65564
16 February 2011 SAN JUAN FIELD	C2-11	C2-11-1	48° 30.442'	122° 57.263' W	75.4	25	0.860999	1.250156	−0.28495	1.574764
		C2-11-2	48° 30.499'	122° 57.164' W	74.2	3	0.92817	1.019095	−0.288662	2.074186
		C2-11-3	48° 30.555'	122° 56.978' W	80	0	2.011659	1.081999	2.226016	6.968468
		C2-11-4	48° 30.545'	122°57.203' W	73.9	4	0.62691	0.757549	−0.870548	4.053025
		C2-11-5	48° 30.544'	122° 57.166' W	73.2	42	0.494276	0.683121	−1.161662	4.993115
		C2-11-6	48° 30.654'	122° 57.174' W	72.2	3	0.679431	0.612113	−0.518584	4.527016
		C2-11-7	48° 30.577'	122° 57.159' W	74.5	1	0.612304	0.923948	−1.028589	3.454976
		C2-11-8	48° 30.710'	122° 57.177' W	73.4	8	0.705638	0.583944	−0.370977	4.333422
		C2-11-9	48° 30.719'	122° 57.221' W	73.1	16	0.824162	0.809048	−0.415873	2.868291
		C2-11-10	48° 30.764'	122° 57.265' W	73.3	4	0.85928	0.836736	−0.541015	2.84772
		C2-11-11	48° 30.779'	122° 57.103' W	70.5	0	0.729409	0.711894	−0.725082	4.236952
		C2-11-12	48° 31.029'	122° 57.238' W	66.1	20	0.866397	0.851112	−0.466089	2.824674
		C2-11-13	48° 31.050'	122° 57.249' W	64.7	0	0.828745	0.861835	−0.560687	3.059643
		C2-11-14	48° 30.941'	122° 57.106' W	69.2	68	0.604378	0.710881	−0.821931	3.979731
		C2-11-15	48° 31.268'	122° 57.070' W	65.3	47	1.733515	0.859093	0.912746	2.817612
		C2-11-16	48° 31.310'	122° 57.143' W	61.1	47	0.835146	1.13316	−0.363947	1.910039
		C2-11-17	48° 31.282'	122° 57.104' W	62.8	76	0.83024	0.88981	−0.502843	2.634484
		C2-11-19	48° 31.313'	122° 57.172' W			1.953552	0.786798	1.426041	4.344054
		C2-11-20	48° 30.511'	122° 57.295' W			2.105554	0.742492	1.796042	5.629863
15 April 2011 SAN JUAN FIELD	C3-11	C3-11-1	48° 25.807'	122° 54.614' W	27.7	0	2.105554	0.742492	1.796042	5.629863
		C3-11-2	48° 25.798'	122° 54.814' W	36.2	0	0.301528	0.663569	0.146784	6.782817
		C3-11-3	48° 25.748'	122° 54.770' W	25.8	0	0.430075	1.388299	−0.517157	2.712989
		C3-11-4	48° 25.688'	122° 54.955' W	31.2	0	0.202147	0.793127	−0.502173	4.218526
		C3-11-5	48° 30.433'	122° 57.181' W	37.2	1	0.744994	0.816597	−0.712606	3.333105
		C3-11-6	48° 30.525'	122° 57.263' W	74.1	6	0.866079	0.942287	−0.480375	2.502384
		C3-11-7	48° 30.583'	122° 57.166' W	74.1	7	1.354396	0.926554	0.31857	2.139121
		C3-11-8	48° 30.660'	122° 57.192' W	72.8	4	0.770751	0.750876	−0.583451	3.589048
		C3-11-9	48° 30.950'	122° 57.138' W	73.2	5	0.731688	0.873996	−0.639592	2.984335
		C3-11-10	48° 31.063'	122° 57.274' W	68.1	3	0.988551	1.044987	−0.155617	1.919282

Table 1. *Cont.*

Date	Cruise ID	Sample ID	Latitude	Longitutude	Water Depth (Meters)	Fish Count	Mean Grain Size (mm)	Standard Deviation (mm)	Skewness	Kurtosis
		C3-11-11	48° 31.236′	122° 57.077′ W	63.2	4	1.11899	1.077244	0.050721	1.754375
		C3-11-12	48° 31.283′	122° 57.125′ W	64.2	2	1.178479	1.046315	0.160777	1.840769
24 June 2011 SAN JUAN FIELD	C4-11	C4-11-1	48° 30.529′	122° 57.205′ W	71.7	0	0.691116	0.655523	−0.602719	4.551284
		C4-11-2	48° 30.478′	122° 57.254′ W	74.2	3	0.782412	0.775554	−0.688209	3.662782
		C4-11-3	48° 30.452′	122° 57.207′ W	72.8	11	0.670635	0.731412	−0.646751	3.712857
		C4-11-4	48° 30.522′	122° 57.259′ W	73.1	5	0.807638	0.924192	−0.538696	2.714451
		C4-11-5	48° 30.538′	122° 57.073′ W	75.3	0	0.674468	1.148823	−0.637554	2.243671
		C4-11-6	48° 30.611′	122° 57.176′ W	72.1	0	0.651117	0.587258	−0.449387	4.890624
		C4-11-7	48° 30.671′	122° 57.266′ W	72.4	0	0.717823	0.747013	−0.596759	3.626665
		C4-11-8	48° 30.652′	122° 57.065′ W	73.1	2	1.11643	1.058659	0.062831	1.8403
		C4-11-9	48° 30.374′	122° 57.194′ W	70.2	0	0.787094	0.722204	−0.403949	3.496516
		C4-11-10	48° 30.766′	122° 57.269′ W	72.7	0	0.662044	0.715804	−0.560269	3.759054
		C4-11-11	48° 30.947′	122° 57.100′ W	78.3	0	0.631338	0.742173	−0.829452	4.010213
		C4-11-12	48° 30.974′	122° 57.206′ W	66.3	0	0.721466	0.737844	−0.587603	3.867885
		C4-11-13	48° 31.032′	122° 57.277′ W	64.8	19	1.165707	0.977239	0.042418	2.003103
		C4-11-14	48° 31.222′	122° 57.049′ W	66	4	1.097705	0.905919	−0.157222	2.187567
		C4-11-15	48° 31.274′	122° 57.150′ W	59.3	11	0.811521	0.909076	−0.554293	2.731016
		C4-11-16	48° 31.324′	122° 57.070′ W	66.8	4	1.069695	1.078217	0.023948	1.802094

3.4. Current Profiles

Seven ADCP current transects including four N–S and three E–W runs within the sand wave field were made in July 2012 on *R/V Centennial* (Table 2). The profiles were all obtained during flood current as shown on the tidal height curve in Figure 5.

Table 2. Summary of ACDP Current Profiler Information.

Transect	SPD, cm/s	DIR, deg	BOT SPD, cm/s	HAB, m
1 N/S	47.04	43.10	35.51	8.01
2 N/S	72.88	8.05	74.57	7.93
3 E/W	67.00	15.66	73.68	6.66
4 E/W	50.15	44.06	79.57	8.96
5 N/S	74.39	5.05	87.74	6.56
6 E/W	75.94	18.11	76.99	7.11
7 N/S	53.17	66.36	56.77	6.53

SPD = Depth Averaged Current Speed; DIR = Direction of Depth Averaged Speeds; BOT Speed = Average Current Speed at Deepest Measurement; HAB = Height above Bottom of Deepest Current Measurement.

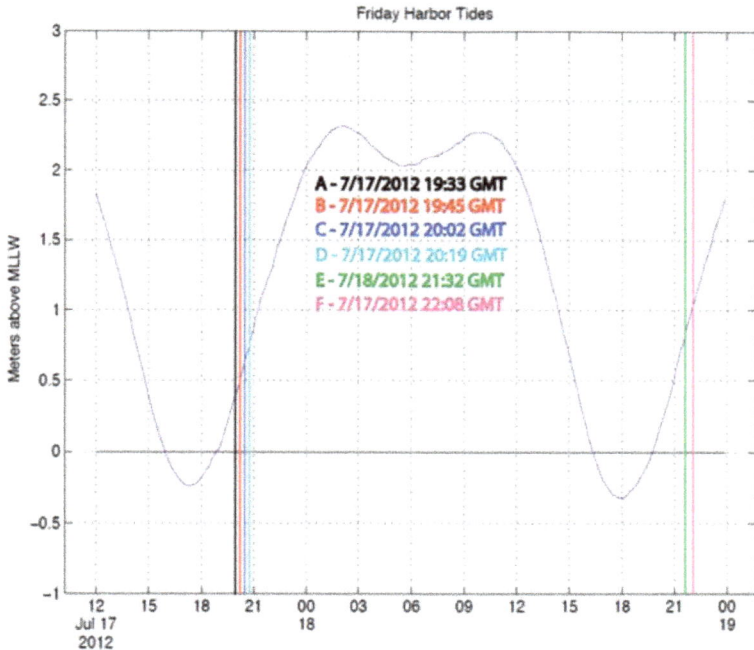

Figure 5. Tidal curves showing tidal heights in meters above Mean Low Low Water (MLLW) within the San Juan Channel for the time of the ADCP survey on 17 July 2012. Data from NOAA tide gauge at Friday Harbor, WA (NOAA, 6–7 October 2010).

A summary of the depth-averaged current speeds and directions for each transect along with the speeds at the deepest reliable measurement in the profiles are shown in Table 2. Bottom current speeds exceeded depth averaged speeds in four of the seven profiles, indicating that the surface current speeds for these profiles were relatively lower than at the bottom. One N–S oriented current transect is plotted in Figure 6. Both surface and bottom currents were directed northward along the entire run. Maximum bottom current speed for this transect was 75 cm/s.

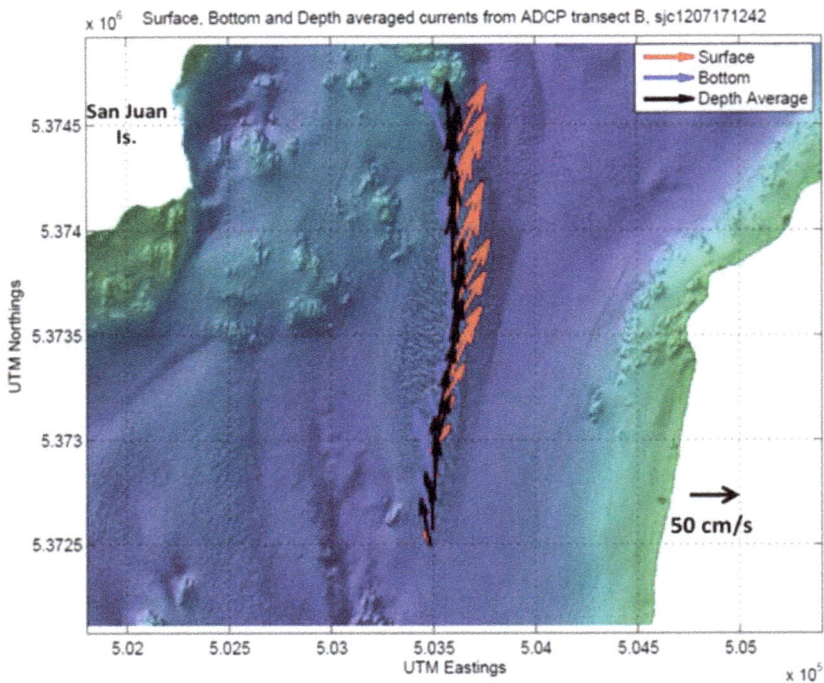

Figure 6. Surface, bottom, and depth-averaged currents observed during ADCP transect B aboard RV *Centennial* at flood tide, 17 July 2012.

The three E–W oriented ADCP current transects had interesting spatial variability in bottom current directions. Bottom currents were directed toward the sand body on either side. This convergence toward the sand body might suggest a means to keep the boundaries of the sand sharply defined (i.e., sand transport has a convergent component toward the feature). Of course, given the limited data set, this idea needs further study. From the current profile data bed load transport was calculated using bed load equations of Van Rijin [34]. Estimates of transport were made for sediment sizes of 0.5 mm and 2 mm, characteristic sizes for bed sediment along the crests and troughs, respectively, of the large sand waves.

To estimate potential migration of the sand waves, we calculated their asymmetry and used the technique of Knaapen [23]. The migration rate C of sand waves was found by Knaapen [23] to be related to bedform asymmetry A, wavelength L, and sand-wave height H. This empirical relationship was developed and validated using 12 different sites in the North Sea where repeated MBES surveys were completed [23]. The technique has been applied in other geographic areas with tidal sand waves [22,23].

4. Results

The seafloor map of the San Juan Channel sand wave field derived from MBES bathymetric data reveals multiple scales of bedforms (Figure 7). Large sand waves with heights H = 1 to 4 m and wavelengths L = 20 to 100 m have two distinct crest orientations, possibly related to steering by ebb and flood tidal currents influenced by the irregular rocky inland sea morphology. One set of sand waves oriented about 30 degrees counterclockwise from North appears to have longer crests and longer wavelengths than the more E-W set (Figure 7). Smaller megaripples and sand waves occur along the edges of the sand body. An enlargement of the sand wave field also shows small- to medium-scale sand waves (1 to 2 m wavelengths) between some of the larger waves, with crest orientations approximately E-W (Figure 7). In general, the large sand waves are nearly straight-crested and two-dimensional. Some bifurcations of the crest lines can be observed. Notice that some of the sand waves are contiguous with one another, with the lee side of one wave encroaching on the stoss side of the adjacent one, whereas others are separated by rippled flat areas that appear to be the glacial rubble surface upon which the waves were constructed (Figure 7).

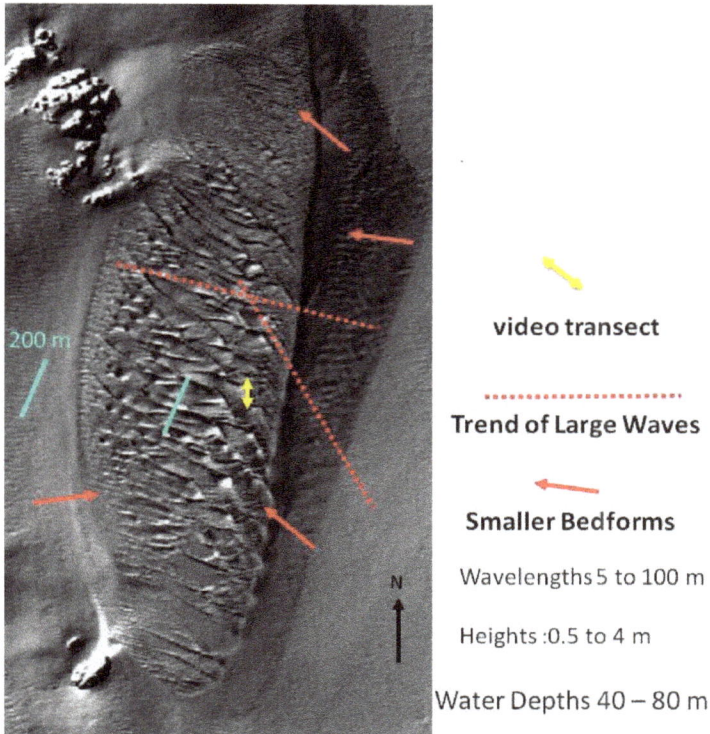

Figure 7. MBES image of the San Juan Channel sand wave field that shows two different sizes of sediment waves from long high amplitude to short low-amplitude waves. Red solid arrows show where ripples and small amplitude sediment waves exist along the edges of the field, dashed red lines show orientation of large sand waves, yellow line with arrows shows video camera transect made to validate sediment type and morphology, and green lines are scale marks used to show wave lengths.

4.1. Sediment Analyses

Grain size analyses previously made by Blaine [18] were repeated from samples collected during this study (Figure 8), confirming that the mean grain size within the San Juan Channel sand wave field is 0.5 mm with a fairly constrained range of 2 mm to 0.125 mm. The sea-floor photographs show that grains on the crests of sand waves have a median size of 0.5 mm (Figure 4b), which is significantly finer than in the troughs, where the median is 2.0 mm (Figure 4c). Also, irregularly shaped shell-like debris is common in the troughs. However, in areas where mean grain size is near 0.5 mm in the larger amplitude sand waves, a greater number of fish were captured (Figure 9, Table 1). The fish appear to generally concentrate in grain sizes that range between 0.25 to 0.71 mm with the highest number of fish (from 30 to 62 per sample) collected in sediment samples with mean grain sizes of approximately 0.35 to 0.5 mm. These results are consistent with that found by Blain [18].

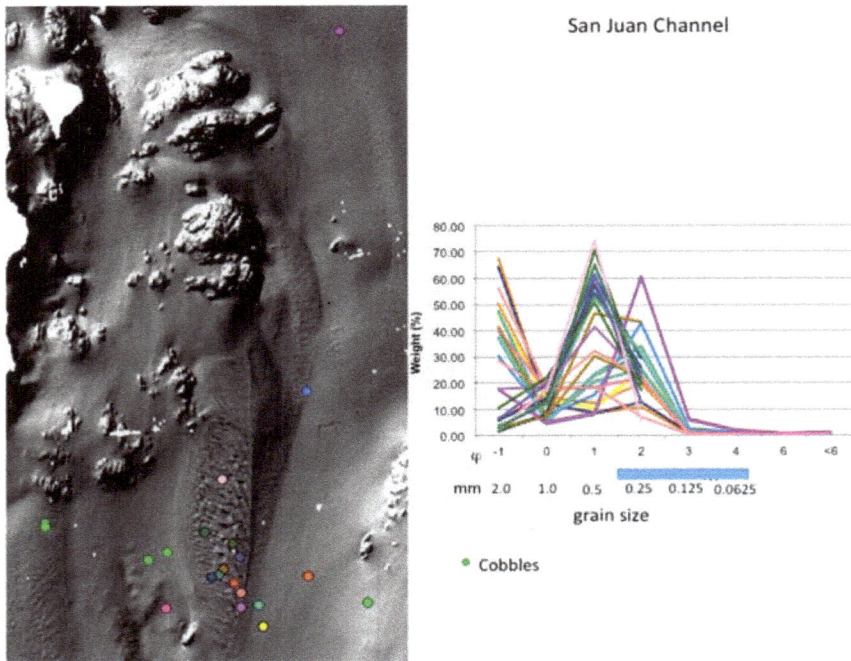

Figure 8. Sediment samples [22] collected and analyzed for this study; (on left, MBES image) location of samples; color dots represent sample sites, and (right) graph of grain sizes from sieve analyses showing grain size distribution.

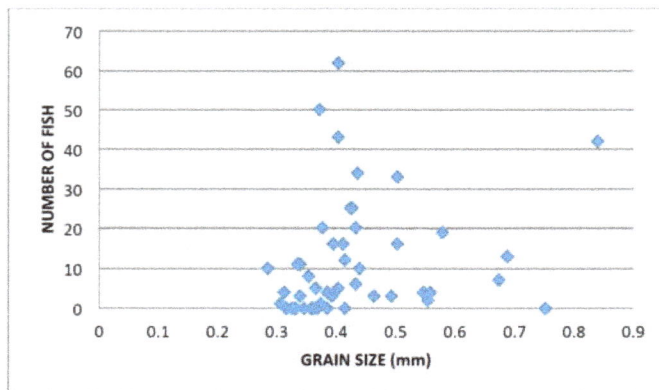

Figure 9. Numbers of fish captured in sediment samples from the San Juan Channel sand wave field. The highest numbers of fish (from 30 to 62 per sample) were collected in sediment samples with mean grain sizes of approximately 0.35 to 0.5 mm.

The photos show loosely consolidated, well sorted, poorly packed sub-rounded to sub-angular grains separated by porous i))nterstices. The apparent good porosity of the sediment facilitates aeration that attracts large concentrations of PSL.

4.2. Sand Transport and Bedform Migration

Bedload transport estimates q_b are plotted in Figure 10 against bottom shear velocities, $u*_b$ for the two dominant bottom sediment sizes. The coarser sediment (1–2 mm) has rather low transport rates (<1 kg/m/s) for $u*_b$ < 0.1 m/s. However, the finer sand (0.5 mm) that was sampled along the flanks and crests of the large sand waves is more mobile at intermediate bottom stresses. For example, at $u*_b$ = 0.08 m/s, q_b = 2 kg/m/s.

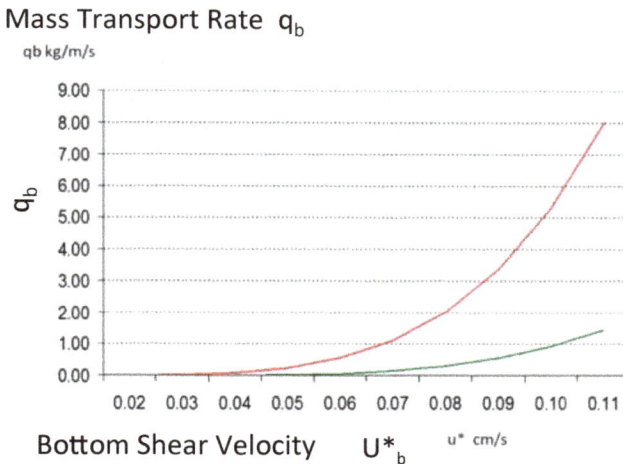

Figure 10. Graph showing mass transport rate of bottom sediment for the San Juan Channel sand wave field based on methods of Van Rijin [34] for grain sizes of 0.5 mm (red line) and 2.0 mm (green line).

Assuming a conservative value of $u*_b$ = 0.05 m/s for tidal currents, based on converting depth-averaged current speeds in Table 2 to bottom shear in tidal flow over a rough bed [35], an estimate of sand wave migration can be made [36].

$$V = \frac{q_b}{0.5\rho_b H} \tag{1}$$

V is bedform migration speed, ρ_b sediment bulk density, and H is bed-form height.

Using Equation (1) for quartz sand (our visual inspection indicates that the sand is arkosic, quartz-rich), a sand wave with H = 3 m would migrate about 0.15 m/h. Since San Juan Channel tides are mixed, tidal current reversals occur nearly every six hours. Also, the limited current profile data from this study, and from that of Ewing's [21] suggest that ebb and flood currents are of similar magnitudes (with flood slightly higher in the limited measurements). These results indicate that the large sand waves would have a small net northward migration (<0.15 m/h.) suggesting that no large supply of sediment is necessary to maintain the field and the field is primarily composed of relic glacial deposits. Because of the limited data and lacking time-series measurements of currents closer to the seabed, this result is only suggestive of small to nil sand wave migration.

4.3. Sand Wave Asymmetry

The sand waves in the San Juan Channel sand wave field can be grouped into two classes (zones) based on size and shape. Nearly symmetric smaller sand waves were found toward the shallower parts of the sand body (Figures 11 and 12). These sand waves in zone 1 to 2 appear sharp and well defined with slightly steeper flanks facing north (Figure 12). The sand wave heights are 1 to 2 m, and wavelengths 30 to 40 m. The sand waves in the deeper section of the sand body in zone 3 to 4 are larger and appear more irregular. The overall impression based on shape and water depth is that these deeper bedforms are inactive and possibly relict from a time of lower sea level. This appears to be consistent with multiple years of MBES surveys over the sediment wave field that suggests little or no change in sediment wave morphology or overall configuration of the sand body (Figure 11).

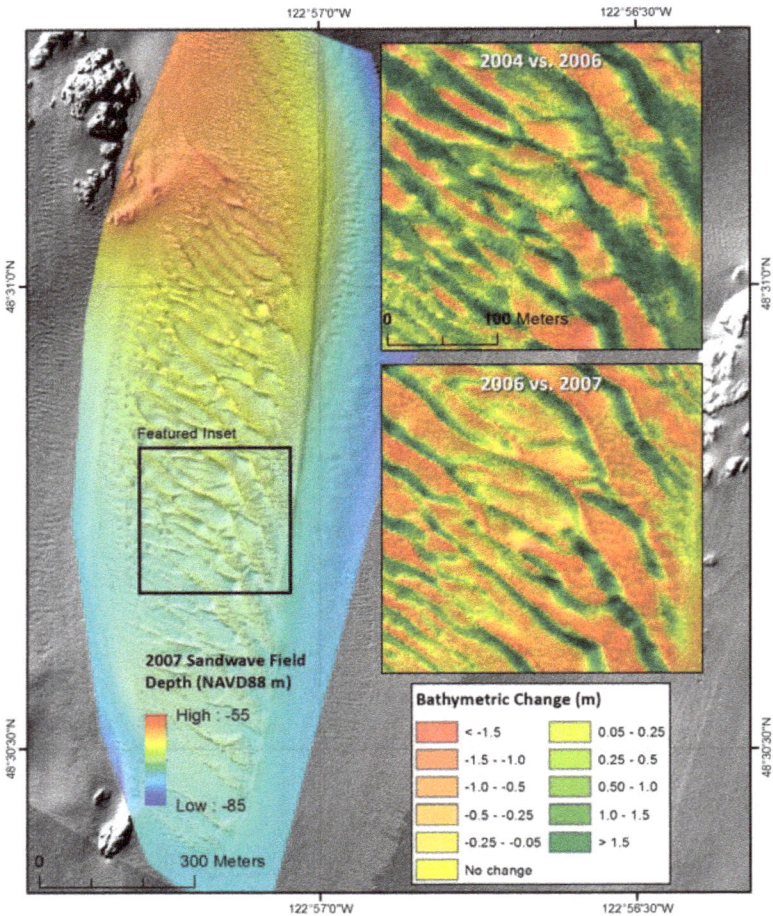

Figure 11. MBES bathymetric image of San Juan Channel sand wave field collected in 2007 with comparisons of changes between years 2004 and 2006 and between 2006 and 2007. Note how the general wave morphology is consistent from one year to the next but that slight sediment erosion (warm colors) and accumulation (green colors), probably the result of shifting back and forth of the sand waves, have occurred.

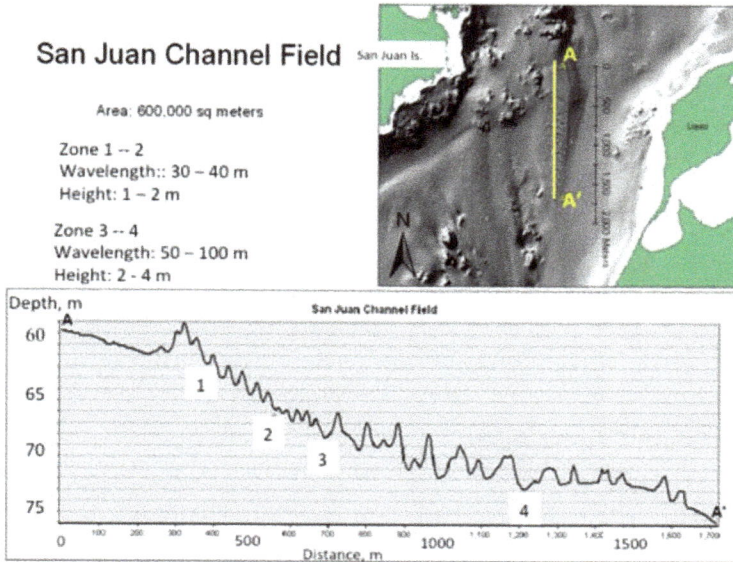

Figure 12. MBES bathymetric imagery and cross-section (A-A') of the San Juan Channel sand wave field showing two different sections (1–2 and 3–4) of sediment waves of differing sizes.

Figure 12 summarizes the results for sand wave migration in the central San Juan Channel using the most recent MBES survey data. Sand waves are predicted to migrate at low rates for all classes (<6 cm/year). This low rate agrees qualitatively with the estimates based on sediment transport calculations (Table 3).

Table 3. Estimate of Migration Rates from Bedform Shape.

	L, m	h_{avg}, m	h_{avg}/L	A	C, cm/year
Small	32.2	1.4	0.043	0.06	1.3
Large	67.5	2.8	0.041	0.12	6.0

Knappen [23] found migration rates of 5 to 800 cm/year based on 12 sites in Southern North Sea and English Channel. Figure 12 estimated sediment migration rates based on bedform shapes in the San Juan Channel sand wave field.

4.4. Bed Phase Diagrams

Current, grain size, and water depth data collected during this study were plotted on bed phase diagrams that had been developed for San Francisco Bay [27]. The results shown in Figure 13 include samples from water depths of 20 to 80 m within the sand wave field in the central San Juan Channel. Depth averaged current speeds ranged from 40 to 80 cm/s (Table 2). The sediment sizes that were used to develop the two bed phase diagrams for San Francisco Bay include the two major bottom sediment size classes for the central San Juan Channel sand wave field (0.5 and 2.0 mm).

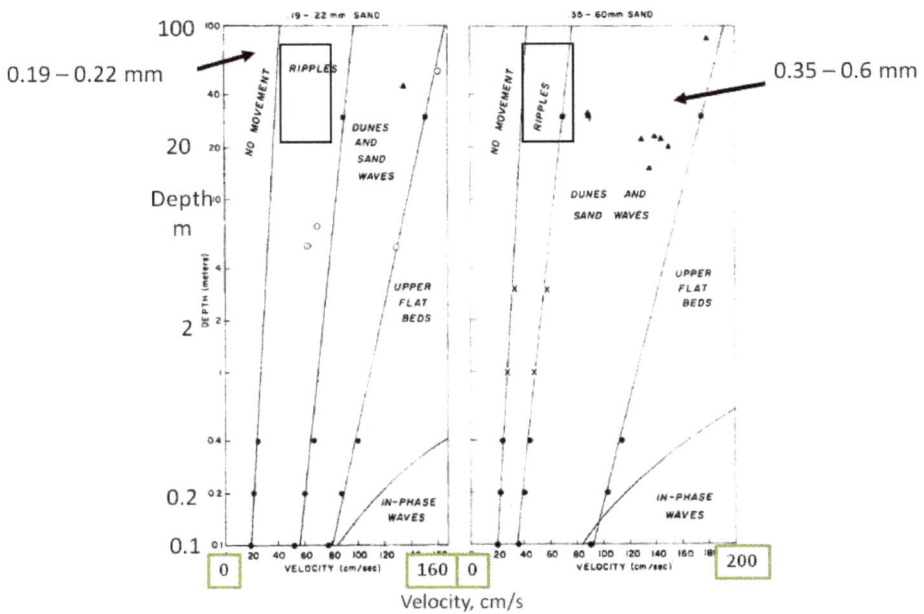

Figure 13. Bed-phase diagram for San Francisco Bay developed by Rubin and McCulloch [27] whose observations were combined with those of Southard [25], Boothroyd and Hubbard [37], and Dalrymple and others [26]. Black boxes are for data collected by this study in the San Juan Channel.

The plots show that the size-depth-speed data from the San Juan Channel sand wave field fall into the ripples-sand wave portion of the diagrams. If depth-averaged currents increase substantially beyond the maximum measured depth-averaged speeds the stability of the sand waves might decrease, leading to significant change to the sandy habitat.

5. Discussion

Several studies of PSL and sand-eel habitats show that these fish prefer sand habitats consisting of medium- to coarse-grained sand of 0.25 to 1 mm in size [8,11,14,18,38], and our results show that PSL select bottom sediment between 0.25 to 0.71 mm in the San Juan Channel sand wave field. Experimental studies (e.g., [38,39]) also show that PSL prefer this range of sediment grain sizes. Furthermore, our results and others (as cited above) show an absence of PSL if the sediment contains more than a few percent (~5%) of fines (fine sand + silt + clay-size material; Figure 14). The preferred sediment is clean, unconsolidated sand that can be easily penetrated, and has an adequate and readily replenished supply of oxygen [5,11,14,39]. If fines such as muds are present, even in small amounts, the gills of the fish can be clogged [39] and the pores of the sediment filled, reducing permeability.

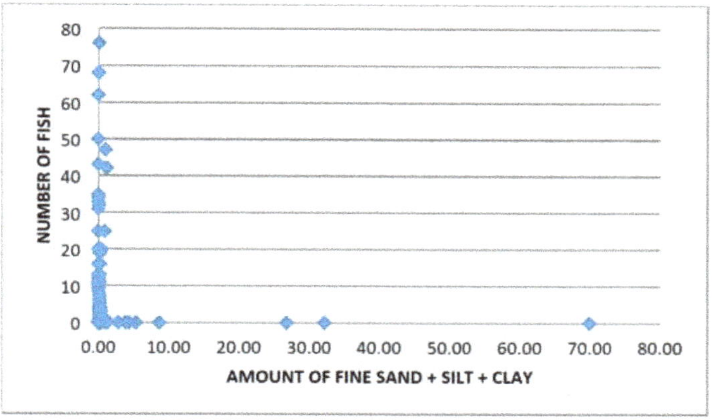

Figure 14. Number of fish in grab sample versus amount of fine sediment (grain size finer than 0.25 mm).

Our results are consistent with that found by Blaine [18] indicating that the San Juan Channel sand wave field is not significantly altered through time. In addition, the field appears to be consistently attractive to the concentration of PSL, as generally no large changes in density of fish have occurred in the past 10 years of study.

Currents were previously measured during ebb tide in the same area, and were found to have velocities similar to those shown here for flood currents [21]. The ebb current profiles were taken with the same ADCP system during November 2011, and cross-sections of currents showed bottom current speeds as high as 70 cm/s [21].

The large sand waves in the central San Juan Channel sand wave field likely have low to nil net migration at present time, based on two independent estimates (i.e., sediment transport calculations and bedform asymmetry). This is further supported based on observations of the resultant bed phase diagrams that relate bottom sediment size, water depth, and depth-averaged current speed to bed phase, indicating that the sand wave field in the central San Juan Channel is stable. However, at substantially higher current speeds the sand waves could be eroded, and the sand body geometry changed to planar bed.

As sea level rises, the volume of tidal flows through the straits at the northern and southern ends of the San Juan Channel might produce higher currents within the channel and on the seafloor. These higher currents could lead to erosion and loss of sand (to upper planar bed) in the channel, and endanger the sub-tidal habitats for PSL.

The strong ebb and flood flow currents appear to generate and maintain sand wave fields with sharp, well-defined boundaries as illustrated in the San Juan Channel sand wave field and other fields such as found off Ice Berg Point of southern Lopez Island and offshore of northern Sucia Island [8] from which we have collected PSL. In contrast to the study of Wright [11] on North Sea sand-eels, we found no evidence that an abundant supply of sand was necessary for maintaining well-aerated sediment for the occupation of PSL. However, similar to the studies of Mercer [9], Meyer [40], and Pinto [39] that found that ripples within the sand field were tangential to the tidal stream and tended to be symmetrical indicating bedforms produced by oscillatory flow, we too found that tidal oscillatory motion produced small scale symmetrical bedforms in the San Juan Channel sand wave field.

No evidence for net sediment transport was found, and we propose that the sand wave field is composed of glacially-derived sediment winnowed through strong current flow, which leaves a well-sorted sand lag deposit that continues to be well-aerated through tidal oscillatory flow. Wright [11] found that sediment permeability in addition to ripple geometry and water percolation better characterized sandeel habitat preferences. Based on eyeball camera data we found that the most desirable sediment characteristics preferred by PSL are concentrated near the crests of the sand waves, which are regularly refreshed by the strong tidal flows.

6. Conclusions

The central San Juan Channel sediment wave field represents an extensive sub-tidal (deep water) habitat feature for PSL, an important forage fish in the Salish Sea. This field contains bedforms with wavelengths up to 100 m and heights of approximately 1 to 4 m within its central area. Smaller sand waves and megaripples are superimposed upon the large sand waves and fringe the sharp edges of the field. Furthermore, two major crest orientations were detected, one with nearly E–W (~280° or W10° N) trend and another with a NW–SE (~315° or W45° N) trend (Figure 7). We used a unique shipboard digital underwater camera (referred to as the "eyeball" camera) to take close-up high-resolution photographs of the seafloor that enabled calculation of median grain sizes [27] and facilitated the observations of consolidation, packing, lithology, and interstitial spaces of the sand waves. Two of these photographs, one located in a sand-wave crest and the other on the adjacent trough (Figure 4b,c) show that the grain size at the crest of a wave is significantly finer (median sand ~0.5 mm diameter) than that in the trough (~2.0 mm median diameter). In addition, the eyeball camera data show finer, better-sorted sediment on the flanks and crests of the larger sand waves. Furthermore, biogenic shell material dominates in the troughs whereas siliciclastic grains dominate on the flanks and crests, thus we can determine the position on a waveform (crest or trough) from which a sample was taken. We used these grain sizes to evaluate the stability of the sand wave field and validate PSL occupancy. Our results suggest that the field is stable, suggesting that other dynamic bedforms exhibiting similar characteristics need to be identified to conserve and sustain these critical forage fish habitats.

It is important to note that sediment grain size is not the only control on the presence/absence of fish in grab samples. For instance, PSL abundance within the sand field varies with time of day, season, and year, irrespective of the suitability of the sediment texture (temporal control). Chemical contamination (e.g., petroleum) and presence of certain seafloor flora (e.g., eelgrass) are other documented potential controls [5,39]. For example, in their North Sea studies on sand-eels Holland and others [14] discuss how grab samples without fish can be portioned into two groups: (1) unsuitable habitat, and (2) suitable habitat but unoccupied by fish. These authors go on to state that only "(2)" should be considered in measuring mean density of fish, and the estimate obtained is the estimate of fish density in suitable habitat. When combined with seabed habitat mapping grab sampling can be directed to those areas where fish habitats are located [14]. Therefore, our study of bedform morphology and processes forming these bedforms is useful in identifying areas where fish might be located and sampled.

Although we focus on sediment size and transport mechanisms to determine the stability of the sand wave field, we realize that other factors such as nutrient supply, temperature and tidal flow also play a role in PSL survival.

Additionally, using median grain size alone to identify sand lance presence or absence can be misleading. For example, examination of Table 1 shows that several samples that did not capture fish have a mean grain size that falls within the range of medium to coarse-grained sand, a range that includes all samples that contained fish. Thus, realizing that absence is never very strong evidence in itself, a clear distinction of sediment preference is not made. A justifiable conclusion would be that PSL were not present in some samples due to other reasons, such as the time of day when the samples were collected or simply out of population size/density dependence. Freeman and others [12] pointed out that in the North Sea other factors such as seawater temperature, physical conditions, and available food in the water column influence the emergence of fish from the sediment and thus may not be present at the time of sampling. However, Figure 14 better defines the sand-lance habitat by the absence or near-absence of fine-grained sediment. All samples that contained fish have less than 1.15% of grains finer than 0.25 mm.

Based on MBES surveys and sediment transport analyses using sediment size samples and ADCP surveys of the central San Juan Channel sand wave field we conclude that surficial sediment within the field are relic glacial materials that have been winnowed and molded into a dynamic bedforms from post-glacial tidal currents. It appears that the field is near stable and could be maintained for some time without significant sediment input. The two survey data sets (MBES and ACDP) compliment each other and indicate that little net sediment transport is taking place and that the larger bedforms are stable. The stability regime could be altered with sea level rise and increased tidal flows, but the reality and rate of such changes are not known. Other dynamic bedforms within the Salish Sea, as well as elsewhere where *Ammodytes* spp. are found, could be stable or marginally stable as well, and the potential effects of sea level rise may diminish this forage fish habitat in sand wave fields.

Acknowledgments: We wish to thank Hank Chezar of the USGS for his assistance in collecting and processing the sediment "eyeball" camera images, and to David Rubin for his help in consultations and analysis of the sediment images. We thank Kurt Rosenberger of USGS and Liz Ewing for their help with decoding, plotting, and analyzing the ADCP current meter data. We appreciate the support given to us by the Pacific Science Center, USGS, to undertake this project. We thank Don Gunderson, Joseph Bizzarro and anonymous reviewers for their critical review of the manuscript and constructive comments. John Aschoff and Charlie Endris assisted in the preparation of the figures and we sincerely appreciate their help. We thank the Dickinson Foundation, Northwest Straits Commission, Geological Survey of Canada and the SeaDoc Society for support in the collection of the MBES and sediment data.

Author Contributions: All authors contributed equally to this study and manuscript.

Conflicts of Interest: The authors declare no conflict of interest.

References

1. Robards, M.D.; Piatt, J.F.; Rose, G.A. Maturation, fecundity, and intertidal spawning of Pacific sand lance in the northern Gulf of Alaska. *J. Fish Biol.* **1999**, *54*, 1050–1068. [CrossRef]

2. Robards, M.D.; Willson, M.F.; Armstrong, R.H.; Piatt, J.F. *Sand Lance: A Review of Biology and Predator Relations and Annotated Bibliography*; Exxon Valdez Oil Spill Restoration Project 99346 Final Report; U.S. Department of Agriculture, Forest Service, Pacific Northwest Research Station: Portland, OR, USA, 1999; p. 327.

3. Robards, M.D.; Rose, G.A.; Piatt, J.F. Growth and abundance of Pacific sand lance, *Ammodytes hexapterus*, under differing oceanographic regimes. *Environ. Biol. Fishes* **2002**, *64*, 429–441. [CrossRef]

4. Tribble, S.C. Sensory and Feeding Ecology of Larval and Juvenile Pacific Sand Lance, *Ammodytes hexapterus*. Master's Thesis, University of Washington, Washington, DC, USA, 2000; p. 98.

5. Haynes, T.B.; Ronconi, R.A.; Burger, A.E. Habitat use and behavior of the Pacific sand lance (Ammodytes hexapterus) in the shallow subtidal region of southwestern Vancouver Island. *Northwest. Nat.* **2007**, *88*, 155–167. [CrossRef]

6. Haynes, T.B.; Robinson, C.I.K.; Dearden, P. Modeling habitat use of young-of-the-year Pacific Sand Land (*Ammodytes hexapterus*) in the nearshore region of Barkley Sound, British Columbia, Canada. *Environ. Biol. Fishes* **2008**, 83473–83484.

7. Johnson, S.W.; Tehdinga, J.F.; Munk, K.M. Distribution and use of shallow-water habitats by Pacific sand lances in Southeastern Alaska. *Trans. Am. Fish. Soc.* **2008**, *137*, 1455–1463. [CrossRef]

8. Greene, H.G.; Wyllie-Echeverria, T.; Gunderson, D.; Bizzarro, J.; Barrie, V.; Fresh, K.; Robinson, C.; Cacchione, D.; Penttila, D.; Hampton, M.; et al. *Deep-Water Pacific Sand Lance (Ammodytes hexapterus) Habitat Evaluation and Prediction for the Northwest Straits Region*; Final Report to Northwest Straits Commission; SeaDoc/Tombolo Mapping Lab and Friday Harbor Labs: Orcas Island, WA, USA, 2011; p. 21.

9. Mercer, C.T. *Sand Eels (Ammodytidae) in the South-Western North Sea: Their Biology and Fishery*; Great Britain Minist. Agic. Fish. Food, Fish. Invest. Ser. II Mar. Fish. 24; Her Majesty's Stationary Office: London, UK, 1966.

10. Reay, P.J. Synopsis of biological data on north Atlantic sandeels of the genus *Ammodytes*. *FAO Fish. Synop.* **1970**, *82*. Available online: http://www.fao.org/3/a-a8703e.pdf (accessed on 22 October 2017).

11. Wright, P.J.; Jensen, H.; Tuck, I. The influence of sediment type on the distribution of the lesser sendeel, *Ammodytes marinus*. *J. Sea Res.* **2000**, *44*, 243–256. [CrossRef]

12. Freeman, S.; Mackinson, S.; Flatt, R. Diel patterns in the habitat utilization of sandeels revealed using integrated acoustic surveys. *J. Exp. Mar. Biol. Ecol.* **2004**, *305*, 141–154. [CrossRef]

13. Wright, P.J.; Pedersen, S.A.; Donald, L.; Anderson, C.; Lewy, P.; Proctor, R. *The Influence of Physical Factors on the Distribution of Lesser Sandeels and Its Relevance to Fishing Pressure in the North Sea*; C.M.-International Council for the Exploration of the Sea: Copenhagen, Denmark, 1998.

14. Holland, G.J.; Greenstreet, S.P.R.; Gibb, I.M.; Fraser, H.M.; Robertson, M.R. Identifying sandeel *Ammodytes marinus* sediment habitat preferences in the marine environment. *Mar. Ecol. Prog. Ser.* **2005**, *303*, 269–282. [CrossRef]

15. Greenstreet, S.P.R.; Tuck, I.D.; Grewar, G.N.; Armstrong, E.; Reid, D.G.; Wright, P.J. An assessment of the acoustic survey technique, RoxAnn, as a means of mapping seabed habitat. *ICES J. Mar. Sci.* **1997**, *54*, 939–959. [CrossRef]

16. Stride, A.H. (Ed.) *Offshore Tidal Sands, Processes and Deposits*; Chapman and Hall: London, UK, 1982; p. 222.

17. Barrie, J.V.; Conway, K.W.; Picard, K.; Greene, H.G. Large scale sedimentary bedforms and sediment dynamics on a glaciated tectonic continental shelf: Examples from the Pacific margin of Canada. *Cont. Shelf Res.* **2009**, *29*, 796–806. [CrossRef]

18. Blaine, J. *Pacific Sand Lance (Ammodytes hexapterus) Present in the Sandwave Field of Central San Juan Channel, WA: Abundance, Density, Maturity, and Sediment Association*; Unpublished: Class Paper, Fish 492 Research Apprentice; Friday Harbor Labs: Friday Harbor, WA, USA, 2006; p. 24.

19. Rood, M. *Length Distribution, Condition Factor, and Feeding Ecology of Pacific Sand Lance in the San Juan Archipelago, Fall 2010*; Student Paper, Pelagic Ecosystem Function Apprenticeship; Friday Harbor Laboratories: Friday Harbor, WA, USA, 2010; 16p.

20. Boyd, S.; Wyllie-Echeverria, T. *Individual Burrowing Behavior of Pacific Sand Lance (Ammodytes hexaperus): A Laboratory Experiment*; Student Paper, BLINKS Research Fellowship; Friday Harbor Laboratories: Friday Harbor, WA, USA, 2010; 14p.

21. Ewing, L. *Physical Influences on Spatial and Temporal Distribution of Pacific Sand Lance (Ammodytes hexapterus) in the Sand Wave Fields of San Juan Channel*; Class Paper; Friday Harbor Marine Labs, Univ. Washington: Friday Harbor, WA, USA, 2000; 28p.

22. Barnard, P.L.; Erikson, L.H.; Rubin, D.M. Analyzing bedforms mapped using multibeam sonar to determine regions bedload sediment transport patterns in the San Francisco costal system. *Int. Assoc. Sedimentol. Spec. Publ.* **2012**, *44*, 273–294.

23. Knaapen, M.A.F.; van Bergen Henegouw, C.N.; Hu, Y.Y. Quantifying bedform migration using multi-beam sonar. *Geo-Mar. Lett.* **2005**, *25*, 306–314. [CrossRef]

24. Ashley, G.M. Classification of large-scale sub-aqueous bedforms: A new look at an old problem. *J. Sediment. Petrol.* **1990**, *60*, 160–172.

25. Southard, J.B. Representation of bed configurations in depth-velocity-size diagrams. *J. Sediment. Petrol.* **1971**, *41*, 903–915.

26. Dalrymple, R.W.; Knight, R.J.; Lambiase, J.J. Bedforms and their hydraulic stability relationships in a tidal environment, Bay of Fundy, Canada. *Nature* **1978**, *275*, 100–104. [CrossRef]

27. Rubin, D.; McCulloch, D. Single and Superimposed Bedforms: A Synthesis of San Francisco Bay and Flume Observations. *Sediment. Geol.* **1980**, *26*, 207–231. [CrossRef]

28. Orr, E.L.; Orr, W.N. *Geology of the Pacific Northwest*; The McGraw-Hill Companies, Inc.: San Francisco, CA, USA, 1996.

29. Barrie, J.V.; Conway, K.W. Late Quaternary glaciation and postglacial stratigraphy of the northern Pacific margin of Canada. *Quat. Res.* **1999**, *51*, 113–123. [CrossRef]

30. Greene, H.G.; Barrie, V. *Potential Marine Benthic Habitats of the San Juan Archipelago. Geological Survey of Canada Marine Map Series, 4 Quadrants, 12 Sheets, Scale 1:50,000*; Geological Survey of Canada: Sidney, BC, Canada, 2011.

31. Greene, H.G.; Bizzarro, J.J.; O'Connell, V.M.; Brylinsky, C.K. Construction of digital potential marine benthic habitat maps using a coded classification scheme and its application. *Mapp. Seafloor Habitat Charact. Can. Geol. Assoc. Spec. Pap.* **2007**, *47*, 141–155.

32. Rubin, D.M. A simple autocorrelation algorithm for determining grain size from digital images of sediment. *J. Sediment. Res.* **2004**, *74*, 160–165. [CrossRef]

33. Wentworth, C.K. A scale of grade and class terms for clastic sediments. *J. Geol.* **1922**, *30*, 377–392. [CrossRef]

34. Van Rijn, L.C. *Principles of Sediment Transport in Rivers, Estuaries, and Coastal Seas*; Aqua Publications: Blokzijl, The Netherlands, 1993; p. 631.

35. Soulsby, R. *Dynamics of Marine Sands*; Thomas Telforb Pubications: London, UK, 2000; p. 249.

36. Rubin, D.M.; Hunter, R.E. Bedform alignment in directionally varying flows. *Science* **1987**, *237*, 276–278.

37. Bothroyd, J.C.; Hubbard, D.K. Genesis of bedforms in mesotidal estuaries. *Esturine Res.* **1975**, *2*, 217–234.

38. Bazzarro, J.J.; Peterson, A.N.; Blaine, J.N.; Balaban, J.P.; Greene, H.G.; Summers, A.P. Burrowing behavior, habitat, and functional morphology of the Pacific sand lance (*Ammodytes personatus*). *Fish. Bull.* **2016**, *114*, 445–460. [CrossRef]

39. Pinto, J.M. Laboratory spawning of *Ammodytes hexapterus* from the Pacific Coast of North America with a description of its eggs and early larvae. *Copeia* **1984**, *1984*, 242–244. [CrossRef]

40. Meyer, T.L.; Cooper, R.A.; Langstone, R.W. Relative abundance, behavior and food habits of the American sand lance, *Ammodytes americanus*, from the Gulf of Maine. *Fish. Bull.* **1979**, *77*, 243–254.

geosciences

MDPI

Article

Multiscale and Hierarchical Classification for Benthic Habitat Mapping

Peter Porskamp (ID) **, Alex Rattray, Mary Young and Daniel Ierodiaconou** * (ID)

Centre for Integrative Ecology, School of Life and Environmental Sciences, Deakin University, P.O. Box 423, Warrnambool, VIC 3280, Australia; pporskam@deakin.edu.au (P.P.); alex.r@deakin.edu.au (A.R.); mary.young@deakin.edu.au (M.Y.)
* Correspondence: iero@deakin.edu.au; Tel.: +61-409-502-980

Received: 28 February 2018; Accepted: 30 March 2018; Published: 2 April 2018

Abstract: Developing quantitative and objective approaches to integrate multibeam echosounder (MBES) data with ground observations for predictive modelling is essential for ensuring repeatability and providing confidence measures for benthic habitat mapping. The scale of predictors within predictive models directly influences habitat distribution maps, therefore matching the scale of predictors to the scale of environmental drivers is key to improving model accuracy. This study uses a multi-scalar and hierarchical classification approach to improve the accuracy of benthic habitat maps. We used a 700-km^2 region surrounding Cape Otway in Southeast Australia with full MBES data coverage to conduct this study. Additionally, over 180 linear kilometers of towed video data collected in this area were classified using a hierarchical classification approach. Using a machine learning approach, Random Forests, we combined MBES bathymetry, backscatter, towed video and wave exposure to model the distribution of biotic classes at three hierarchical levels. Confusion matrix results indicated that greater numbers of classes within the hierarchy led to lower model accuracy. Broader scale predictors were generally favored across all three hierarchical levels. This study demonstrates the benefits of testing predictor scales across multiple hierarchies for benthic habitat characterization.

Keywords: Multibeam bathymetry; benthic habitat mapping; multiscale; Random Forests

1. Introduction

Development of hydrographic survey technologies over the past few decades has provided an ever more focused lens through which scientists can study the relationships between the physical nature of the seafloor and the benthic communities found there. Multibeam echosounders (MBES) provide the ability to collect detailed full-coverage information on fine-scale features that are required for the development of benthic habitat maps [1,2]. However, the spatial scale of drivers of habitat distribution are often mismatched and in many cases there is a need to explore seafloor structure at multiple spatial scales in order to match local drivers of habitat distribution [3]. Understanding resource and habitat distribution is crucial for effective management. As a result, maps that accurately reflect distribution of key biological resources are powerful tools for marine spatial management and planning [1].

Processes that drive the distribution and composition of marine benthic communities operate from global to local scales [4]. For example, changes in marine communities along latitudinal temperature gradients have been well documented [5–7]. In temperate coastal oceans, light availability dictates the primary division of photosynthetic algae and filter feeding organisms [8,9]. This pattern can be modified by factors such as local light attenuation (turbidity), productivity, wave energy and current energy [8,9]. In addition to temperature and light, local seabed characteristics can also drive species distribution. Previous studies have demonstrated the importance of seafloor geology in supporting macroalgae and invertebrate communities [10,11] (e.g., availability of hard substrate for attachment), while complex seafloor features support high diversity for benthic [12] and mobile species [13]. Since key environmental drivers may operate over a range of spatial scales, matching the scale of predictors used to model habitat distribution is key to attaining accurate habitat maps [2,14,15]. Many different modelling approaches have been used to develop habitat maps (e.g., frequentist, machine learning), however, users of those models often pre-determine the analysis scale rather than allowing the model to determine the best scale for each predictor [16–22].

Collecting biological data over broad spatial scales is often time consuming and expensive, resulting in the use of abiotic surrogates, such as terrain attributes, to associate with habitat classes for extrapolation [12,23]. Bathymetry and backscatter are the primary products of acoustic sampling providing full coverage data of the seafloor. Bathymetry provides information on seafloor depth, and through post processing can be used to derive measures of seafloor complexity [3,24], while backscatter provides information on acoustic energy scattered by the seafloor [25]. Ierodiaconou et al. [17] found that combining both bathymetry and backscatter can result in more accurate habitat maps. Derivatives from bathymetry and backscatter are often calculated using a default focal window size of 3×3 [4]. It is important to note that a single fixed scale cannot capture all features of interest [26], and testing at multiple scales is considered best practice [3,4]. In marine studies, predictor scale has often been chosen based on the resolution of bathymetry and backscatter datasets (see reviews [1,2,14]). Wilson et al. [3] conducted a multiscale analysis using derivatives from a bathymetric surface, and found a combination of terrain attribute scales resulted in a more accurate model when compared to using a fixed scale for each. Investigations of attribute scale importance in benthic habitat mapping studies are still limited, but choosing the most useful scale via model statistics is expected to result in more accurate habitat models [3].

In addition to deriving terrain attributes, high resolution bathymetry is essential for downscaling regional wave models in the coastal zone, allowing for the inclusion of wave energy as a predictor in habitat models [18]. Wave energy impacts water circulation and nutrient delivery within coastal ecosystems [27,28], such as those along Australia's southern coastline, which are exposed and experience some of the highest wave energy in the world [29]. Rattray et al. [18] conducted a study along Victoria's coastline and found that distribution of reef biotopes is strongly mediated by variation in hydrodynamic energy.

The marine environment is commonly classified in a variety of ways with a focus on abiotic features [30], biotic features [31], or combinations of both [32]. Focusing on the interactions between the abiotic (terrain features) and biotic (communities) environments result in a habitat map representing biotopes, that are ecologically relevant [31,33,34]. Since ecosystems are structured hierarchically, using a hierarchical approach when classifying thematic habitat maps is a logical path. Previous studies have discussed the benefits of habitat modelling using hierarchical approaches [33,35,36] (e.g., aggregating smaller classes into higher levels, mapping at various levels based on the scale and size of the map to suit user needs), but few studies have done so [33], especially in the marine environment. Bock et al. [37] investigated two terrestrial case studies that compared classification performance at different hierarchical levels. They found that progression down the hierarchy (increasing hierarchy resolution) produced less accurate models. Using satellite sensor technology to classify coral habitat, Mumby et al. [38] found high variability in accuracy within habitat classes, and hypothesized that this was due to the poor delineation between classes. Understanding how hierarchical levels within a classification scheme impact model uncertainty for the marine environment is important, especially when such data are used to inform management.

Assessing the performance of habitat classifications at different hierarchical levels provides a unique opportunity to evaluate the influence of spatial scale in predicting benthic habitat classes. The aim of this study is to investigate a multi-scalar and hierarchical classification approach to improve the accuracy of benthic habitat maps using a machine learning modelling approach.

2. Materials and Methods

2.1. Study Area

The study area encompasses 700 square kilometers surrounding Cape Otway in Southeast Australia with full MBES data coverage in depths ranging from 10–80 m and is located along the Bass Strait and Western Bass Strait ecoregions [39]. The Cape Otway coastline is highly exposed due to its orientation to prevailing southwesterly ocean swells (Figure 1). In the east, the site is characterized by large sandy embayments with headland reef systems extending offshore. To the south and west, complex rocky reef systems extend offshore from an erosional shoreline of cliffs and limestone stacks [40]. Shallow reefs support diverse communities dominated by canopy forming kelps or open areas of red seaweeds. Deeper reefs in the region are generally populated by diverse sessile invertebrates or crustose coralline algae [40].

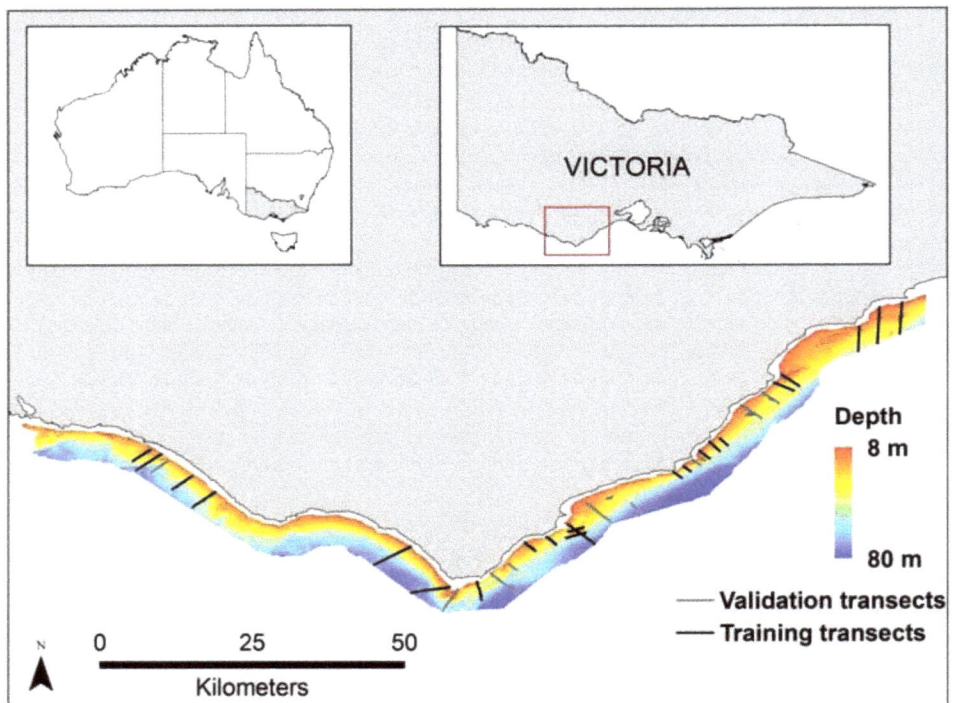

Figure 1. Location of Cape Otway in Victoria, Australia, and the multibeam echosounder (MBES) bathymetry overlaid on a bathymetric hillshade of the study site. The western side of the cape is exposed to prevailing south-westerly ocean swells, while the eastern side is moderately protected by comparison. Black lines represent the transects selected for training and the gray lines represent transects selected for validation.

2.2. MBES Data Collection and Processing

MBES data were collected from 2006–2008 using a Reson Seabat 8101 MBES operating at a frequency of 240 kHz aboard the Australian Maritime College vessel *R.V. Bluefin*. Vessel positon was determined by differential GPS (DGPS) and motion values were captured by a POS-MV motion sensor recording heave, pitch, roll and yaw with an accuracy of ±0.02 degrees. Bathymetry was collected with a horizontal accuracy of ±0.3 m. Bathymetry and backscatter values were processed using Starfix suite 7.1 and University of New Brunswick (UNB1) algorithm. The processed bathymetry and backscatter data were gridded to 2.5 m resolution.

For analysis, we merged the rasters from each survey into a single bathymetry and a single backscatter surface using ENVI 5.3.1 [41]. ENVI was used to stitch and fill any holes in the dataset using Delaunay triangulation of values from the surrounding surface. Once the two surfaces were created a mask was used to exclude any islands in the data and clip both surfaces to common extents. We extracted derivatives using ENVI and ArcMap 10.4.1 [42] that represent variability in seafloor structure (Table 1).

Table 1. List of variables and their respective equations. See Figure 2 for examples of derivative surfaces.

Derivative	Software	Description
Bathymetry Mean	ArcMap 10.4 (Spatial Analyst)	Local mean value of pixel to neighborhood $\bar{x} = (\Sigma Xi)/N$
Bathymetry Standard Deviation	ArcMap 10.4 (Spatial Analyst)	Local standard deviation value of pixel to neighborhood $\sigma_X = \sqrt{(\Sigma(Xi - X)^2/N)}$
Backscatter Mean	ArcMap 10.4 (Spatial Analyst)	Local mean value of pixel to neighborhood $\bar{x} = (\Sigma Xi)/N$
Backscatter Standard Deviation	ArcMap 10.4 (Spatial Analyst)	Local standard deviation value of pixel to neighborhood $\sigma_X = \sqrt{(\Sigma(Xi - X)^2/N)}$
Backscatter Rugosity (VRM)	ArcMap 10.4 (Benthic Terrain Mapper)	Incorporates the heterogeneity of both slope and aspect using three-dimensional dispersion of vectors. See [43] for more details.
Bathymetry Rugosity (VRM)	ArcMap 10.4 (Benthic Terrain Mapper)	Incorporates the heterogeneity of both slope and aspect using three-dimensional dispersion of vectors. See [43] for more details.
Bathymetry Slope	ENVI 5.3.1	Change in elevation over designated neighborhood size $\tan^{-1}(\text{Rise/run})$ [3]
Bathymetry Complexity	ENVI 5.3.1	Rate of change of slope over designated neighborhood size $\tan^{-1}(\text{rise(slope)/run(slope)})$ [3]
Maximum Curvature	ENVI 5.3.1	Steepest curve of convexity for a pixel over designated neighborhood size $K(x) = \mid e^x \mid /(1 + e^{2x})^{3/2}$
Gray-Level Co-Occurrence Matrix (GLCM) Mean Backscatter	ENVI 5.3.1	Uses a co-occurrence matrix to represent the number of occurrences between a pixel and its neighbor Local mean value of pixel to neighborhood $P(i)$ = probability of each pixel value N_g = Number of distinct gray levels $\sum_{i=1}^{N_g} \sum_{j=1}^{N_g} i * P(i,j)$ [44]
GLCM Standard Deviation Backscatter	ENVI 5.3.1	As described above Local standard deviation of pixel to neighborhood $\sum_{i=1}^{N_g} \sum_{j=1}^{N_g} (i - u)^2 P(i,j)$ [44]
GLCM Entropy Backscatter	ENVI 5.3.1	As described above. Statistical measure of randomness of pixel to neighborhood $-\sum_{i=1}^{N_g} \sum_{j=1}^{N_g} P(i,j) \log(P(i,j))$ [44]
Eastness	ENVI 5.3.1	The sine of the angle of slope in the analysis window. Equation: $\sin(\text{aspect})$ [3]
Northness	ENVI 5.3.1	The cosine of the angle of slope in the analysis window. Equation: $\cos(\text{aspect})$ [3]

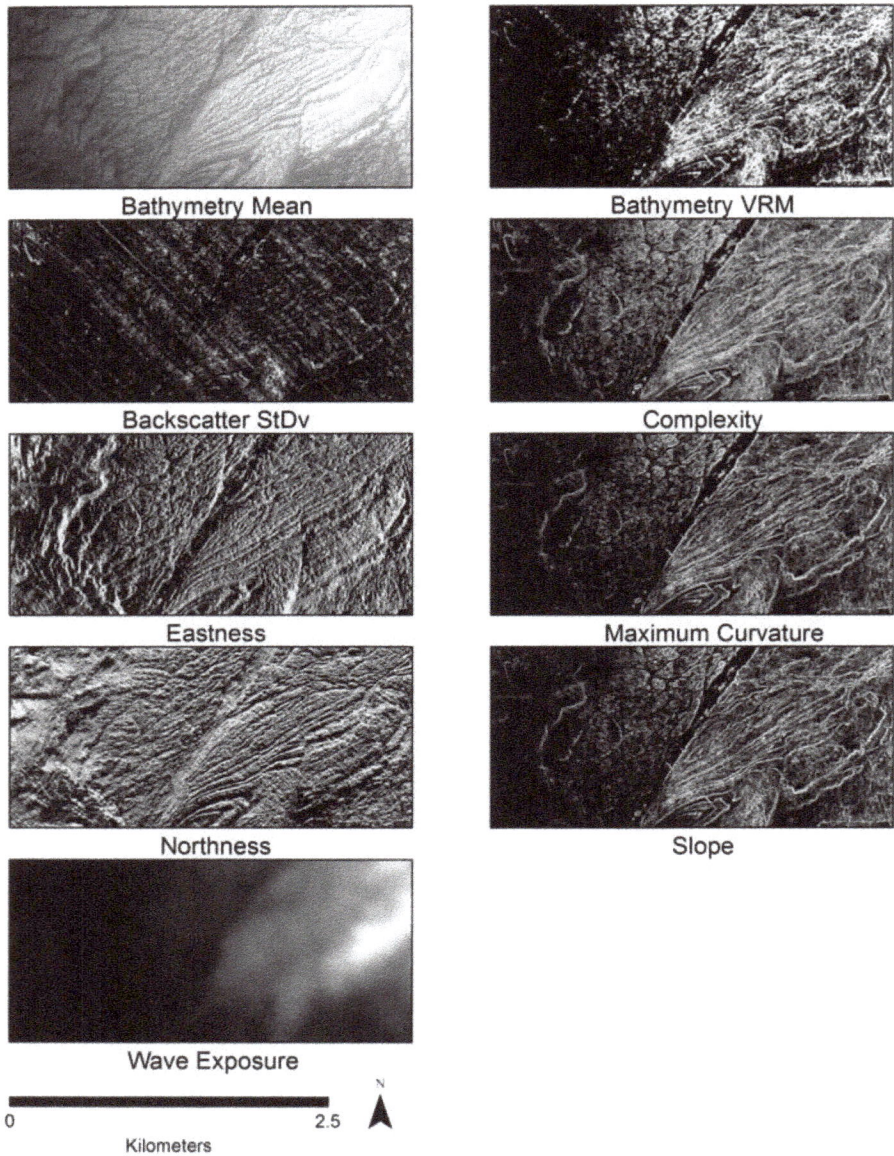

Figure 2. Example of derivative surfaces selected for use in hierarchical models. Scale of analysis for examples is 3 × 3.

Derivatives were created from the bathymetry and backscatter products based on methods from previous habitat mapping studies [3,19] (Table 1). To gain a better understanding of the influence of scale on model results, each derivative was created at an analysis kernel size ranging 3 × 3 to 21 × 21 pixels. This resulted in products for each of the 14 seafloor derivatives at analysis scales of 3, 5, 7, 9, 11, 13, 15, 17, 19 and 21 pixels, a total of 140 derived products.

2.3. Groundtruth Collection and Classification

Over 180 linear kilometers of towed video groundtruth data were collected by the Deakin Marine Mapping Group using a custom towed-video system (for details see [19]). The Cape Otway dataset is extensive and combines data collected over two years across 30 shore-normal transects. During tows, the camera system was piloted via live-stream video from an umbilical and kept ~1 m off the seafloor resulting in a field of view of roughly 2 m × 2 m. Camera position was recorded via an ultra-short baseline transponder and the vessel's DGPS.

The towed video dataset was classified using the Combined Biotope Classification Scheme (CBiCS) which has been adopted by the Department of Environment, Land, Water and Planning (Victorian Government, Australia) [45]. The video data was segmented into multiple categories of habitat type for each of the CBiCS hierarchies, based on observations of biota, substrata, geoform, exposure, and biogeographic region [45]. CBiCS uses components of the Joint Nature Conservation Committee-European Nature Information System (JNCC-EUNIS [46]), and United States Coastal and Marine Ecological Classification System (CMECS [47]), with a total of six hierarchical levels representing the biotic component [48]. Three levels of the scheme were used in this study; broad habitat classes (BC2), habitat complexes (BC3), and biotope complexes (BC4) (Table 2). CBiCS also provides classifications for biotopes (BC5) and morphospecies (BC6), however, they were not explored due to the large number of rare classes with limited observations for training and validation data sets within the study area.

In order to limit spatial autocorrelation between the training and validation sets, transects were chosen at random, 20 for training and 10 for validation. There were 14,785 observations in the training data set and 7835 observations in the validation data set. To ensure only a single towed video observation represented each raster cell the observation data were thinned with a buffer of 4 m between observations.

Table 2. Combined Biotope Classification Scheme (CBiCS) hierarchies (BC2, BC3, BC4) used to train and validate the three hierarchical models in the present study with depth range, depth mean, depth standard deviation, and number of observations (N) for each level at BC4.

BC2	BC3	BC4	Range (Mean, Standard Deviation)	N
Infralittoral rock and other hard substrata (IR)	High energy infralittoral rock (HIR)	High energy *Durvillaea potatorum* communities (DUR)	13–16 m (14 m, 1 m)	10
		High energy *Ecklonia radiate* communities (HECK)	5–23 m (12 m, 4 m)	600
		High energy *Ecklonia-Phyllospora comosa* communities (EP)	12–30 m (18 m, 1 m)	10
		High energy lower infralittoral zone (HLI)	18–42 m (31 m, 5 m)	1617
Circalittoral rock and other hard substrata (CR)	High energy open-coast circalittoral rock (HCR)	Bushy bryozoan-dominated communities (BBR)	32–37 m (36 m, 1 m)	49
		Crustose coralline algal communities with combinations of thallose red algae and scattered sponges on high energy circalittoral rock (CCA)	33–55 m (42 m, 3 m)	535
		High energy circalittoral rock with seabed covering sponges (SpoCov)	27–45 m (36 m, 5 m)	516
		Low complexity circalittoral rock with non-crowded erect sponges (LxSml)	41–43 m (42 m, 1 m)	38
		Moderate to high complexity circalittoral rock with seabed covering sponges (CxSml)	39–45 m (42 m, 2 m)	102
		Sandy low profile reef wave surge communities with sand trapped around sponges (SLO)	39–45 m (42 m, 11 m)	558
Sublittoral sediment (SS)	Non-reef sediment epibenthos (EPI)	Erect octocorals on sediment (OCT)	27–71 m (48 m, 13 m)	239
	Sublittoral sand and muddy sand (SSa)	Circalittoral fine sand (CLFiSa)	29–70 m (48 m, 9 m)	4078
		Infralittoral fine sand (ILFiSa)	10–40 m (30 m, 10 m)	6433

2.4. Wave Exposure Model

As a surrogate for wave exposure, we used a fine-scale (60 m) estimation of wave induced orbital velocities of the seabed. The model was created using the WaveWatch III global hindcast model downscaled to a regional spectral wave model using the high-resolution bathymetric surface [18]. Wave-induced orbital velocities were transferred to the seafloor using linear wave theory (for more detailed information see [18]). Bayesian kriging was used to resample the wave exposure surface to a resolution of 2.5 m and mean kernel size analysis was performed to produce surfaces at identical scales to those defined in Section 2.2.

2.5. Statistical Approaches

2.5.1. Data Analysis

The Random Forests (RF) ensemble classification approach was used to derive rule-based relationships between geophysical derivatives and corresponding observational data using the "randomForest" package in R [49]. RF is a machine learning modelling technique that uses tree-type classifiers and bootstrap aggregation based on subsets of the input data [50]. The benefits of using the RF classification approach are that it reduces the chance of overfitting the model by including the results of multiple trees from bootstrap samples of the training data, and measures of variable importance to model accuracy can be derived [51]. RF also keeps bias low via random predictor selection. RF have been shown to perform well when compared with other rule-based classification approaches, as demonstrated by Stephen and Diesing [52], who reported that tree-based methods, including RF, performed best when predicting sediment classes from acoustic and groundtruth data sets.

Parameters for the optimal number of predictors randomly selected at each split (mtry), and the optimal number of trees within models (ntree) were assessed for each of the three hierarchical models using a repeated K-fold cross-validation routine in the R package 'caret' [53]. Model parameters mtry and ntree were subsequently set at values of 5 and 300 respectively for the three hierarchical models used in the study.

2.5.2. Variable Importance

Variable importance was obtained by randomly permuting each predictor value in the "out of bag" (OOB) observations for each tree. The error rate for classification is calculated for each tree in the model, and the same is done for each predictor variable. The differences between the two errors are averaged over all trees and normalized using the standard deviation from the differences. Predictors for each of the three hierarchical models were chosen based on variable importance without repetition (i.e., if two variables with analysis kernels of 21 and 19 pixels were ranked equally only one was included). Predictors with a Pearson product-moment correlation value greater than 0.8 were not retained for model runs (Figure 3). See Table 3 for predictors used in RF modelling for the three hierarchical models.

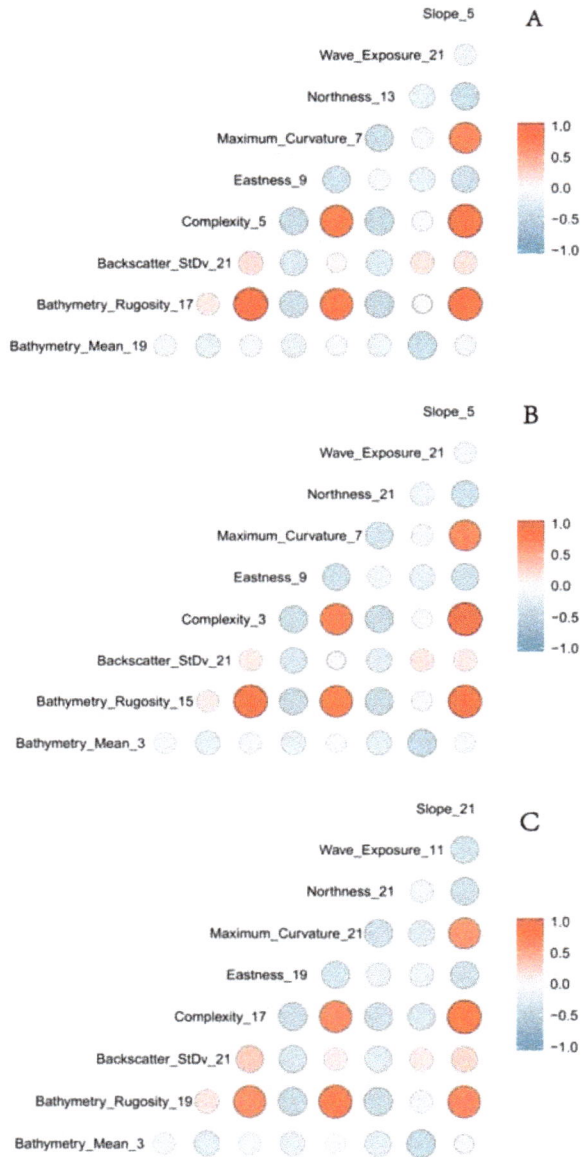

Figure 3. Correlation matrix containing all derivatives used for the three hierarchical models (**A**) BC2, (**B**) BC3, (**C**) BC4. Color and size represent the magnitude and relationship between derivatives. The numbers after the labels represent the scale of analysis (in pixels).

Table 3. Model predictors and spatial scale retained for each model.

Model	Derivatives	Kernel Size
BC2	Bathymetry Mean	19
	Bathymetry Rugosity (VRM)	17
	Backscatter Standard Deviation	21
	Complexity	5
	Eastness	9
	Maximum Curvature	7
	Northness	13
	Slope	5
	Wave exposure	21
BC3	Bathymetry mean	3
	Bathymetry Rugosity (VRM)	15
	Backscatter Standard Deviation	21
	Complexity	3
	Eastness	9
	Maximum curvature	7
	Northness	21
	Slope	5
	Wave exposure	21
BC4	Bathymetry Mean	3
	Bathymetry Rugosity (VRM)	19
	Backscatter Standard Deviation	21
	Complexity	17
	Eastness	19
	Maximum Curvature	21
	Northness	21
	Slope	21
	Wave Exposure	11

2.5.3. Predictive Mapping

RF models were used to predict the final classified habitat maps using the "ModelMap" package in R [54]. Model accuracies were determined by creating confusion matrices comparing the predicted classifications with the validation dataset providing overall accuracies, kappa statistics and measures of within class model performance (sensitivity and specificity) for each model using the "caret" package in R. The Kappa statistic (K) measures the agreement between an observed accuracy and an expected (values ranging from −1 to 1), but differs from a simple percent agreement calculation as it considers the agreement occurring by chance [55].

3. Results

3.1. Habitat Suitability Maps

The results of the benthic habitat maps indicate at the BC2 hierarchy the Cape Otway site is characterized by sublittoral sediments with extensive circalittoral rock and infralittoral rock more dominant west of the cape (Figure 4). The habitat map for BC3 is similar to BC2, but has an additional class, non-reef sediment epibenthos, which is present in small patches west of the cape.

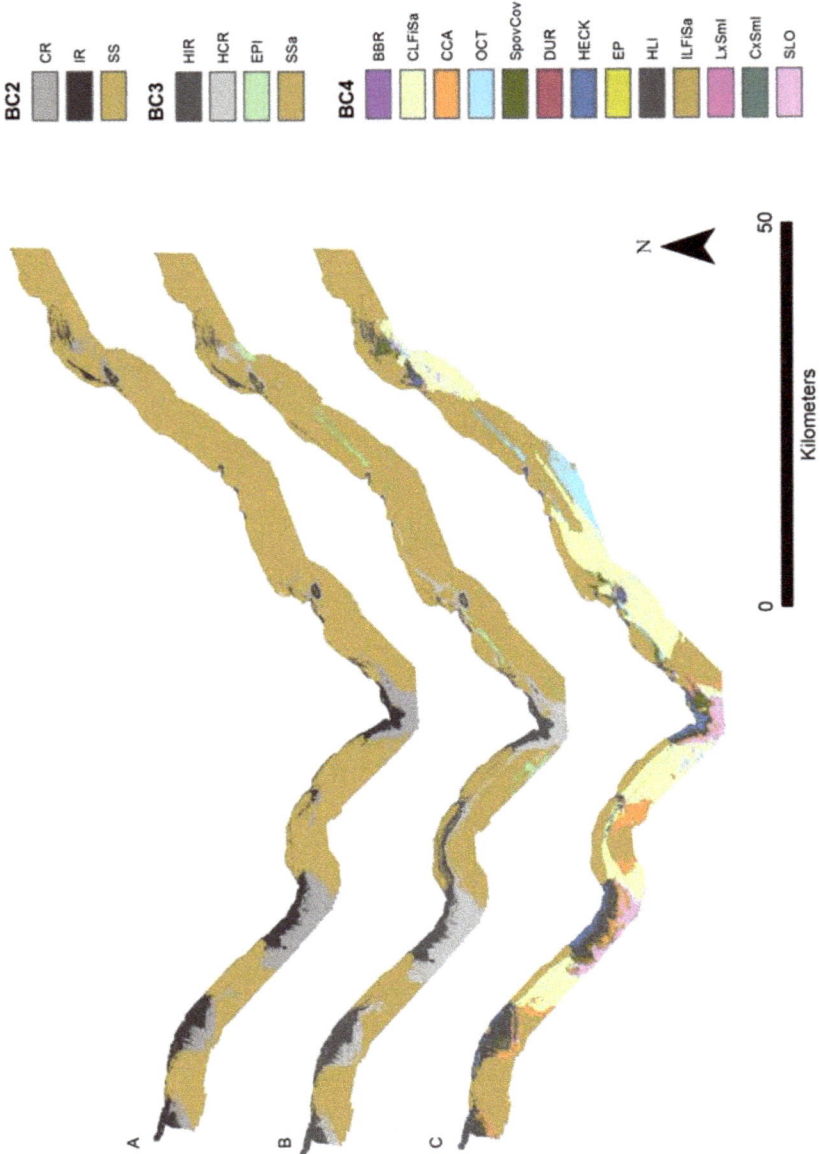

Figure 4. Predictive classification maps across three hierarchies BC2 (**A**), BC3 (**B**), and BC4 (**C**). Produced using Random Forests (RF) models in R. Refer to Table 2 for class descriptions.

The smallest class observed in groundtruth for BC4 was high energy *Durvillaea* communities located in the south of the cape at depths less than 16 m. High energy *Ecklonia-Phyllospora* biotope complexes were distributed in patches at a mean depth of 19 m ± 2 m. High energy *Ecklonia* biotope complexes had a mean depth of 22 m ± 4 m within reefs along the western side of Cape Otway heads. The most dominant class, accounting for 71% of the infralittoral hierarchy, was the class high energy lower infralittoral zone. Biotopes included in the biotope complex high energy lower infralittoral zone for this study area are *Ecklonia radiata* park, and foliose red algae with sessile invertebrates.

The circalittoral zone, classified at the level of BC4, contained the highest number of biotope complexes. Bushy bryozoan-dominated communities were primarily distributed south of the headlands with a mean depth of 35 m ± 2 m. The biotope complexes of high energy circalittoral rock with seabed covering sponges, low complexity circalittoral rock with non-crowded erect sponges, and moderate to high energy circalittoral rock with seabed covering sponges were found at a mean depth of 42 m ± 6 m. Sandy low-profile reef wave surge communities with sand trapped around sponges were found throughout the site at a mean depth of 47 m ± 8 m. The most dominant biotope complex by area, accounting for 45% of the circalittoral habitat in the study area, was crustose coralline algal communities with combinations of thallose red algae and scattered sponges on high energy circalittoral rock at a mean depth of 48 m ± 8 m.

Two of the three classes, infralittoral and circalittoral fine sand, within the sublittoral sediment hierarchy represented 72% (by area) of the BC4 benthic habitat map. The remaining class was "erect octocorals on sediment" and was present east of the heads at a mean depth of 65 m ± 15 m.

3.2. Hierarchical Comparison

Three benthic habitat maps, one for each hierarchical level, were assessed comparing predicted classes to a validation dataset comprised of spatially independent transects not used in model development. The classification for BC2 performed best with an overall accuracy of 87.4% and a Kappa statistic of 0.59 (Table 4). Hierarchical models for BC3 and BC4 had overall accuracies of 69.7% (K = 0.31) and 39.9% (K = 0.16), respectively.

Class specific classification accuracies for BC2 were generally good with all three classes predicting at 60% or higher. Within BC2 sublittoral sediment (SS) performed the best at 91.9% while circalittoral rock (CR) and infralittoral rock (IR) performed moderately at 63.5% and 72.7% respectively (Table 4). The next level in the hierarchy, BC3, predicted three classes accurately with high-energy infralittoral rock (HIR), high-energy open-coast circalittoral rock (HCR) and sublittoral sand and muddy sand (SSa) performing well at 70.6%, 63.4% and 86.3%, respectively. The model failed to accurately differentiate any of the non-reef sediment epibenthos which was incorrectly classified predominately as sublittoral sand and muddy sand (Table 5). As the number of hierarchy levels increased the class specific accuracies typically decreased, with seven of the 13 classes of BC4 consistently incorrectly classified. Five of the seven classes consistently misclassified were from the high-energy circalittoral rock habitat complex: high-energy circalittoral rock with seabed covering sponges, bushy bryozoan-dominated communities, crustose coralline algal communities with combinations of thallose red algae and scattered sponges on high-energy circalittoral rock, low complexity circalittoral rock with non-crowded erect sponges, and moderate to high complexity circalittoral rock with seabed covering sponges. The two other misclassified biotope complexes were from the high-energy infralittoral rock habitat complex: high-energy *Durvillaea* communities, and high-energy *Ecklonia-Phyllospora* communities (Table 6). The remaining classes varied in their accuracies with high-energy *Ecklonia* communities (HECK) preforming the best at 97.1%.

Table 4. Error matrix for BC2. Class codes as follows: Infralittoral rock and other hard substrata (IR); Circalittoral rock and other hard substrata (CR); Sublittoral sediment (SS). Italicized values indicate correctly classified observations.

	Reference	Overall Accuracy = 87.4%, K = 0.59			
		CR	**IR**	**SS**	**Sensitivity**
Predicted	CR	*621*	20	279	63.6
	IR	39	*213*	218	72.7
	SS	317	60	*5636*	91.9
	Specificity	95.4	96.4	70.3	

Table 5. Error matrix for BC3. Class codes as follows: High-energy infralittoral rock (HIR); High-energy open-coast circalittoral rock (HCR); Non-reef sediment epibenthos (EPI); Sublittoral sand and muddy sand (SSa). Italicized values indicate correctly classified observations.

	Reference	Overall Accuracy = 69.7%, K = 0.31				
		HIR	**HCR**	**EPI**	**SSa**	**Sensitivity**
Predicted	HIR	*207*	33	0	261	70.6
	HCR	20	*619*	10	404	63.4
	EPI	0	6	*0*	7	0.0
	SSa	66	319	1180	*4234*	86.3
	Specificity	95.8	93.2	99.8	36.4	

Table 6. Error matrix for BC4. Class codes are as follows: High-energy Durvillaea communities (DUR); High-energy Ecklonia-Phyllospora communities (EP); High-energy *Ecklonia* communities (HECK); High-energy lower infralittoral zone (HLI); Bushy bryozoan-dominated communities (BBR); Crustose coralline algal communities with combinations of thallose red algae and scattered sponges on high-energy circalittoral rock (CCA); High-energy circalittoral rock with seabed covering sponges (SpoCov); Low complexity circalittoral rock with non-crowded erect sponges (LxSml); Moderate-to-high complexity circalittoral rock with prominent sea plumes, sea tulips and hydroid fans (PSH); Moderate-to-high complexity circalittoral rock with seabed covering sponges (SpoCov); Sandy low-profile reef wave surge communities with sand trapped around sponges (SLO); Erect octocorals on sediment (OCT); Circalittoral fine sand (CLFiSa); Infralittoral fine sand (ILFiSa). Italicized values indicate correctly classified observations.

| | Reference | | | | | | | | | | | | | Overall Accuracy = 39.94%, K = 0.16 |
		BBR	CLFiSa	CCA	OCT	SpoCov	DUR	EP	HECK	HLI	ILFiSa	LxSml	CxSml	SLO	Sensitivity
Predicted	BBR	0	8	0	0	0	0	0	0	1	0	0	0	1	0
	CLFiSa	3	617	0	792	0	0	0	0	0	343	34	9	56	22.7
	CCA	0	192	0	0	0	0	0	0	0	5	0	0	45	0
	OCT	0	0	0	102	0	0	0	0	0	3	0	0	0	8.6
	SpoCov	0	46	0	0	0	0	0	0	0	42	0	0	0	0
	DUR	0	0	0	0	0	1	0	0	0	0	0	0	0	0
	EP	0	0	0	0	0	11	19	0	0	0	0	0	0	0
	HECK	0	0	0	0	0	1	0	136	0	56	0	0	0	97.1
	HLI	0	10	0	0	0	0	0	4	71	83	0	0	3	58.2
	ILFiSa	0	1842	21	296	46	0	0	0	50	1625	23	7	143	74.4
	LxSml	0	0	0	0	0	0	0	0	0	0	0	0	0	0
	CxSml	0	3	0	0	0	0	0	0	0	0	0	0	0	0
	SLO	0	5	0	0	0	0	0	0	0	26	20	4	277	52.8
	Specificity	99.9	72.4	96.1	99.9	98.7	100	99.9	98.8	98.6	50.4	100	99.9	99.2	

3.3. Variable Importance

For each hierarchical model, variable importance was ranked using the mean decrease in accuracy metric (Figure 5). A larger value for mean decrease in accuracy indicates that the environmental predictor was more important during the classification process. Wave exposure performed the best for hierarchical level BC4. Bathymetry rugosity (VRM) performed best for hierarchal levels BC2 and BC3. Backscatter mean, backscatter rugosity (VRM), backscatter standard deviation, GLCM entropy, GLCM mean, GLCM variance and maximum curvature shared positive linear trends across predictor scale and between each hierarchy (Figures A1–A3 and 6–8). Eastness and northness had no patterns in predictor scale or between hierarchical scale. Bathymetry mean performed well across all predictor scales and hierarchies, favoring finer scales for BC4. Bathymetry standard deviation for BC2 and BC3 favored predictor scales towards the mid-range of pixel kernel size (11–15) while BC4 displayed a positive linear relationship (Figures A1–A3). At coarse hierarchical levels slope favored finer scales, while at finer hierarchical levels slope favored broader scales. In BC2 the relationship between complexity and variable scale favored finer scales with BC3 showing no pattern, and BC4 favoring broader scales.

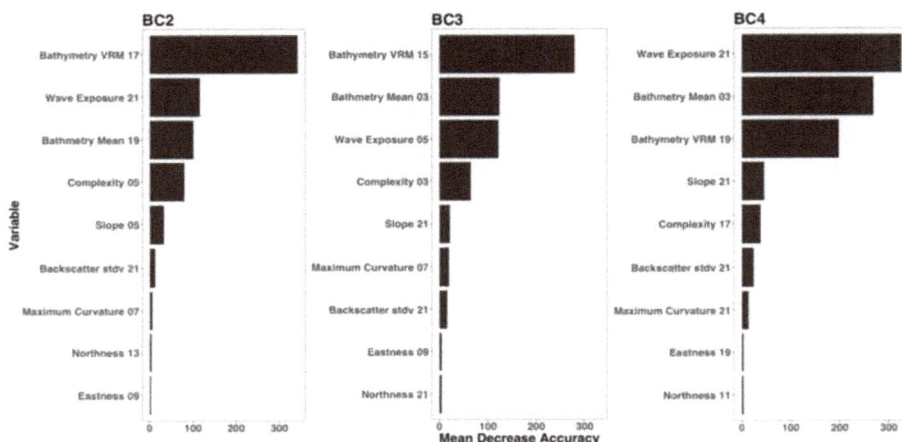

Figure 5. Variable importance for retained model variables. Mean decrease in accuracy represents the RF model decrease in accuracy when that variable is removed, therefore a larger value for mean decrease in accuracy indicates that the environmental predictor was more important.

BC2 Importance

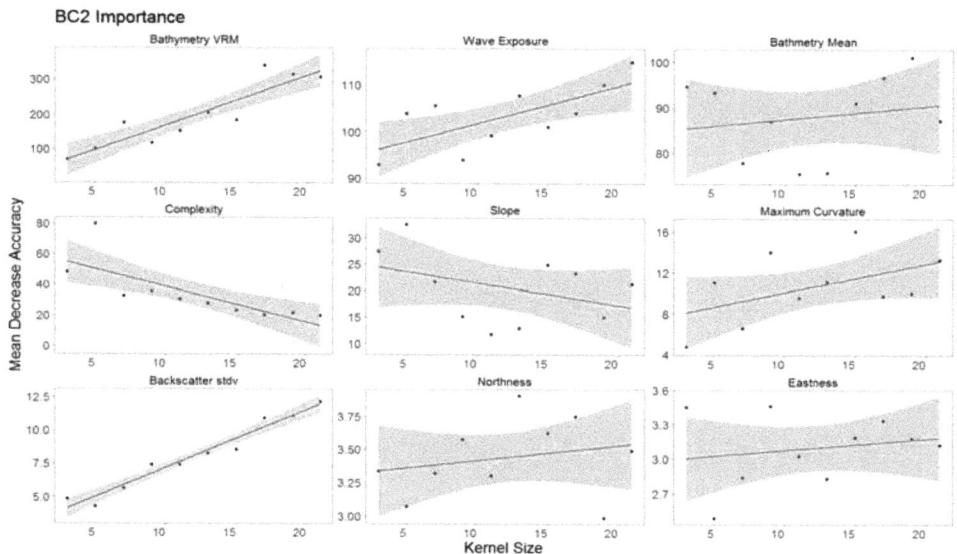

Figure 6. Variable importance for BC2, for variables retained. Sorted by derivatives in order of importance. Generalized linear model (GLM) trend represented in blue. Gray area represents 95% confidence interval for predictions.

BC3 Importance

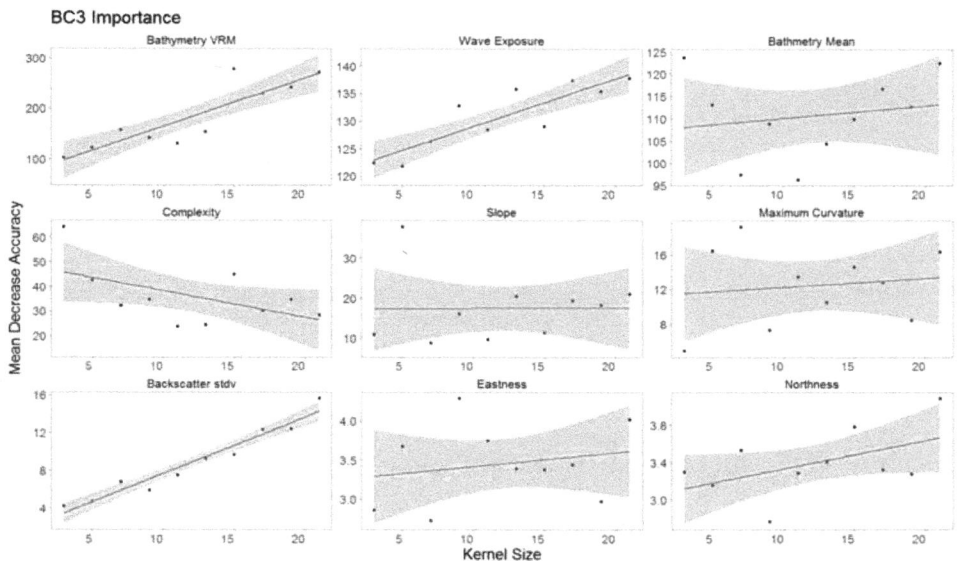

Figure 7. Variable importance for BC3, for variables retained. Sorted by derivatives in order of importance. Generalized linear model (GLM) trend represented in blue. Gray area represents 95% confidence interval for predictions.

BC4 Importance

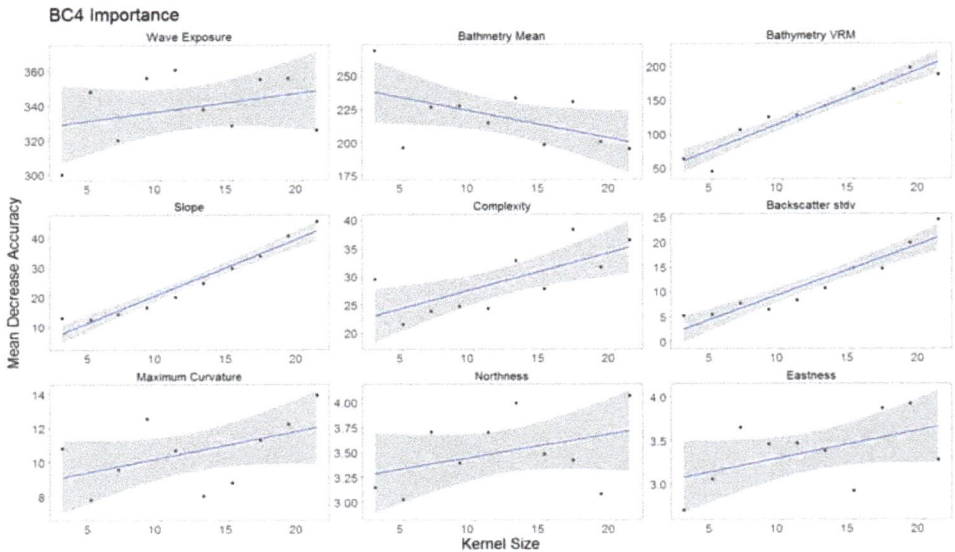

Figure 8. Variable importance for BC4, for variables retained. Sorted by derivatives in order of importance. Generalized linear model (GLM) trend represented in blue. Gray area represents 95% confidence interval for predictions.

4. Discussion

Numerous habitat mapping studies have tested the importance of environmental predictor scale when creating habitat maps [2,3,56]. There have also been a number of studies, primarily terrestrial, that have examined the effects of hierarchical classifications on habitat prediction [33,35,36]. In this study, we tested both hierarchical and environmental variable scale in a marine environment, using MBES and terrain derivatives, and demonstrate their importance in developing statistical relationships between observations and geophysical products to create benthic habitat maps.

4.1. Impact of Classification Hierarchy on Model Performance

We observed a decrease in map accuracy at increasing levels of complexity of a hierarchical classification scheme. The lower accuracy of the hierarchical models as more classes were added in BC3 and BC4 is likely due to the decrease in samples per class and increased complexity of the classification task, a common challenge experienced across many studies (see review [57]). Additionally, as class complexity increased, some rarer habitats were under represented, with too few samples for rigorous model training and validation. Mumby et al. [38] found similar results using a hierarchical cluster analysis to predict coral reefs habitats according to three levels (broad, intermediate, and fine) of hierarchy across five different satellite sensors [38]. They found a significant decrease in accuracy between broad to fine descriptive resolutions [38]. Capolsini et al. [58] found a similar pattern in decreased accuracy when using more classes to define coral reef habitats in South Pacific Islands.

According to benchmarks established by Landis et al. [59] kappa values in the present study indicate that BC2 had moderate strength of agreement, BC3 had fair strength of agreement and BC4 displayed slight strength of agreement. The moderate and fair agreement for BC2 and BC3 provide confident results. The slight strength in agreement for BC4 is likely due to multiple rare and similar classes, with two of the classes having 10 observations and the remainder having similar characteristics; all located on circalittoral rock, with either crustose coralline algal communities, sponges, or bryozoans.

Sediment classes with no visible epibiota; SS, SSa, and ILFiSa were consistently accurately classified, likely because sediment morphology displays patterns that are readily discernible from reef using MBES derivatives [1,60]. In the hierarchical level BC3, the model failed to accurately classify any of the sediment classes containing visible epibenthos. This may reflect poor class separability using acoustic methods, as demonstrated by Freitas et al. [61], who found that acoustic methods had difficulty separating grain sizes between fine and very fine sand. Reef classes had, in general, lower accuracies. This is possibly due to low profile reefs in the study area having sand veneers making them difficult to discern using MBES derivatives. Classes with similar biotic components, namely crustose coralline algal communities, and seabed/erect sponges, were misclassified throughout BC4. This finding is similar to a predictive modelling study conducted by Che Hasan et al. [62], which found that misclassification occurred between similar biota classes (i.e., mixed brown algae, mixed brown and invertebrates) because they shared many acoustic properties due to similar species compositions and substratum types. Some classes are rare, with low density biota making it difficult to discern from surrounding acoustic signatures.

4.2. Impact of Variable Scale

A common finding across all three hierarchies was the importance of wave exposure in reducing classification uncertainty. Wave exposure was the highest ranked variable in the BC4 model and was the second highest variable in the BC2 and BC3 hierarchical models, which is unsurprising given the strong east-west exposure gradient at the site. Characteristics used to determine biotope classes often include exposure level, reinforcing the desirability of incorporating variables describing local exposure gradients in habitat maps [63]. Our results align with findings from previous studies where oceanographic features and seafloor habitat variables were incorporated into habitat models [18,64]. Rattray et al. [18] found that incorporating wave exposure as a variable in habitat mapping significantly increased model accuracy when compared to models derived using MBES derivatives alone. Downscaling the original hindcast model using Bayesian kriging likely weakened the mean decrease in accuracy relationship between each scale because the original scale of the model is coarser than the coarsest scale extracted.

Seafloor habitat variables derived from bathymetry rugosity (VRM) and backscatter standard deviation shared common trends as both predictors were important across hierarchies. Another pattern observed across hierarchies was the increasing importance of environmental predictors derived from the backscatter layer as scale increased (Figures A1–A3), while predictors derived from bathymetry displayed weaker trends. An explanation for this pattern may be the larger number of artefacts in the backscatter layer in comparison to bathymetry. Deriving backscatter derivatives at broader scales potentially reduces speckle and nadir noise common in backscatter mosaics [19].

Within each classification hierarchy most environmental predictors showed increased importance at broader scales of analysis, with a few exceptions; for example, eastness, northness, bathymetry mean and complexity at hierarchical models BC2 and BC3 were more important at finer scales. In similar multiscale studies, a combination of fine and broad-scale variables were found to be the best predictors of bird habitat [65] and macrofaunal habitat [22]. This reflects findings in the marine environment indicating that species/habitat associations may be based on a combination of both broad-scale and fine-scale variables [66]. However, when habitats are best represented by broad-scale variables a previous study by Wilson et al. [3] found that there may be an upper limit to the scale of analysis before variable importance decreases. Scale importance in our study did not appear to reach any upper limits for most variables; suggesting future studies should examine derivatives at even broader scales for model input.

Importance of scale is likely site dependent and driven by the heterogeneity of the seabed and size of the study area [67,68]. Results from this study provide an example from a highly exposed, temperate coastline, which may be representative of other temperate sites. Using the methodology described here, the effect of scale on classification can be explored and applied to other study areas.

Predictive habitat maps based on small class sample sizes are unlikely to be as accurate as those based on classes with large samples, especially when class characteristics share similarities with surrounding classes [67,68], or when predictors do not adequately represent processes driving biotic distribution. Groundtruth surveys are typically carried out after the collection and processing of MBES bathymetry and backscatter. Using unsupervised classification techniques to guide collection of observation data is a logical first step to ensuring adequate sample sizes across classes of interest. For example, Przeslawski and Foster [69], suggest that transect length should be dependent on the spatial properties of the target biota, using short transects where the biota has large spatial autocorrelation and using long transects where the biota has small spatial autocorrelation. In an effort to reduce spatial autocorrelation we split our observations into training and validation by randomly selecting transects. This was not ideal and likely had a negative impact on model accuracy. Several classes from the BC3 and BC4 levels were only present in a single transect, resulting in either no training or validation points for some classes, thus were excluded from the model.

5. Conclusions

Advances in MBES technology have provided products at increasing resolution, however, studies exploring the impact of variable scale on the habitat mapping process have been limited. When assessing variable importance across multiple scales, our study generally shows positive trends towards broader scale variables. We demonstrate that predictor scale is important for improving map accuracy and suggest multiscale analysis to limit user bias when selecting variable scales for analysis as they are likely to be site specific, which is driven by the heterogeneity of the seascape.

The number of predictive classes and the ability to differentiate between them accurately is a key consideration when deriving benthic habitat maps. Using a hierarchical classification approach for model prediction, we found as the number of predictive classes increased model accuracy decreased. Classification schemes are not necessarily designed with predictive mapping in mind and can result in classes sharing similar abiotic characteristics that can be difficult to differentiate using acoustic approaches. Therefore, classification schemes that incorporate appropriate levels for accurate predictive mapping that aligns with management objectives are desirable.

Rare classes pose a challenge for predictive habitat modelling, because capturing an adequate number of samples when collecting groundtruth data is difficult. For future studies we recommend considering the tradeoff between statistical rigor and useful habitat maps by using a combination of automated and manual digitization of rare classes. In conclusion, this study brings attention to hierarchical classifications and the importance of using multiscale approaches when classifying benthic habitats.

Acknowledgments: This work was supported by the National Heritage Trust and Caring for Country as part of the Victorian Marine Habitat Mapping Project with project partners Glenelg Hopkins Catchment Management Authority, Department of Environment and Primary Industries, Parks Victoria, University of Western Australia and Fugro Survey. We thank the crew from the Australian Maritime College research vessel Bluefin, which was used for the multibeam data collection. We also thank the crew from Deakin University research vessel Courageous II for assisting Daniel Ierodiaconou & Alex Rattray collecting the towed video data used in this project. Towed video was classified by Australian Marine Ecology and Fathom Pacific using CBiCS under a project funded by the Department of Environment, Land, Water and Planning. Multibeam bathymetry and classified towed video data was accessed via the Victoria Marine Data Portal, available online: https://vmdp.deakin.edu.au/ (accessed 2 April 2018). We would like to thank Matt Edmunds and Adrian Flynn for providing comments on the manuscripts first draft and for their contributions in the creation of CBiCS. Alex Rattray and Mary Young were supported by the Victorian Marine Habitat Mapping Program with funds through Department of Environment, Lands Water and Planning, Parks Victoria and Australian National Data Services (ANDS) through funding from the Australian Government's National Environmental Science Programme.

Author Contributions: Daniel Ierodiaconou conceived the project and led the fieldwork. Peter Porskamp led the analyses and writing with contributions from Daniel Ierodiaconou, Alex Rattray and Mary Young.

Conflicts of Interest: The authors declare no conflict of interest.

Appendix A

BC2 Importance

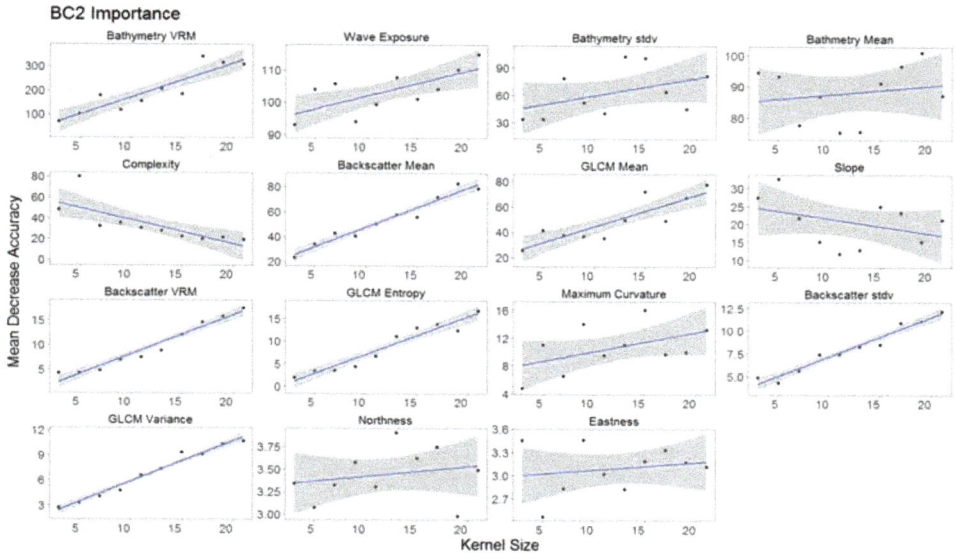

Figure A1. Variable importance BC2 sorted by derivatives in order of importance. GLM trend represented in blue. Gray area represents 95% confidence interval for predictions.

BC3 Importance

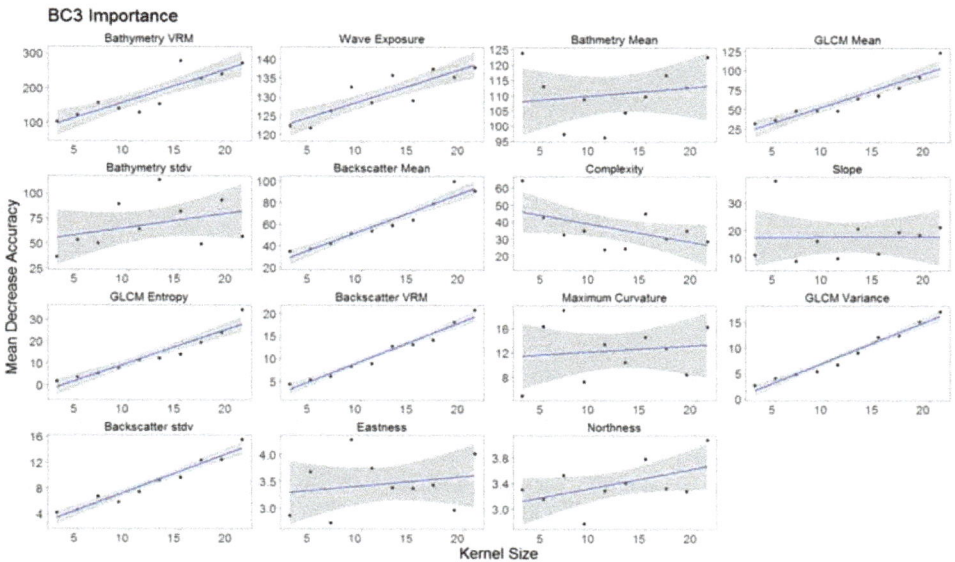

Figure A2. Variable importance BC3 sorted by derivatives in order of importance. GLM trend represented in blue. Gray area represents 95% confidence interval for predictions.

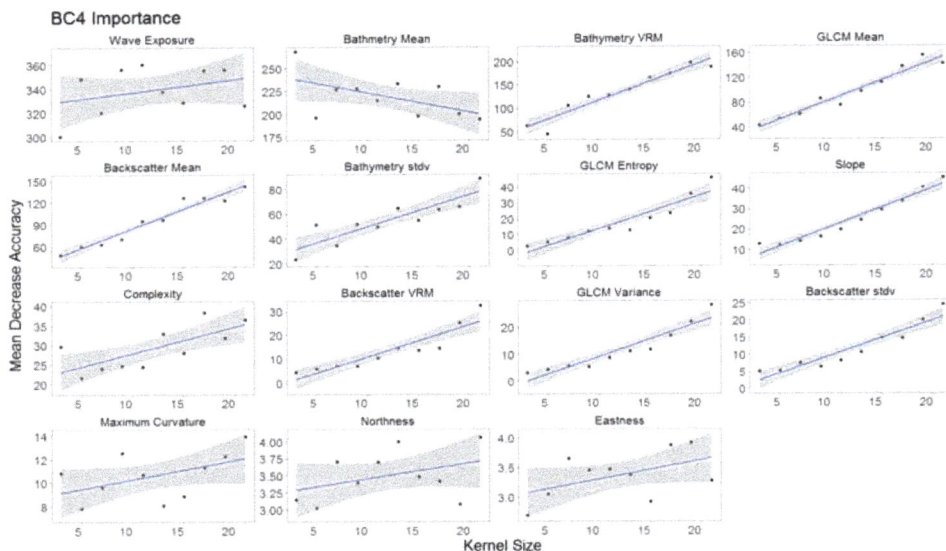

Figure A3. Variable importance BC4 sorted by derivatives in order of importance. GLM trend represented in blue. Gray area represents 95% confidence interval for predictions.

References

1. Brown, C.J.; Smith, S.J.; Lawton, P.; Anderson, J.T. Benthic habitat mapping: A review of progress towards improved understanding of the spatial ecology of the seafloor using acoustic techniques. *Estuar. Coast. Shelf Sci.* **2011**, *92*, 502–520. [CrossRef]
2. Lecours, V.; Devillers, R.; Schneider, D.; Lucieer, V.; Brown, C.; Edinger, E. Spatial scale and geographic context in benthic habitat mapping: Review and future directions. *Mar. Ecol. Prog. Ser.* **2015**, *535*, 259–284. [CrossRef]
3. Wilson, M.F.J.; O'Connell, B.; Brown, C.; Guinan, J.C.; Grehan, A.J. Multiscale Terrain Analysis of Multibeam Bathymetry Data for Habitat Mapping on the Continental Slope. *Mar. Geodesy* **2007**, *30*, 3–35. [CrossRef]
4. Diesing, M.; Mitchell, P.; Stephens, D. Image-based seabed classification: What can we learn from terrestrial remote sensing? *ICES J. Mar. Sci. J. Cons.* **2016**, *73*, 2425–2441. [CrossRef]
5. Lemme, M.C.; Koppens, F.H.L.; Falk, A.L.; Rudner, M.S.; Park, H.; Levitov, L.S.; Marcus, C.M. Gate-Activated Photoresponse in a Graphene p–n Junction. *Nano Lett.* **2011**, *11*, 4134–4137. [CrossRef] [PubMed]
6. Engle, V.D.; Summers, J.K. Latitudinal gradients in benthic community composition in Western Atlantic estuaries. *J. Biogeogr.* **1999**, *26*, 1007–1023. [CrossRef]
7. Wernberg, T.; Smale, D.A.; Tuya, F.; Thomsen, M.S.; Langlois, T.J.; de Bettignies, T.; Bennett, S.; Rousseaux, C.S. An extreme climatic event alters marine ecosystem structure in a global biodiversity hotspot. *Nat. Clim. Chang.* **2013**, *3*, 78–82. [CrossRef]
8. Gattuso, J.-P.; Gentili, B.; Duarte, C.M.; Kleypas, J.A.; Middelburg, J.J.; Antoine, D. Light availability in the coastal ocean: Impact on the distribution of benthic photosynthetic organisms and contribution to primary production. *Biogeosci. Discuss.* **2006**, *3*, 895–959. [CrossRef]
9. Anthony, K.; Ridd, P.V.; Orpin, A.R.; Larcombe, P.; Lough, J. Temporal variation of light availability in coastal benthic habitats: Effects of clouds, turbidity, and tides. *Limnol. Oceanogr.* **2004**, *49*, 2201–2211. [CrossRef]
10. Bax, N.; Kloser, R.; Williams, A.; Gowlett-Holmes, K.; Ryan, T. Seafloor habitat definition for spatial management in fisheries: A case study on the continental shelf of southeast Australia. *Oceanol. Acta* **1999**, *22*, 705–720. [CrossRef]

11. Roff, J.C.; Taylor, M.E.; Laughren, J. Geophysical approaches to the classification, delineation and monitoring of marine habitats and their communities. *Aquat. Conserv. Mar. Freshw. Ecosyst.* **2003**, *13*, 77–90. [CrossRef]

12. McArthur, M.A.; Brooke, B.P.; Przeslawski, R.; Ryan, D.A.; Lucieer, V.L.; Nichol, S.; McCallum, A.W.; Mellin, C.; Cresswell, I.D.; Radke, L.C. On the use of abiotic surrogates to describe marine benthic biodiversity. *Estuar. Coast. Shelf Sci.* **2010**, *88*, 21–32. [CrossRef]

13. Kostylev, V.E.; Erlandsson, J.; Ming, M.Y.; Williams, G.A. The relative importance of habitat complexity and surface area in assessing biodiversity: Fractal application on rocky shores. *Ecol. Complex.* **2005**, *2*, 272–286. [CrossRef]

14. Lecours, V.; Devillers, R.; Simms, A.E.; Lucieer, V.L.; Brown, C.J. Towards a framework for terrain attribute selection in environmental studies. *Environ. Model. Softw.* **2017**, *89*, 19–30. [CrossRef]

15. Mitchell, P.J.; Monk, J.; Laurenson, L. Sensitivity of fine-scale species distribution models to locational uncertainty in occurrence data across multiple sample sizes. *Methods Ecol. Evol.* **2017**, *8*, 12–21. [CrossRef]

16. Ierodiaconou, D.; Monk, J.; Rattray, A.; Laurenson, L.; Versace, V.L. Comparison of automated classification techniques for predicting benthic biological communities using hydroacoustics and video observations. *Cont. Shelf Res.* **2011**, *31*, S28–S38. [CrossRef]

17. Ierodiaconou, D.; Laurenson, L.; Burq, S.; Reston, M. Marine benthic habitat mapping using Multibeam data, georeferencedvideo and image classification techniques in Victoria, Australia. *J. Spat. Sci.* **2007**, *52*, 93–104. [CrossRef]

18. Rattray, A.; Ierodiaconou, D.; Womersley, T. Wave exposure as a predictor of benthic habitat distribution on high energy temperate reefs. *Front. Mar. Sci.* **2015**, *2*. [CrossRef]

19. Ierodiaconou, D.; Schimel, A.C.G.; Kennedy, D.; Monk, J.; Gaylard, G.; Young, M.; Diesing, M.; Rattray, A. Combining pixel and object based image analysis of ultra-high resolution multibeam bathymetry and backscatter for habitat mapping in shallow marine waters. *Mar. Geophys. Res.* **2018**. [CrossRef]

20. Hasan, R.; Ierodiaconou, D.; Monk, J. Evaluation of Four Supervised Learning Methods for Benthic Habitat Mapping Using Backscatter from Multi-Beam Sonar. *Remote Sens.* **2012**, *4*, 3427–3443. [CrossRef]

21. Siwabessy, P.J.W.; Tran, M.; Picard, K.; Brooke, B.P.; Huang, Z.; Smit, N.; Williams, D.K.; Nicholas, W.A.; Nichol, S.L.; Atkinson, I. Modelling the distribution of hard seabed using calibrated multibeam acoustic backscatter data in a tropical, macrotidal embayment: Darwin Harbour, Australia. *Mar. Geophys. Res.* **2017**. [CrossRef]

22. De Leo, F.C.; Vetter, E.W.; Smith, C.R.; Rowden, A.A.; McGranaghan, M. Spatial scale-dependent habitat heterogeneity influences submarine canyon macrofaunal abundance and diversity off the Main and Northwest Hawaiian Islands. *Deep Sea Res. Part II Top. Stud. Oceanogr.* **2014**, *104*, 267–290. [CrossRef]

23. Bouchet, P.J.; Meeuwig, J.J.; Salgado Kent, C.P.; Letessier, T.B.; Jenner, C.K. Topographic determinants of mobile vertebrate predator hotspots: Current knowledge and future directions: Landscape models of mobile predator hotspots. *Biol. Rev.* **2015**, *90*, 699–728. [CrossRef] [PubMed]

24. Walbridge, S.; Slocum, N.; Pobuda, M.; Wright, D.J. Unified Geomorphological Analysis Workflows with Benthic Terrain Modeler. *Geosciences* **2018**, *8*, 94. [CrossRef]

25. Le Bas, T.P.; Huvenne, V.A.I. Acquisition and processing of backscatter data for habitat mapping—Comparison of multibeam and sidescan systems. *Appl. Acoust.* **2009**, *70*, 1248–1257. [CrossRef]

26. MacMillan, R.A.; Shary, P.A. Chapter 9 Landforms and Landform Elements in Geomorphometry. In *Developments in Soil Science*; Elsevier: Amsterdam, The Netherlands, 2009; Volume 33, pp. 227–254. ISBN 978-0-12-374345-9.

27. Lindegarth, M.; Gamfeldt, L. Comparing Categorical and Continuous Ecological Analyses: Effects of "Wave Exposure" on Rocky Shores. *Ecology* **2005**, *86*, 1346–1357. [CrossRef]

28. Bustamante, R.H.; Branch, G.M. Large Scale Patterns and Trophic Structure of Southern African Rocky Shores: The Roles of Geographic Variation and Wave Exposure. *J. Biogeogr.* **1996**, *23*, 339–351. [CrossRef]

29. Hughes, M.G.; Heap, A.D. National-scale wave energy resource assessment for Australia. *Renew. Energy* **2010**, *35*, 1783–1791. [CrossRef]

30. Dartnell, P.; Gardner, J.V. Predicting Seafloor Facies from Multibeam Bathymetry and Backscatter Data. *Photogramm. Eng. Remote Sens.* **2004**, *70*, 1081–1091. [CrossRef]

31. Costello, M. Distinguishing marine habitat classification concepts for ecological data management. *Mar. Ecol. Prog. Ser.* **2009**, *397*, 253–268. [CrossRef]

32. Simboura, N.; Zenetos, A. Benthic indicators to use in Ecological Quality classification of Mediterranean soft bottom marine ecosystems, including a new Biotic Index. *Mediterr. Mar. Sci.* **2002**, *3*, 77. [CrossRef]

33. Guarinello, M.L.; Shumchenia, E.J.; King, J.W. Marine Habitat Classification for Ecosystem-Based Management: A Proposed Hierarchical Framework. *Environ. Manag.* **2010**, *45*, 793–806. [CrossRef] [PubMed]

34. Shumchenia, E.J.; King, J.W. Comparison of methods for integrating biological and physical data for marine habitat mapping and classification. *Cont. Shelf Res.* **2010**, *30*, 1717–1729. [CrossRef]

35. Klijn, F.; de Haes, H.A.U. A hierarchical approach to ecosystems and its implications for ecological land classification. *Landsc. Ecol.* **1994**, *9*, 89–104. [CrossRef]

36. Frissell, C.A.; Liss, W.J.; Warren, C.E.; Hurley, M.D. A hierarchical framework for stream habitat classification: Viewing streams in a watershed context. *Environ. Manag.* **1986**, *10*, 199–214. [CrossRef]

37. Bock, M.; Xofis, P.; Mitchley, J.; Rossner, G.; Wissen, M. Object-oriented methods for habitat mapping at multiple scales—Case studies from Northern Germany and Wye Downs, UK. *J. Nat. Conserv.* **2005**, *13*, 75–89. [CrossRef]

38. Mumby, P.J.; Green, E.P.; Edwards, A.J.; Clark, C.D. Coral reef habitat mapping: How much detail can remote sensing provide? *Mar. Biol.* **1997**, *130*, 193–202. [CrossRef]

39. Department of the Environment and Heritage. *A Guide to the Integrated Marine and Coastal Regionalisation of Australia: IMCRA Version 4.0*; Australian Government, Department of the Environment and Heritage: Canberra, Australia, 2006; ISBN 978-0-642-55227-3.

40. Bezore, R.; Kennedy, D.M.; Ierodiaconou, D. The Drowned Apostles: The Longevity of Sea Stacks over Eustatic Cycles. *J. Coast. Res.* **2016**, *75*, 592–596. [CrossRef]

41. *ENVI*; Exelis Visual Information Solutions: Boulder, CO, USA, 2010.

42. *ArcGIS*; Environmental Systems Research Institute (ESRI): Redlands, CA, USA, 2015.

43. Sappington, J.M.; Longshore, K.M.; Thompson, D.B. Quantifying Landscape Ruggedness for Animal Habitat Analysis: A Case Study Using Bighorn Sheep in the Mojave Desert. *J. Wildl. Manag.* **2007**, *71*, 1419–1426. [CrossRef]

44. Haralick, R.M.; Shanmugam, K.; Dinstein, I. Textural Features for Image Classification. *IEEE Trans. Syst. Man Cybern.* **1973**, *3*, 610–621. [CrossRef]

45. Edmunds, M.; Flynn, A. *A Victorian Marine Biotope Classification Scheme*; Report to Deakin University and Parks Victoria; Australian Marine Ecology Report No. 545; Deakin University: Melbourne, Australia, 2015.

46. Davies, C.E.; Moss, D.; Hill, M.O. *EUNIS Habitat Classification Revised 2004*; European Topic Centre on Nature Protection and Biodiversity: Paris, France, 2004; pp. 127–143.

47. Federal Geographic Data Committee. *Coastal and Marine Ecological Classification Standard*; FGDC-STD-018-2012; Federal Geographic Data Committee, Marine and Coastal Spatial Data Subcommittee: Reston, VA, USA, 2012.

48. Eigenraam, M.; McCormick, F.; Contreras, Z. *Marine and Coastal Ecosystem Accounting: Port Phillip Bay*; State of Victoria Department of Environment, Land, Water and Planning: Victoria, Australia, 2016.

49. Liaw, A.; Wiener, M. Classification and Regression by randomForest. *R News* **2002**, *2*, 18–22.

50. Breiman, L. Random forests. *Mach. Learn.* **2001**, *45*, 5–32. [CrossRef]

51. Cutler, D.R.; Edwards, T.C.; Beard, K.H.; Cutler, A.; Hess, K.T.; Gibson, J.; Lawler, J.J. Random forests for classification in ecology. *Ecology* **2007**, *88*, 2783–2792. [CrossRef] [PubMed]

52. Stephens, D.; Diesing, M. A Comparison of Supervised Classification Methods for the Prediction of Substrate Type Using Multibeam Acoustic and Legacy Grain-Size Data. *PLoS ONE* **2014**, *9*, e93950. [CrossRef] [PubMed]

53. Kuhn, M.; Wing, J.; Weston, S.; Williams, A.; Keefer, C.; Engelhardt, A.; Cooper, T.; Mayer, Z.; Kenkel, B.; The R Core Team; et al. *Caret: Classification and Regression Training*; GitHub, Inc.: San Francisco, CA, USA, 2017.

54. Freeman, E.; Frescino, T. *ModelMap: Modeling and Map Production Using Random Forest and Stochastic Gradient Boosting*; USDA Forest Service, Rocky Mountain Research Station: Ogden, UT, USA, 2009.

55. Hallgren, K.A. Computing Inter-Rater Reliability for Observational Data: An Overview and Tutorial. *Tutor. Quant. Methods Psychol.* **2012**, *8*, 23–34. [CrossRef] [PubMed]

56. Lecours, V.; Devillers, R.; Edinger, E.N.; Brown, C.J.; Lucieer, V.L. Influence of artefacts in marine digital terrain models on habitat maps and species distribution models: A multiscale assessment. *Remote Sens. Ecol. Conserv.* **2017**. [CrossRef]

57. Lecours, V.; Dolan, M.F.J.; Micallef, A.; Lucieer, V.L. A review of marine geomorphometry, the quantitative study of the seafloor. *Hydrol. Earth Syst. Sci.* **2016**, *20*, 3207–3244. [CrossRef]

58. Capolsini, P.; Andréfouët, S.; Rion, C.; Payri, C. A comparison of Landsat ETM+, SPOT HRV, Ikonos, ASTER, and airborne MASTER data for coral reef habitat mapping in South Pacific islands. *Can. J. Remote Sens.* **2003**, *29*, 187–200. [CrossRef]

59. Landis, J.R.; Koch, G.G. The Measurement of Observer Agreement for Categorical Data. *Biometrics* **1977**, *33*, 159. [CrossRef] [PubMed]

60. Brown, C.J.; Collier, J.S. Mapping benthic habitat in regions of gradational substrata: An automated approach utilising geophysical, geological, and biological relationships. *Estuar. Coast. Shelf Sci.* **2008**, *78*, 203–214. [CrossRef]

61. Freitas, R.; Rodrigues, A.M.; Quintino, V. Benthic biotopes remote sensing using acoustics. *J. Exp. Mar. Biol. Ecol.* **2003**, *285*, 339–353. [CrossRef]

62. Che Hasan, R.; Ierodiaconou, D.; Laurenson, L. Combining angular response classification and backscatter imagery segmentation for benthic biological habitat mapping. *Estuar. Coast. Shelf Sci.* **2012**, *97*, 1–9. [CrossRef]

63. Galparsoro, I.; Connor, D.W.; Borja, Á.; Aish, A.; Amorim, P.; Bajjouk, T.; Chambers, C.; Coggan, R.; Dirberg, G.; Ellwood, H.; et al. Using EUNIS habitat classification for benthic mapping in European seas: Present concerns and future needs. *Mar. Pollut. Bull.* **2012**, *64*, 2630–2638. [CrossRef] [PubMed]

64. Young, M.; Ierodiaconou, D.; Womersley, T. Forests of the sea: Predictive habitat modelling to assess the abundance of canopy forming kelp forests on temperate reefs. *Remote Sens. Environ.* **2015**, *170*, 178–187. [CrossRef]

65. Graf, R.F.; Bollmann, K.; Suter, W.; Bugmann, H. The Importance of Spatial Scale in Habitat Models: Capercaillie in the Swiss Alps. *Landsc. Ecol.* **2005**, *20*, 703–717. [CrossRef]

66. Kendall, M.; Miller, T.; Pittman, S. Patterns of scale-dependency and the influence of map resolution on the seascape ecology of reef fish. *Mar. Ecol. Prog. Ser.* **2011**, *427*, 259–274. [CrossRef]

67. Hernandez, P.A.; Graham, C.H.; Master, L.L.; Albert, D.L. The Effect of Sample Size and Species Characteristics on Performance of Different Species Distribution Modeling Methods. *Ecography* **2006**, *29*, 773–785. [CrossRef]

68. Kadmon, R.; Farber, O.; Danin, A. A systematic analysis of factors affecting the performance of climatic envelope models. *Ecol. Appl.* **2003**, *13*, 853–867. [CrossRef]

69. Przeslawski, R.; Foster, S. *Field Manuals for Marine Sampling to Monitor Australian Waters*; National Environmental Science Programme, Marine Biodiversity Hub: Tasmania, Australia, 2018.

MDPI

St. Alban-Anlage 66

4052 Basel

Switzerland

Tel. +41 61 683 77 34

Fax +41 61 302 89 18

www.mdpi.com

Geosciences Editorial Office

E-mail: geosciences@mdpi.com

www.mdpi.com/journal/geosciences